Natural Resources
in
Tropical Countries

Natural Resources
in
Tropical Countries

Edited by
OOI JIN BEE

Published by Singapore University Press with a grant
from the Commonwealth Foundation
and the United Nations University

ISBN 9971–69–063–2 (cloth)

Typeset and Printed by Fong & Sons Printers Pte Ltd

This volume is dedicated to
Professor R.W. Steel
Director
Commonwealth Geographical Bureau
1972–81

Contents

List of Figures

List of Contributors

Abdul Hamid Abdullah, B.A., M.A., Ph.D.
Associate Professor
Department of Geography
Universiti Kebangsaan Malaysia

D.O. Adefolalu, B.Sc., M.Sc., Ph.D.
Department of Geography & Regional Planning
University of Calabar

R.M. Auty, B.A., M.A., Ph.D.
Department of Geography
University of Lancaster

C.J. Barrow, B.Sc., Ph.D.
Centre for Development Studies
University of Wales
University College of Swansea

Deryck Bernard, B.A., M.Phil.
Department of Geography
University of Guyana

Edison Dayal, M.A., Ph.D.
Senior Lecturer
Department of Geography
The University of Wollongong

Chief Adetoye Faniran, B.A., Ph.D.
Professor of Geography
University of Ibadan

A.N. Gillison, Q.D.A., B.Agr.Sc., M.Sc., Ph.D.
Principal Research Scientist
CSIRO Institute of Biological Resources
Australia

J. Dennis Huckabay, B.A., M.A., Dip.Ed.
Department of Geography
The University of Zambia

Richard, Jackson, M.A., D.Phil.
Professor of Geography
University of Papua New Guinea

Mustafa M. Khogali, Ph.D.
Department of Geography
University of Khartoum

H.M. Mushala, B.A., M.A.
Department of Geography
University of Dar-es-Salaam

Richard S. Odingo, B.A., Ph.D.
Professor of Geography
University of Nairobi

Ooi Jin Bee, B.A., M.A., D.Phil.
Professor of Geography
National University of Singapore

Maheshwari Prasad, B.A., M.A., D.Litt.
Reader
Department of Geography
Ranchi University
Bihar

In this first major publication of the Commonwealth Geographical Bureau, it is proper to note some of the attainments of the Bureau since its foundation during the International Geographical Congress held in India in December 1968. As Chairman of the Working Party set up in New Delhi to discuss the possible creation of such an organization, I and the others who worked with me were very conscious of the vision of some who had long dreamt of such a Bureau (notably the late Professor K. Kularatnam of Sri Lanka who became, very appropriately, the first Director of the Bureau). We were also aware of the likely organizational difficulties and financial commitments of a newly created Commonwealth body that wanted to be active in many different ways and also truly international in its operations and in its choice of officers and committee members.

The Bureau came into being with a Sri Lankan Director, a Chairman in Britain, and a Secretary/Treasurer in Canada, and within two years, thanks to the generous and understanding financial support of the Commonwealth Foundation, was able to mount a very successful conference for West African geographers in the University of Ghana in September 1970. This set the pattern for a series of Commonwealth gatherings of which the workshop held in Swansea in April 1981 is the latest and, in at least one respect, the most productive in that this volume is made up of the papers read and discussed there.

At the International Geographical Congress in Montreal in 1972, a well-attended and representative meeting of Commonwealth geographers voiced some criticisms of the Bureau's activities and made some valuable suggestions for its future development. The late Professor Charles A. Fisher agreed to be the Chairman of the Bureau, on the understanding that his colleague in the School of Oriental and African Studies in the University of London, Dr R.W. Bradnock, would take on the important work of the Secretary/Treasurer; and at its first meeting the Board invited me to become its Director.

I remained Director for nine years — much longer than I should have stayed in that office, though I made an effort to hand my role over to someone else on several occasions. Regional conferences were organized and held — in Singapore in 1975 for South and East Asia, in Dar-es-Salaam in 1976 for East and Central Africa, and in Guyana in 1980 for Latin America including the Caribbean. When Charles Fisher indicated that he wished to give up the chairmanship, I was persuaded to combine the two offices of Director and Chairman; and for some time I had to play the central role in the organization of the Bureau when not only were nearly all the officers located in Britain but it seemed that nearly all of them were doing what is eminently right and proper for a good geographer, taking study leave from their institutions in the United Kingdom and visiting different parts of the Commonwealth including India, Malaysia, and Hong Kong.

Despite the success of the Bureau's regional conferences, the Committee of Management felt that a different kind of Bureau activity needed to be encouraged and stimulated. With the appointment of a very energetic Secretary, Garth Cant of Christchurch, New Zealand, plans were developed for a workshop on "Natural Resources of the Third World", and Professor Ooi Jin Bee of Singapore agreed to act as a coordinator and to undertake the editing of the volume that it was hoped would emerge from the workshop. Since at the time I was not only Principal of the University College of Swansea but also Vice-Chancellor of the University of Wales, I suggested that the workshop be held in my own College in Swansea.

There is no doubt that the Swansea Workshop was a great success, both academically and socially. Geographers when they meet together usually enjoy themselves, and where the group is small, as it was, the pleasure is especially marked, particularly when there is a shared interest within the traditionally, and even notoriously, wide range of geography. It is unnecessary to say much more about the workshop since this volume is its record and it speaks far more eloquently than I can about the papers discussed and the range of subjects covered. But I want to pay special tribute to the organizational work done (much of it from a distance of 13,000 or more miles) by the Secretary, Garth Cant, in New Zealand, and by the Local Secretary, Chris Barrow of the Centre for Development Studies in the University College of Swansea, and others in the College who assisted him. The Treasurer, Mike Eden, of Bedford College, University of London, dealt not only with the finance made available through the Commonwealth Foundation but also with the very generous grant made to the workshop by the United Nations University. Two members of that University were especially helpful, and both were able to attend part of the Workshop — Professor W. Manshard, formerly Vice-Rector, Programme on the Use and Management of Natural Resources of the University (and now once again Professor of Geography at the University of Freiburg, in the Federal Republic of Germany), and his colleague, Professor R.S. Odingo, who has since returned to his post at the University of Nairobi in Kenya. Professor Ooi Jin Bee acted as chairman of the workshop, and his work in editing this volume underlines the wisdom of the Bureau in inviting him in 1981 to be my successor in the Directorship.

The valuable contribution that the Commonwealth Geographical Bureau can make to the progress of geography will be obvious from the contents of this volume, which looks at the natural resources of the tropics from many different points of view and provides, I venture to suggest, some new insights into many topics that are of vital concern not only to many specialists besides geographers but also to the people of the Third World generally. My hope is that the nature of this work will inspire in all its readers a confidence in the future development of the Bureau with its broadly based and widely representative Committee of Management and with its present officers — Professor Ooi Jin Bee in Singapore, Dr Garth Cant in New Zealand, and Mr Michael Eden in London, the latter with a special responsibility for maintaining a close link with the Commonwealth Foundation, without whose help and encouragement the Bureau could never have come into existence and made the progress that it has since its beginnings in 1968. This permanent record of the Swansea Workshop of April 1981 will, I trust, be seen as an indication of the valuable work that the Bureau has already done and will, I firmly believe, continue to do in the future.

March 1982

Robert W. Steel
Principal, University College of Swansea,
University of Wales;
Director, Commonwealth Geographical
Bureau, 1972–81

Editor's Preface

The idea of a workshop on the theme of natural resources in tropical countries was first put forward at a meeting of the Commonwealth Geographical Bureau Committee of Management held in Dar-es-Salaam. An invitation to contribute papers to the workshop was published in July 1979 in the Bureau's newsletter. More than thirty Commonwealth geographers responded to the invitation, submitting proposals which covered a broad spectrum of topics spanning the tropical world. The Commonwealth Geographical Bureau's original budget allowed only for a workshop of about ten paper writers. This was subsequently expanded to sixteen when, through the good offices of Professor R.W. Steel, the United Nations University provided the Bureau with a grant sufficient to cover the cost of an enlarged workshop.

It would be appropriate to place on record here the Bureau's appreciation and thanks to the Commonwealth Foundation and to the United Nations University, particularly Professor Walther Manshard, Vice-Rector and Professor R.S. Odingo, Senior Programme Officer, who were then with the Natural Resources Programme of the United Nations University, for their support for this workshop.

Even with an expanded budget it was not possible to include many of the proposals, good though they were, and it is the Bureau's hope that the authors will have an opportunity to publish their research findings in journals or other academic publications. The fifteen papers which were finally selected were presented at the workshop held in the Centre for Development Studies, University College of Swansea, between 7 and 14 April 1981. The facilities of the Centre and at Neuadd Martin, where the delegates were housed, were made available through the kindness of Professor Steel, the Principal of the University College. To him and the University College of Swansea, we owe our thanks. Dr C.J. Barrow, the local organizer, and Dr H.R.J. Davis were both extremely helpful to all the delegates and did much to make the workshop a success. We are grateful to Dr Graeme Humphrys for taking us on an excellent field trip to the Swansea Valley. We owe our thanks, too, to Professor F.T. Banner of the Department of Oceanography for giving us a thought-provoking talk on "The Resources of Shelf Seas and Coasts". After the workshop most of the delegates visited the Department of Geography at the University of Liverpool on the kind invitation of Professor Mansell Prothero. The Commonwealth Geographical Bureau and the delegates deeply appreciate the thoughtfulness of the invitation and the hospitality extended to them by Professor Prothero and members of his department.

The papers in this book are, in form and content, substantially the same as those presented at the workshop. They have been organized in a straightforward manner and put in one of the three major sections. There is, inevitably, a certain degree of arbitrariness involved in the placement of the material, in that one or two papers may sit somewhat awkwardly in their section, and in that the material in others may straddle sections. The papers themselves cover a wide range of topics in the field of natural resources. As is the case with any collection from a workshop, individual papers vary widely in length, depth of subject treatment, and mode of presentation. I have attempted to place them in some kind of perspective in the introductory overview which forms the first chapter of this book.

Many people have helped me in the organization of the workshop and in the preparation of this book. Apart from those mentioned earlier, I would particularly wish to thank Dr Garth Cant, the Commonwealth Geographical Bureau's Secretary, and Mr Michael Eden, the Treasurer. Mrs Irene Chee and Mrs Lim Kim Leng of the Department of Geography, National University of Singapore, provided me with all the secretarial and typing assistance I needed while Mr Poon Puay Kee, Senior Cartographer, was responsible for the maps and diagrams.

The Bureau is deeply grateful to the Commonwealth Foundation and the United Nations University for helping to defray the cost of publishing this book.

Singapore 1982   Ooi Jin Bee

1
Natural Resources in Tropical Countries: An Examination

OOI JIN BEE

Tropical countries have long been recognized as being "underdeveloped" or less developed. The standard term today for such countries is "developing". A developing country exhibits characteristics which are considered undesirable from the standpoint of "development". It usually has a non-diversified economy producing a low per capita volume of exchangeable goods and services, a subsistence sector of some importance, a high proportion of the labour force in agriculture and other forms of primary production, an embryonic manufacturing sector, poor credit and marketing facilities, poorly developed transport and communications facilities especially in the rural areas, a low volume of trade per capita, and a low income per capita. These are some of what Myrdal (1970:34) terms the "constellation of numerous undesirable conditions for work and life" common in such a country.

The index most frequently used to differentiate a developed (rich) country from a developing (poor) country is the gross national product (GNP) per capita. This index suffers from a number of weaknesses. It does not, for example, take into account the actual distribution of incomes among classes or groups in a country. Moreover, its computation is subject to a considerable margin of error in those countries where the exchange economy is not fully developed and where a significant part of the national product is derived from subsistence production, as in many of the countries of Africa south of the Sahara. As a measure of real income and standards of living, it does not take into consideration differences in the requirements for food, shelter, and clothing (and heating) arising from geographical location and from differences in climatic conditions. The population data from which the GNP per capita is computed are often of questionable validity. Any discussion of a country's level of development as measured by its GNP per capita should therefore have these reservations in mind.

About thirty years ago, economist John Kenneth Galbraith noted: "(If) one marks off a belt a couple of thousand miles in width encircling the equator one finds within it no developed countries . . ." (1951:693). The correlation between the factor of tropicality and the low level of development is as close today as it was then. Figure 1.1 shows the distribution of the developing countries of the world in 1979, based on World Bank data. They are divided into two groups made up of 36 low income countries with a per capita GNP of US$370 or less and 60 middle-income countries with a per capita GNP ranging from US$380 to US$4,380. The figure indicates clearly that though there are some countries (notably China) outside the tropics which are poor, nearly all the tropical countries share this common fact of poverty.

However, the developing countries should not be regarded as an undifferentiated whole, and many writers have been at pains to stress the heterogeneity of the less developed world (see, for example, Bauer and Yamey 1967). Differences in rates of economic growth, in territorial and population sizes and therefore man-land ratios, in resource endowments, and in cultural and social attributes — to mention only a few variables — are quite marked among these countries. Similarly, the other feature which most of them share — the tropical climate — will be seen to be, on close examination, distinctly complex, involving a large variety of environmental conditions.

There is no exact definition of the tropics, although many attempts have been made to delimit them (see Gourou 1968; Garnier 1958; Fosberg *et al.* 1961; Miller 1971; Oliver 1979). It is possible to distinguish three main types of tropical climates. First, there is the wet equatorial climate, characterized by high average temperatures (27 °C) for every month of the year; a small seasonal temperature range; and annual rainfall in excess of 2,000 mm, but with marked differences in monthly averages. Areas lying within the belt 10° on either side of the equator experience this climate. At the other extreme are the tropical desert and steppe climates, centred roughly on the Tropics of Cancer and Capricorn. Here the climates are controlled by dry air masses subsiding and moving away from continental high pressure cells. The deserts of north Africa, the Kalahari, Arabia, Iran, West Pakistan, and Australia typify areas which experience a tropical desert climate. Temperatures are very high during the period of high sun; the annual range is 17–22 °C, and the daily range may also be just as high. Aridity is the dominant feature of such a climate, and annual rainfall is usually less than 40 mm, with the steppe areas receiving between 40 and 120 mm of rain. Third, there are areas lying between these two extreme climatic types that experience a tropical wet-dry climate, with moist equatorial and maritime air masses bringing in rain during the period of high sun and a dry season controlled by dry continental tropical air masses during the period of low sun. In Central and South America, Africa, and Australia the latitude belts which have such a climate lie between 5° and 25°, but in Asia the configuration of the land mass is such that this belt is located between latitudes 10° and 30°.

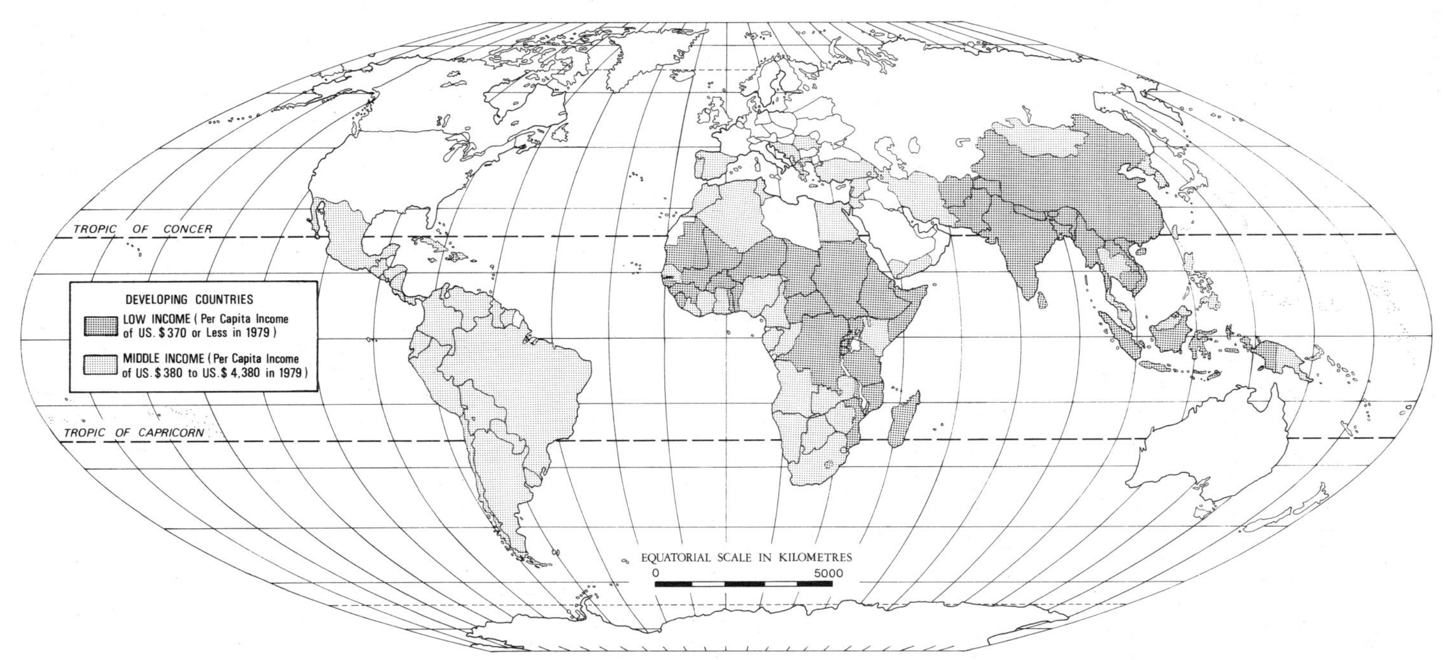

Fig. 1.1 The developing countries of the world
(Source: based on data in World Bank, 1981)

Natural Resources

Natural resources are physically a part of that segment of the physical world ("nature") that has a present or potential use for the survival and physical well-being of man, to be developed through the application of technological knowledge. They can be divided into renewable or non-renewable resources, their renewability being relative to human but not geologic or astronomic time. Renewable resources — climate, land, soils, water, forests, wild life, fisheries, and the non-mineral energy sources of solar, wind, tidal, and geothermal systems — if properly managed, can provide a regular production or flow for an indefinitely long period of time. The non-renewable resources are the finite stocks of minerals which, like petroleum, have accumulated over a long period of geological time and which, once used and dispersed, are thereafter unavailable.

Natural resources, as Carl Sauer (1969:3) has said, are "cultural appraisals". Every group will interpret its natural resources within the framework of its own social structure. The variety of cultures in the pre-industrial societies of the developing countries of the tropical world is as great as is the variety of techniques applied by these societies in the conversion of resources in their natural state into raw materials and into goods. Such societies tend to be tied to a local habitat and to the resources of that habitat, unlike the technologically more complex industrial societies whose links with natural resources are not local but global.

A further difference needs to be noted: modern man conceptually isolates and assesses natural resources as that part of the physical world which is of material use to him, whereas pre-industrial man looks upon them as part of nature, within a single order system which includes man, nature, and the gods (Spoehr 1965).

These and other differences in the cultural configurations of societies must be borne in mind when considering the role of natural resources in economic development in the developing world. Economic development demands that the goals and objectives of the society (or at least the leaders) seeking such development be oriented towards the regular production of exchangeable goods and services beyond basic subsistence needs. This may mean the relegation of established socio-religious values to a subsidiary position in favour of material considerations.

The value which they place on natural resources, and indeed their appraisal of what constitutes a natural resource, is based on a set of criteria appropriate to and evolved from a pre-industrial setting. Within a developing tropical country may lie natural resources which are non-utilizable to the indigenous people (strictly speaking, these are "neutral matter" and not resources to them) but which have value to an industrial country with the technology to use them. Their exploitation, as Ritter (1975) has noted, has given rise to the many examples of dualistic economic systems in the Third World.

Any assessment of the role of natural resources in economic development must be based on the value attached to these resources or to their products in the international market-place. A developing tropical country has to assess its resource endowment by this criterion, as such resources or their products, if abundant and in demand, can act as a means of capital accumulation and economic growth.

It should be noted that the economic value of natural resources is relative to the places in which they are found, to time, and to their technological and cultural setting. As Sundrum (1977) has pointed out, one of the determinants of the per capita level of exchangeable goods and services in a country is the attitude of the people to their resources — how they view them, and what they want to do with them. Another is the technology they use to transform them into goods and services. Since these may change over time, the economic meaning and valuation of natural resources would also shift.

It would be apparent that one of the major requirements for development planning in a developing country is knowledge of its resource base, that is, an inventory of those naturally occurring materials that are required in some degree by the national economy responding to internal consumption demands and its position in international trade (Perloff and Wingo 1965). However, there is a general paucity of resource inventory data in most developing countries. The generation of information on the nature, location, and other characteristics of natural resources is itself costly in terms of two items which are in short supply in these countries — money and skilled manpower. Nevertheless, resource development projects cannot proceed on the basis of guesswork, and the necessary investment must be made in collecting resources information for project evaluation.[1]

In nearly all cases, the assembling of complete information is impossible, so that decisions will have to be made on the kinds of data to be gathered, in what detail, order of priority, and by what methods. The strategy of resources information gathering poses a difficult problem in management, as the pressures for information needed to reduce risk and uncertainty in natural resource development may result in some information being developed prematurely and some being gathered which may never be used at all. The guidelines should be to find out as much about the natural resources of a country as is necessary to make efficient use of them in production, and to be reasonably certain that expenditure incurred in generating resources

---

[1] The problems of resource inventory — determination of the extent of the resource and identification of the primary environmental determinants of the resource — in respect of tropical vegetation are analysed by Gillison (Chapter 11). In his paper Gillison also presents a case for an improved approach to field survey and vegetation classification methods.

information will result in additional gain to economic development (see Herfindahl 1969).

In many parts of the tropical world, the execution of a natural resource inventory is constrained by difficult terrain, scarcity of trained personnel, low levels of infrastructure development, wet season inaccessibility, and problems relating to inventory design and data processing. From the technical and cost-effective point of view remote sensing from space satellites such as the NASA Landsat satellites offers a practical solution to some of the problems of resource inventory information gathering (CENTO 1971; U.N. 1975). Information can be generated at a fraction of the time and cost of traditional methods. For example, a complete inventory of land use in Costa Rica was completed for less than one-seventh the cost of conventional mapping, and provided more information than was available from the latest agricultural census (NASA 1979). However, most of the developing countries have yet to use this new technological tool, for reasons which range from the purely technical to the political (Ambrosetti and Goward 1981).

Organizations engaged in natural resource surveys in developing countries may be international institutions such as the Food and Agriculture Organization or, depending on the level of expertise available, national governmental departments and universities and research institutions (Young 1968). International funding agencies may, however, be reluctant to finance survey operations covering resources whose value and productivity cannot be accurately assessed, as for example, the vegetation and water resources of the Sahel which are under stress because of population growth and uncontrolled increases in the number of cattle and other ruminants (Panzer 1981).

Estimates of the availability of natural resources require that a careful distinction be made between resource potentialities and resource endowments. Much confusion can be avoided if these estimates are expressed within a clearly defined framework such as that drawn up by Lovejoy and Homan (1965) and modified by House and Williams (1977):

> *Total stock* is the sum of all components of the environment that would be resources if they could be fully extracted from it. Assessment of the total stock is largely the concern of earth and life scientists, and the state of knowledge concerning it depends on the adequacy of prevailing theory, the state of exploration and survey technology and the extent of its application. Applied to resources that are conventionally referred to as nonrenewable resources, the total stock is finite and thus eventually exhaustible. Applied to renewable resources, the total stock consists of highly complex systems in a state of dynamic, delicately balanced, and only partially understood, equilibrium.
>
> *Resources* comprise that proportion of the total stock that man can make available under technological and economic conditions different from those that prevail. The assumed state of technology will set the limits within which different economic and social variables (will) determine what proportion of the total stock can become available. Assessment of a resource (level) involves

not only physical and biological scientists but applied and social scientists as well. They must make judgments: about the directions and rate of change of technological developments (e.g., gradual increase of efficiency of extraction of a mineral deposit or yield of a crop as against dramatic, order-of-magnitude, breakthrough changes); about the impact of changed economic conditions and new alignments in international relations; and about public attitudes on such varied matters as birth and population control, transportation preferences, and clean air. In essence the question is one of judging man's potential for creating resources out of the total stock; of selecting the chief agent of change from among technological, economic, or other societal forces; and of determining the relevant time and space dimensions.

Reserve refers to that proportion of a resource that is known with reasonable certainty to be available under prevailing technological, economic, and other societal conditions. This term embraces current extraction rates, yield, management practices, legal frameworks, and social attitudes. It is, therefore, the least speculative, shortest term, most place-specific, and smallest of the three types of estimates.

The framework makes provision for resources moving from one threshold of availability to another in response to changes in the physical, human, and economic variables of the country concerned.

Natural Resources and Economic Development

Economic development may be defined as "a process which makes people in general better off by increasing their command over goods and services and by increasing the choices open to them" (Elkan 1978:15–16). As seen earlier, the per capita income of the developing countries is low. The central problem in these countries is to bring about a rise in income levels through a sustained increase in the production of exchangeable goods and services. Economic development has also been defined as the process whereby the people of a country use the available resources to achieve this sustained increase in the per capita production of goods and services (Williamson 1965).

Sundrum (1977) has compiled the following list of factors which economists have put forward as determinants of the per capita level of exchangeable goods and services in a country:

1. The endowment of natural resources — the iron, the coal, the rivers and the coast, and the climate — not just in the way it helps to grow rubber or coffee but also the effect on people — all taken in relation to the size of the population.

2. Man-made capital — both the capital used for directly productive activities and that involved in infrastructure or social overhead capital.

3. Technology — the methods used by people to transform their resources into exchangeable goods and services, as determined by their knowledge and willingness to apply this knowledge.

    4. Relations with other countries — the movement of people and of goods and of finance from country to country.

    5. The attitudes and aptitudes of people — how they want to use their time, what they want to do with their resources, and how they respond to economic incentives.

    6. The role of institutions — of law and order, of the right of property, and of markets in goods and services, especially of financial assets.

    7. The role of the government — the extent of its control in economic affairs and the ways it exercises this control.

The division of the earth's surface into a large number of political units of uneven sizes has inevitably resulted in great variations in the natural resource endowments of these units. It has been observed that there is little correlation between the level of economic development of a country, whether tropical or temperate, and its natural resource endowment. The economic history of many nations has shown that differences in development and prosperity were not connected with the discovery or exhaustion of natural resources within their territories, so that the chance distribution of these resources among them cannot be used as an explanation of such differences in development (Bauer and Yamey 1957). The situation is summed up aptly by Herfindahl (1969:4).

> . . . no particular type of natural resource is essential to a high level of national income or to economic progress. We can assert with confidence that a country's endowment of natural resources need not exercise a determining influence on the course of its national income over time — if it is able to trade. "All" that is necessary for economic progress is the availability of a substantial quantity of services of capital and labour — with a considerable part of the capital embodied in persons — plus a social system with certain characteristics favorable to systematic improvement of production practices. And the more capital per person the better. The truth of this observation is evident from the economic success of countries with limited natural resources and the success of countries with greatly different natural endowments. To enjoy economic success, a country must have access to natural resources, but this can be had through trade with other nations with a different or better natural endowment. It is not necessary to economic progress that a nation have iron ore, good soil, forests, or any particular one of the natural resources.

Economic historians have traced the stages through which most developed countries passed in attaining their present economic positions. In the very early pre-industrial stage, natural resources contributed directly to meeting their basic needs of food, shelter, and clothing. As population increased and markets widened, there was greater specialization of labour, larger accumulations of capital, and a more complex economic organization. This was followed by the establishment of light and then heavy manufacturing industries. In the final stages the industrial pattern became (and is today) highly diversified. The availability of and *access* to natural resources was essential to these development processes, but as the countries achieved a sustained increase in their per capita income, the contribution of natural resources in producing that income decreased relative to the contribution of other factors such as capital and technology (Fisher 1964; Schultz 1965).

Natural resources play a more important role in the productive services of developing countries than they do in developed industrialized countries. The contribution of natural resources to all resources employed to produce the national income in developing countries today may be as high as 20 to 25 per cent. In contrast, the lower limit of the contribution in developed countries may be only about 5 per cent except in countries such as Canada, Australia, and South Africa (Schultz 1965).

Although natural resources within territorial boundaries are not crucial to or a prerequisite to economic development in a country, possession of such resources can have a favourable influence on the level of income and its growth. This is especially so in a developing country where the variable resources of capital and labour skills tend to be limited. In such a situation its fixed or nature resources may provide a ready-made nature-given stock of capital which it can draw upon and which may otherwise take a long time to accumulate out of income. Clearly, the richer and more varied the natural resource endowment the better it is for that country (Ginsburg 1957).

Such natural resources, however, are not finished capital goods. If they are to provide productive services, additional investment outlays (including investment on information on the resource endowment) will have to be made. Such investment would include the development of infrastructural facilities. Natural resources, from this economic perspective, can therefore be regarded as unfinished capital goods (Herfindahl 1969).

The additional investment outlay necessary to develop natural resources depends on their location and accessibility. Being natural phenomena, the distribution of natural resources is always localized (except for air), and is the outcome of geological accident, physical processes, or location in relation to the earth's surface. Such resources are movable only to a limited extent, compared with other forms of capital, although many of the products which they yield can be moved freely and over long distances. Some resources, such as amenity resources, are fixed and immovable. In the developing tropical countries, accessibility to natural resources is of particular significance, and a major objective in the opening up of new land through the development of transport is to make their resources more accessible.

The discussion in this paper is concerned with only one of the factors — natural resources — which Sundrum (1977) has put forward as determinants of the per capita level of exchangeable goods and services in a country. No attempt will be made here to examine the other factors, but it is patently neither accurate nor helpful to single out any one factor and attribute to it a special causative role in determining the wealth of nations. All of them — natural resources, population, capital technology and skills, trade relations, the attitudes and aptitudes of the people, and the roles of institutions and of the government — must be combined effectively for a country to increase its production of exchangeable goods and services and to raise its per capita income levels. The role of natural resources in the economic development of the developing countries of the tropical world must thus be viewed within this perspective.[2]

Climate and Economic Development

The correlation between the factor of tropicality and the fact of under-development in tropical countries has been indicated earlier. This correlation is as good as most correlations are between non-economic factors and economic development. Development economists have largely neglected the climatic factor in their studies on the theory of economic development although some, such as Gunnar Myrdal (1970) and Benjamin Higgins (1968) have devoted some attention to it (see also Kamarck 1976). Geographers addressing themselves to the problems of development in the tropical world have adopted a cautious stance in advancing explanations for the correlation between climate and economic development, partly as a reaction against the deterministic theories of Ellsworth Huntington and his school.

Huntington postulated that though man could survive under a wide range of environmental conditions, people who lived under the least favourable climatic conditions, such as the hot, wet tropics, would be at a disadvantage compared with those who lived near the climatic optima. He then attempted to draw up optima for physical energy, mental efficiency, civilization, and health (Huntington 1924; 1930). His critics maintained that he exaggerated the correlation between climate and human progress, vitality, energy, health, social organization, and other factors, and that he reached inaccurate conclusions from insufficient data, misinterpreted statistics, and confused the primary factor of climate with secondary controls such as disease and diet (see, for example, Sorokin 1928; Vance 1932; Wicken 1930). But, as Kindleberger (1965:78) notes: "The arguments against Huntington are telling, but the fact remains that no tropical country in modern times has

---

[2]In his paper on natural resources and economic development in developing countries, Odingo (Chapter 16) ranges over a wider spectrum of issues and topics than that of the role of natural resources in economic development.

achieved a high state of economic development. This establishes some sort of presumptive case — for the end result, if not for the means."

Popularization of Huntington's theories provided the support needed as a rationalization of Western colonial policy, that the colonial peoples of the tropics were somehow inferior and nothing could be done to improve the productivity of the colonies. They also served to perpetuate the myth of the lazy native.

Modern-day reactions against such attitudes have tended to cause the pendulum to swing the other way, to the extent that in the examinations of the causes of underdevelopment, climate is either ignored or regarded as being of little consequence. As expressed by an economist: "Because economic growth is currently most rapid in the temperate zones, it is fashionable to assert that economic growth requires a temperate climate, but the association between growth and temperate climates is a very recent phenomenon in human history. . . . The climate hypothesis does not take us very far." (Lewis 1955:53, 416) Another reason for not taking climate into account in development analysis is the belief that nothing can be done about the effects of climate (see Kamarck 1976:8).

Gilfillan (1920) has traced the "coldward course of progress" in human history. The earliest civilizations were located in the tropics and sub-tropics — in the fertile valleys of the Tigris and Euphrates, the Nile Valley, the Indus Delta, and tropical America. These tropical regions were the countries of primitive economic development, but in the fifty centuries that followed the tropical civilizations withered and gave way to new centres of development located in colder localities on both sides of the Atlantic (Fig. 1.2). The geographic coverage of these centres may be expanded today to include Japan, Australia, and New Zealand.

It would be facile to claim on this evidence that there is therefore a one-for-one link between climate and development. Logical reasoning would indicate that such a premise would be faulty. The evidence does not prove anything; it does not reveal anything about the essential relationships of cause and effect. If, for example, a Huntington had lived during the period when the centres of development and growth were in the tropics and sub-tropics he could well have arrived at quite different conclusions on the relationships between climate and human progress. Much would depend on which part of the historical spectrum shown in Figure 1.2 one chooses to focus on; much would also depend on what is meant by human progress.

As pointed out by Sir Arthur Lewis, the association between growth and temperate climates is a very recent phenomenon in human history. Growth had its starting point in Europe after 1500 when there was a marked acceleration in economic development which culminated in the Industrial Revolution of the eighteenth and nineteenth centuries. Looked at from the historical

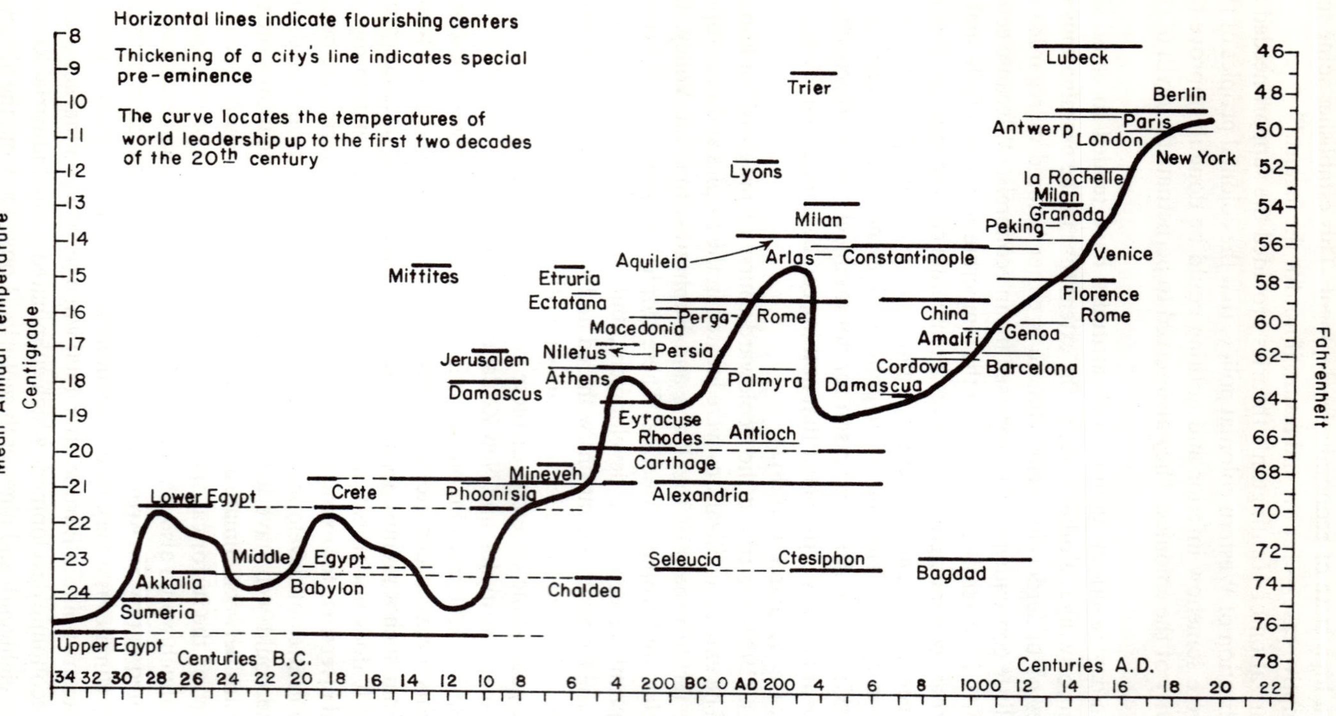

Fig. 1.2 The coldward march of progress
(Source: after Gilfillan 1920)

dimension, the development of the Western (European) economies was a unique event:

> It is a combination of a whole series of scientific and technical events (printing, metallurgy, mechanical time-keeping, improved draft tillage, unprecedented geographical discoveries in the Americas, round the Cape, in Brazil, etc., and cultural and religious upheavals such as the Renaissance and the Reformation), each of them deeply rooted in the past, which explains that unique phenomenon, the emergence of the European economy. It is, like it or not, fortuitous, and a scandal in the eyes of some, but it is this combination of events which gradually altered man's attitude to nature and to his fellow beings and it is this which gave him a taste for progress, for fulfilment and for a deliberate control over his future in general (David 1982:90).

To these might be added the point made by Higgins (1968:187) — that the Black Death gave the European countries relief from population pressure. The resultant labour scarcity caused a shift to more land- and capital-intensive techniques in agriculture, the products of which found ready markets through the expansion of world trade. By a remarkably fortuitous turn of events, the growth of manufacturing provided ready employment for labour forced out of farming by enclosures.

Economic historians looking at these and related aspects of past Western history have derived certain generalizations about this experience which, put together, have come to be known as the Western model for development. Viewed from such a perspective, the neo-classical prescription for economic growth in the developing countries of the world today, based on the Western model, would be industrialization, which would move labour away from traditional agriculture, livestock rearing, and other forms of primary production into manufacturing. Industrialization, dependent on high capital investment and inanimate energy, would lead to higher levels of productivity and higher incomes for the workers, which in turn would lead to capital accumulation through savings. Such capital, supplemented by imported capital where necessary, would be reinvested in factories, machines, and infrastructural facilities. The result should be a spiralling of the economy to increasingly higher levels of development.

A necessary requirement of this prescription would be social engineering on a massive scale, whereby social organizations, attitudes, and values not compatible or supportive of modern industrial growth would be replaced, through education, by new sets of values and attributes based on discipline, hard work, time consciousness, and responsiveness to monetary incentives (Abu Lughod 1977).

Both the Western and the socialist (Marxist-Leninist) models for development appear to be independent of climate and geographic space. Neither places any emphasis on environmental factors in general or on climate in particular. Most economists writing on development, as mentioned by

Kamarck (1976), have paid no regard to the role of climate. A notable exception is Gunnar Myrdal who, struck by the fact that nearly all the developing countries are in the tropics, has said, "This cannot be entirely an accident of history but must have to do with some special handicaps, directly or indirectly related to climate, faced by countries in the tropical or sub-tropical zones." (Myrdal 1970:143–44). Sundrum, as seen earlier, has included climate (including its effects on people) in his list of proximate factors affecting the per capita level of exchangeable goods and services in a country.

As part of the physical endowment of tropical countries, climate can be considered from the viewpoint of its effects, direct or indirect, on the productiveness of the resources of these countries. The difficulty is to isolate these climatic effects from the effects of other factors and conditions and to arrive at valid generalizations about the relationships between climate and economic development.

Climate also affects the productivity of labour. The output of a worker in the developing tropical countries is generally very low. The question that needs to be asked is whether this is causally linked to the tropical climate. Early laboratory studies on the effect of climate on human activity have concluded that "the leisurely habits of those who live in the tropics have a sound basis in physiological necessity" (Dill and Edwards, quoted by Price 1939:216). Evidence cited by Lee (1957) indicates that while there seem to be no permanent deleterious direct climatic effects in healthy persons living under natural conditions in the tropics, there remains an increased disinclination for work which tends to reduce output.

Surveys conducted among farmers and fishermen in many parts of the tropics — Malaysia, the West Indies, Zambia — have shown that their working day was on an average only four or five hours. The short working day was linked to low energy levels, defective or deficient diets, poor health, and the nature of their economic activities (see Ooi 1959:33–36). Kamarck (1976:58) has reported that a threshold temperature exists for different jobs and, above this, increases in temperatures bring about a drop in production (see also Haas 1975).

But the physiological effects of the tropical climate are only one of the many factors affecting labour productivity. Work attitudes may override the effects of climate, as Higgins (1968:214) observes:

> The attitude toward work, leisure, and income in Australia seems much the same from subtropical Darwin, where summer heat is more intense than in most equatorial countries, to chilly Hobart, with its ten-month-long winter and cool summers. Nor is there any significant difference in attitudes or productivity between Indonesians living at sea level and those living in the invigorating climate 4,000 feet up in the mountains. In the Philippines, the mountain people have remained the most primitive in the country.

The indirect effects of climate on economic development include those on plants, soils, crops, animals and pests, and diseases. Many of these manifest themselves in agricultural production which remains the dominant form of land use in the tropics, as seen in the fact that 71 per cent of the labour force in the 36 low-income countries (of which 35 are wholly or partly tropical) were engaged in it in 1979 (World Bank 1981).

Tropical soils are generally poor, except those derived from recent alluvium and from basic volcanic matter. The fertility of "8 or 9 square miles of average land south of the Sahara and north of the Union of South Africa is equivalent to that of one acre of Iowa soil" (Kindleberger 1966:64). Gourou (1953:13) has also pointed out that tropical soils are poorer and more fragile than those of temperate lands.

The low fertility of most soils in tropical countries can be traced to a common causative factor — the climate. Temperatures are higher by 10 °C to 20 °C, and chemical reactions take place at two to four times the speed usual in temperate climates, and without the break associated with winter. In the hot, wet tropics where the annual precipitation exceeds the annual evaporation, the characteristic movement of water is downwards, which leads to a continuous leaching of the soils. The easily soluble bases are removed so that the soil becomes highly acid. Leaching also washes out organic matter and plant nutrients which cannot be replaced by capillary action but only through addition from above, that is, by the decomposition of plant and animal organic matter in a rainforest situation, or by the application of manure or fertilizers by farmers.

The rate of decomposition of organic matter increases with temperature so that the higher the temperature the more rapid the rate of oxidation. The slender nutrient capital which may be built up in a rainforest area will be quickly oxidized away on exposure to the sun on forest clearing. Similarly, it has been found that the high annual temperatures of the tropics are not conducive to nitrogen accumulation in soil (Ooi 1959:79–84).

Compounding the problem posed by temperature is that posed by rainfall. In the hot wet tropics, the frequent and heavy rainfall leads not only to leaching of the soils but also to a loss of top soil through sheet erosion on exposed land, and to the compaction of the soil structure to the extent that the soil provides too few contacts for plants to take in water and nutrients. In those parts of the tropics which experience a dry season, such as in many parts of tropical Africa, periods of drought and flood are common, and pose special problems in the timing of farming operations.

The replacement of plant nutrients by the application of fertilizers is the universal method used by Western farmers to raise the productivity of the soil. But in the developing tropical countries this seemingly simple solution is not always possible. First, not enough is known about nutrient and trace element deficiencies in specific soil types and about crop responses to various fertilizer compounds in the tropics. Second, most tropical farmers have little experience in the use of fertilizers. The third and perhaps the most important reason is that oil-derived factory-produced fertilizers are too expensive for the developing countries to import in the quantities necessary to raise agricultural productivity substantially.

One other factor which affects agricultural output in the tropics needs mention. Crops grown under monoculture conditions, often in large plantations, are vulnerable to attacks by disease and pests. Under tropical conditions, disease pathogens and insect pests and vectors multiply throughout the year, while the rapid rate of insect reproduction results in several periods of injurious activities by a single species in a year. Pure stands are susceptible to infestations which can rapidly attain epidemic proportions. Among the crops thus affected in the past were cotton, rubber, coffee, cocoa, and banana. The depredations from disease and pests are not confined to crops — the livestock of the tropical pastoralist are equally exposed to a variety of diseases spread by micro-organisms and insect vectors which grow and multiply freely in the tropical environment. Man himself has to fight a constant battle against infection from various pathogenic agents and insect vectors such as the mosquito (Jusatz 1973).

The developing countries are mainly those that produce and export raw materials. They are the main producers of tropical food products and the only producers of industrial raw materials based on agriculture, with some exceptions such as wool and cotton. They supply all or nearly all of the world's coffee, bananas, palm kernels, cocoa beans, jute, and natural rubber, and more than 70 per cent of the tea and groundnuts (Kebschull and Schoap 1975).

This orientation towards raw materials production and export, which is partly a result of their colonial history, has put the developing countries in a dependent position vis-á-vis the industrialized, mainly Western countries. Clearly, such dependency is greater where these exports contribute significantly to the national income. The economic position of these countries is greatly affected by fluctuations in the demand for the export commodities. To this is added the double jeopardy of costly imported petroleum supplies and higher prices for imports of food and manufactured goods from the developed countries resulting from sustained worldwide inflation. Producers in the developing countries have also had their economies disrupted by technological advances in the industrial countries which resulted in the production of synthetic counterparts to natural products, for example, synthetic fibres, rubbers, resins, and dyestuffs.

Producers of tropical food products and agricultural raw materials, in adjusting to world market fluctuations for such products, may allocate their scarce arable land resources towards the production of crops and livestock in regions which may be unsuitable in terms of climate and/or soil, to such crops and livestock. Thus the rainforest regions of the tropics are unfavourable to livestock production, owing to the absence of natural grassland, the high incidence of animal diseases and parasites, and the likelihood for the livestock to be dwarfed. Such areas are also environmentally unfavourable to the cultivation of shallow-rooted annuals because of the risks of soil compaction and erosion arising from the turning over the soil after each harvest, the invasion of weeds, and the tendency towards luxuriant leaf growth. Perennials are ecologically more suitable for the hot, wet tropics because they are in the ground every day of the year to take advantage of the conditions for continuous growth. Not surprisingly, the perennials are the highest yielder of calories per unit area, with sugarcane, oil palm, and banana leading the group (Masefield 1970). But perennials suffer from both short-run and long-run price inelasticity of supply. Moreover, the long-term commitment involved in perennials results in land resource immobility, so that once a decision is taken to plant cocoa, coconut, rubber, or other perennials there is little prospect for changes in the use of the land and other fixed capital during the economic life-span of the crops (Wharton 1963).

The need to intensify the rate of development in the developing tropical countries may result in the ill-conceived allocation of scarce resources for short-term apparent gains, when the goal should be to develop sustainable productive systems whereby certain areas of land and water are assigned to specific long-term uses based on ecological realities, including the effects of climate on the productivity of land (Tosi and Voertman 1964).

The adverse impacts of climate on crops, animals, and man in the tropics should not be regarded as being permanent and unalterable. It is possible to a small extent to alter the local climate by air-conditioning just as it is to alter the wintry climate of temperate countries by heating. Similarly, the other adverse effects of the tropical climate can be avoided or eliminated by planned policies, except that to overcome them would require large inputs of capital and trained manpower, both of which are scarce in the developing countries.

However, as Kamarck (1976:12) remarks, as a tropical country develops and becomes richer

> . . . the adverse effects of climate become less economically important. As individual incomes rise, according to Engel's law, a smaller and smaller proportion of income is spent on food. The proportion of GNP produced in agriculture will therefore tend to drop and, consequently, climatic factors affecting agriculture will become less important to national economic development. The relative importance in the development process of possessing natural resources varies inversely with the level of development. As GNP rises, the possession of natural resources becomes progressively less important.

Finally, growth in GNP enables greater expenditure on health problems, attenuating this impact of climate.

Exhaustible Natural Resources

The exhaustible natural resources include mineral ores and fossil fuels, the occurrence of which is the outcome of geological accident. Their distribution will necessarily be uneven, with some countries favoured over others, and some regions within countries better endowed with these resources than others. A point to bear in mind is that the location and quantities of mineral and fossil fuel resources are not fully known, and additional outlays must be incurred by a country or a company to find out where a resource is, how much of it is there, and what characteristics it has. From such geological exploration and delineation of underground deposits, those that are regarded as economically exploitable at current prices and available technology constitute the proven reserves. Those that are not proven reserves may be classified, according to the degree of certainty of existence and cost of recovery, into conditional, hypothetical, and speculative resources, the last two categories being undiscovered resources.

Estimates of the fossil fuel and mineral resources of the developing countries must take into account the fact that reserves are a function of exploration. Since large areas in these countries have not been explored for minerals (the ratio of reserves to land is low — see Table 1.1), it is likely that the reserves position shown in Table 1.1 would be revised upwards with

TABLE 1.1 MINERAL RESERVES OF THE WORLD

Region	Land Area (percentage)	Estimated Mineral Reserves (percentage of world total in volume)	Reserves: Area
Developed market economies	26	35	1.35:1
Less developed market economies	49	38	0.77:1
Centrally planned economies	25	27	1.05:1
Total world	100	100	

Source: Based on Bosson and Varon 1977:60.

future exploration. Most of the mineral reserves of the world are located in the countries in which these minerals are being produced. The discovery of large reserves serves to attract development capital, and the establishment of production and infrastructural facilities would in turn act as an incentive to further exploration. The fact that, on an average, more than ten years will lapse between the start of exploration and commercial production implies that the existing geographic patterns of reserves and production for many minerals will change slowly.

The developing countries' share of the world's reserves of non-fuel minerals is also shown in Table 1.1. These estimates must be viewed with some caution, since they pertain to the position in the early 1970s and because reserves estimates of minerals vary as a result of measurement and conceptual differences. Figure 1.3 indicates that the developing countries, with 49 per cent of the world's land area, have half or more than half of the world's reserves in respect of only six minerals: tin, sulphur, aluminium, columbium, phosphate rock, and copper. They possess significant (20 to 48 per cent) reserves of seven other minerals: nickel, manganese, iron ore, silver, titanium, diatomite chromium, and lead.

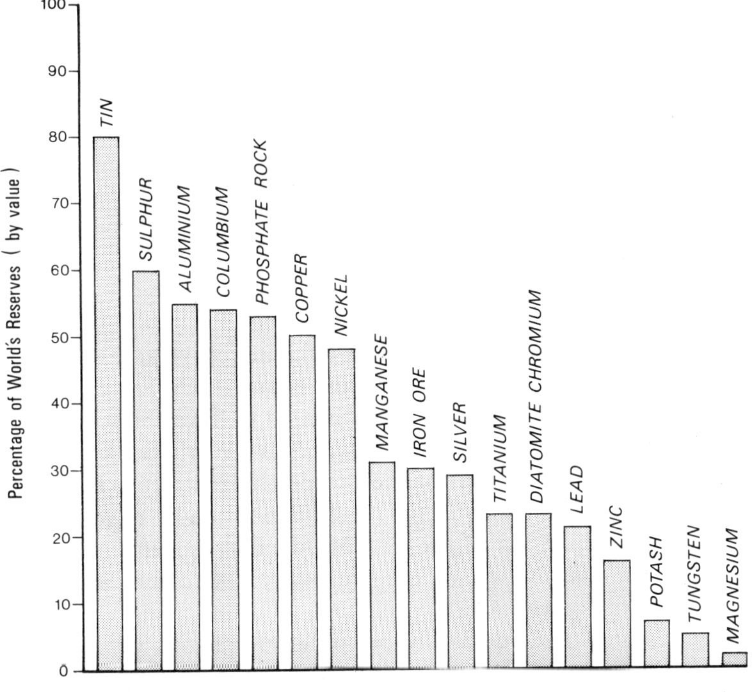

Fig. 1.3 Mineral reserves in the developing countries
(Source: data from Bosson & Varon, 1977, Table G.1)

The outlook for petroleum (oil and natural gas) in the oil-importing developing countries is poor in the short term in that the proven reserves represent only about 2 per cent of the world's total reserves (World Bank 1981). For the developing countries as a whole, apart from the members of the Organization of Petroleum Exporting Countries, the proved reserves were 9 per cent of the world's reserves for oil and 7 per cent for gas (*Oil and Gas Journal*, Dec. 1979).

A 1978 survey of the petroleum prospects of 70 developing countries has shown that 23 of them have a "high" to "very high" potential (Table 1.2).

TABLE 1.2 PETROLEUM PROSPECTS OF 70 DEVELOPING COUNTRIES

| | | Size of Potential Resources | | | |
| | | Low | Fair | High | Very High |
Type of Country	No. of Countries	>100 m/b	100–750 m/b	750–1,500 m/b	<1,500 m/b
Oil producer/net importer	12	1	2	3	6
Non-producer/known reserves	10	1	3	2	4
Non-producer/no discoveries	45	30	10	4	1
Non-OPEC producer/ exporter	3	0	0	1	2
Total	70	32	15	10	13

Source: World Bank 1979a.

An estimated 20 per cent of the world's ultimate recoverable resources of oil and gas are in the non-OPEC developing countries (Nehring 1978). The oil-importing developing countries are estimated to have about 15 per cent of the ultimate recoverable resources of oil and gas (World Bank 1979a).

About 34 million km² or half of the prospective petroleum areas of the world are in the developing countries. The distribution by region is: Latin America 36 per cent, Africa 28 per cent, Middle East 21 per cent, and Asia 15 per cent. The distribution of the prospective petroleum areas in the tropics is shown in Figure 1.4A.

Large sections of the potentially petroliferous regions of the tropical world are either unexplored or only partially explored (Fig. 1.4B). The 1978 World Bank survey indicated that only 7 of the 23 developing countries with high or very high petroleum prospects had been "adequately" explored.

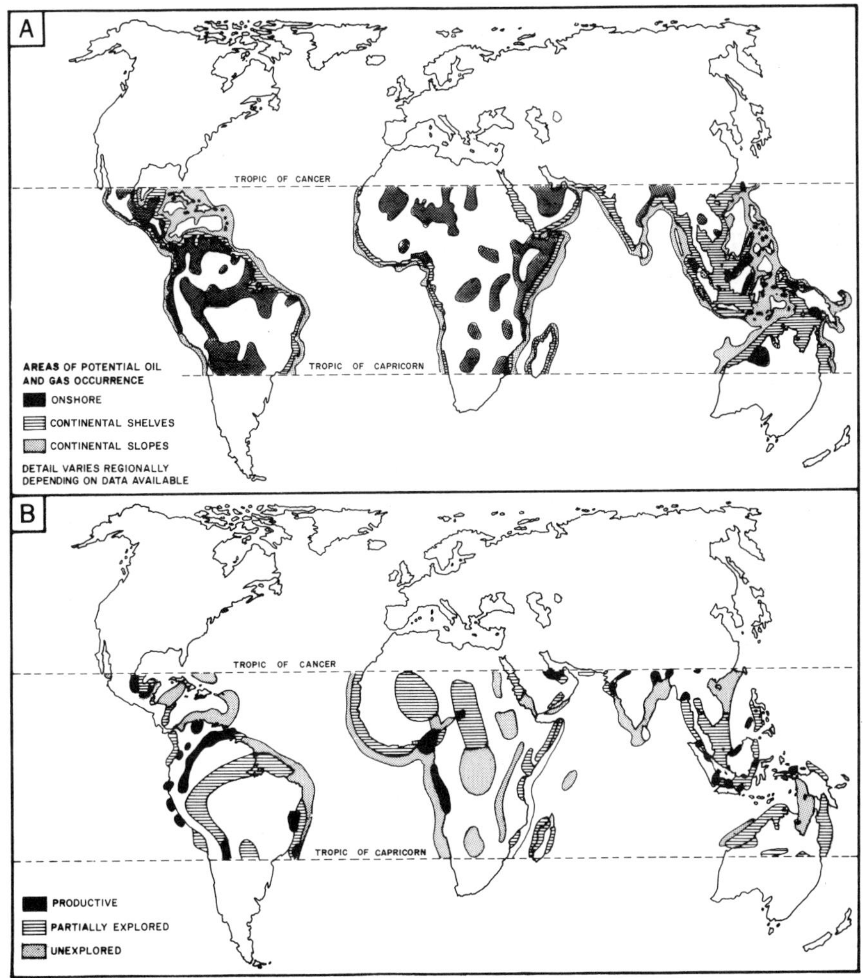

Fig. 1.4 Oil and gas areas in the tropics
A. Areas of potential occurrence (Source: Odell, 1981)
B. Productive areas, and exploration status (Source: World Bank, 1979)

Thirty-eight countries were inadequately explored, and 22 countries were moderately explored. One measure of the intensity of exploration is the drilling density, in terms of the number of wells drilled per thousand square miles (2,590 km²). The drilling density in the developing countries was only one per cent of the U.S. drilling density (Africa and Middle East 0.1 per cent, Asia 0.7 per cent, Latin America 2 per cent) (OPEC Secretariat 1980). The drilling density in the oil-importing developing countries amounted to only about 5 per cent of the world total (World Bank 1979b).

The problems of petroleum exploration in the developing countries range from those posed by the physical and geological conditions to those of technology, capital, and the uncertain political climate prevailing in these countries. The 1978 World Bank survey noted that 54 of the 70 developing countries required assistance in updating existing data from geological and geophysical surveys and/or in implementing new surveys. Since each project could cost as much as $5 million (1978 figures) in a country where no exploration work has been carried out, and more than one project may be needed to cover all potential areas, the costs involved in the first stage of exploration are quite substantial.

Aside from petroleum exploration, exploration for minerals in the developing tropical countries is hindered by a number of problems. First, in the hot, wet tropics the formulation of a thick mantle of soil and weathered material on which a luxuriant rainforest develops makes for difficulty of movement and difficulty in locating and assessing the mineral potential of the underlying rocks. Rock formations in the temperate lands and semi-arid and arid zones of the tropics are, in contrast, well exposed. Furthermore, the high rainfall conditions of the hot, wet tropics pose special problems in locating the more easily soluble minerals such as potassium and sodium salts, dolomite, limestone, and gypsum. As against this, the tropical countries derive a comparative advantage from the surface concentrations of such minerals as alluvial tin, bauxite, some iron ores, manganese, and nickel, derived from the intensity of the weathering processes in these countries (Singh 1980).

Second, the geophysical and geochemical techniques of mineral exploration were largely developed for and applied to the non-tropical developed parts of the world. Instruments, for example, may have to be modified for use under conditions of high heat and humidity. Even the interpretation of data may require a different frame of reference in the tropics because their physical and chemical parameters are different (see U.S. Agency for International Development 1972; Karmarck 1976).

While developing countries of some size will have resource endowments which include mineral resources (and, for the more fortunate few, petroleum resources), there are some where the mineral sector is dominant in the economy. The World Bank has identified a group of 28 mineral economies, of which 15 are petroleum exporters and 13 mineral exporters, in which the mineral sector contributed 10 per cent or more to the gross domestic product and/or 40 per cent to the export earnings (Table 1.3).

The need in these mineral economies is to maximize the returns from the exploitation of their mineral resource endowment and to use such returns to earn foreign exchange, and to set in motion the development process; in particular, to make the transition from a mineral dependent to a diversified modern economy.[3] As the resources are finite and exhaustible, the returns should, ideally, be reinvested in some form of reproducible capital in order to maintain earnings on the capitalized value of the resources. On exhaustion of its own resources, the country will have to depend on external sources to satisfy its demand for such resources.

After the Industrial Revolution, the industrializing countries had looked for minerals mainly within their own borders, although many of them had exploited the opportunities available to them in their colonies abroad, which today are the independent developing countries. Mineral exploitation then and now involved the participation of large, vertically integrated mining companies. Up to the Second World War, the mining companies concentrated their efforts in the developed countries. But after the war the gradual depletion of high-grade ores as well as the sharp rise in demand for minerals in these countries stimulated a move towards the exploitation of the mineral resources of the developing countries. Infrastructural developments, particularly in transport, facilitated the move. High-grade ores and lower labour costs resulted in above-average profits to the multinational mining companies. As a result, mineral production in the Third World (developing) countries grew at an average rate of 7 per cent a year during 1948–70, as against only 2.5 per cent in the developed countries (Bairoch 1975:50, 53; Tanzer 1980).

But the nationalist movements in the developing countries in the 1960s have caused an erosion in the profits of the multinational mining companies as the host countries sought to increase their share of the gains from mineral exploitation through nationalization (as has happened in Algeria, Bolivia, Burma, Chile, Ghana, Guyana, Peru, Zaire, and Zambia in the recent past), increasing the tax burdens of the companies, or through the formation of producer associations such as OPEC.

Fears about security of tenure in many of the developing countries, weak markets for many mineral commodities, rising costs of exploration and mine development, and physical isolation from major markets are among

---

[3]In his paper on Caribbean bauxite, Auty (Chapter 2) analyses the factors that have caused a significant delay in the establishment of forward linkages, via Jamaican refining and Guinanese refining and smelting, in the aluminium production chain in the region. Such forward linkages would have increased the proportion of value added within the bauxite producing countries and thereby increased the returns from the exploitation of their mineral resource.

TABLE 1.3 THE MINERAL ECONOMIES

Country	Major Mineral	Share of Mining[1] in GDP 1967–75 (%)	Share of Mineral Exports[2] in Total Merchandise Exports (%) 1974–76
Non-Fuel Mineral Economies			
1. Bolivia	Tin	11.9	73.6
2. Chile	Copper	8.5	65.8
3. Guinea	Bauxite	n.a.	70.2
4. Guyana	Bauxite	16.5	26.0
5. Jamaica	Bauxite	11.0	18.4
6. Liberia	Iron Ore	31.9	70.1
7. Mauritania	Iron Ore	27.9	88.6
8. Morocco	Phosphate	7.5	55.5
9. Peru	Copper	5.6	36.7
10. Sierra Leone	Iron Ore	17.0	10.9
11. Togo	Phosphate	9.8	66.3
12. Zaire	Copper	18.5	66.6
13. Zambia	Copper	32.0	96.5
Petroleum Economies			
1. Algeria		22.4	88.9
2. Angola		11.0	43.0
3. Congo People's Republic		9.5	66.0
4. Ecuador		11.4	56.1
5. Gabon		34.5	86.3
6. Indonesia		14.0	69.2
7. Iran		34.2	85.3
8. Iraq		45.7	97.6
9. Kuwait		69.2	80.2
10. Libya		55.5	93.4
11. Nigeria		24.9	93.3
12. Saudi Arabia		70.5	91.0
13. Syrian, A.R.		9.0	66.2
14. Trinidad and Tobago		23.9	31.0
15. Venezuela		25.0	63.7

Notes: [1]Includes quarrying.

[2]Only includes ores and concentrates, unwrought and unrefined (except copper which may be refined but is unwrought).

Source: World Bank, 1979b.

some of the factors that have brought about a recent shift, on the part of the mining companies, away from these countries to "safer" areas. In the past two decades, about 80 per cent of world non-oil mineral exploration expenditures had been channelled to five developed countries: Australia, Canada, Ireland, South Africa, and the United States (U.N. Secretariat 1981).

However, minerals will continue to play a crucial role in the development efforts of many developing countries, especially those with mineral economies. The complexities arising from the physical nature of mineral resources (heterogeneity, non-renewability, uncertain occurrence, often in underground locations), the gains that can be realized through economies of scale, the large infrastructure requirements, and the capital-intensiveness of modern mining and processing methods often provide the host countries with little choice but to depend on foreign mining companies to develop their resources. Here there is often a conflict of interest between the host country, which seeks to maximize its share of revenues, and the mining company, which is concerned with corporate profits.[4] The performance record of the mining companies has been less than exemplary. As Bosson and Varon (1977:11) have stated:

> The international companies pattern of project selection, investment, location of processing facilities, and revenue-sharing arrangements — while defensible on narrow microeconomic and political grounds — has not always been consistent with the development needs, legitimate interests, aspirations, and sensitivities of the developing countries. The leverage of developing countries over the mining industry has been kept low by the industry's history of guarding its knowledge of reserves and resources, technology and markets, overstressing its complexity to prevent easy intervention. In a number of instances the multinationals have been less than candid with regard to the results of exploration, the parameters of investment and expansion decisions, processing and marketing properties of the product, technological and economic trends, and market value. Host countries have thus been deprived of knowledge about their own resources and of the basic tools with which to evaluate decisions deeply affecting their self-interest. In a number of areas the multinationals have themselves been the major source of the information needed for independent evaluation of their performance, fostering suspicion and placing them at odds with host countries and with legislators in their own countries.

This performance record is one of the reasons why some have advocated leaving the mineral resources in the ground. Two other reasons are also cited in support of this policy: first, that the resources would be more valuable in the future and, second, that under the economic and political systems prevailing in most developing countries today the revenues obtained

---

[4]The issues and problems associated with such a conflict of interest are discussed by Jackson (Chapter 3) in his paper on the exploitation of mineral resources in Papua New Guinea.

from minerals are largely wasted, and would be more useful if left unexploited until some future date (Tanzer 1980). Others, such as Kay and Minnlees (1975) believe that future generations would be better served if they are left production equipment rather than minerals in the ground.

Those developing countries which choose to exploit their mineral resources now rather than in the future would have to face a complexity of issues concerned with (1) the depletion rate of the minerals, (2) their economic relationships with the foreign mining companies (for those who do not or cannot mine the resources themselves), and (3) producer cooperation in the mining, marketing, and processing of the minerals.

The economics of mineral resource depletion have been treated exhaustively in the publications of Hotelling (1931), Herfindahl (1967), Solow (1974), Pearce and Rose (1975), among others. Briefly, if the present and future prices of the resource and the marginal costs of extraction are known, the resource should be extracted or exploited at a rate such that the resource price, net of costs, increases at the same rate as the prevailing rate of interest. A country should therefore extract its minerals and re-invest the revenues so earned into, say, factories, only when the returns from the factories are higher than the appreciation of the minerals in the ground.

For the developing countries with mineral economies (including oil and gas), their stocks of minerals are the main sources of development capital. In theory, their depletion policy should be based on the continuous exchange of their (wasting) mineral resources for real capital assets and investing these in areas which yield the highest rate of economic development. The time frame within which the mineral resources are depleted would necessarily vary with the circumstances of each country, especially the size of its mineral reserves and the size of its population. The objective should be that when the mineral resources are finally depleted the country would have attained the best possible level of economic development (for a discussion of this topic see *Review of Economic Studies* 1974; Al-Janabi 1979; Motamen 1979; Saad 1979–80).

In the real world however, many uncertainties prevail. There is, first, the geological uncertainty, whereby inventories of minerals by quantity, quality, and location are subject to often wide margins of estimating error. Second, there is uncertainty with respect to the trend of future prices. Third, there is the uncertainty regarding technological advances which, in the past, have lowered costs and have made it possible to exploit lower grade resources at costs comparable with past costs for higher grades. Fourth is the uncertainty pertaining to the substitution of one category of resources for another. In the past there was a shift from renewable to nonrenewable resources. For example, firewood was replaced by coal and later by oil and natural gas, natural manures by chemical fertilizers, and wood by plastics. In the case of minerals, there is the substitution of aluminium and stronger steels for copper and of light metals for stainless steel.

Such uncertainty, coupled with the urgency of the need to convert the mineral resources from their physical state into financial resources for development purposes, among other reasons, have worked towards most of the developing countries with mineral economies depleting their minerals resources at maximum rates, except when production cutbacks were necessary to maintain prices (World Bank 1979a).

The second issue connected with the exploitation of mineral resources in developing countries is that of the role of the multinational mining companies. There are some such as Bosson and Varon (1977) who hold that as the companies control the capital, technology, and markets, the developing countries would have to depend on them for resource exploitation. Tanzer (1980), on the other hand, argues that the profit rates the mining companies seek are much higher than the real cost of capital, and that by allowing them to exploit its minerals, a country would forfeit a crucial opportunity of accumulating its own capital from potential profits. Tanzer also cites other reasons, such as possible intervention in the affairs of a country, for minimum reliance on the multinationals. One way to doing this, in so far as technology is concerned, is for the country to pay the going market rate for it from specialized companies. It is possible, for example, to engage the services of such companies to conduct oil exploration for a flat fee (Tanzer 1978).

The alternative to dependence on the multinational mining companies is for the developing country to run its own mines and, by so doing, accrue to itself all the scarcity, differential, monopolistic, and windfall rents associated with mineral resource development. An early example of a state-run mining enterprise is PEMEX, the state oil company of Mexico, formed in 1938 when the Mexican government nationalized the oil industry. PEMEX discovered 7 billion barrels of oil through its exploration efforts conducted between 1938 and 1966, and another 14 billion barrels in more recent years. Similar steps towards state control in oil activities have been taken by India and Vietnam (Tanzer 1978).

There is a noticeable tendency among the developing countries to exercise sovereignty over their exhaustible mineral and fossil fuel resources at all stages of their exploration, exploitation, and marketing. The degree of state control in mining varies: Zambia and the Congo People's Republic own more than 90 per cent of their non-fuel mines; Bolivia, Chile, and Indonesia about 50 per cent; and Jamaica about 10 per cent (Gillis 1978). But for various reasons — lack of managerial and technical inputs, the very demanding nature of mining activity, high entry barriers in marketing, to cite a few — multinational mining companies are likely to continue to play a significant role in the mining activities of the developing countries (World Bank 1979c).

The third set of issues in the conversion of mineral resources from their physical state to a flow of financial resources concerns increasing the revenues through producer alliances or cartels such as OPEC. A similar organization of copper producers (CIPEC) was formed in 1967, comprising the four major copper-producing developing countries — Chile, Zaire, Peru, and Zambia. In March 1974, seven major bauxite producers — Australia, Guinea, Guyana, Jamaica, Sierra Leone, Surinam, and Yugoslavia — grouped themselves together to form the International Bauxite Association (IBA). The trend continued in 1975 with the formation of the Association of Iron-Ore Exporting Countries made up of Algeria, Australia, Chile, India, Mauritania, Peru, and Venezuela, and the International Association of Mercury Producers, comprising Algeria, Italy, Mexico, Spain, and Turkey.

These producer alliances attempt to gain more equitable returns from the exploitation of their scarce natural resources, and to increase their bargaining power over the industrialized consuming countries which exert dominance over capital, technology, processing, marketing, and shipping. Several conditions must be met if these alliances are to succeed in attaining their objectives. First, a few countries should control as large a proportion of the world production of the mineral as possible. It would be an added advantage if they also possess a high percentage of the reserves and resources, and have a large annual-production-to-reserves ratio. Second, the supply of the mineral from non-alliance sources should be high cost and inelastic. Third, the economies of the members should not be tied too closely to those of the consuming nations. Fourth, members should be politically compatible and share common interests.

Factors which militate against the success of producer alliances are substitution and recycling possibilities, stockpiles in consumer countries (an example is the tin stockpile in the United States), the possession of small reserves and resources among the producer countries, and in the longer term the changes in the supply situation that will result from mining the mineral resources of the oceans (Krasner 1974; Varon and Takenchi 1974; Tanzer 1980).

An analysis of the gains which producer alliances have obtained from cartelization in respect of oil (OPEC), bauxite (IBA), and copper (CIPEC) has shown that relative gains were made mainly in the first five years after cartelization and that such gains were highest for bauxite (60 to 500 per cent over the competitive solution), next highest for oil (50 to 90 per cent), and lowest for copper (8 to 30 per cent) (Pindyck 1978). Cartelization and price increases will stimulate the development of substitutes which may, however, require heavy initial investments and start-up costs. The rise in oil prices has brought some substitutes such as tar sands on the verge of profitability. Other oil-substitution possibilities are oil shales, coal, nuclear fission, and, perhaps in the future, nuclear fusion. It would appear that, as the gains from cartelization are likely to be short-run (Robinson 1975), such gains must be large enough to permit diversification of the producer's economy and to increase its flexibility, more so in the case of the mineral economies of the developing world.

The debate about cartelization is summed up by the World Bank (1979c:95).

> . . . given some market dominance, the potential gains from cartelization, if any, require detailed knowledge of the industry and of the market. Thus although it is difficult to predict what other minerals may successfully join the ranks of oil and bauxite, it is clear that a prerequisite of such action is the intensive study of the mining industry and mineral markets by the producer countries. The objectives of such a study would be to determine the price levels that would stay clear of "limit prices" at which substitutes become attractive, and to reveal the potential gains from cartelization as a means of encouraging group cohesion. Until such knowledge becomes available to the producer countries, the exaction of monopolistic rents must remain a highly uncertain proposition.

Renewable Natural Resources

Farm land, soils, forests, fisheries, air, surface water, and solar, wind, and tidal energy are usually regarded as renewable natural resources. Natural resources as stocks may be augmented by discovery or reduced by natural processes such as death, decay, or predation. In the case of non-renewable resources stock reduction would be through extraction. The objective in the management of renewable resources should be sustained yield production, whereby the resources continue to provide an optimum flow of services over an indefinite length of time.

Some resource stocks, as parts of larger natural resource systems, are renewable by natural processes. Others may be renewable by human-assisted processes. The renewability of resources such as farm lands and soils, forests, reserves of ground water, and fisheries often depends on non-destructive methods of use and management. Farm lands and soils can be destroyed by the "mining" of soils, and the history of agricultural land use in the tropics is replete with examples of such mismanagement. Forests can be cut down without replacement, and fish stocks can be depleted through overfishing and uncontrolled exploitation.

There was a time when, because of low population numbers, the farm lands, forests, fisheries, and other renewable resources of the tropical world were considered inexhaustible. Sparse settlement was accompanied by attitudes to resource exploitation in which the concept of sustained yield held little attraction. Instead, the productivity of the resources was seriously impaired by profligate practices which, in some cases, made large areas barren. The situation was not unique to the tropical world; it also prevailed in other countries such as the United States, Australia, and the Soviet Union during their early history when resources were abundant in relation to population numbers (Ackerman 1965).

But the population/resource situation has changed significantly since then and, as Table 1.4 shows, present-day rates of population increase in most of the tropical countries are very rapid, and will result in a doubling of the population within an average of about twenty-eight years. Furthermore, higher levels of expectations would mean even more rapid increases in

TABLE 1.4 POPULATION NUMBERS AND MAN/LAND RATIOS IN TROPICAL
            COUNTRIES

	Estimated Population (millions) Mid-1981	Rate of Natural Increase (annual, %) Mid-1981	No. of Years to Double Population, Mid-1981	Persons per km² of Arable and Pastureland Mid-1981
LOW INCOME COUNTRIES GNP per capita, 1979 (listed in increasing order of GNP per capita)				
Kampuchea, Dem. Rep.	5.5	1.8	38	152
Lao, PDR	3.6	2.4	29	217
Bangladesh	92.8	2.6	27	958
Chad	4.6	2.3	30	10
Ethopia	33.5	2.5	28	43
Somalia	3.8	2.8	25	13
Mali	6.8	2.8	24	21
Burma	35.2	2.4	29	340
Guinea Bissau	0.8	1.8	39	52
Vietnam	54.9	2.8	25	512
Burundi	4.2	2.7	25	245
Upper Volta	7.1	2.6	27	37
India	688.6	2.1	33	381
Malawi	6.2	3.2	22	150
Rwanda	5.3	3.0	23	361
Sri Lanka	15.3	2.2	32	592
Benin	3.8	3.0	23	372
Mozambique	10.7	2.6	27	23
Sierra Leone	3.6	2.6	26	129
Tanzania	19.2	3.0	23	38
Haiti	6.0	2.6	26	425
Zaire	30.1	2.8	25	97
Gambia	0.6	2.5	28	105
Niger	5.7	2.9	24	46
Guinea	5.1	2.5	27	72
Central Africa Rep.	2.4	2.2	31	40
Madagascar	8.8	2.6	27	24
Uganda	14.1	3.0	23	132

TABLE 1.4 *(cont.)*

	Estimated Population (millions) Mid-1981	Rate of Natural Increase (annual, %) Mid-1981	No. of Years to Double Population, Mid-1981	Persons per km² of Arable and Pastureland Mid-1981
Mauritania	1.7	2.8	25	4
Togo	2.6	3.0	23	162
Indonesia	148.8	2.0	35	524
Sudan	19.6	3.1	22	62
MIDDLE INCOME COUNTRIES				
Kenya	16.5	3.9	18	273
Ghana	12.0	3.1	22	90
Yemen Arab Rep.	5.4	2.3	30	63
Senegal	5.8	2.6	27	72
Angola	6.7	2.4	28	22
Zimbabwe	7.6	3.4	21	104
Yemen, PDR	2.0	2.7	26	21
Liberia	1.9	3.3	21	315
Zambia	6.0	3.2	22	17
Honduras	3.9	3.5	20	104
Bolivia	5.5	2.5	28	18
Cameroon	8.7	2.3	30	55
Guyana	0.8	2.1	33	61
Thailand	48.6	2.0	35	273
Philippines	48.9	2.4	29	538
Congo, People's Rep.	1.6	2.6	27	11
Nicaragua	2.5	3.4	20	52
Papua New Guinea	3.3	2.8	25	700
El Salvador	4.9	3.2	22	383
Nigeria	79.7	3.2	22	178
Botswana	0.8	3.4	21	2
Peru	18.1	2.7	26	59
Columbia	27.8	2.1	33	120
Guatemala	7.5	3.1	22	282
Ivory Coast	8.5	3.1	22	72
Ecuador	8.2	3.1	22	159
Paraguay	3.3	2.6	26	20
Namibia	1.0	2.8	24	2
Malaysia	14.3	2.3	30	220
Panama	1.9	2.2	31	109
Mexico	69.3	2.5	28	71
Chile	11.2	1.5	47	63
Brazil	121.4	2.4	29	58
Costa Rica	2.3	2.8	25	113
Suriname	0.4	2.3	30	753
Venezuela	15.5	3.0	23	70
Gabon	0.7	1.1	62	13
Hong Kong	5.0	1.2	60	62,675
Singapore	2.4	1.2	58	30,250

Sources: World Bank, 1981
            Population Reference Bureau, 1981

demand for food, energy supplies, fresh water, and other material goods and services. The combination of more people with larger requirements per person would clearly place a great strain on available resources.[5]

In 1981 nearly three-quarters of the labour force in the low-income tropical countries and half of the labour force in the middle-income tropical countries were engaged in agriculture (World Bank 1981). If livestock rearing, forestry, and fishing are included, the percentage of the population dependent directly on the renewable natural resources of soil and arable land, pasture land, water, forests, and fisheries would be even higher.

In such a situation, the more a country has of these resources, the more varied the resources are, and the higher their quality, the better off it will be, other things being equal. The probability of a country possessing large, varied, and high quality natural resources increases with its territorial size. The probability of variety in natural resource endowments increases further if, in addition to size, a country also straddles several climatic zones.

Countries such as Chile, Mexico, Columbia, Kenya, and even small countries such as Taiwan, Sri Lanka, and Haiti have differentiated climates because of either latitudinal extent or altitudinal variations, and have sizeable areas or at least pockets of well-watered arable land, forest, and other renewable resources. In contrast, the countries bordering the Sahara such as Mauritania, Mali, Chad, Niger, and Sudan, despite their size, are short of water and forests as they are in a single, climatically difficult zone.

A primary concern among countries of the tropical world today is whether, given a country's resource endowment, it is able to increase its output of goods and services to a level which will provide it with a socially desirable standard of living. As seen earlier, the natural resource endowment is only one of many determinants of the per capita level of goods and services in a country. Possession of a poor natural resource endowment need not exert a negative influence on a country's economic progress if it can have access to natural resources through trade. In the developing countries of the tropics where two of the other important determinants of the per capita level of goods and services — capital and technology — are in markedly short supply, and where a very high percentage of the economically active population is engaged in primary production based on agriculture, pastoralism, forestry, and fishing, the natural resource endowment is evidently of greater importance to their economic health than in the industrial countries where only a small percentage of the working population is in primary production and where there is a concentration of capital and technological skills, among other factors.

---

[5]These and other issues are covered in detail by Faniran (Chapter 15) in his wide-ranging paper on population change and natural resources in Africa south of the Sahara.

The renewable natural resources are part of larger complex ecosystems which are interlinked to form the natural environment. Within an ecosystem the availability of resources and the manner they are utilized will influence its carrying capacity, defined here as the level of human activity that can be sustained in a region at acceptable "quality of the life" levels (House and Williams 1977).

The manner and rate of use of the renewable natural resources in a region should be such as to ensure the maintenance of the long-term productivity of the resources. In sum, the flow of goods and services derived from the resources should be so regulated as to preserve the ecological integrity and dynamic equilibrium of the area and allow for the continued flow of such goods and services.

In any natural environment there is a balance between the rates in which it can provide resources and the rates in which it can assimilate the resultant waste products and residuals. Beyond this, the environment will be under stress, as happens when man increases his pressure on the natural resources of the environment. Exceeding the stress limits and finite assimilative capacity of the environment leads to conditions of instability and degradation, one manifestation of which is pollution, defined as "too much substance or matter in an unintended place". Prolonged and increasing stress will lead to a final breakdown.

The variety of technological systems extant in the world today is very great, ranging from simple food-collecting and hunting to the highly complex technology of the industrial nations. In the tropical world the technical systems are predominantly agricultural and pastoral, with hunting and gathering in localized areas. Some, as seen earlier, are centred on mineral exploitation. Very few are evolving towards urban/industrial complexity.

Regardless of the degree of complexity, every technology is based on a knowledge of the natural resources which are exploited through the working of that technology. In a society bound by tradition, as many societies in the tropics are, the same procedures are followed from year to year, extending through many generations. Interest in natural resources tends to be directed only towards those on which they depend, and exploitation techniques will be set along familiar patterns.

In the pre-colonial era, the traditional societies were closely surrounded by barriers, cultural as well as economic, and have thus been described as "closed" societies. Their coherence was derived from their social unity. Each man had a fixed place in life around which his culture was integrated and his wants satisfied. The pattern of life was a pre-established order, and the aim of life was the maintenance of that order. Production, based on the utilization of locally available natural resources, was for subsistence, and there was little incentive to produce regular surpluses over and above the limited needs of the community. There was little functional specialization.

Population was kept in check by Malthusian processes, and in a situation of abundant natural resources in relation to population numbers, the carrying capacity of these resources was seldom exceeded. Such strain on resources and ecological disruption as occurred with, say, shifting cultivation was temporary. Migration from the area of cultivation would relieve the strain and allow nature to restore the ecological equilibrium. In areas of sedentary agriculture in the Asian tropics, wet padi cultivation allowed for high population densities. Relief from ecosystem stress was readily available through the extension of wet padi cultivation into undeveloped level land, or in some localized areas in Java and the Philippines, through the terracing of hill slopes.

Ackerman (1965) has expressed the relationship between natural resources, population and standard of living in the form of an equation:

$$PS = RQ(TAS_t) + E_s + T_r \pm F - W$$

where

P numbers of people
S standard of living
R amount of resources
Q factor for natural quality of resources
T physical technology factor
A administrative techniques factor
S_t resource stability factor
W frugality element (wastage or intensity of use)
F institution advantage and "friction" loss element consequent upon institutional characteristics of the society
E_s scale economies element (size of territory, etc.)
T_r resources added in trade.

Among the developing countries of the tropical world today, the population parameter has changed considerably in that the numbers of people have increased and are increasing at a rapid rate (Table 1.4). At the same time, with the revolution of rising expectations, these increased numbers seek improved standards of living. But the natural resources component of the equation remains fixed (unless augmented by international trade or territorial acquisition).

To accommodate the twin demands of intensified population densities and higher standards of living, the traditional patterns of resource use, originally geared towards self-sufficiency and low standards of living, would have to be modified to increase the flow of goods and services derived from them. But there is little reason to expect such modifications if the society concerned is ignorant, indifferent, or even antagonistic to them:

In other words, interest in developing natural resources comes only where there is dissatisfaction with current ways of supplying economic needs and an awareness or belief that improvements can be made. While the initiative for change may come from a relatively small percentage of the total population, it is not enough that they alone have this awareness or belief. They must be working in a social environment that accepts change and encourages the adoption of new processes and new products. The pace of discovery and exploitation will be conditioned by the level of technical knowledge and experience. Moreover, within a favorable social environment, advances in technology and experience form the most dynamic element in bringing about changes in the organization of economic life, including the utilization of natural resources (Fisher 1964:35).

At the same time, an awareness of the potential contributions of available resources has to be fostered, as well as an understanding of how these resources may be exploited, together with the acquisition of the appropriate techniques of exploiting them.

An approach to the use of the renewable natural resources of the tropics must be adopted in which not only economic but ecological considerations are taken into account. This is of course easier said than done. Some difficulties that come to mind are: first, resources such as amenity or recreation resources are difficult to quantify and are not easily subject to cost/benefit analysis. Second, common property resources such as water resources, fisheries, and forests pose problems such as the determination of the optimum rates of use in order to maximize returns and minimize waste, the monitoring of ecosystem changes that result from exploitation, and the establishment of controls on free, competitive access to them so as to prevent the extinction of the fish or forest stocks.

Further, natural scientists, ecologists, and geographers who are concerned with physical and biological changes arising from resource use do not usually view these in economic terms, and resource economists understandably seldom examine resource management problems in other than economic terms. As Herfindahl has said:

> The "conventional" economist who ventures to work on resource problems will sometimes err because he does not sufficiently understand the physical and biological relationships involved and because he is not accustomed to working with problems in which benefits and costs external to the decision-making unit occupy so prominent a place. The remedy is not to abandon conventional economic principles but rather to learn enough about the problem at hand to apply them usefully (Brooks 1974:6).

In point of fact, as resource management and policy issues include those connected with environmental quality, equity, and social stability, they will require inputs from many other non-economic disciplines such as law, sociology, engineering, hydrology, and the like (Howe 1979).

The following sections will review briefly some aspects of the development of the renewable natural resources in the tropics, and the consequences arising from the pressure on resources resulting from increases in the human and livestock populations in tropical countries.

Water resources. The development of a country's water resources is important from both the economic (as an input to the economic growth process) and the social (as part of basic needs — health, sanitation — package) points of view. Those sectors of the economies of the developing countries of the tropics most likely to be affected by water resources availability are agriculture, power, and transportation. The industrial demand for water is low, about the same order as the demand for domestic consumption or from 20 to 40 m^3 per person a year.

Water as an economic good and as a factor of production in the developing countries is most important in agriculture as this is the dominant sector in the economies of many countries. In comparison, water does not play as critical a role in power generation and for transportation as it does in agriculture because of the availability of substitutes, in the form of thermal power generation and other forms of transport.

The demand for water for agriculture in the developing tropical countries is met mainly from rainfall. More than 85 per cent of the cultivated land depends on rainfall (Ambroggi 1980). But rainfall is variable even in the humid tropics, so that rainfed agriculture is subject to uncertainties which in turn will affect yields. Controlling the supply of water through irrigation and combining it with the planting of improved plant varieties and the application of fertilizers and pesticides can raise crop yields quite significantly. Rice yields, for example, are substantially higher in countries where there is water control than where the crop depends on the rainfall (Fig. 1.5).

Apart from increasing yields, irrigation can provide other benefits to agriculture in the tropics. In semi-arid or arid lands, it allows for the cultivation of crops on land which otherwise could not be farmed at all. It also makes possible the cultivation of more than one crop a year on the same piece of land. But perhaps the greatest benefit it confers on the farmer is that it provides him with a secure supply of water for his crops.

But efforts to control the hydrologic cycle through the impounding of water, whether for irrigation, flood control, or power generation, also result in environmental problems.[6] In many tropical countries, such impound-

---

[6]The environmental impacts and consequences of water resources development in tropical countries are discussed by Barrow in his comprehensive paper (Chapter 13).

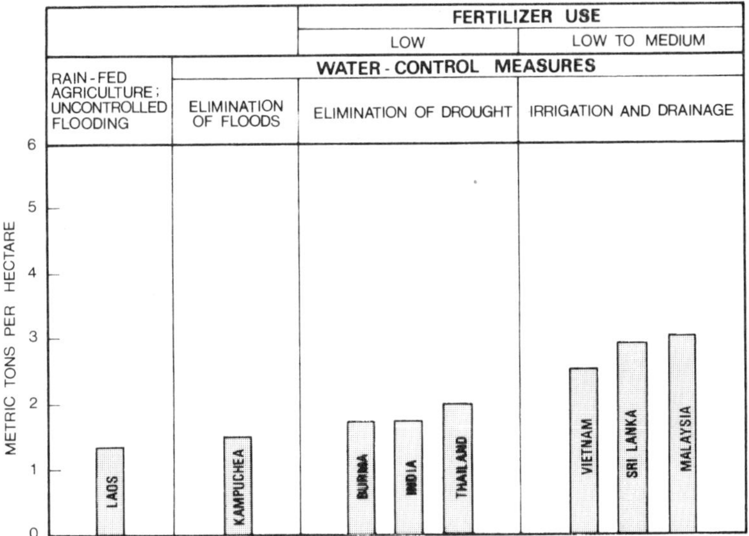

Fig. 1.5 Rice yields per hectare in seven tropical countries in Asia
(Source: after Ambroggi, 1980)

ments create ideal conditions for the multiplication of disease vectors, hosts, and parasites. Malaria and filariasis-transmitting mosquitoes breed in the standing water of irrigation canals. Human contact with the water can also lead to infestation with the blood flukes which cause schistosomiasis, a debilitating disease affecting about 200 million people in tropical Africa, South America, and Asia (Kinckley 1980). Another problem relates to the waterlogging of soils which results when irrigation water is applied to poorly drained flat land over a long period, thereby raising the water-table to root- and crop-killing levels. Furthermore, large areas of land can be lost to agriculture when the water evaporates and causes a build-up of salts.

Water is not only an economic good but also a social necessity. It has been estimated that about 30 m³ per person per annum of water for domestic consumption are necessary to sustain an acceptable quality of life; one cubic metre of this total is needed for drinking. The urban population in developed countries consumes 180 cubic metres per person per annum. In contrast, the consumption of the population in the developing countries is low; in an extreme case, among the rural population of southwest Malagasy Republic, it can be as low as under 2 cubic metres per person per annum (Ambroggi 1980). In the developing countries, only 29 per cent of the rural population and 75 per cent of the urban population have water supply services through house connections or standpipes. The inadequacy of water supply services in these countries was the main cause of the deaths in 1979 of about 13 million children under the age of five (Fano 1981).

Human health depends on the availability of water in sufficient quantity and quality. Many tropical countries face problems of providing enough water for drinking, cooking, and the maintenance of personal cleanliness (Fig. 1.6). In East Africa, for example, some of the population have as little as 2 litres of water a day, with much time spent by women and children in drawing water from distant sources (White *et al.* 1972).

Many countries, especially those in the arid zone, are looking to ground water as an important source of supply, supplementing surface water. It has been estimated that there are thirty times more fresh water in the ground than in all the world's lakes and rivers (Howe 1979). In arid countries ground water may be the only reliable source of water. In countries such as Bangladesh where the water table is high, ground water may be tapped for domestic consumption. About 70 per cent of the rural population of Bangladesh now obtains its drinking water from this source, which is generally free of disease pathogens if adequately protected.

In the fifteen years between 1962 and 1977, the United Nations has launched 52 ground water survey and exploration projects and 21 ground water development projects, mainly in the arid and semi-arid countries. Among the projects are those concerned with the development of ground water for urban and industrial use and for tourism in Senegal, Togo, El Salvador, Guatemala, Bahamas, and Surinam; to meet industrial, mining, and tourism needs in non-urban areas in Somalia, Upper Volta, and Northern Chile; and to meet rural and livestock needs in Benin, Cameroon, Mali, Mauritania, Somalia, Togo, Uganda, and Upper Volta (Taylor 1979).

Ground water is a renewable natural resource if the rate of withdrawal is less than the rate of recharge. Some aquifers such as those in some parts of the Sahara contain fossil water deposited in ancient times (Weistroffer *et al.* 1977). They receive no recharge now because of sparse rainfall, the nature of the soil structures, or because they lie below the zone of surface recharge. In such cases, the ground water is a finite non-renewable stock, and the principles of management should be the same as for any finite stock of non-renewables.

Given the average per capita consumption of water, the global water supply can in theory support a world population of between 20 and 25 billion (Ambroggi 1980). The actual distribution of this supply is however, geographically uneven and governed by physical processes over which man has so far not been able to control. In most parts of the world, tropical or non-tropical, the only way to obtain a reliable supply of water is by the active management of the water resources, whereby water is collected, stored, allocated, and distributed. Such intervention in the water cycle requires, among other factors, capital and technological skills, both of

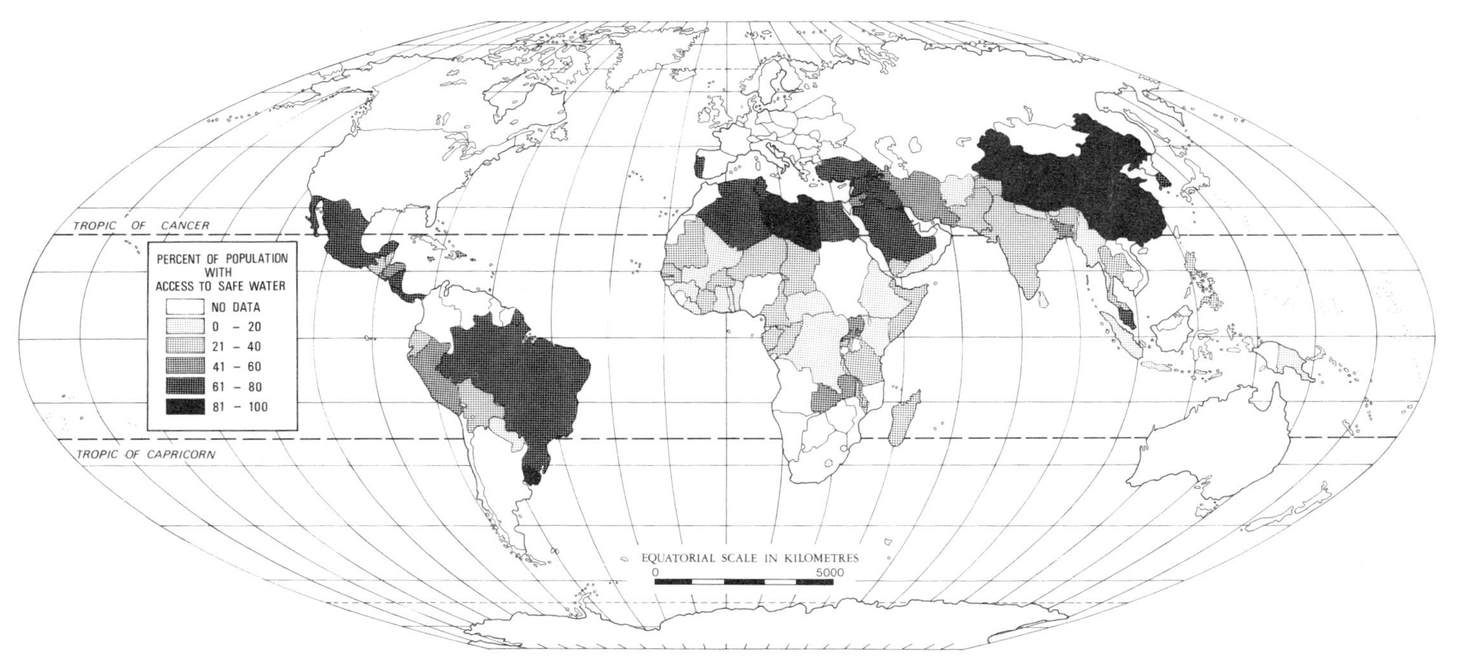

Fig. 1.6 Access to safe water
(Source: based on Ambroggi, 1980, pp. 44–45)

which are scarce supply in the developing countries. Structures such as dams and canals are expensive to construct and may also require lead-times as long as twenty years for large projects.

The high cost of water resource development and scarcity of capital funds make it all the more important that investment in such development should be made only after very careful analysis. Decisions on the balance between development for economic growth and development for social ends are made difficult by the fact that many of the benefits cannot be meaningfully quantified. In either case, the basic economic laws of diminishing returns would apply. Thus in areas of water scarcity in tropical, especially semi-arid or arid countries, the initial provision of water for, say, agriculture would lead to significant increases in output, with further inputs of water showing diminishing returns. Similarly, the returns in terms of health benefits are very high from the provisions of the first 25 to 40 litres of good quality water per person per day, but drop progressively with further inputs until 80 to 100 litres per day, beyond which there is very little incremental health benefit (Julius and Buky 1980).

Soil resources.   The soil is a perceived and important natural resource in all the tropical countries. It is the product of a whole range of factors, a mix of organic and mineral matter occurring naturally on the land surface. It is also the thinnest natural resource and one which is particularly vulnerable to alteration and destruction by man. From the resource perspective, it may be regarded as a stock of nutrients and organic matter that is utilized for the production of food and fibre. The rate of depletion of this stock depends on farming and soil conservation practices. In the case of shifting cultivation in the tropics, the stock of nutrients and organic matter is depleted within a very short period of farming operations. In contrast, in the case of intensive rice cultivation, the crop can be cultivated on flooded land for year after year without any perceptible drop in yields or in the soils' fertility (Chandler Jr. 1979).

To be considered as a renewable resource, therefore, the soil should maintain its fertility and structure. This is only possible when the land is used rationally, and patterns of land use are determined not solely by economic and social forces but by taking into consideration the constraints imposed by climate, vegetation, slope, and the underlying rock formations. Qualitative degradation of the resource manifests itself in lower productivity due to the loss of humus from the soil, displacement of nutrients, or leaching (Jung 1977).

The contribution of soil resources to the economic development of tropical countries depends on their potential to support crops. In assessing this potential, it is necessary to bear in mind certain characteristic features of tropical soils and soil-forming processes. First, the high temperatures and humidities in the hot, wet tropics cause weathering processes to be two to four times as rapid as those in temperate latitudes. Second, climates in the humid tropics have remained stable for several million years (Lockwood 1976). The prolonged exposure of soils in the tropics to intense weathering has resulted in old, climatically impoverished soils which cannot support crops on a sustained yield basis without skilled agricultural management. However, not all tropical soils are poor. Newly formed soils derived from volcanic material and humus-rich soils deposited along rivers can be quite fertile.

Soils cannot be considered in isolation but must be related to crops, local agricultural practices, and other factors such as water availability. Many tropical soils are not as infertile as generally believed, and are capable of greater productivity with proper management techniques. But agricultural development is hampered by the variability of tropical soils over small areas, and by the fact that, in many of the developing countries, not much is known about the nature and distribution of the local soils.

In many of these countries, the pressure of population on the soil resources may manifest itself in accelerated rates of soil erosion. Soil erosion is the physical detachment and transport of soil particles which result in a change in the place and/or form of the soil resource. Under stable, natural conditions the slow removal of soil is part of the normal geological process of land mass denudation. Soil erosion becomes a problem only when its rates are accelerated, usually through human interference causing a change in the conditions of the vegetative cover and the physical state of the ground surface. Table 1.5 shows the rates of soil erosion under different rainfall conditions and types of land use in four areas in tropical Africa. The high

TABLE 1.5 RATES OF SOIL EROSION IN PARTS OF TROPICAL AFRICA

Locality	Average Annual Rainfall (mm)	Slope (%)	Erosion (tons/hectare/year)		
			Forest	Crop	Bare Soil
Ouagadongou (Upper Volta)	850	0.5	0.1	0.6–0.8	10–20
Sefa (Senegal)	1,300	1.2	0.2	7.3	21.3
Bouake (Ivory Coast)	1,200	4.0	0.1	1.26	18–30
Abidjan (Ivory Coast)	2,100	7.0	0.03	0.1–90	108–170

Source: Based on Charreau, Table 5.5, in Greenland and Lal 1977.

rates of erosion when the soil is under crops, and even higher rates when it is denuded of its protective vegetation cover provide evidence of the vulnerability of tropical soils when exposed to the erosive forces of moving water and, in the drier parts of the tropics, of wind.

Soil erosion is a persistent problem which threatens the fragile economies of many tropical countries, especially those countries where the density of population is high and the carrying capacity of the land is being stretched to its limits. In tropical Africa, for example, nearly every report on agriculture, animal husbandry, or forestry refers to the seriousness of the soil erosion problem. The underlying causes are the increase in the human and cattle population in recent times, leading to the intensification of cultivation due to land hunger and to overgrazing of pastoral land.

Shifting cultivation, which is still widely practised in tropical Africa, is an ecologically well-adapted system of crop production, but its carrying capacity is limited and it breaks down when the population expands beyond the equilibrium. Emigration is one manifestation of this breakdown. In the forested areas of Ghana, for example, the carrying capacity of the land under shifting cultivation is between 363 and 388 per km^2; emigration occurs when densities exceed these levels (Hunter 1966). The soil exhaustion which results when plots are re-cultivated too soon also leads to soil erosion — another manifestation that the agricultural system is unable to cope with the demands of a rapidly increasing population.

In many parts of the tropics the most common cause of soil erosion is overgrazing. The problem is particularly serious and widespread in tropical Africa. As Hodder (1968) says: "Africa is perhaps the classic instance of a continent suffering from soil erosion caused by overgrazing, and defective livestock management there is in general responsible for greater losses from erosion than is defective cultivation." Overstocking and overgrazing are due mainly to the pastoralists' desire to own large herds as their status in society are often gauged by the size of their flocks and herds (Worthington 1958). Apart from the social significance attached to them, the animals are also a form of insurance against the uncertainties of drought or disease, a capital investment, a means of transport and traction, a source of organic manure and occasionally of domestic fuel, as well as source of meat and animal products.[7]

It is not surprising that, given the many roles which they play and their social significance, there are as many livestock in tropical Africa as there are people (300 million). But their productivity is very low, and African herds

---

[7]The problems of overstocking and overgrazing and their effects on the grazing resources of the Sudan are examined by Khogali (Chapter 9). Many of these problems stem from the free access nature of the pastures and the value the Sudanese nomads place on their livestock as a form of wealth.

produce less than 20 per cent of the meat and milk they are capable of producing with proper management techniques (International Livestock Centre 1981).

Overstocking may be defined as "the maintenance of animals on a piece of land to the detriment of its carrying capacity" (Hornby, quoted in Worthington 1958:317). In tropical Africa, overstocking is due, among other factors, to the cessation of intertribal warfare, reduction in the toll taken by wild animals, the control of many animal diseases, and low annual offtake arising from social considerations.

The carrying capacity of pastures range from one beast per 0.5 ha in the case of the Kikuyu grass country of the East African highlands to one beast per 12 ha or more in the arid parts of Africa. Overstocking, overgrazing, and their consequential effects on the soil had been recorded in localized areas even before the Second World War: in the Ukamba and Kumasi reserves of Kenya, for example, overgrazing had caused more than one-third of the land to be eroded down to the subsoil in 1938 (Hailey 1957). The situation has worsened since then, and "overgrazing in some areas (in Africa) has produced erosion so extensive that the time required for recuperation of the soil and pasture has been lengthened almost to infinity" (Worthington 1958:318).

Overgrazing and overcultivation are among the main causes of wind erosion, and creeping deserts, in arid areas. Continual grazing and trampling by large herds destroy soil-holding vegetation. The exposed top-soil is blown away by prevailing winds. This is one of the factors contributing to desertification, defined as "the intensification or extension of desert conditions, this being a process leading to reduced biological productivity, with consequent reduction in plant bio-mass, in the land's carrying capacity for livestock, in crop yields and human well-being" (Fano 1978:178).

Desertification is in fact an extreme form of soil erosion, and a result of human pressure on fragile ecosystems in adverse climatic conditions. In Africa about 650,000 km² of arable and grazing land on the southern margin or the Sahara have been affected during the past fifty years. The desert boundaries of the Sahara are extending westwards into Senegal and eastwards into the Sudan at an average rate of 5 km each year. In Morocco, Algeria, Tunisia, and Libya, overgrazing, the extension of farming onto fragile land, and firewood gathering are contributing to the loss of more than 100,000 ha of range and cropland to the desert each year (Eckholm and Brown 1977). About 6.6 million km² of land in Africa, 8 million km² in Asia, and 1.7 million km² in South America are exposed to a high to very high degree of desertification hazards ("Desertification in Africa" 1978).[8]

---

[8] In his paper Adefolalu (Chapter 12) examines the various factors and causes of desertification of the Sahel, and concludes that the most important of these is the destruction of the natural plant cover caused by the rapid increase of people and livestock in the region.

Total soil loss from desertification and other forms of erosion, from mining, transport development, or housing and urban development, reduces the natural resource endowment of a country. But partial soil loss as well as the qualitative degradation of the soil arising from inappropriate land use practices can have equally negative, if less immediately obvious, effects on a tropical country's efforts to develop economically, by affecting the renewability of the soil resources, its capacity to support crops and livestock, and its ability to continue to do so over time.

Forest resources.   Estimates of the tropical forest cover vary considerably. In the mid-1970s the total forested land in the humid tropics was estimated to be 17.15 million km², or 54.5 per cent of the total forested land, and only 25 per cent of the total land area of the humid tropics (Sommer 1976).

Table 1.6 shows the results of a more recent joint survey conducted by the Food and Agriculture Organization and the United Nations Environment Programme involving 76 tropical countries and covering 97 per cent of the total area of countries lying wholly or mainly within the Tropics of Cancer and Capricorn. The area under forest (excluding forest fallows and shrublands) amounted to 19.3 million km² in 1980.

TABLE 1.6 TROPICAL FOREST FORMATIONS, 1980 (million km²)

Tropical Regions	Forests			Forest Fallows	Shrub-lands	Total
	Closed	Open	All			
Africa (37 countries)	2.2	4.8	7.0	1.7	4.4	13.1
America (23 countries)	6.8	2.2	9.0	1.7	1.4	12.1
Asia (16 countries)	3.0	0.3	3.3	0.8	0.4	4.5
Total (76 countries)	12.0	7.3	19.3	4.2	6.2	29.7

Source: FAO/UNEP, 1982.

The tropical forest can be regarded as a stock of trees and other vegetation that is available for the production of lumber, fuelwood, and other forest products. If properly managed, it will continue to produce such goods without any appreciable decrease in the amount or quality. But it is subject to forces which can lead to its depletion. The rate at which this stock is depleted depends on the socio-economic circumstances of the country concerned. At any one time, there is pressure being exerted on the forest, not only for the products which it provides but also for the land on which it stands. In tropical countries with high population growth rates, demographic pressure is compelling the governments to release more and more forested land for permanent settlement, agriculture, and other forms of non-forest land use (see, e.g., Furtado 1979). Table 1.7 shows the rate of depletion in thirteen selected countries. The annual depletion rate of 2.16 million ha, if extrapolated, would be 11 million ha for the humid tropics as a whole (Sommer 1976).

TABLE 1.7 RATES OF DEPLETION OF FORESTED
LAND IN 13 SELECTED COUNTRIES

Country	Area of Forest Reported To Be Lost Per Year (million ha)
Bangladesh	0.01
Colombia	0.25
Costa Rica	0.06
Ghana	0.05
Ivory Coast	0.4
Lao	0.3
Madagascar	0.3
Papua New Guinea	0.02
Philippines	0.26
Thailand	0.3
Venezuela	0.05 (?)
North Vietnam	0.01
Malaysia	0.15[1]
Total	2.16

Note: [1]The depletion rate for Malaysia in 1978 was 370,000 ha (Ministry of Finance, Malaysia 1978).
Source: Sommer 1976:19.

Such forests, cleared for agriculture and permanent settlement, evidently do not belong any longer to the renewable natural resource category, but must be regarded as finite stocks which are driven to extinction through total clearance. In such instances, the returns from the forests must be maximized through the recovery of the usable forest products, and waste must be reduced to minimum levels. However, for various reasons such as transport inadequacies, distance from markets, ignorance of the value of some forest products, or attitudinal barriers, much of the forest endowment of the developing countries is wasted, usually by burning, on clearance.

It has been estimated that more than half of the assumed world reserve of 1.8 billion ha of potentially arable land lies in the tropics (Brunig 1979). The carrying capacity of land under unmanaged rainforest is less than one person per km^2, as compared with 30–50 persons per km^2 under some forms of shifting cultivation and 400–700 persons per km^2 under intensive wet-padi cultivation (Ooi 1959; Bene *et al.* 1977). The rapid increases in population in the tropical countries, with its obvious effects on the demand for land, make it inevitable that the conversion of forest land into higher carrying capacity agricultural land will continue, in all likelihood at an increasing pace. Such conversions to permanent agriculture are an economically rational shift to permit a more productive use of the land.

Tropical forest ecosystems may be regarded as a multi-purpose (or output) system. They may be managed for (1) their economic products, (2) for conservation (where the forests are protected as watersheds and wildlife habitats), and (3) for recreation (where they are an amenity resource). The forests, if properly managed, are a renewable natural resource which can yield a wide variety of forest products, including timber, fuelwood, pulp, and minor products such as gums and latex, oil-bearing nuts, medicines, dyes, essential oils, resins, tannins, bark products, rattans, honey, and beeswax.

Although the mature rainforest has an average density of 250–300 m^3/ha, the volume of commercial wood obtained from it is low, ranging from 5 to 30 m^3/ha ("African Wood in World Trade" 1976). The yield per unit area from the forests of the Malay Peninsula is only one-tenth that of the coniferous forests of Europe and America (Ooi 1976). A comparative study by the FAO of the Republic of the Congo and Finland, with similar land areas and forest estates, revealed that the value of forest products exported from Finland was sixty times greater than that from the Congo (quoted in Bene *et al.* 1977).

There are several reasons for these low yields and low returns. First, the rainforest has an extremely rich flora (in the Malay Peninsula there are 8,000 species of flowering plants, of which 2,500 species are trees), but they are distributed in a heterogeneous manner, and only a few of them produce timber of commercial value. Thus patches of forest from 8 to 40 ha in area may have only two or three heavy hardwoods of marketable size in them (Ooi 1976).

Second, few species have a market value: in West and Central Africa, the main sources of forest products in Africa, 10 species made up 87 per cent of timber production and 97 per cent of timber exports, although more than 100 types of wood are actually exploited (Morallet 1976). This appears to substantiate the Ricardian hypothesis that states that, among several pools of a given resource, the pools of highest quality (or the lowest extraction cost) tend to be used first. Harvesting of timber in the tropics has been concentrated on the most valuable trees located in the most accessible areas. The rest of the forest will be composed of trees of lesser quality and a full stock of species of no commercial value. This method of extraction would result in a progressive lowering of the timber quality.

Third, tropical timbers are susceptible to decay and insect depredations in living trees, so that the final recovery of saleable timber is low. Fourth, a high percentage of the forest products are exported in an unprocessed form, so that the value added from further processing and manufacture does not accrue locally. More than half of the export earnings from forest products in the less developed countries are from unprocessed logs, and only 45 per cent are from sawn wood and wood-based panels, with a negligible percentage from pulp and paper (World Bank 1978).

Tropical forests have traditionally been the sources of fuelwood, the demand for which has been reinforced by the energy crisis that has raised the cost of alternative domestic fuels such as kerosene, bottled gas, and heating oil.[9] Wood is the preferred fuel for the low income groups in the rural areas, as it can be gathered at little or no cost and, as charcoal, is smokeless, light, and easily transportable. In 1974 fuelwood accounted for 85 per cent of total wood use and nearly one-quarter of total energy use in the developing countries (Arnold 1979). In Africa about 90 per cent of the wood from forests is consumed as fuel, 6 per cent as poles and posts, and 3 per cent as sawn wood (Sanger *et al.* 1977).

Depletion of the forest resources does not occur where fuelwood collection is from dead wood and fallen branches and twigs, as when there is a balance between tree growth and population numbers. But where demand exceeds dead wood supply, because of either a local increase in population numbers or demand generated from urban centres, living trees from nearby forests are cut, often illegally, to meet the shortfall. The physical shortage of fuelwood and charcoal in many towns has led to the systematic clearance of woodland and forest surrounding them. Niamey and other West African towns are surrounded by circles of treeless desert as a result of such deforestation for fuel (Eckholm 1979).

In the rural areas, fuelwood is obtained from sources within walking distance. Households within 10 km of a forest would obtain about 70 per

---

[9]Mushala (Chapter 5), in his paper on rural energy resources in tropical countries, traces the rural energy situation in some African countries and the impact of the oil crisis and rapid population growth on the forest resources from which fuelwood and charcoal are obtained. The Tanzanian situation is examined in the second part of the paper.

cent of their fuelwood from it; the percentage decreases beyond 10 km until it is almost nil beyond 15 km (Arnold 1979).

The effect of the demand for fuelwood on the forest varies among the developing countries. A forest under heavy pressure as a source of fuelwood will not attain enough stature for timber or other forest products. Where there is a very heavy demand, the forest may be destroyed, and all the tree and shrub cover may be removed, as has happened in the densely peopled Gangetic plain of India. Such a practice is tantamount to mining the forest for fuelwood. The demand for fuelwood in the developing countries is projected to grow at a rate faster than could be met by the annual increment in supply (Openshaw 1978). This, together with the conversion of forest land to agriculture, would lead to a rapid reduction in the forest capital.

Management of the forests on a sustained yield basis is difficult because of the demographic circumstances of the developing countries. It is made even more difficult because of the long time which the most valuable forest product — timber derived from hardwood trees — takes to reach maturity. Where most crops would yield a harvest in a few months or, at most, a few years, tropical rainforests may require seventy years or more to produce marketable timber on a sustained yield basis (Marshall 1958). Therefore, forestry projects not only take a long time to begin production but may take an even longer time to run to completion, if they do so at all. Few developing countries can plan for forest production on a seventy-year rotation cycle, if only because their population would have increased several-fold by then. Unless they are land-affluent, or the nature of the economies changes, the consequent demand for agricultural and pastoral land will lead to the conversion on land set aside for forestry.

The area that will be retained under forest will depend largely on political decisions that determine land use in each tropical country. The allocation of land according to its capacity of each potential use (such as agriculture, forestry, wild life, and recreation) is a highly complex matter, made even more complicated by local land laws, customs, and ownership patterns. In many parts of the less developed world, for example, private ownership of land is a foreign concept, and forest land may be tribally owned. Where forest land is publicly owned, control by the state may be tenuous because of the lack of forest officers. Both these affect the willingness and ability to protect the forest and to make the long-term investment necessary for forest development (see Bene *et al.* 1977).

Estimates of the rate of depletion and conversion to other uses vary. The World Bank estimated the rate to be 150,000 to 200,000 km^2 a year. At such a rate, and even with no growth in demand, the forest stock of the developing countries would last for 60 to 80 years (World Bank 1978). The U.S. National

Academy of Sciences has put the rate of depletion at 245,000 km² a year, with total depletion occurring within 50 years (*New Scientist*, 17 Apr. 1980).

The FAO/UNEP survey shows that the closed and open tropical forests are being depleted at annual rates of 0.62 and 0.52 per cent, respectively (Table 1.8). The overall annual rate of 0.58 per cent or 113,000 km² corresponds closely with the earlier estimate of Sommer (1976).

TABLE 1.8 RATES OF DEFORESTATION OF TROPICAL FORESTS ('000 km²)

Tropical Regions	Closed Forests		Open Forests		Total	
	Area	%	Area	%	Area	%
Africa	13.3	0.61	23.5	0.48	36.8	0.52
America	43.4	0.64	12.7	0.59	56.1	0.63
Asia	18.2	0.60	1.9	0.61	20.1	0.60
Total	74.9	0.62	38.1	0.52	113.0	0.58

Note: Deforestation is defined as total clearing of natural forest formations for any purpose, including shifting cultivation.

Source: FAO/UNEP, 1982.

Although available statistics indicate that forest products contribute only about 5 per cent of the value of all renewable resources produced in developing countries (Bene *et al.* 1977), the loss of these forests may have more far-reaching consequences than the solely economic one, as two million of Earth's five million plant and animal species are in these forests, and they play an important role in terrestrial photosynthesis, and in the oxygen and carbon cycle of the plants.

Free access natural resources: fisheries. The term "free or open access resources" is given to a group of natural resources — ambient air and water systems, the oceans, fisheries, wild life, wilderness areas, public lands, natural beaches, and parks — which share a common characteristic: in normal circumstances they are available to all those who care to use them. In legal terminology, common property (*res communes*) refers to property rights to a resource exercised by a group of users to which all other users have no access, whereas unowned resources (*res nullius*) are resources to which no property rights have been assigned and to which there is free access. In the case of free access resources within national boundaries (e.g., parks, beaches, wilderness areas, coastal and inland fisheries) the government generally act as owners, but in the case of a resource such as the oceans (including the fisheries and the minerals of the deep sea bed), which until recently have been regarded as the property of all the nation-states of the world, the question of unfettered access is being debated in the various Law of the Sea conferences.

Many renewable natural resource systems such as grazing land, forests, and wild life are common property resources and have been managed on a sustained yield basis for a long time by groups of users with usufructuary rights and access to such resources, for example, pastoralists and hunters and gatherers in the tropics. Several factors had operated to maintain these societies in equilibrium with their resources. First, common ownership of the resources and the rules of sharing minimized over-use and depletion of the resources for individual gain. Second, in the pre-colonial era, there was no incentive to accumulate surpluses as there were no markets to sell such surpluses. Third, control over the rate of depletion was exercised whenever necessary through such measures as closed hunting seasons, as practised among the Acholi of Uganda. The introduction of taxes and of a market economy during the colonial era resulted in an over-use of resources to obtain the cash to pay such taxes and to acquire the goods available on the market (see Ciriacy-Wantrup and Bishop 1975).

Each of the natural resources to which there is free access generates its own set of problems — from air, which is a ubiquitous resource, to fish, which is a fugitive (mobile) resource that has to be captured. Some of the problems which the tropical countries face in respect of their free access resources are exemplified in the case of fisheries. Consideration will be limited here to fisheries within national jurisdiction, as those in the high seas, though they are also fugitive resources, have different management and institutional problems (Ciriacy-Wantrup 1971).

In general the condition of free access and of undefined or unassigned property rights to a natural resource has led, in the context of increasing commercial demands and expanding populations, to such serious present-day problems as overgrazing, overfishing, the permanent depletion of ground water, diminishing stocks of wild life and the extinction of some species, and air and water pollution through the use of air and water bodies as waste sinks. Without restrictions there is a tendency to over-exploit the resource, or to exploit it too rapidly for regeneration to take place, or to exploit it to destruction and extinction. The ills of such mis-management are summed up in the phrase "everybody's property is no one's property" (Haveman 1973; Plourde 1975).

To obtain a perspective of fishery development in the developing tropical countries, it is helpful to look at the current status of such development and the potential contribution of the fish resources to the economic well-being of these countries.

Although water covers 71 per cent of the earth, only 3 per cent of the world's food supplies are derived from the sea. Of the total production of vegetable matter from the world's seas and oceans, only 0.03 per cent is ultimately harvested in the form of fish — as the final link in a long food chain. Most fishing is carried out at depths of 500 m or less, that is, in the waters of the continental shelves. About half of the world fish production is from the developing countries. Fishing in the oceans of the northern hemisphere, carried out by the fishing fleets of the industrialized countries, appears to have reached its limits, but there is still a considerable untapped potential in the waters of Southeast Asia, the Indian Ocean, and the coastal areas off West Africa and South America. Nearly all the fishing grounds here are within the 200-mile (322 km) exclusive economic zone of the developing countries.

The other major source of fish is from inland waters — lakes, lagoons, rivers, swamps, canals, and brackish water areas. However, the multiple uses to which the inland waters are often put may pose problems in the maintenance of adequate stocks of fish for the fishing industry. Nevertheless, some tropical countries still have a significantly large potential from their inland waters. The major lakes of Africa, for example, are undeveloped (Table 1.9).[10]

The fish resources of the tropical world are important in at least two respects: they are a source of food (and high-value protein) and they are also a source of income and employment and of foreign exchange earnings when exported. Fish is a major source of protein in a large number of African countries (Table 1.10). In all the countries of Southeast Asia except Indonesia, the per capita consumption of fish is several times the world average of 11 kg per capita. The value of the fish catch is also high compared with that in the developed countries. In Southeast Asia it is 3 to 4 per cent of the GNP, compared with 1 per cent or less in developed countries (Marr 1976).

One of the most important developments in recent years is the adoption of the concept of the exclusive economic zone (EEZ), formulated in the U.N. Law of the Sea conferences. The EEZ extends beyond a territorial sea of 19.3 km up to a total distance of 322 km from the baselines from which the territorial sea is measured, within which the coastal state has sovereign

---

[10]Much of the potential of the fisheries of northern Zambia appears to be more apparent than real, as there are indications of overfishing in some areas. The situation in northern Zambia (discussed at length by Huckabay in Chapter 7) provides an example of the problems that an African developing country faces in the exploitation of a common property resource.

TABLE 1.9 AFRICA: POTENTIAL FISH PRODUCTION
OF THE MAJOR LAKES

Site	Area (in km²)	Average Annual Production 1970–75 ('000 t)	Estimated Potential ('000 t)
Lake Victoria	67,000	115	420
Lake Tanganyika	34,000	79	310
Lake Nyasa (Malawi)	28,500	30	100
Lake Turkana (Rudolph)	7,000	2	7
Lake Kyoga	2,280	62	100
Lake Edward	2,220	16	25
Lake Mobutu Sese Seko (Albert)	5,500	14	30
Lake George	270	4	6
Lake Kivu	2,700	3	1
Lake Rukwa	2,000	5	8
Lake Chilwa	1,000	2	3
Lake Meru/Ntipa	5,200	18	30
Lake Bangweulu	12,000	8	10
Lake Kariba	5,300	3	14
Lake Chad	14,800	135	150
Lake Kainji	1,250	8	10
Lake Volta	8,727	35	40
Lake Nasser/Nubia	5,000	9	20
Lake Tsana	3,500	5	15
Lake Kossou	1,750	2	8
Rift Valley Lakes — Abays, Zuveys, etc.	3,100	—	17
Total	213,097	548	1,324

Source: The Courier, Jan./Feb. 1977, p. 61.

rights for the purpose of exploring and exploiting, conserving and managing
the natural resources, including the living resources of fish and other marine
life.

Since the seventeenth century, the basic principle of the law of the sea was
that a narrow strip of offshore waters should be under the exclusive sovereignty of the coastal state and that the high seas beyond should be free.
However, in so far as fisheries are concerned, the traditional assumption of
unlimited harvests from the seas could no longer be maintained as improvements and innovations in fishing equipment resulted in overfishing and

TABLE 1.10 RELATIVE IMPORTANCE OF FISH IN
            FOOD SUPPLIES IN AFRICAN
            COUNTRIES
            (% of supply of animal protein)

More than 40%	Between 40 and 20%	Less than 20%
Congo (72%)	Zaire	Upper Volta
Sierra Leone	Burundi	Mauritania
Gambia	Chad	Tunisia
Benin	Guinea	Egypt
Ghana	Gabon	Morocco
Senegal	Nigeria	Kenya
Cameroon	Mali	Zimbabwe
Zambia	Mauritius	Algeria
Togo	Uganda	Sudan
Ivory Coast	Tanzania	Niger
Liberia	Malawi	Somalia
Angola	Madagascar	Ethiopia
	Mozambique	Rwanda

Source: Forecasts on agricultural products, 1970–80, vols. I and II,
         Rome (1971). Classification in order of the importance of fish
         in overall supply of animal protein.

depletion in many areas. Many, mostly developed, countries began to
exploit the fish resources of seas far beyond their own coasts. Developing
countries which did not have the financial and technical means to exploit
the fish resources of their coasts, and apprehensive that these would be
depleted by the fishing fleets of the richer nations, were instrumental in
formulating the concept of the EEZ, and having it accepted in the U.N.
Law of the Sea conferences.

But the benefits arising from the establishment of the EEZ are not confined mainly to the less developed countries; several developed countries
with large territorial expanses, notably the United States, Canada, Australia,
and the Soviet Union, are in fact major beneficiaries. While the developing
countries with long coastlines or which are archipelagoes will gain considerably from the EEZ, a large number which are land-locked, shelf-locked, with
short coastlines or with poor fish and other natural resources offshore, will
be geographically disadvantaged.

Although the 322-km EEZs will cover only some 35 per cent of the oceans they will include 90 per cent of the living resources now being exploited, including most fisheries (Carroz 1977). Article 56 of the 1980 Law of the Sea Draft Convention gives coastal nations sovereign rights over all natural resources within their 322-km economic zones. Many countries have already enacted legislation, or are in the process of doing so, to establish their EEZs. Among those that have enacted their EEZs are most of the countries of South and Southeast Asia and most of the South Pacific countries. The EEZs of the South Pacific island nations have brought some 6 million miles^2 (15.5 million km^2) of the Pacific under their national jurisdiction. The EEZ of Fiji, for example, covers 330,900 miles^2 (856,700 km^2), and that of Papua New Guinea covers 690,000 miles^2 (1,786,000 km^2). In the Indian Ocean the EEZ of the Seychelles is 393,500 miles^2 (1,019,000 km^2).

The advantages that could be obtained by these coastal and archipelagic states from the establishment of the EEZs will depend on the abundance of living resources in the EEZs and on the capacity of these states to exploit and manage them. The problems facing many developing countries in the development of the fish resources in both the inland waters and in the seas within the EEZs are complex, partly because of the biological characteristics of these resources, partly for technical and manpower reasons, and partly because of the open access nature of the resources. Some of these are discussed below.

First, in accepting that national control of the fish resource is necessary to the conservation of stocks for sustained yields, it is implicit that there should be surveillance and enforcement facilities available to protect these resources from encroachment by third parties. However, many developing countries do not have such facilities, especially the small island nations which have very large ocean areas within their EEZs. Many South Pacific island nations have expressed their concern over the problem of poaching by foreign countries as a result of their inability to police their large EEZs (Van Dyke and Heftel 1981). They cannot afford to provide adequate protection measures.

Second, to manage the fish resources effectively, data must be available on the extent, distribution, and dynamics of the resources. Such data are often lacking or inadequate. The developing countries will have to mount extensive and costly research and assessment studies to plan and give effect to management policies.

Third, the migratory nature of fish poses special problems in management. All species of fish are mobile for at least part of their lives, although the extent of their movements vary markedly from one species to another. Most confine their movements to the waters of the continental shelf, but some, such as the tuna and the whale, migrate across entire oceans. Stocks of fish, migrating parallel to coastlines or across regional seas, may move through the waters of two or more coastal states. Exploitation of the stocks by these states may, in aggregate, be larger than the annual yield of the stocks. From the biological perspective, therefore, migratory species, whether such migration is across EEZs or from EEZs to open seas and vice versa, would have to be managed on a regional basis.

Fourth, many of the developing countries do not have the infrastructural facilities, such as ports, processing and handling installations, fishing fleets with modern fish detection and catching equipment, and trained manpower, to exploit their fish resources to the optimum level. It will take a long time, given the restrictions caused by lack of capital and technical manpower, for the countries to build up their fishing capacities. In the interim some of them generate revenue from licensing fees and catch taxes paid by foreign boats fishing in their waters. Another source of revenue is bilateral agreements with major fishing countries such as Japan. Senegal, Guyana, Ghana, and Mauritius are examples of countries with well-developed joint-venture programmes (Stoneman and Disney 1980). However, in countries such as the South Pacific island states where negotiations on access and licensing are conducted on an individual basis with foreign countries, the returns to the island states may be reduced because of competitive undercutting among themselves (Van Dyke and Heftel 1981). Some of the developing countries with large fishing fleets have in recent years also entered into joint-venture agreements with other countries with poorer fishing capacities. One example is the agreement between Thailand and Oman signed in July 1981, whereby Thailand will receive 30 per cent of the total profit derived from fishing in Oman's territorial waters (*Asia Research Bulletin*, 28 Feb. 1982).

Fifth, certain biological characteristics of the fish population pose constraints to their exploitation. For example, there is no method available for predicting the environmental factors that favour high recruitment, namely, aggregation, spawning, and fry survival. This makes it difficult to apply the sustained yield concept to fish resources and to adjust the catch rate to the fish population. Species representation in many tropical seas is very high — in the South China Sea, for example, there are 2,500 fish species of which 324 are commercially important. But there is an almost complete lack of knowledge on the distribution and range of the self-perpetuating fish population (Marr 1976). The exploitation methods developed by the industrial countries for catching large quantities of certain preferred species of fish, practically unmixed, are of little use in the tropical seas where such a variety of fish species are found.

Sixth, the open access nature of the fisheries (referred to earlier) generate special management problems which are complicated by the fugitive nature of the resource and the fact that fish swim across boundaries set by man — from national to international waters and vice versa, from one part of the lake to another, and from one part of a river to another. Exploitation of such a resource involves capturing it while it is available.

In the absence of restrictions any individual may catch the fish without charge. He would have no incentive to restrain his fishing effort in anticipation of future returns, as there is no certainty that he will obtain his due share of such returns. This results in overcapitalization of the industry, economic waste, overuse of the resource, and depletion of the stocks to the point where the annual yield is less than the maximum that can be sustained. This is "the tragedy of the commons" which Hardin refers to in his famous essay of 1968 (Hardin 1968).[11]

Attempts to resolve this problem generally involves removing the open access nature of the fish resources. Specific management techniques include (1) regulations to limit the catch from a fishery, (2) regulations which relate to the kind of technology that can be used, and (3) regulations which seek to impose a limit on the fishing effort through special licensing of trawlers, or through taxes and fees. Both the developed and developing countries may employ all or a combination of these techniques to modify the open access nature of their fisheries (see Christy Jr. 1975; Marr 1976; Howe 1979).

Resource Management

Natural resources are those components of the physical environment that are of use to man. Because the world is divided into a multiplicity of states of varying sizes located in different environments, the resources which any one state has within its national boundaries are finite (except for air). The finite nature of resources gives them their scarcity (or economic) attribute: most resources are scarce to varying degrees, and such scarcity, whether local or global, is usually reflected in prices. Resource planning and resource management would not be necessary if resources are freely available in unlimited amounts to all and sundry. But because they are scarce, their present and future availability is a matter of concern to policy analysts and decision-makers. The identification and monitoring of existing scarcity, the prediction of future scarcity, and the allocation and regulation of these scarce resources to meet societal or national objectives are among some of the problems which resource managers have to face.

Countries in the early stages of economic development, whether located in the tropics or in non-tropical regions, tend to view the maximization of economic benefits as the major objective of natural resource management

---

[11]The problems of overcapitalization, biological overfishing, and economic waste are traced by Abdul Hamid Abdullah (Chapter 6) in his paper on the marine fish resources of Southeast Asia.

and development. Resource-exporting countries generally depend on a small number of commodities for foreign exchange receipts, and until recently the exploitation of their resources was largely in the hands of foreign enterprises. In their efforts to overcome poverty and promote economic development, they have sought to assert control over their natural resources, to expand their economic base through industrialization, to expand foreign trade, and to gain access to markets for their raw materials and semi-processed or processed goods.

The question of control over natural resources was raised in the United Natons as early as 1952 during the debate on the Draft International Covernants on Human Rights. The doctrine of permanent sovereignty over natural resources was accepted a decade later (U.N. resolution 1803 of Dec. 1962), and reaffirmed on a number of occasions in subsequent years, by U.N. bodies as well as by the General Assembly itself. To a few developing countries, notably the members of OPEC, sovereignty over natural resources has been translated from concept to reality, and has brought about a more equitable sharing of the benefits that accrue from the exploitation of these resources.

But to many of the other developing countries, especially the least developed countries, sovereignty over their natural resources remains a theoretical goal. Lack of the financial and technological means to locate and develop these resources often places them in a position of dependency — on the international public sector and on the international private sector.[12] The former, represented by the U.N. Development Programme, the regional development banks such as the Asian Development Bank and the Inter-American Development Bank, and by the World Bank, has not been able to render effective assistance in such areas as petroleum exploration (Zakariya 1980).

The latter is represented by the multinational corporations. To the least developed countries and many other developing countries, the multinationals are often the main sources of capital, organizational ability, and technological expertise — all necessary to the location and development of their natural resources. In offering this assistance the multinationals are concerned with the returns they can expect, and are often able to maximize these returns because the least developed countries lack alternative sources of aid and are therefore in a weak bargaining position. Their choice of policy would depend on the options and their leaders' perception of net gain — while they would understandably opt for national ownership and control of their natural resources, those without capital, organizational capacities, and/or technological expertise but nevertheless needing to produce as much

---

[12]Guyana is one country that has sought to reduce its dependency status and move towards a convergence policy in the development of its natural resources. Bernard (Chapter 14) traces the background to this policy and considers the problems of its implementation in Guyana.

of their resources as possible now would have to accept multinational participation and perhaps move towards joint-venture relationships later.

A strategy favoured among developing countries is to adopt a collective or cooperative approach to improve their economic position, an approach which has proved effective in the case of oil. As stated in a 1973 U.N. resolution on permanent sovereignty over natural resources (Resolution 1737-LIV), "one of the most effective ways in which the developing countries can protect their natural resources is to promote or strengthen machinery for cooperation among them, having as its main purpose to concert pricing policies, to improve conditions of access to markets, to coordinate production policies and thus to guarantee the full exercise of sovereignty by developing countries over their natural resources". The issues and problems of forming and maintaining such producer alliances have been discussed earlier.

In the developing countries, the aim today is to increase the per capita flow of goods and services to levels sufficient to raise standards of living. In so far as natural resources are concerned, this would involve changing traditional patterns of resource use geared to self-subsistence and adopting new patterns of use which will allow for such an increased flow. But the objective is easier to state than to achieve, not least because patterns of resource use are culture-bound, and societies which had not thought of resource use in economic and money terms may find it difficult to orientate their perceptions to accommodate the change. To many this would involve abandoning the commons in hunting and gathering, the enclosure of agricultural and grazing land, and restrictions on access to fishing grounds. In many African countries, changes in the conditions of access to land through changes in land tenure have yet to be achieved or, if achieved, to be effectively implemented (Adeyoju 1976).

Questions involving the management of natural resources would hinge upon, among other things, the value societies place on the future. Decisions on resource utilization over time can and do vary from one society to another. But in the developing countries generally, rapid rates of population growth are narrowing the options on the time distribution of resource use. More and more such countries are being confronted with the problems of resource depletion because of the inability to expand their resources to keep pace with population growth. In situations of emergency, which may in time become chronic with uncontrolled population increase, societies may have no choice but to allow for the irreversible depletion of resources in order to survive.

But where there is a choice, decisions would have to be made about resource use and resource management. These are plainly decisions which involve value judgements — about the forms which economic development should take and the role of natural resources in such development, about the rates of depletion of stock resources, about the management of renewable resources, and about the impact of resource exploitation on local and global ecosystems.

There are some who maintain that economic development in the developing countries cannot and should not follow the model of the industrialized countries, on the grounds that (1) to build an economic infrastructure dependent on high inputs of non-renewable energy resources in a situation of diminishing stocks of such resources is illogical and potentially disastrous, and (2) to favour a highly capital-intensive pattern of industrialization over labour-intensive production in a situation of capital scarcity and abundant labour is equally hazardous (Rifkin 1980; May 1981).

Whatever form of development it chooses, a country with large reserves of non-renewable resources, of which petroleum is currently the most important, and able to obtain equitable prices for them (again, as in the case of petroleum), would have a comparative advantage over those countries without such resources. The problem in resource-affluent countries is to optimize the depletion rates so that when the resources — as wasting assets — are depleted the countries would have attained the highest possible level of economic development and, ideally, self-sustaining economic growth. The objective is the continuous exchange of the wasting assets for productive social and material capital over the time horizon designated "the development horizon" (Jafar Mausur Saad 1979/80). The problem is a highly involved one because of several uncertainties — uncertainty about the actual amounts of resources available at a given time, uncertainty about future technology and the discovery of substitutes, and uncertainty about the population variable.

Faced with such uncertainties, some countries may decide to exploit their resources more fully now rather than conserve them for the future. As Johnson (1967:155) puts it: "Given that the less developed countries are anxious in industrialize as rapidly as possible, and in so doing expect to increase the flexibility of their economic structures, it might well be an optimum strategy for them to attempt to maximize their profits from primary production over the short run, at the expense of future earnings, in order to secure their development objectives."

The developing countries which do not have stocks of minerals and fossil fuels to draw upon for development capital will have to continue to depend on their renewable resources for material support. The management of renewable resources is at best a chancy business, mainly because it involves dealing with complex natural ecosystems in tropical environments which are yet little understood. A single management act may have multiple effects, some foreseen and others not. There have been many instances of costly mistakes resulting from failure or inability to predict the ecological effects of management decisions.

Indigenous tropical societies have, over time, evolved systems of resource management in which the flow of goods and services from their resource base (usually fixed and finite) is in balance with their needs. Such systems are now breaking down because of the pressure of population and also because the "revolution of rising expectations" is causing many developing countries to exploit their renewable resources more rapidly than can be replaced by natural processes. This is so with water, soils, forests, fisheries, and wild life.[13] Such overexploitation runs counter to what is generally recognized as the best policy for renewable resources — that of sustained yield, whereby resources are harvested at rates equal to the rate of replacement. Over-rapid harvesting may also result from natural fluctuations in resource availability, in the case of water and fish resources.

As a result, the outlook for the developing countries is not promising. To take some examples: by the year 2000, population growth alone will double the demand for water in nearly half of the countries of the world, with the greatest pressure occurring in the developing countries with low per capita water availability and high population growth. The per capita water availability everywhere will decline (Table 1.11).

A similar scenario is predicted for the forest and arable land resources of the developing tropical countries. The stock of mature closed forest in these countries is being depleted at the rate of several hundred thousand ha a year, and the two-thirds of the tropical forests which are accessible would be totally depleted by the turn of the century if current rates of deforestation and population growth remain unchanged.

Projections of arable land resources availability indicate that pressure on these resources due to population growth will increase to the point where many of the developing countries, especially the Sudan-Sahel countries of Africa, East Africa, parts of South Asia, and Latin America, will encounter major problems arising from soil degradation, soil loss and sedimentation, desertification, and salinization. Population growth and physical constraints, in terms of both the scarcity of good cultivable land and the absolute lack of arable land for agricultural expansion, will lead to a progressive decline in the arable area per capita in the developing countries, as shown in Table 1.12 (*Global 2000 Report, 1982*, vol. 2, pp. 96–99).[14]

The increasing demands that the developing countries will make on their natural resources (renewable as well as non-renewable) in striving for a higher GNP per capita will inevitably have impacts on the environment. The

---

[13]Pullan (Chapter 8) examines in detail the role of wildlife in the development of an African country, and attempts to answer the questions: Is wildlife a valuable resource? If so, what are the costs involved in protecting and managing it?

[14]Dayal (Chapter 10) provides a country example of the decline in the area of arable land per capita and the consequent deterioration of the man-land ratio as a result of population increase.

TABLE 1.11 PER CAPITA WATER AVAILABILITY IN SELECTED TROPICAL
COUNTRIES, 1971 AND 2000

	'000s of cu. m. per capita per year		Percentage Change
	1971	2000	
ASIA			
Sri Lanka	4.7	2.8	− 40.4
Burma	24.6	13.2	− 46.3
Thailand	4.8	2.0	− 58.3
Vietnam	19.2	10.6	− 44.8
Philippines	10.1	4.2	− 58.4
Malaysia, Singapore, Brunei	33.3	14.4	− 56.8
Indonesia (excluding Irian Jaya & Timor)	13.0	6.7	− 48.5
AFRICA			
Nigeria	4.7	2.4	− 46.9
Ghana	7.8	3.1	− 60.3
Zaire	58.4	34.6	− 40.8
Kenya	3.4	1.2	− 64.7
Tanzania	5.7	2.4	− 57.9
Malawi	2.0	1.0	− 50.0
TROPICAL AMERICA			
Guatemala	15.1	7.0	− 53.6
Honduras	38.0	15.3	− 59.7
Cuba	3.1	1.8	− 41.9
Jamaica	1.1	0.6	− 45.5
Venezuela	73.2	28.5	− 61.1
Guyana	317.0	164.2	− 48.2

Source: Based on *The Global 2000 Report to the President*, 1982, vol. 2, Table 9.10.

quickened pace of economic activity in the mining, refining, and energy
sectors and the added pressures on the biological systems (fisheries, forests,
grasslands, and crop lands) which supply the non-mineral and non-petrol-
eum-derived raw materials for industry will produce correspondingly larger
quantities of waste products. Without measures to protect the environment,
most of these wastes will enter the environment as pollutants, especially the
atmosphere, the streams, and the oceans, which until recently were regarded
as convenient waste sinks with a practically inexhaustible capacity to absorb
whatever was put into them.

TABLE 1.12 ACTUAL AND PROJECTED ARABLE AREA PER CAPITA IN THE
DEVELOPING COUNTRIES

Region	Arable Land Per Capita in Hectares				
	1951–55	1961–65	1971–75	1985	2000
Latin America	.56	.51	.47	.35	.28
North Africa/Middle East	.68	.58	.47	.33	.22
Other African LDCs	.72	.73	.62	.49	.32
South Asia	.38	.32	.26	.19	.13
Southeast Asia	.38	.41	.35	.28	.20
East Asia	.15	.15	.13	.11	.08
Less developed countries	.45	.40	.35	.27	.19

Source: *Global 2000 Report, 1982*, vol. 2, Table 6.13.

Keeping such wastes out of the environment would involve increased
costs, which in most cases are not borne by the originators unless there is
public sector or state intervention.[15] The common property nature of the
global environmental systems means that unilateral action by the industrial
countries to control wastes discharge is by itself insufficient to prevent
changes to these systems arising from other countries' failure to keep wastes
and pollutants out of the environment. To many of the developing countries,
however, the environmental and ecological issues raised by the industrial
nations appear to be another attempt to curtail their economic growth. At
the 1972 Stockholm Conference, the developing countries had advanced the
argument that it was the industrial nations that were the main polluters, and
they should pay the costs of cleaning up the environment.

It is not uncommon for a developing country to accept environmental
pollution as the price of economic advancement. Such countries would
provide the "pollution havens" for the industrial nations to site factories
with manufacturing processes which do not meet their home environmental
regulations. Multinational corporations engaged in highly polluting
processes and in the production of toxic substances may find the developing

[15]In his paper, Prasad (Chapter 4) examines the environmental consequences of open-pit
mining, under largely uncontrolled conditions, in the mineral belts of Chotanagpur, India. He
considers land dereliction and the other adverse environmental impacts of mining as inevitable,
as neither India nor the other less developed countries of the tropics with mineral economies
can afford the costs of waste prevention and control.

countries increasingly attractive as factory sites (Marlay 1979; *Global 2000 Report, 1982*, vol. 2:424).

Associated with the environmental pollution question is the issue of resource conservation and the "limits-to-growth" thesis. Many in the developing countries see this as being another attempt by the industrial countries to keep them under economic subservience. In the words of C.T. Kurien:

> It is a small affluent minority of the world's population that whips up a hysteria about the finite resources of the world and pleads for a conservationist ethic in the interests of those yet to be born; it is the same group that makes an organized effort to prevent those who happen to be outside the gates of their affluence from coming to have even a tolerable level of living. It does not call for a divine's insight to see what the real intentions are (cited in Rifkin 1980:190).

Such a view is not really on the issue of resource conservation per se but on the developing countries' adverse terms of trade with the industrial countries which prevent them from realizing the economic benefits they expect from the exploitation of their natural resources (usually exported in the form of raw materials). If such present-day exploitation does not provide them with equitable returns, why then conserve their resources for the continued future benefit of the industrial countries?

The above is essentially an argument in equity, applicable in situations involving the export of raw materials derived from the exploitation of natural resources. But leaving aside the matter of exports, developing countries also derive goods and services from their natural resources for domestic use, and therefore need to consider what conservation policy they should adopt towards such resources.

Much would depend on what is meant by conservation. As defined by Fisher (1964:53–54), it means the wise and economic use of resources, involving an attitude of caution and husbandry, especially towards the non-renewable resources, and the adoption of policies of controlled use combined with measures to avoid waste and increase yield. Herfindahl (1965:232–33), however, sees a conservative act as one which saves something for future use instead of present use, or which saves something for use instead of non-use. The conservation of one resource may be at the expense of another resource. The most important question is whether the gains outweigh the costs: if they do, then the resource should be conserved; if not, conservation may not be necessary.

Much would also depend on the options each country has. Often such options are interrelated and competing. With respect to the non-renewable resources, resource-dependent countries are also faced with the dilemma of whether to produce as much as possible now or to conserve them for the future. Decisions regarding these resources are in many ways easier to arrive at as the stocks of such resources are finite, measurable, and do not have common property or free access attributes. Decisions regarding the renewable resources are complex and difficult to make because the resources are part of the physical and biological environment, and may be subject to multiple use. Often these uses are either partially or totally incompatible. Multiple use raises many conceptual and practical problems in decision-making, especially when two groups of users seek to maximize returns from the same resource. In many cases a solution may be sought through a policy of partitioning the resources of terrestrial and marine ecosystems into best-use categories based on their carrying capacities and limiting factors operating in these ecosystems (Hinckley 1980).

The other major class of problems in resource management pertains to the common property nature of renewable resources, mentioned earlier in regard to environmental externalities. In most of the developing countries, water, forests, soil, and fisheries are resources which, in the absence of restrictions, can be exploited by any individual without charge. Such free access may in turn give rise to overexploitation of the resources to the point where they can no longer be renewed by natural processes.

The interrelated or interdependent nature of productive ecosystems compounds the problems of resource management in that action on one part of an ecosystem will in turn affect all the other parts of the ecosystem to varying degrees. Thus hydrologic and soil problems are affected by how the watershed is managed, which in turn is tied up with forestry management. Forestry, in its turn, is affected by encroachments on forest land for agriculture, for grazing and other forms of land use. Furthermore, the consequences of actions affecting ecosystems may take a long time to develop. The resource manager would thus need to be aware of the complexities and interconnections of ecological processes. His task is not made easier by the fact that administrative boundaries of political units seldom coincide with the geographical boundaries of ecological systems, such as a river basin or a watershed.

The issues connected with the development and management of natural resources in the tropical, developing world are mainly political issues as they involve the weighting of criteria accorded to their resources policies and the effects of such weighting on various groups in society. Measures to implement these policies may, in fact, disrupt the accustomed ways of life of many people.

Countries with low incomes would be expected to opt for a policy which seeks to maximize net economic benefits in the development and management of their natural resources. However, some may base decisions to develop resources in a region on considerations of equity among regions rather than those of maximum economic advantage. Similarly, governments may be motivated by values of a non-economic nature in deciding on the manner in which restrictions on common property resources should be imposed or, indeed, whether they should be imposed at all. Judgements on the proper relationship between privately-provided and state-provided natural resources-derived goods and services would depend on the ideology the society subscribes to.

In the final analysis, decisions on natural resources development and management will hinge on the options available to each country, and on how its leaders perceive the problems and decide on priorities.

REFERENCES

Abu-Lughod, J. 1977. "Development and Urbanization". *Habitat International* 2, nos. 5/6:417–26.

Ackerman, E.A. 1965. "Population and Natural Resources". In Burton and Kates (eds.), pp. 127–52.

Adeyoju, S.K. 1976. "Land Use and Tenure in the Tropics". *Unasylva* 28, nos. 112–13: 26–41.

Adnan Al-Janabi. 1979a. "Production and Depletion Policies in OPEC". *OPEC Review*, Mar., pp. 34–44.

_______. 1979b. "The Concept of Conservation in OPEC Member Countries". *OPEC Review*, Autumn, pp. 16–26.

"African Wood in World Trade". 1976. *The Courier*. Nov./Dec., pp. 38–40.

Ambroggi, R.P. 1980. "Water". In *Economic Development* (Scientific American, U.S.A.), pp. 39–49.

Ambrosetti, E., and Goward, S. 1981. "Resource Assessment for National Development". Professional Paper no. 13, Department of Geography and Geology, Indiana State University, pp. 11–17.

Arnold, J.E.M. 1979. "Wood Energy and Rural Communities". *Natural Resources Forum*, Apr., pp. 229–52.

Bairoch, P. 1975. *The Economic Development of the Third World since 1900*. Berkeley, Calif.

Bauer, P.T., and Yamey, B.S. 1957. *The Economics of Underdeveloped Countries*. Cambridge.

Bene, J.G., *et al*. 1977, *Trees, Food and People*. Ottawa: IDRC.

Bosson, R., and Varon, B. 1977. *The Mining Industry and the Developing Countries*. Washington, D.C.: World Bank.

Brooks, D.B., ed. 1974. *Resource Economics: Selected Works of Orris C. Herfindahl*. Baltimore, Md.

Brunig, E.F. 1979. "Utilization of the World's Forests: Possibilities and Limitations". *Applied Sciences and Development* 14:15–25.

Burton, I., and Kates, R.W. 1965. *Readings in Resource Management and Conservation*. Chicago.

Carroz, J.E. 1977. "The Conference on the Law of the Sea: Its Implications for ACP Fisheries". *The Courier*, Jan./Feb., pp. 76–78.

CENTO. 1971. *Seminar on the Application of Remote Sensors in the Determination of Natural Resources*. Ankara.

Chandler, R.F., Jr. 1979. *Rice in the Tropics: A Guide to the Development of National Programs*. Boulder, Colorado.

Christy, F.T., Jr. 1975. "Property Rights in the World Ocean". *Natural Resources Journal*, Oct., pp. 695–712.

Ciriacy-Wantrup, S.V. 1971. "The Economics of Environmental Policy". *Land Economics*, Feb., pp. 36–45.

_______, and Bishop, R.C. 1975. "Common Property as a Concept in Natural Resource Policy". *Natural Resources Journal*., Oct., pp. 713–28.

Claiborne, R. 1970. *Climate, Man and History*. New York.

David, Jacques. 1982. "The Economist's Purgatory". *The Courier*, Jan./Feb., pp. 89–94.

"Desertification in Africa'. 1978. *The Courier*, Jan./Feb., pp. 38–39.

Eckholm, E. 1979. *Planting for the Future: Forestry for Human Needs*. Worldwatch Paper 26. Washington, D.C.

______, and Brown, L.R. 1977. *Spreading Deserts: The Hand of Man*. Worldwatch Paper 13. Washington, D.C.

Eklan, W. 1978. *An Introduction to Development Economics*. London.

Fano, E. 1978. "The United Nations Desertification Conference". *Natural Resources Forum*, Jan., pp. 177–84.

______. 1981. "The Role of the International Community in the Drinking Water Supply and Sanitation Decade". *Natural Resources Forum*, July, pp. 261–69.

Fisher, J.L. 1964. "The Role of Natural Resources". In Williamson and Buttrick, eds., pp. 22–26.

Fosberg, F.R., *et al.* 1961. "Delimitation of the Humid Tropics". *Geographical Review*, 51:333–47.

Furtado, J.I. 1979. "The Status and Future of the Tropical Moist Forest in Southeast Asia". In C. MacAndrews and Chia Lin Sien, eds., *Developing Economies and the Environment*. Singapore.

Galbrath, J.K. 1953. "Conditions for Economic Change in Underdeveloped Countries". *Journal of Farm Economics*, Nov.

Garnier, B.J. 1958. "Some Comments on Defining the Humid Tropics". *Research Notes* Ibadan 11:9–25.

Gilfillan, S.C. 1920. "The Coldward Course of Progress". *Political Science Quarterly*, Sept., pp. 393–410.

Gillis, M., *et al.* 1978. *Taxation and Mining: Non-Fuel Minerals in Bolivia and Other Countries*. Cambridge, Mass.

Ginsburg, N. 1957. "Natural Resources and Economic Development". *Annals, Association of American Geographers*, Sept., pp. 197–212.

Global 2000 Report to the President, 1982. London.

Gourou, P. 1968. *The Tropical World*. London.

Greenland, D.J., and Lal, R. 1977. *Soil Conservation and Management in the Humid Tropics*. Chichester.

Grossling, B.F. 1976. *Window on Oil*. Financial Times, London.

Hass, J. 1975. "Working Capacity and Total Metabolic Rate of Man Performing Industrial and Traditional Tasks in a Tropical Climate". *Applied Sciences and Development* 5:78–94.

Hailey, Lord. 1957. *Africa Survey*. London.

Hardin, G. 1968. "The Tragedy of the Commons". *Science*, Dec., pp. 1243–48.

Haveman, R.H. 1973. "Common Property, Congestion and Environmental Pollution". *Quarterly Journal of Economics*, May, pp. 278–87.

Herfindahl, O.C. 1965. "What Is Conservation?" In Burton and Kates, eds., pp. 229–36.

______. 1967. "Depletion and Economic Theory". In M. Gaffney, ed., *Extractive Resources and Taxation*, pp. 68–90.

______. 1969. *Natural Resource Information for Economic Development*. Baltimore, Md.

Hotelling, H. 1931. "The Economics of Exhaustible Resources". *Journal of Political Economy*, Apr., pp. 137–75.

Hunter, J.M. 1966. "Ascertaining Population Carrying Capacity under Traditional Systems of Agriculture in Developing Countries: Notes on a Method Employed in Ghana". *Professional Geographer* 18:151–54.

Higgins, B. 1968. *Economic Development: Principles, Problems and Policies*. New York.

Hinckley, A.D. 1980. *Renewable Resources in our Future*. Oxford.

House, P.W., and Williams, E.R. 1977(?). *The Carrying Capacity of a Nation*. Lexington.

Huntington, E. 1924. *Civilization and Climate*. New Haven.

_______. 1930. "The Effect of Climate and Weather". In E.V. Cowdry, ed., *Human Biology and Racial Welfare*. New York, pp. 295–330.

Jafar Mansur Saad. 1979-80. "Conservation: Towards a Comprehensive Study". *OPEC Review*, pp. 153–57.

Johnson, H.G. 1967. *Economic Policies towards Less Developed Countries*. Washington.

Julius, D.S., and Buky, J.B. 1980. "Assessment of the Economic Contribution of Water Resources to National Development". *Natural Resources Forum*. April, pp. 212–19.

Jung, L. 1977. "The Behaviour of the Soil in Response to Changes in Land Use". *Natural Resources and Development* 5:46–51.

Justaz, H.J. 1973. "The Significance of the Situation Relating to Infectious Diseases for the Development of Tropical Countries". *Applied Sciences and Development* 1:87–101.

Kamarck, A.M. 1976. *The Tropics and Economic Development*. World Bank, Washington, D.C.

Kay, J.A., and Miarlees, J.A. 1975. "The Desirability of Natural Resources Depletion". In Pearce and Rose, eds., pp. 140; 76.

Kebschull, D., and Schoop, H.G. 1975. "The Importance of Raw Materials in Economic Development". *Natural Resources and Development* 2:7–19.

Kindleberger, C.P. 1966. *Economic Development*. New York.

Krasner, S.D. 1974. "Oil Is the Exception". *Foreign Policy*. Spring.

Lee, D.H.K. 1957. *Climate and Economic Development in the Tropics*. New York.

Lewis, W.A. 1955. *Theory of Economic Growth*. London.

Lovejoy, W.F., and Homan, P.T. 1965. *Methods of Estimating Resources of Crude Oil, Natural Gas and Natural Gas Liquids*. Baltimore.

Marley, R. 1979. "Environment and Resource Policy in Developing Countries". *Natural Resources Forum* 3:179–86.

Marr, J.C. 1976. *Fishery and Resource Management in Southeast Asia*. Resources for the Future, Washington, D.C.

Marshall, C. 1958. "Land Utilization in the Humid Tropics: Forestry". *Proceedings of the 9th Pacific Science Congress*, 1957 (Bangkok), vol. 20, pp. 148–56.

Masefield, G.B. 1970. "Food Resources and Production". In J.P. Garlick and R.W.J. Keay, eds., *Human Ecology in the Tropics*.

May, B. 1981. *The Third World Calamity*. London.

Miller, A.A. 1971. *Climatology*. London.

Ministry of Finance, Malaysia. 1978. *Economic Report 1978/79*. Kuala Lumpur.

Morallet, J. 1976. "Trends in Tropical Timber Production". *The Courier*, Nov./Dec., pp. 47–48.

Motamen, H. 1979. "Economic Policy and Exhaustible Resources". *OPEC Review*, Mar., pp. 45–75.

Myrdal, Gunnar. 1970. *An Approach to the Asian Drama*. New York.

NASA. 1979. *Significant Foreign Landsat Results*, Doc. No. 0238A, Jan. 18.

Nehring, R. 1978. *Giant Oil Fields and World Oil Resources*. Rand, Santa Monica.

Oliver, J. 1979. "A Study of Geographical Imprecision: The Tropics". *Australian Geographical Studies*, Apr., pp. 3–17.

Ooi Jin Bee. 1959. "Rural Development in Tropical Areas, with Special Reference to Malaya". *Journal of Tropical Geography* 1:222.

_______. 1976. *Peninsular Malaysia*. London.

Openshaw, K. 1978. "Woodfuel: A Time for Re-assessment". *Natural Resources Forum*, Oct., pp. 35–51.

Pauzer, K.F. 1981. "Inventory and Monitoring of Renewable Resources in the Sahel".

Natural Resources and Development 14:78–88.

Pearce, D.W., and Rose, J. eds. 1975. *The Economics of Natural Resources Depletion.* London.

Perloff, H.S., and Wingo, L., Jr. 1965. "Natural Resource Endowment and Regional Economic Growth". In Burton and Kates, eds., pp. 427–42.

Pindyck, R.S. 1978. "Gains to Producers from the Cartelization of Exhaustible Resources". *Review of Economics and Statistics*, May, pp. 238–51.

Plourde, C. 1975. "Conservation of Extinguishable Species". *Natural Resources Journal*, Oct., pp. 791–97.

Population Reference Bureau, Inc. 1981. *1981 World Population Data Sheet.* Washington, D.C.

Price, A. Grenfell. 1939. *White Settlers in the Tropics.* New York.

Rifkin, J. 1980. *Entropy: A New World View.* New York.

Ritter, Wigand. 1975. "Natural Resources in Developing Countries". *Natural Resources and Development* 1:44–58.

Robinson, C. 1975. "The Depletion of Energy Resources". In Pearce and Rose, eds., pp. 21–55.

Sanger, C., *et al.* 1977. *Trees for People.* Ottawa.

Sauer, Carl. 1969. *Agricultural Origins and Dispersals.* 2nd ed. Cambridge, Mass.

Schacher, O. 1977. *Sharing the World's Resources.* New York.

Schultz, T.W. 1965. "Connections between Natural Resources and Economic Growth". In Burton and Kates, eds., pp. 397–403.

Sobharam, Singh. 1980. "Mineral Exploration in Tropical Rain Forests: Practical Problems and Directions". *Natural Resources Forum*, Oct., pp. 437–47.

Solow, R.M. 1974. "The Economics of Resources or the Resources of Economics", *American Economic Review*, May, pp. 1–14.

Sommer, A. 1976. "Attempt at an Assessment of the World's Tropical Moist Forests". *Unasyla* 28:5–26.

Sorokin, P. 1928. *Contemporary Sociological Theories.* New York.

Spoehr, A. 1965. "Cultural Differences in the Interpretation of Natural Resources". In Burton and Kates, eds., pp. 110–18.

Stoneman, J., and Disney, J. 1980. "Fisheries Development in the ACP Countries". *The Courier*, Nov./Dec., pp. 65–68.

Sundrum, R.M. 1977. *The Limits of Economics for the Study of Development.* Occasional Paper no. 7. Development Studies Centre, Australian National University, Canberra.

Tanzer, M. 1978. "Oil Exploration Strategies for Developing Countries". *Natural Resources Forum*, July, pp. 319–26.

———. 1980. *The Race for Resources.* London.

Taylor, G.C., Jr. 1979. "The U.N. Groundwater Exploration and Development Programme: A Fifteen-Year Perspective". *Natural Resources Forum*, Jan., pp. 147–66.

"The International Livestock Centre for Africa". 1981. *The Courier.* Jan./Feb., pp. 75–77.

Tosi, J.A., Jr., and Voertman, R.F. 1964. "Some Environmental Factors in the Economic Development of the Tropics". *Economic Geography*, July, pp. 189–205.

U.N. 1975a. *Predicted Costs and Benefits Involved in the Practical Application of Remote Sensing Technology.* Doc.A/AC. 105/153.

U.N. 1975b. *Summary of Studies on Cost Effectiveness in Remote Sensing.* Doc. A/AC. 105/139 27 Jan., A/AC. 105/139 Add. 1, Apr. 17.

U.N. Secretariat. 1981. "Two Decades of Mineral Resources Development: The Role of the United Nations". *Natural Resources Forum*, Jan., pp. 15–30.

U.S. Agency for International Development. 1972. *The Application of Geochemical, Botanical, Geophysical, and Remote Sensing Mineral Prospecting Techniques to Tropical Areas.* Report TA/OST 72-13, Washington, D.C.

Vance, R.B. 1932. *Human Geography of the South.* Chapel Hill, U.S.

Varon, B., and Takeuchi, K. 1974. "Developing Countries and Non-Fuel Minerals". *Foreign Affairs*, Apr.

Weistroffer, E.K.K., *et al.* 1977. "Fossil Reserves of Groundwater in the Central Sahara". *Natural Resources and Development* 5:19–45.

Wharton, C.R. 1963. "Monocultural Perennial Export Dominance: the Inelasticity of Southeast Asian Agricultural Trade". Mimeo.

White, G.F., *et al.* 1972. *Drawers of Water: Domestic Water Use in East Africa.* Chicago.

Wilkens, C.H. 1930. "Dr. Huntington and Low Latitudes". *Economic Record* 6:123–27.

Williamson, H.F., 1964. "Introduction". In Williamson and Buttrick, eds., pp. 3–21.

———, and Buttrick, J.A. eds. 1964. *Economic Development: Principles and Patterns.* New Delhi.

Winckler. G. 1977. "Fishing: Making More of a Major Natural Resource". *The Courier*, Jan./Feb., pp. 48–50.

World Bank. 1978. *Forestry.* Washington, D.C.

World Bank. 1979a. *A Program to Accelerate Petroleum Production in the Developing Countries.* Washington, D.C.

World Bank. 1979b. *Energy Options and Policy Issues in Developing Countries.* Washington, D.C.

World Bank. 1979c. *Development Problems of Mineral-Exporting Countries.* Washington, D.C.

Worthington, E.B. 1958. *Science in the Development of Africa.* Hertford, U.K.

Young, A. 1968. "Natural Resource Surveys for Land Development in the Tropics". *Geography*, July, pp. 229–48.

Zakariya, H.S. 1980. "Sovereignty over Natural Resources and the Search for a New International Economic Order". *Natural Resources Forum*, Jan., pp. 75–84.

Zuleta, B. 1980. "The Third UN Conference on the Law of the Sea and Its Impact on Third World Countries". *The Courier*, Nov./Dec., pp. 63–64.

I

MINERAL AND ENERGY RESOURCES

2
MNC Strategy and Regional Growth Potential of Mining: Caribbean Bauxite's Forward Linkage

R.M. AUTY

The rapid expansion of non-communist primary aluminium production from 0.73 million tonnes in 1940 to 10.96 million tonnes in 1974 on the eve of the first OPEC-deepened recession was largely based on bauxite mining in the tropics (U.S. Bureau of Mines 1944; 1975). Yet only a small fraction of the total value added in the aluminium production chain took place in the tropical mining regions, ranging from about fifth in Oceania to less than one-tenth in the tropical Americas.[1] In this respect, aluminium compares unfavourably with other major metals: the World Bank reported that in 1970 the proportion of value added within the host country was 79 per cent for tin, 78 per cent for copper, 69 per cent for lead, 42 per cent for zinc, 25 per cent for iron ore, and 10 per cent for bauxite (Radetzki 1977:327).

Caribbean economists have blamed multinational corporate strategy for severely reducing the regional economic impact of Caribbean bauxite (Huggins 1965:46; Girvan 1971:15–98). Reformulated in terms of export base theory, their argument states that corporate strategy minimized the Caribbean's forward linkage (alumina refining, aluminium smelting and fabrication) from bauxite mining and curtailed backward, final demand and fiscal linkages (local revenue retained). Since the magnitude of backward, final demand, and fiscal linkage is largely conditional on securing forward linkage (Auty 1980:169–72), this paper will focus on forward linkage. The issue of potential revenue retention is reserved for full examination elsewhere.

The alleged sub-optimal location of forward linkage outside the Caribbean and the developing countries as a whole (Radetzki 1977:328–29) has been ascribed to corporate reluctance to make large investments outside the OECD countries. The Caribbean economists' case implies that a strategy of vertical integration engendered oligopolistic competition, rendered the

---

[1] Brazil and India are excluded from this generalization as they functioned as small, self-contained industries until the current phase of capacity expansion.

corporations insensitive to spatial variations in costs, and deflected refining, smelting, and fabrication away from cost-minimizing locations in the Caribbean (Girvan 1971:75–98). The need to distort data to comply with dependency theory is evident in Girvan's work (1970): his case rests on incorrect conclusions about the nature of competition under oligopoly and highly selective use of available cost data. The present work synthesizes a wide range of published and unpublished data drawn from industry consultants, the aluminium corporations, and government agencies in the Caribbean and North America to provide a strong empirical base for the conclusions reached.

The link between corporate strategy and the location of forward linkage is examined in two stages. First, the implications of the aluminium industry's technological characteristics and recent emergence for corporate strategy are examined. While the existence of economies of scale, large capital requirements, and imperfect markets for both company-specific resources and certain corporate inputs fostered vertical integration, oligopoly, and uniform target pricing, it is shown that none of these characteristics is incompatible with intense locational cost competition. The second part of the paper uses cost estimation (Auty 1975) to trace the evolving locational economics of each stage in the production chain, with particular reference to the competitiveness of the Caribbean. The sensitivity of the corporations to locational costs is explored for each of three distinct periods following the termination of Alcoa's U.S. monopoly in 1940. Such a historical perspective is necessary, first, because much of the case against the corporations rests on studies made of this period and, second, because there is a strong element of spatial inertia in an industry characterized by heavy investments with lead-times of up to ten years and life-times of twenty years or more. The paper concludes by restating the determinants of the location of forward linkage and assessing Caribbean prospects for attracting such forward linkage.

Corporate Strategy and Market Structure, 1940–80: Declining Oligopoly

The North American aluminium industry has passed through three differing periods of operating conditions since the termination of Alcoa's U.S. monopoly in 1940. The first period, lasting until the late 1950s, was dominated by government incentives aimed not only at the rapid expansion of capacity for the Second World War and the Korean War but also at increasing the number of domestic aluminium producers to weaken Alcoa (Table 2.1). Each corporation operating in the first period eventually adopted a strategy of full vertical integration for a single product. There followed a period of excess capacity exacerbated by retaliatory action from foreign aluminium producers and North American copper and steel companies as the large U.S. aluminium producers sought to diversify both their geographical markets and product range. As both demand and the number of producers expanded (Table 2.1), it became increasingly possible to depart

TABLE 2.1 ENTRY OF PRODUCERS INTO THE NORTH AMERICAN PRIMARY ALUMINIUM INDUSTRY

Year	Company	North American Smelting Capacity, 1975 (million tonnes)	Prime Motive for Entry
1888	Alcoa	1.429	Process invention.
1928	Aluminium Ltd./ Alcan	0.907	Pre-1928 Alcoa subsidiary. Ownership connections totally cut 1950 by U.S. Court.
1940	Reynolds	1.060	Backward integration from foil fabrication.
1946	Kaiser	0.629	Light metals maker for U.S. government in WW2.
1955	Anaconda	0.272	Market defence by copper corporation.
1955	Ormet[1]	0.103	Revere copper corporation and Ohlin (WW2 U.S. Government plant operator) joint venture.
1958	Harvey	0.191	Backward integration. 1969 taken over by Martin Marietta defence contractor.
1963	Conalco[2]	0.161	60 per cent Alusuisse: retaliatory entry to U.S.
1966	Howmet[3]	0.396	Pechiney retaliatory entry to U.S.
1966	Amax		Diversification by U.S. metal corporation.
1969	National-Southwire	0.163	Market defence by two steel corporations.
1971	Noranda	0.064	Canadian metal corporation diversification.

Notes: [1]Ohlin's share was acquired by Conalco and Revere entered the industry in its own name. Capacity figure represents latter alone.

[2]Alusuisse also owns 40 per cent of original Ormet, Revere owns 34 per cent, and Phelps Dodge (Alusuisse' Conalco partner) owns 26 per cent. Ormet 1975 capacity 0.236.

[3]Howmet has 50 per cent equity in two smelters with Alumax.

Sources: U.S. Bureau of Mines 1975; President's Council on Wage and Price Stability 1976:126.

from the fully vertically integrated, single product strategy of the previous period. Besides geographical and product diversification, increasing recourse was made to joint ventures at the smelting stage and consortia for mining and refining, thereby spreading the risks of large new investments (Meyer 1977) and easing the barriers to entry and expansion posed by continued increases in the minimum viable scale of production at all four stages in the chain. In the third period, OPEC-induced upheavals in factor markets paralysed investment in the mid-1970s, eventually eliminating excess capacity and triggering a new round of expansion. Although oligopoly declined in North America as the number of producers increased to twelve, 71 per cent of the region's smelting capacity was controlled by four fully-integrated dominant-product companies, and uniform target pricing and price leadership persisted, provoking a major investigation in the mid-1970s by the Council on Wage and Price Stability (1976). By the late 1970s, six large corporations (the big four North American companies together with Pechiney and Alusuisse) owned half of world smelter capacity, fifty private companies owned one quarter, and twenty-four governments owned the other quarter (Stamper and Kurtz 1978:1).

Economies of scale (notably in refining), the early superiority of vertical integration, and the need to reduce the risks on large investments characterized by long lead-times and long payback periods, explain evolving North American corporate strategy and market structure. Alcoa successfully defended its monopoly of the U.S. market against domestic entrants from its establishment in Pittsburgh in 1888 until 1940, through a combination of patents and control of alumina refining (Lanzillotti 1961). On the eve of the Second World War, the corporation supplied the entire U.S. market with four smelters fed by a single refinery which was nevertheless still operating below the minimum optimum size. During the massive war-time expansion, three new refineries were built at a cost of about $30 million each, ten times the cost of a smelter of minimum viable size and five times such a smelter's requirements (Klagsbrunn 1945). Although continued U.S. market expansion since 1939 has reduced the minimum viable size of a refinery from 150 per cent of the total domestic market to less than 7 per cent, both present size (800,000 tonnes) and cost (about $500 million) cause the refinery to persist as a formidable barrier to entry for a corporation. The total capital costs of establishing the entire vertical chain have compounded the problem of entry. In the absence of special concessions such as a government subsidy, the total cost of entry was estimated at $400 million in the mid-1950s (Peck 1961), $700 million in the late 1960s (Younger 1971), and $1 billion in the late 1970s. The two latter figures exclude investment in smelter-owned power facilities and assume consortia membership at the mining and refining stages in order to match alumina supply with ingot capacity, thereby avoiding the investment requirements of a totally self-sufficient system estimated at over $2 billion in 1979.

The preference for vertical integration in spite of the formidable capital requirements, especially among early entrants, reflects the vulnerability of an unintegrated smelter, with its high fixed costs of capital and contract to purchase large blocks of power, to manipulation of imperfect markets by material supplier, ingot purchaser, or government taxation bureau. For example, during the first period of expansion, new entrants such as Kaiser (1946) and Anaconda (1955), which chose initially to rely on alumina purchases from established corporations, were charged rates which added about 6 per cent to their total costs at time when alumina normally accounted for no more than 15 per cent of total revenue (Loeb 1950; Lanzillotti 1961). Since bauxite, unlike alumina, varies in chemical composition and refineries are therefore designed to handle specific types of ore, there is an even stronger incentive to integrate backwards from refinery to mine than from smelter to refinery. Although third-party sales of bauxite have been made throughout the period under study, especially in the Far East, they still comprise less than one-seventh of all sales.

At the other end of the production chain, Aluminium Limited, which had specialized as a low-cost exporter of ingot, was forced to invest more than $600 million during the period of excess capacity in unprofitable fabricating outlets to secure markets for its own ingot, as the U.S. corporations terminated long-term purchase agreements with the Canadian company to maximize capacity utilization throughout their fully integrated production chains. Those fabricating companies which had not integrated backwards into smelting at this time also came under intense pressure as the three largest U.S. corporations (Alcoa, Reynolds, and Kaiser) adjusted prices to squeeze the profit margin on fabrication and raise it on the earlier stages in the production chain where entry was more difficult and the recent grant of Western Hemisphere Trade Corporation status reduced tax liability. Between 1957 and 1965, the contribution of fabrication to operating profits of the three largest U.S. corporations is estimated to have fallen from 27 to 17 per cent (Donaldson *et al.* 1966). Therefore, in imperfect markets, the integrated production chain gives considerable flexibility in the short-run absorption of higher costs or lower margins at any particular stage in response to aggressive moves by competitors or governments; or when a new plant comes onstream and initially operates below optimum capacity.

Stigler (1951) has hypothesized that increasing market size leads to increased specialization of companies in differing facets of production and the replacement of monolithic companies by numerous interacting units. One explanation for the North American aluminium companies' continued preference for a vertically integrated dominant product strategy, despite evidence that it was one of the least successful strategies in post-war U.S. industry as a whole (Rumelt 1974), is that unlike most other industries pursuing that product strategy, aluminium was still in the youthful and dynamic section of the generalized S-shaped growth curve. Newer and less fully integrated entrants such as Alumax and Noranda may prove more successful as the industry enters what most observers regard as the mature phase of its development. A second explanation for the persistence of vertical integration has been outlined by Dunning (1977) and rests on the existence of imperfect markets for company-specific resources as well as location-specific ones. To capture the benefits of unpriced expertise and to eliminate uncertainties over the continuing flow of materials arising from inadequate futures markets and government policies, vertical integration may not only persist but be efficiently accompanied by horizontal integration.

The prevailing corporate strategy led to an industry dominated by companies with large sums at risk and new investments that are frequently lumpy and characterized by lead-times of four to ten years. The caution bred by these risks is reflected in the continuance of target pricing from Alcoa's monopoly days. The target price is based on long-run costs plus a specified percentage, usually 15 to 20 per cent, reflecting the anticipated pre-tax rate of return considered necessary to generate an adequate cash flow for the industry at an assumed level of capacity utilization of about 70 per cent (Peck 1961). With the emergence of excess capacity in the late 1950s, the corporate goals of growth and market share were downgraded in favour of the return on capital. The current expansion shows that alternative schemes for system expansion are consistently monitored and ranked, after bargaining with governments and prospective partners, according to the price needed to secure the target rate of return. All else being equal, there is a preference for brownfield over greenfield expansion since less capital is required and fewer risks are incurred. Price movements automatically trigger capacity expansion, regulated by the constraint imposed by cash flow and the need of less promising projects for a higher target price.

During the first period of expansion, the established corporations, Alcoa and its former subsidiary Aluminium Limited, acted as a "core" pursuing a conservative expansion strategy, fearful of anti-trust dismemberment, for the more dynamic "fringe" group of new entrants. Alcoa functioned as the barometric price leader during most of this period, moving prices with considerable sensitivity to competitor response. Under such circumstances, the core price leader would favour a lower price than the fringe as a result of lower costs arising from scale economies, lower (historic, sunk) capital costs, and greater ease in generating either equity or long-term loans because its equity base is much larger than those of its more highly levered fringe rivals.

Kaiser and Reynolds appear to have functioned as part of the core during the period of excess capacity from the late 1950s to the early 1970s (Langton 1978), with the smaller entrants acting the role of the fringe. Three important consequences of this partial oligopoly for the fringe are: first, the core allows them to reduce prices in times of overcapacity and so operate closer to full capacity (Council on Wage and Price Stability 1976). Second, starting from an initial position close to capacity, the fringe tends to expand sooner as supply tightens. Thus, both Reynolds and Kaiser expanded faster than Alcoa when they were fringe members in the first period (Table 2.2), while the smaller, newer entrants did this in the second period (Table 2.3), and a fringe company, Alumax expanded early and rapidly in the present phase of growth (*Business Week*, 1978). The third consequence for the fringe is that the core sets an ultimate limit on market share erosion by lowering prices itself at that limit. The core moved in this manner around 1963–64 (*Corporate Annual Reports and Accounts*).

Partial oligopoly with barometric price leadership and target pricing does not necessarily result in excess profits within the industry. Returns were certainly high in the 1950s prior to the onset of oversupply, with the highly levered fringe companies of that period, Kaiser and Reynolds, recording rates of return on shareholders' equity of 15–20 per cent (*Corporate Annual Reports and Accounts*). Nevertheless, this occurred in response to government war-time incentives with little certainty that peace-time uses could be found for the extra plant. Moreover, returns subsequently slumped: the Council on Wage and Price Stability (1976:202) calculated the rate of return on total assets employed in 1965–74 at 6.7 per cent for the aluminium industry, compared with 6.9 per cent for steel, 10.1 per cent for copper,

TABLE 2.2 NORTH AMERICAN INGOT CAPACITY OF FOUR LARGEST
PRODUCERS AND FRINGE, 1945–74

	Company						Total (million tonnes)
Year	Alcan	Alcoa	Reynolds	Kaiser	Fringe	Four Largest (%)	
1945 (million tonnes)	0.454	0.376	0.176	0.118	—	100	1.124
1945 (%)	40	33	16	10	—	100	1.124
1954 (%)	32	28	20	20	—	100	1.839
1964 (%)	24	24	23	18	10	90	3.236
1974 (%)	17	26	20	12	25	75	5.405
1974 (million tonnes)	0.907	1.429	1.060	0.659	1.350	75	5.405

Sources: Reynolds 1947:25; Aluminium Association 1979:34–35; U.S. Bureau of Mines 1975.

TABLE 2.3 NORTH AMERICAN CORE AND FRINGE SMELTER EXPANSION
BY REGION, 1959–74

Group	Region	1959 million tonnes	1959 %	1964 million tonnes	1964 %	1969 million tonnes	1969 %	1974 million tonnes	1974 %
Core	Northeast	0.756	29	0.842	30	0.965	27	1.034	26
	South	1.055	40	1.089	39	1.338	37	1.482	37
	Northwest	0.695	26	0.717	25	1.020	28	1.116	28
	Midwest	0.132	5	0.164	6	0.307	8	0.381	9
	Total	2.638		2.812		3.630		4.013	
	Number of plants	24		25		25		25	
Fringe	Northeast	—		—		—		0.160	10
	South	—		0.029	9	0.127	15	0.264	17
	Northwest	0.103	38	0.140	42	0.482	55	0.587	37
	Midwest	0.163	61	0.153	49	0.258	30	0.573	36
	Total	0.266		0.332		0.887		1.582	
	Number of plants	3		4		6		12	

Source: U.S. Bureau of Mines 1960; 1965; 1970; 1975.

15.1 per cent for lead and zine, and 10.5 per cent for Standard and Poors'
425 industrials. With limited price competition on largely undifferentiated
products, cost efficiency becomes the critical area of competition between
companies. Throughout the entire period under study, the corporations
were keenly aware of the role of location in cost minimization, especially as
they diversified from hemispheric to global markets in the late 1950s and
encountered longer hauls, wider ranges in input costs, and alternative
bauxite sources to the Caribbean.

Market-Orientation of Processing to the Late 1950s

Since bauxite varies considerably in chemical composition, refineries are
designed to draw from specific mines and this, together with freight savings
from the weight-loss in transforming two or three tonnes of bauxite into one
tonne of alumina, exerts a strong pull on the refinery to the mine. Smelting
also results in a weight loss of about 2:1 as alumina is converted to aluminium, but historically cheap power has been the critical location factor
for this stage. Since the Guianese Highlands offered such cheap power
throughout the period under study, and from the late 1950s nuclear power
held the prospect of cheap energy in the Northern Antilles, forward integration from mine to smelter appeared viable within the region. Although
fabrication involves little weight loss, savings on reheating for semi-fabrica-
tions such as sheet, coil, and rod have encouraged close proximity between

smelters and mills, if not between the former and extrusion plants where the product is characterized by a high volume:weight ratio. If the above generalizations hold, then the mine-site location of the refinery becomes the lynch-pin in securing virtually complete forward integration within the Caribbean region, drawing on hydro-electric power in the south and nuclear power in the north. That the pull of the refinery to the mine was, with one exception, ineffective in the Caribbean during the first expansionary period appears to support the thesis of the sub-optimal location of refinery capacity in North America. The case becomes even stronger if Stern's figures, cited by Huggins (1965:46), are accurate in suggesting that in the mid-1950s a new Jamaican refinery would have been significantly cheaper than a similar plant on the U.S. Gulf Coast. Moreover, since brownfield expansion has been preferred to greenfield expansion not only for its lower risks but also for its frequently lower unit costs, the implied spatial inertia makes the initial location of mines, refineries, and smelters significant for subsequent growth patterns.

Refinery Location (Fig. 2.1)

In the first phase of expansion, strategic considerations during the Second World War briefly retarded the growing reliance of North America

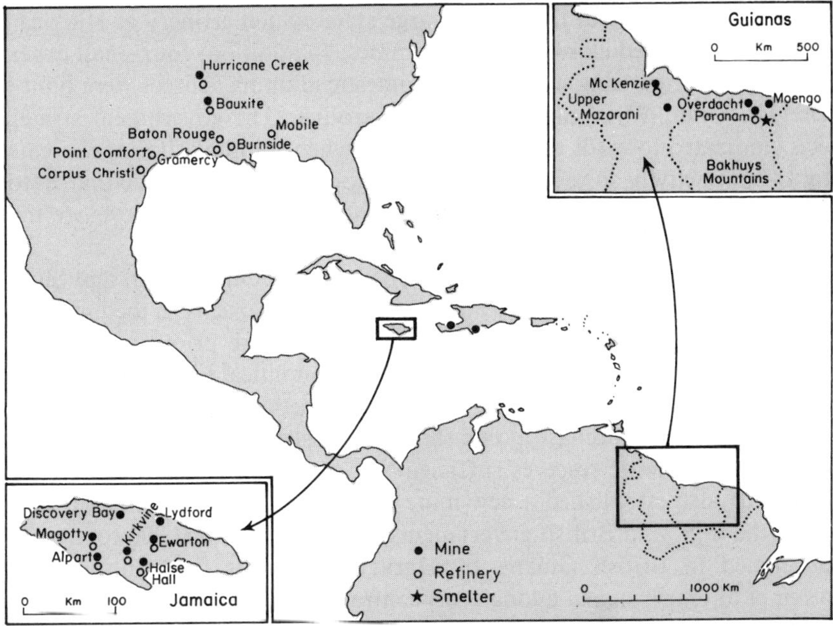

Fig. 2.1 *Mines and refineries in the United States and Caribbean, late 1970s*

on imported bauxite from the Guianas and later the Northern Antilles. Proven U.S. domestic reserves were small and of low quality (requiring over 2.6 tonnes/tonne of alumina compared with 1.9 tonnes for Guianese ore). Overland freight costs were also extremely high. However, since in the mid-1940s f.o.b. mine costs of imported ore at $12.94/tonne alumina were only fractionally cheaper than domestic ore at $13.72/tonne alumina (U.S. Bureau of Mines 1944), limited long-term supply prospects was the critical shortcoming of domestic ore. Thus, although its pre-war refinery at East St. Louis (established in 1903) drew almost equally on domestic and imported ores, Alcoa like Aluminium Limited, based its war-time expansion programme on bauxite from the Guianas (Suralco 1955; Roach 1957). While the Canadian company expanded its sole refinery in Quebec to secure greater economies of scale, Alcoa built a greenfield plant at Mobile to handle Surinamese ore. The strategically determined U.S. alumina import duty of $11/tonne exceeded the potential savings on transfer costs of a refinery located in Suriname.

Seven additional refineries were built during the war, but these were government-backed schemes and cost considerations were secondary to the strategic imperative of a rapid, secure expansion of supply. Although one sizeable government refinery followed Alcoa's lead and was built at Baton Rouge, the small Listerhill facility built by the government to assist the entry of Reynolds and the second large state-owned refinery at Hurricane Creek were located close to domestic mines. In addition, four small experimental refineries using non-bauxite domestic alumina sources were built in Oregon, Utah, Wyoming, and South Carolina. The government strategy was vindicated towards the end of the war when shipping shortages created by U-boat activity so severely curtailed imports from Suriname that Baton Rouge had to rail in domestic ore, pushing its bauxite costs to more than double those of Hurricane Creek (Klagsbrunn 1945).

Although the Guianese mines of Alcoa, Aluminium Limited, and Shell's Surinamese mining subsidiary, Billiton, had excess capacity in the immediate post-war years, the U.S. fringe companies set a high priority on securing their own mines. Reynolds shut its small uneconomical Listerhill refinery in 1946 and leased the government's Hurricane Creek plant until 1949, when it made an outright purchase at one-third the construction cost, having gained control of domestic reserves sufficient for at least ten years of operation. Reynolds also established a new mine in northern Jamaica with soft loans from the U.S. and British governments, took over the 200,000 tonne mine developed in British Guiana by Harvey during its first (unsuccessful) attempt at entry, signed a long-term contract with the unintegrated Billiton firm, and considered but rejected the possibility of constructing a refinery in the Pacific Northwest using Indonesian bauxite (Reynolds 1947). In first leasing, and then purchasing cheaply, the Baton Rouge refinery from the U.S. government, Kaiser inherited a twenty-year contract for up to 500,000 tonnes bauxite a year from Alcoa's Surinamese mine. But this option was phased out from the mid-1950s as Kaiser brought its own new mine into production in southwest Jamaica.

Table 2.4 shows that the four largest North American producers drew most of their incremental bauxite supplies from the Northern Antilles in the 1950s. Although the new ores had 5 to 10 per cent less alumina content than those of the Guianas, this was partly offset by the smaller volume of silica requiring removal. However, the Northern Antilles bauxite was covered by only a few centimetres of overburden compared with the tens of metres in the non-plateau Guianese mines which threatened the viability of production there (Suralco 1955). Per tonne of aluminium, extraction costs in Jamaica were below $14 compared with $18 for bauxite from the Guianas, both figures f.o.b. Being more than 2,000 km closer to North America and accessible to vessels with over five times the draft of those plying the shallow rivers of the Guianas, Jamaican bauxite had transfer costs of $1.75/tonne to the Texas Gulf Coast compared with $7/tonne from Suriname and $9/tonne between British Guiana and Quebec (Brubaker 1967). The

TABLE 2.4 BAUXITE SOURCES OF FOUR NORTH AMERICAN CORPORATIONS, 1950-74

Corporation	Year	Source						
		Arkansas	Suriname	Guyana	Jamaica	Haiti	Dominican	Guinea
		(million tonnes)						
Aluminium Ltd[1]	1950	—	—	1.485	—	—	—	—
	1959	—	—	1.273	1,070[4]	—	—	0.301
Alcoa[2]	1950	0.457	1.603	—	—	—	—	—
	1960	0.610	1.963	—	—	—	0.734	—
Kaiser[3]	1950	—	n.a.	—	—	—	—	—
	1959	—	n.a.	—	2.335	—	—	—
Reynolds	1950	0.806	n.a.	—	—	—	—	—
	1959	0.816	—	0.161	1.853	0.312	—	—

Regional Total	Year	North America %	Guianas %	Northern Antilles %	Africa %	Australia %	Total %
	1950	29	71	0	0	0	100
	1959/60	12	30	55	3	0	100
	1974	6	18	45	9	22	100

Notes: [1]Excludes French subsidiary (0.3 million tonnes, 1960), Brazil, India, Malaysia.
[2]Shipped as 0.476 million tonnes of alumina.
[3]Data not available for 1959.
[4]Kaiser received shipments from Alcoa which declined in volume through the decade. In 1950 Reynolds imported bauxite from Billiton in Suriname.
Sources: Girvan (1971); Moment (1962); Corporate 10-K Reports (1974).

three U.S. fringe companies which entered the industry in the 1950s were less successful in quickly securing captive bauxite/alumina supplies than Reynolds or Kaiser: Anaconda purchased alumina first from Reynolds and then, in return for helping finance a new refinery, from Kaiser at a substantial premium; Ormet resorted to multiple sourcing of bauxite from third parties for its Burnside refinery; and Harvey, having lost its British Guiana mine to Reynolds, imported alumina from Japan to feed its Oregon smelter.

Under the stimulus of the Korean War expansion programme's rapid amortization allowances, Alcoa built a new inland refinery at Bauxite using domestic ore and Reynolds expanded Hurricane Creek (though drawing on some imported ore). However, the fact that the small refinery at Bauxite was not expanded and concentrates on non-metal grade aluminas underscores the transfer cost disadvantage of inland refineries where the railway companies could charge high freight rates on both bauxite and alumina. North Caribbean mines incurred cheap shipping charges, and the option of shipping alumina up the Mississippi or around the coast from coastal refineries effectively restrained the prices which the railways could charge on hauls from such refineries to the Midwest, Northeast, or Pacific Northwest (Badger 1956:14). Whereas Pacific Northwest refineries built with accelerated amortization and using local non-bauxite ores or imported Southeast Asian bauxite might have been competitive with alumina delivered from Gulf refineries, none of the corporations had sufficient smelting capacity in the region to justify the erection of a plant of minimum viable size. Moreover, while such potential plants could not deliver alumina at competitive prices east of the Rockies, alumina railed or shipped from the Gulf to the Pacific Northwest cost only twice as much ($12/tonne) as journeys one-fifth that distance to smelters in the South (Badger 1956:49).

With the termination of rapid amortization allowances by the mid-1950s, the critical question was no longer whether to locate a refinery in the Pacific Northwest, a region of diminishing significance for smelter expansion, but rather, whether the U.S. corporations should follow Aluminium Limited's pioneering venture in Jamaica and locate incremental capacity at the Caribbean mines instead of on the Gulf Coast. Although by the mid-1950s Kaiser had sufficient capacity in the Pacific Northwest to support an alumina refinery of adequate size there, the much lower transfer costs from the Northern Antilles to the Gulf, together with an expansion strategy still geared to the markets of the Industrial Northeast favoured a Gulf or offshore location (Krutilla 1955:276). Table 2.5 shows that the bulk of U.S. alumina expansion undertaken or planned in the 1950s opted for the Gulf Coast, both through extensions of war-time refineries (Mobile and Baton Rouge) and through greenfield units at Point Comfort (Alcoa), Corpus Christi (Reynolds), Gramercy (Kaiser), and Burnside (Ormet). The second

TABLE 2.5 NORTH AMERICAN ALUMINA CAPACITY, BY LOCATION, 1945–74

Location	Capacity									
	1945[1]		1950[2]		1958[2]		1958[2] Planned		1974[3]	
	million tonnes	%	million tonnes	%	million tonnes	%	million tonnes	%	million tonnes	%
Interior U.S.	1.178	37	0.639	25	1.043	20	1.043	15	1.102	9
Gulf Coast	1.044	32	0.945	37	2.475	49	3.680	49	5.550	47
Quebec[4]	0.994	31	0.994	38	1.088	21	1.134	17	1.066	9
Caribbean	—		—		0.490	10	0.957	14	1.719	15
Australia	—		—		—		—		2.383	20
Total	3.216	100	2.578	100	5.096	100	6.817	100	11.820	100

Sources: [1] Reynolds (1947).
[2] U.S. Bureau of Mines (1955 & 1959).
[3] Excludes capacity in western hemisphere supplying markets in Europe and Ghana. U.S. Bureau of Mines (1975) and corporate reports.
[4] Annual Reports of Alcan (formerly Aluminium Ltd.).

significant trend shown in Table 2.5 is for the Canadian corporation to base the bulk of its expansion on mine-site refineries in the Caribbean.

The Caribbean economist Huggins (1965:46) cites Stern's estimates for the mid-1950s, which suggest that a Jamaican refinery could operate with unit costs about 14 per cent below those of a refinery on the Gulf Coast (Table 2.6). However, Stern's figures are both incomplete and inaccurate in that they exclude capital costs and contain incorrect transfer costs. Capital costs would be higher in Jamaica, reflecting the need for greater expenditure on infrastructure, longer hauls for machinery and more severe start-up problems associated with developing countries. These factors are estimated in Table 2.6 to penalize a Jamaican refinery by $2/tonne on depreciation and $4/tonne on capital return. Figures used by Brubaker (1967:154) suggest a total capital cost penalty against a developing country of $8–12/tonne on refineries.

Stern's bauxite transfer costs are more like those from the Guianas to the Gulf and about three times actual Jamaica-Gulf transfer costs. The close proximity of Jamaica to the U.S. Gulf Coast rendered savings on bauxite transfer costs insufficient to compensate for the higher capital cost of Jamaican refinery. Two additional factors favoured the choice of Gulf locations for the U.S. corporations. First, the external economies of con-structing chemical plants serving non-aluminium domestic markets adjacent

TABLE 2.6 STERN'S ESTIMATED REFINERY COSTS BY LOCATION, 1956

Input	Gulf Coast	Jamaica
	$/tonne	
Bauxite	15.66	15.66
Bauxite transfer costs	15.19	—
Labour	5.95	2.98
Fuel	3.42	5.09
Administration and miscellaneous inputs	9.92	9.92
Alumina transfer costs to the United States	—	4.83
United States' duty on alumina	—	5.51
Total variable cost/tonne alumina	50.14	43.99
Capital cost (5% depreciation, 10% return)	26.91	32.29

Note: Bauxite transfer costs used are over three times those actually paid.
Sources: Huggins 1965:46, except capital charges which are based on data from Kaiser's Annual Report and Accounts 1956.

to refineries; and second, the option of multiple sourcing at a time of intense worldwide exploration for new bauxite deposits.

However, where long shipping hauls were involved, mine-site refinery capacity was constructed in the Caribbean by the North American core corporations in the 1950s. Aluminium Limited based its massive planned smelter expansion in British Columbia on refineries in Jamaica, but delays in the construction of smelting capacity resulted in much of the Jamaican alumina being traded with Scandinavian affiliates in a barter deal for ingot. Three factors favouring the Jamaican location were the prospect of low-cost mining, transfer costs to British Columbia less than three-quarters of those from the Guianas, and the provision of one-sixth of the capital under Marshall Aid repayable in ingot to the U.S. government. By the end of the 1950s the corporation had invested $100 million in installing two greenfield refineries in Jamaica with 675,000 tonnes of alumina capacity, and the prospect remained of future brownfield and greenfield expansion, including the eventual replacement of the aging Canadian refinery at Arvida by Caribbean plant.

The success of Aluminium Limited with Jamaican alumina and Reynolds and Kaiser with Jamaican bauxite shipments to Gulf refineries, lay behind Alcoa's opening of a new mine in the Dominican Republic and its acquisition of mining rights in Jamaica at the close of the first period. The abrupt cessation of mine expansion in the Guianas in the mid-1950s reflected increasing doubts about the competitiveness of bauxite mining there as deeper ores were encountered and the increasing size of ships heightened the freight disadvantage of the upriver mines (Suralco 1955). By effectively halving transfer costs, refineries at these mines not only reduced the risk of uncompetitiveness but also widened the potential markets for bauxite from the southern Caribbean as oversupply in the U.S. aluminium industry was directing the attention of the four largest North American corporations towards new markets outside the western hemisphere.

Smelter Location (Fig. 2.2).

Government incentives reinforced the pull of North American refineries and markets on smelter location during the first period as the relative attraction of cheap power began to wane. Alcoa's pre-war system of four smelters fringing the Industrial Northeast, drawing upon hydro power from Niagara, eastern Appalachia, and the T.V.A., secured both market proximity and cheap power. The typical pattern of operation has been to locate the smelter and mill as close as possible and to concentrate mill production on lines in which the operator holds a competitive edge through a techn-

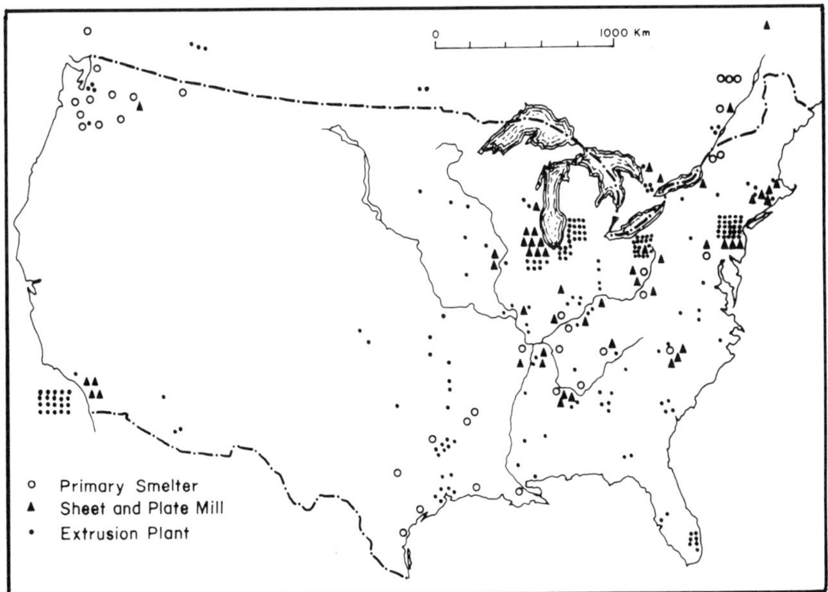

Fig. 2.2 Location of North American smelters, mills, and extrusion plants, mid-1970s
Sources: Aluminum Association (1977) pp. 35–36
Metal Bulletin (1977) pp. 147–49

ological lead, using up surplus capacity on products sold at lower margins, or even at cost if necessary. Market proximity reduced freight costs on both ingot and semi-fabricated products, which have had freight rates typically three and five times, respectively, those for bauxite or alumina (Brubaker 1967:162). The freight advantage has been compounded by the facility with which new scrap can be retrieved from customers and, more important, the greater ease with which orders could be attuned to individual customer specifications of metal composition, size, and delivery. Apart from Aluminium Limited prior to its forward integration drive in the sixties, integrated fabricating plants formed the main markets for the primary ingot producers, with the government stockpile affording an important alternative during the first period. The independent fabricators secured a greater share of sales of extrusions, castings, cable, and finished products than for plate and sheet which were dominated by the mills of the ingot producers. In consequence, the factories of the independents have tended to be more dispersed than those of the integrated companies which clustered near the smelters (Fig. 2.2).

The faster expansion of the fringe companies compared with the core and the areal diversification of both sets of companies are shown in Table 2.7. In fact, the diversification reflects the post-war orientation of the fringe companies towards eastern markets and the large-scale expansion of the Canadian core company — excluded from the emerging alternative power-

TABLE 2.7 NORTH AMERICAN SMELTER EXPANSION, 1945–59

Group	Region	1945		1953		1959	
		million tonnes	%	million tonnes	%	million tonnes	%
Core	Northeast	0.547	66	0.611	62	0.647	45
	South	0.205	25	0.283	28	0.431	30
	Northwest	0.078	9	0.098	10	0.360	25
	Midwest	—		—		—	
	Total	0.830	100	0.992	100	1.438	100
Fringe	Northeast	—		—		0.173	11
	South	0.080	27	0.437	59	0.624	41
	Northwest	0.215	73	0.306	41	0.438	29
	Midwest	—		—		0.296	19
	Total	0.295	100	0.743	100	1.531	100

Source: U.S. Bureau of Mines 1945; 1954; 1960.

market cost combinations available to its southern rivals — in British Columbia. The imperative to locate at hydro-electric sites declined in the immediate post-war years as the costs of alternative power sources fell and the penalty of remoteness from markets weighed more heavily. The downward pressure exerted by cheap oil on other fossil fuels together with the realization of economies of scale in ever-larger coal-fired thermal power stations lay behind this trend (Lester 1979:17). Power costs in North American regions under active consideration as potential smelter sites ranged from 2 mills/kwh in the Pacific Northwest through 3 mills in British Columbia and 3.3 mills in the T.V.A. to between 4 and 5 mills for strip-mined coal incinerated in large power stations in the Tennessee and Ohio valleys (Badger 1956:20). Estimated power costs per tonne of aluminium during the Korean War are shown in Table 2.8. Assuming an ingot price of $397/tonne, power costs ranged from less than 14 per cent of revenue in the Northwest to 24 per cent in Tennessee. Moderately priced power was also available in central Texas and Louisiana based on natural gas or carbonized lignite (Alcoa Annual Reports and Accounts). Offsetting the cheap power costs of the Northwest were the higher costs of alumina, carbon, and labour, effectively equalizing total costs there with those in the Texas-Gulf Coast region. Transfer costs from smelters in these two areas to mills located in the Midwest or Northeast were similar.

TABLE 2.8 ESTIMATED RAPID AMORTIZATION IMPACT, TWO U.S. SMELTER LOCATIONS, EARLY 1950s

Mill Input	Normal Depreciation		Accelerated Depreciation	
	Northwest	Gulf	Northwest	Gulf
		($/tonne)		
Alumina	117.24	98.66	117.24	98.66
Power	64.48	82.01	55.95	53.92
Carbon	25.80	20.50	25.80	20.50
Labour	30.40	25.57	30.40	25.57
Transfer to Northwest	0.72	21.87	0.72	21.87
to Midwest	23.79	20.94	23.79	20.94
Total: Northwest market	238.64	248.61	230.11	220.52
Midwest market	261.71	247.68	253.18	219.59

Note: Accelerated depreciation would only fall on the current rectifiers from public power lines in the Pacific Northwest compared with the entire plant in the Gulf. The figures assume power rates for the privately-built power station in the Gulf are priced to cover variable costs and the 20% of the investment not eligible for rapid amortization.

Source: Krutilla 1955:281.

The operation of the government's rapid amortization scheme during the Korean War, allowable on 80 per cent of total investment over five years, not only precluded consideration of smelter expansion outside the United States but also favoured projects using private, rather than public, power. Table 2.9 shows how the operation of this incentive affected Kaiser's decision to build a privately powered greenfield smelter in Louisiana rather than a smelter using cheap Bonneville power in the Pacific Northwest. Lanzillotti's (1961:155–56) data on net smelter costs for the period 1947–55 for the Alcoa, Reynolds, and Kaiser systems show that the addition of the southern smelter maintained Kaiser's edge in power costs over its rivals and helped to offset Kaiser's erstwhile disadvantage from high alumina costs. The South experienced major expansion of capacity in the late 1940s and early 1950s, with Alcoa pioneering the first non-hydro-electric smelter in the United States, as it had pioneered the opening up of smelting in the Pacific Northwest a decade earlier.

Increasing competition for gas and lignite in the South, together with government discrimination in allocating power to small users in the Pacific Northwest, made the market locations of the Midwest and St. Lawrence with 4 to 5 mill power and cheap water-borne alumina the optimum U.S. greenfield smelter locations by the mid-1950s. Three-quarters of the consumption of semi-finished aluminium was in states in, or adjacent to, the Ohio valley (Isard and Whitney 1952). Kaiser (*Annual Report and Accounts 1956*:7) summarized the new position in announcing its expansion plans for its second decade: "Principal units of this newest expansion are located in the east, a sheet and foil mill and an aluminium reduction plant at Ravenswood, West Virginia, on the Ohio River. These integrated facilities extend Kaiser aluminium production into the very centre of major aluminium markets. Ravenswood is located within a 500 mile radius of more than 70 per cent of

TABLE 2.9 ESTIMATED NORTH AMERICAN POWER COSTS BY REGION, EARLY 1950s

Region	Energy Form			
	Purchased	Private Gas	Private Coal	Comparison
	($/tonne)			
Tennessee Valley	97.22	—	—	97.22
Texas Gulf Coast	—	79.37–86.31	80.36–86.31	79.37–86.31
Pacific Northwest	54.56–64.48	—	—	54.56–64.48

Source: Krutilla 1955:275.

the nation's aluminium usage. Important also is the fact that the new reduction plant being built at Ravenswood will use coal-generated electricity . . . it represents a new freedom for the aluminum industry from its former dependence on hydro-electric power." Table 2.10 shows that a market-oriented location in the Ohio valley could undercut cheap hydro-electric locations in the Pacific Northwest or Latin America by the late 1950s. However, even allowing for the shortcomings in the data for the Latin American plant outlined in the notes to Table 2.10, they suggest that a smelter built at a cheap hydro-electric site near a mine-refinery complex in the area and oriented to Europe might provide a flexible and viable option as the North American corporations turned towards new markets.

TABLE 2.10 ESTIMATED COSTS OF A 100,000-TONNE SMELTER IN 1960, BY LOCATION

Location	Ohio River	Oregon	Peru
	\$/tonne		
Alumina: 1.9 tonnes at Gulf Coast refinery	104.72	104.72	104.72
Barge transport up Ohio River	8.38	—	—
Rail transport to the Pacific Northwest	—	25.46	—
Ship transport to Peru	—	—	12.68
Carbon electrodes and pot lining	41.89	41.89	59.52
Electrolyte	16.53	14.33	18.22
Power: 1,500 kwh (Ohio 4.1 mills, Oregon 2.1, Peru 2.0)	67.79	35.27	33.90
Direct labour and social security	51.16	51.16	14.07
Management and supervision	6.61	6.61	6.61
Local taxes and insurance	6.61	6.61	7.93
Repairs and maintenance	9.92	9.92	9.92
6 per cent interest on long term debt (50% finance) + 5 per cent depreciation	52.91	52.91	63.36
Freight to: Eastern and Midwest markets	4.41	—	—
Western and Midwest markets	—	26.46	27.56
Import duty	—	—	27.56
Total cost c.i.f.	370.93	375.34	386.05

Note: Arthur D. Little data for the Peruvian smelter converted to the Senate Committee format using ratio derived from the Ohio River smelter costs contained in each data source.

Sources: U.S. Senate Committee on Public Relations 1962:155–56; Arthur D. Little 1961:12.

Global Competition and the Growth of Mine-Site Refineries, 1957–74

In the second period of analysis, the North American core is expanded to include Reynolds and Kaiser, which by this time were operating on a scale that had more in common with Aluminium Limited (henceforth referred to as Alcan) and Alcoa than with the other fringe companies. Faced with excess capacity, all three U.S. core corporations quickly initiated spatial and product diversification in the search for new growth markets. Between 1959 and 1964, the core's North American capacity rose a mere 6.5 per cent as the U.S. core companies sought to emulate Alcan's penetration of over-seas markets through fabrication plants and then smelters (Table 2.11). Most of Alcan's investment during this period went into defensive measures: the proportion of its annual investment expended on fabrication plant rose from 6 per cent in the mid-1950s to 60 per cent by the late 1960s, and its dependence on third party ingot sales was reduced from 70 to 25 per cent of corporate revenues (*Alcan Annual Reports and Accounts*). Of the three U.S. core corporations, Kaiser pursued the boldest diversification pro-gramme, expanding overseas smelting to one-third of its total capacity (Table 2.11) and pushing the contribution of its real estate and chemical and commodity trading divisions to one-third of revenue and nearly half of profits by 1972–74 (Kaiser Form 10-K 1974:1). The profit rate replaced growth as the prime corporate goal during the second period. It has been estimated that prices were adjusted along the vertical chain of production between 1957 and 1965 to squeeze the share of fabrication in total operating profit for the U.S. core from 26 to 17 per cent (Donaldson *et al.* 1966:16) and reduce the number of independent fabricators competing with the core's captive fabricating operations (Marcus 1974:20). However, despite diversification, profit manipulation along the vertical chain, and the squeeze on independent fabricators, the real price of aluminium ingot declined and profits languished while five new companies entered the U.S. aluminium industry during the second period (Table 2.1). Using constant 1976 dollars, the price of aluminium ingot fell from $0.56/lb in 1957 to $0.45 in 1965 and $0.39 in 1974, after dipping to $0.32/lb in 1973 (Stamper and Kurtz 1978: 20). Table 2.12 shows that the return on shareholders' equity was sub-stantially below that of the mid-1950s in the second period and it did not recover until the late 1970s.

Smelting Capacity Location (Fig. 2.2)

The North American corporations had five broad options for expanding smelting capacity during the second period, namely, (1) inside existing large markets at brownfield sites, (2) inside existing large markets at greenfield sites, (3) inside small protected markets anticipating large long-term

TABLE 2.11 ESTIMATED ALUMINIUM CAPACITY OF THE SIX MAJORS BY REGION, 1954–74

Corporation	Year	North America	Latin America	West Europe	Asia	Oceania	Africa	Total	Free World (%)	X (%)
		million tonnes								
Alcan	1954	0.579	—	0.043	0.053	—	—	0.659	25.8	88
	1964	0.783	0.015	0.050	0.134	—	—	0.953	18.6	84
	1974	0.941	0.048	0.351	0.243	0.032	—	1.615	13.6	58
Alcoa	1954	0.517	—	—	—	—	—	0.517	20.3	100
	1964	0.779	0.020	0.056	—	0.041	—	0.895	15.1	87
	1974	1.432	0.102	0.076	—	0.051	—	1.660	14.0	86
Reynolds	1954	0.376	—	—	—	—	—	0.376	14.8	100
	1964	0.753	—	0.051	0.030	—	—	0.834	16.6	90
	1974	1.045	0.025	0.167	0.006	—	0.015	1.259	10.6	83
Kaiser	1954	0.366	—	0.001	—	—	—	0.367	14.8	100
	1964	0.590	—	0.005	0.020	0.053	—	0.667	11.3	88
	1974	0.658	0.003	0.067	0.046	0.068	0.138	0.981	8.3	67
Alusuisse	1954	—	—	0.110	0.003	—	—	0.113	4.4	97
	1964	0.029	—	0.185	0.008	—	—	0.222	3.1	83
	1974	0.251	—	0.369	—	—	0.017	0.637	5.4	58
P.U.K.	1954	—	—	0.141	—	—	—	0.141	5.5	100
	1964	—	—	0.334	—	—	0.053	0.387	6.5	86
	1974	0.163	0.003	0.678	0.009	—	0.032	0.862	7.3	79
Free World Total	1954	1.838	0.012	0.599	0.102	—	—	2.552		
	1964	3.184	0.056	1.267	0.372	0.093	0.053	5.025		
	1974	5.569	0.314	3.413	1.807	0.319	0.353	11.838		

Note: X is % capacity in home continent.

Sources: 1954 and 1964: IBRD Memo of 310568, EC-163, *Past and Prospective Trends in the World Aluminium Industry*, pp. 19–20. 1974: President's Council on Wage and Price Stability, *Aluminium Prices 1974–75*, pp. 60–61 and 65.

TABLE 2.12 PERCENTAGE RATE OF RETURN ON SHAREHOLDERS' EQUITY,
BY CORPORATION

Corporation	1953–55	1958–60	1963–65	1968–70	1973–75	1977–79
Alcan	12.76	6.30	8.06	9.76	9.53	17.56
Alcoa	12.19	6.36	7.14	9.37	7.80	15.09
Kaiser	25.40	7.98	8.12	9.77	9.24	15.21
Reynolds	17.73	9.67	7.68	6.69	9.06	11.45

Source: Corporate Annual Reports and Accounts.

demand, (4) at greenfield sites adjacent to cheap power, and (5) at green-field sites adjacent to bauxite and cheap power.

The investment decisions were taken against a background of expected continued rapid growth in demand, especially in Europe and Japan, and the gradual erosion of tariff barriers, notably those which had protected European coal mining from cheaper energy. An additional factor was uncertainty about the political stability of the developing countries as colonial regimes were dismantled. The aluminium industry had direct experience of the growing wave of nationalization (Williams 1975:260) in 1961 when Guinea expropriated Alcan's partially completed $100 million mine and refinery complex at Boke and the corporation lost almost one-tenth of its bauxite supply and $24.5 million (Smith Barney and Co. 1959; *Aluminium Ltd. Annual Report and Accounts 1961:9*). However, offsetting such investment disincentive was the diminishing importance of any one mine as growth brought new mines onstream and increasing recourse was available to consortia financing.

The U.S. market continued to be supplied by domestic ingot supplemented by imports from Canada, through the continental systems of Alcan and Reynolds, with Kaiser's Ghanaian smelter providing a new, but small, input (U.S. Bureau of Mines 1964; 1969; 1974). Table 2.13 indicates that subsidies were required to erode the competitive edge held by U.S. greenfield smelters over the principal alternatives. A combination of a 5 per cent tariff on ingot, high freight rates on ingot compared to bulk-hauled bauxite and alumina, and the lower capital costs afforded by the superior infrastructure and reduced risk associated with domestic sites (Table 2.14) offset the advantages of cheap energy and materials at overseas locations. The fringe companies expanded primarily through the construction of new smelters (Table 2.3): of the nine fringe smelters established in the second period, three were in the Midwest, three in the South, two in the Pacific Northwest,

TABLE 2.13 ESTIMATED SMELTING COSTS AT ACTIVELY CONSIDERED
LOCATIONS, LATE 1960s

Inputs for 120,000-Tonne Smelter	U.S.	Europe	Middle East	Australia
	¢/lb			
Alumina (2 tonnes at $66/tonne c.i.f.)[1]	7.50	7.90	7.00	6.20
Power (8 kwh/lb aluminium)[2]	3.60	4.80	2.00	3.60
Other raw materials	1.50	1.50	2.00	1.50
Labour and overhead	2.75	2.75	2.75	2.75
Depreciation (17 years)[3]	3.10	2.50	3.20	3.20
Interest on Fixed Capital (5% M.E., 8% elsewhere)	2.00	1.70	1.40	2.20
Freight on ingots[4]	0.75	0.50	1.50	1.50
Tariff on ingot	—	—	1.00	1.00
Total	21.20	21.65	20.83	21.95

Notes: [1]U.S. includes $5/tonne internal freight and $10/tonne external; Europe $3 and $10
+ 8.8% tariff; Middle East $10 sea freight; Australia $2 internal freight.
[2]U.S. coal at 4.5 mills/kwh; Europe coal and nuclear at 6 mills; Middle East gas at
2.5 mills; Australia coal at 4.5 mills/kwh.
[3]U.S. on $1,100/tonne capacity; Europe on $950 i.e. subsidy 15%; Middle East and
Australia $1,200, reflecting higher infrastructure costs.
[4]U.S. and Europe only include internal freight; Middle East and Australia include
ocean freight plus U.S. internal coastal freight.

Source: Younger 1971:18.

TABLE 2.14 ESTIMATED IMPACT OF DIFFERENT CAPITAL REQUIREMENTS
ON PRODUCTION COSTS

Plant Capital Costs ($/tonne)	Annual Capital Cost by % Rate of Return ($/tonne)			
	7	10	12	15
772	75.90	86.39	97.56	114.86
849 (10% higher cost)	77.44	95.03	107.32	126.35
887 (15% higher)	80.95	99.35	112.19	132.09
926 (20% higher)	84.48	103.66	117.08	137.83
965 (25% higher)	88.00	107.98	121.95	143.57

Note: Assumes 20-year straight line depreciation, with compounding to yield required rate
of return.

Source: Brubaker 1967:234.

and one in the Northeast. Proximity to the market continued to offset the cheap power cost advantage of the Pacific Northwest for integrated production, as the cost/tonne figures for smelters in the early-1960s demonstrate (Table 2.15).

TABLE 2.15 COST/TONNE FIGURES FOR SMELTERS IN THE EARLY 1960s

Smelter	Power	Alumina transfer	Ingot transfer	Sheet transfer to Detroit	Total ($/tonne)
Mead (Pacific Northwest)	35.14	25.67	2.09	50.71	113.61
Alcoa (Tennessee)	72.15	11.21	—	18.92	102.28
Ravenswood (Midwest)	72.15	11.21	—	10.25	93.61

Source: Bloch and Moment 1967:41.

The West Coast market was so small that five-sixths of the region's output was shipped as ingot at this time (Fulkerson 1965:6), and part of these sales faced intensified competition due to the "promotional policies" of governments in Norway, India, Ghana, and Suriname (Bloch and Moment 1967:259–67). However, the prospect of importing alumina in bulk shipments from Australia at $6.06/tonne, together with the provision of two million kwh at a cost of 2.01 mills in 1965, stimulated further expansion in the region, leaving it the locus for one-third of fringe capacity at the end of the second period. The advantages of expansion at existing sites are clearly demonstrated by the growth of the joint Pechiney-Amax smelter near Bellingham, which was undertaken in three distinct stages of 69,000 tonnes capacity each. Construction of the second and third stages cost $652/tonne compared with $942/tonne for the initial stage (Reimer 1968: 22–23), representing a considerable saving on production costs since capital charges, exclusive of the return on equity, were more than one-fifth of these costs.

The core added only one new smelter in the second period, and that was Alcoa's Midwest unit at Evansville whose construction had been deferred from the mid-1950s in response to excess capacity. Instead, virtually the whole of the core's North American growth was met by the addition of new potlines at existing locations. As a result, the regional distribution of the core's smelting capacity changed only slightly, with the Northeast and South yielding 6 per cent of their continental share to the Midwest and the Pacific Northwest (Table 2.3). Not only were unit costs reduced by brownfield expansion but additions could be made in smaller increments, thereby lowering the probability of a company being caught with substantial excess capacity. In this way, the core was able to retain its low-cost advantage, while Alcoa and Alcan maintained an extra margin over Kaiser and Reynolds. One estimate put the cost of greenfield expansion for a fully integrated production system at $2,400/tonne in 1973, but calculated that the U.S. core companies had built up their systems for half this cost, and carried them in their books at only $850/tonne (Marcus 1974:19).

All three U.S. core corporations, like Alcan before them, invested in heavily protected markets in the larger developing countries such as India, Brazil, Mexico, and Venezuela. However, it was the fragmented European market which attracted most overseas interest from the core corporations during the second period. In each case, smelter investment was preceded by the acquisition of captive fabricating outlets (*Corporate Annual Reports and Accounts*). The initial concentration on hydro-electric-powered smelters in Scandinavia gave way to market locations inside and adjacent to the rapidly expanding European Economic Community as both substantial subsidies became available and power costs dropped from 8 mills towards 5 (*Economist*, 12 June 1976). Thus, in the late-1960s Alcan, Kaiser, and Reynolds were each involved in separate U.K. smelter projects, while Kaiser and Reynolds invested in West Germany and Alcoa brought a smelter onstream in the Dutch colony of Suriname. The British example shows that in addition to assistance with capital costs (Table 2.13), the corporations extracted substantial concessions on energy. According to the IBA (1977: 36), Kaiser and Reynolds secured 25 to 30-year power contracts at below 2 and 7 mills respectively, while Alcan's Lynemouth power station obtained a 25-year contract for coal at a discount from the NCB's average price equivalent to 5 mills by 1977.

In the 1960s, locations with cheap power (about 2 mills), such as existed in some developing countries, would only yield a saving on electricity of $50/tonne against revenue of $600/tonne aluminium (Karim 1968:4–9). Offsetting this advantage non-market locations faced tariffs ranging from $28/tonne in the United States and Benelux, through $50 in France and West Germany to over $70 in Italy and Japan (*Kaiser Annual Report and Accounts 1964*). Annual capital charges, exclusive of the return on equity, were $10–40/tonne greater at remote new sites (Table 2.14), while net freight savings were realizable only where a refinery complemented cheap power (Table 2.13). Thus, for example, twelve years after the formation of the Bahrainian smelter consortium, this pioneering Middle East venture was still unable to compete effectively in world markets (*The Times*, 6 Feb. 1980), while the second Middle East smelter due for completion in Dubai in 1980 also had an inauspicious start (*The Times*, 21 Jan. 1980). Against this background of market attraction in the second period, Alcoa nevertheless invested \$150 million in 1959–65 in establishing a 150 megawatt hydro plant, a 400,000-tonne refinery, and a 50,000-tonne smelter in Suriname (*Forbes* 1 May 1968:52–54). The documents leading to the signing of the Brokopondo Agreement (1959) indicate that the government of the Dutch colony was unable to raise capital for the hydro scheme without incurring a significant risk premium, and that Alcoa was only persuaded to participate by backdoor access to the European Economic Community through the Netherlands, tax concessions from Western Hemisphere Trade Corporation status, and — the corporation's principal objective — a concession to explore for new bauxite reserves in northeastern Suriname (Brokopondo Agreement 1959). Surinamese plans for a second hydro-smelter complex of 180,000 tonnes capacity in the west of the colony were explored by several major corporations, but they foundered in the late 1960s in the face of expansion in Western Europe.

Despite favourable reports on its hydro-electric power resources, Guyana had less success than Suriname in establishing a smelter. The excess capacity at the end of the 1950s forced Alcan, the largest mining corporation in Guyana, to postpone major hydro expansion schemes in Quebec and British Columbia and to withdraw from projects under consideration with Pechiney in Guinea and British Aluminium in Ghana. It was estimated that Alcan could bring new capacity of more than 100,000 tonnes onstream in Quebec and British Columbia for C\$1,075/tonne and C\$358/tonne respectively (Smith Barney 1959): the Corporation therefore had little incentive to invest in the politically troubled South American country, and the opportunity was not taken up by other producers. However, Kaiser, with Reynolds and British Aluminium as minority partners, did undertake the Ghanaian scheme in the mid-1960s. The cheap power of the Volta was supplemented by a ten-year tax holiday, and corporate risk was reduced to a minimum by raising \$96 million from the Export-Import Bank and securing guarantees from the U.S. Development Loan Fund to cover Kaiser's \$32 million equity stake. Elsewhere in West Africa, Pechiney abandoned \$175 million plans for a 145,000-tonne smelter in Guinea in favour of a smaller project in the Cameroons. Contrary to assertions by Girvan, there is no evidence that a nuclear-powered smelter in Jamaica was a practical or viable proposition during the second period (Auty 1980:173–74). In summary, it is clear that during the second period unsubsidized market locations in North America and subsidized ones in western Europe, especially where brownfield expansion could be undertaken immediately or in the near future, afforded low-cost, low-risk smelter locations. But as the North American corporations adopted a more international marketing strategy, their potential bauxite and alumina supply points increased, transforming the role of the Caribbean bauxite producers.

Refinery Location (Fig. 2.1)

Bauxite imports into the United States rose substantially during the second period as the American corporations reaped the combined advantages of reduced taxation through the operation of the Western Hemisphere Trade Corporation (WHTC) loophole, and the lower incremental capital costs of brownfield expansion at the Gulf Coast refineries. The U.S. corporations transferred a significant proportion of their gross profit from fabrication to mining and reduced taxation on these profits by at least one-third through the adoption of WHTC status. For example, Kaiser raised the internal transfer price of Jamaican bauxite from $9.30 to $14.50/tonne between 1961 and 1963 at a time of minimal cost inflation (Moment 1968: 11–14). One source has estimated that the American corporations raised the percentage of their gross profits attributed to mining from 4.6 to 8.6 per cent between 1957 and 1965 (Donaldson *et al.* 1966:15). Brownfield expansion was not undertaken at interior U.S. refineries owing to limited supplies of low-quality bauxite and the high transfer costs to smelters. Data for the Reynolds system (Marcus 1974:40–44) show that the interior Hurricane Creek refinery operated with f.o.b. costs about 15 per cent above those of the efficient coastal Corpus Christi unit. Although Hurricane Creek incurred low freight costs on alumina railed to the nearby Arkansas smelter, rail freight to Listerhill and Massena exceeded charges for alumina transferred by barge or ship from coastal refineries three to five times these distances. Alcoa turned its small inland Bauxite refinery over to chemical grades of alumina, and when its nearby Tennessee smelter was expanded in the late-1960s, the alumina was railed from Mobile.

The North American corporations widened their options for alumina sourcing during the second period by building greenfield refineries in the early 1960s in the southern Caribbean; in Australia in the mid-1960s; and in Jamaica and western Europe from the late 1960s. Each plant was designed to permit expansion to between three and six times its initial capacity, but the option was not always exercised. For example, while the Surinamese refinery was boosted by Alcoa and Billiton to 1.13 million tonnes in 1965–68 to serve smelters in western Europe and the northeastern United States, Alcan did not expand its Guyanese unit. Instead of using the rapid amortization privileges secured in Guyana to expand capacity there, Alcan chose to reduce the book value of its Guyanese investment and, in combination with maintenance of spare capacity in Quebec and Jamaica, minimize the damage of potential nationalization to its system (Hodgson 1965:31).

By the late 1960s, several factors were encouraging the U.S. producers to build greenfield refineries in Jamaica, the only location in the northern Caribbean with bauxite reserves sufficient to sustain large-scale refinery expansion. Tightening environmental controls and a lack of suitable sites in the United States, coupled with small potential cost advantages in Jamaica (Table 2.16) and WHTC tax opportunities, stimulated the construction of three Jamaican refineries. The importance of rapid large-scale expansion for competitiveness is demonstrated by estimates for Alpart's S.W. Jamaican refinery: average costs were projected to fall from $79.89/tonne at the start-up of the 200,000-tonne unit to $61.22/tonne in the tenth year, with production of more than one million tonnes (Moment 1968:iii–8). Alpart filed a transfer price with the Securities Exchange Commission of $70.87/tonne, indicating the intention of the consortium to declare a significant profit in Jamaica where it held WHTC status. Alpart's targeted cost was comparable with f.o.b. costs reported for transfers by Alcoa from Western Australia to the United States and by Pechiney from Guinea to western Europe. However, all three new Jamaican refineries (Alcoa's Halse Hall, Alpart's Nain, and Revere's Maggotty) experienced severe start-up problems and remained unprofitable through the 1970s (Davis 1978:6–8). For example, in 1973 Alpart's costs were reported to be more than $100/tonne, about 25 per cent above the "high-cost" Hurricane Creek plant (Marcus 1974:40–44). However, Alcan's two established Jamaican refineries remained comfortably profitable until the fiscal changes of 1974 (*Alcan Jamaica Annual Report and Accounts 1974*), largely on account of the low capital costs associated with older units: 1973 f.o.b. costs were fractionally below those of the large

TABLE 2.16 ESTIMATED PRODUCTION COSTS FOR U.S. GULF AND
JAMAICAN REFINERIES 1967

Location	U.S. Gulf		Jamaica	
	0.275 million tonnes capacity	0.550 million tonnes capacity	0.275 million tonnes capacity	0.550 million tonnes capacity
Total capital cost ($ million)	52.5	87.0	55.1	91.4
Average capital cost ($/tonne)	191.0	158.2	201.6	166.1
Bauxite (2.5 tonnes)	21.36	21.36	16.35	16.35
Caustic soda, fuel, other materials	12.24	12.24	12.51	12.51
Maintenance materials and supplies	2.75	2.75	2.75	2.75
Direct and indirect labour	8.27	5.51	6.61	4.41
Depreciation (U.S. 10%, Jamaica 12%)	19.28	15.98	24.25	20.12
Overhead and other expenses	1.38	1.10	1.65	1.38
Average cost ($/tonne)	65.29	58.94	64.30	57.70

Note: Bauxite costs are modified from original estimate by Karim who cites figures several times below cost figures for Jamaica at this time.

Source: Karim 1968.

new Gladstone refinery in Australia. Gladstone was technically the sister plant to Alpart's Jamaican refinery, but its 1973 average cost of about $62/tonne was less than two-thirds of Alpart's average cost.

The first large new mine of the second period came onstream in 1963 when Kaiser and a subsidiary of Rio Tinto Zinc began exploiting the 4 billion tonne bauxite reserve at Weipa in northern Queensland. With a conversion ratio of 2.1 tonnes of bauxite per tonne alumina, the unprocessed bauxite could compete on international marktes, unlike the poorer quality ores simultaneously being developed by Alcoa in Western Australia which required refining prior to export. At the close of the 1960s, when Guyana nationalized Alcan's Demba subsidiary, the North American corporations had the options of market expansion in subsidized European greenfield refineries supplied by Australian and West African bauxite, or mine-site expansion in Jamaica, Suriname, Australia, and West Africa. Guyanese nationalization underscored corporate dependence on the region and accelerated the process of diversification. By 1970, Alcoa was demonstrating a distinct preference for incremental expansion from Australia, as were the other corporations which had adopted policies of making substantial third party sales, Kaiser, and Alusuisse. Alcoa's Kwinana plant, which had earlier been passed over in preference for expansion in Suriname, was boosted to 1.3 million tonnes in 1973 when a sister plant as Pinjarra was brought onstream and raised to a similar capacity by 1974 (*Alcoa of Australia Annual Report and Accounts, 1976*). The Queensland refinery of Kaiser, Rio Tinto Zinc, Pechiney, and Alcan started production in 1967 and had reached two million tonnes capacity by 1974, with fixed investment per tonne falling from A$240/tonne to A$145/tonne in the process (*Queensland Alumina Ltd., Report of Directors and Annual Review 1977*). In Northern Australia, Alusuisse's 70 per cent-owned Gove refinery was expanded from zero to one million tonnes between 1972 and 1974. The profitability of the Alcoa and Comalco operations is demonstrated by rates of return on equity about 50 per cent above the average for the parent companies involved. Australian alumina was absorbed partly in the domestic market, partly by Japan, but mainly by west coast American smelters which benefited from its lower costs in comparison with alumina shipped from the Caribbean or railed from Caribbean-sourced Gulf Coast refineries. Alcoa's extremely efficient Western Australia operations also managed to capture third party sales in Argentina, the Middle East, and South Asia.

By 1974 bauxite production at Weipa had been expanded to 10 million tonnes (about 13 per cent of total world output), of which half was refined at Gladstone; one-sixth was exported as bauxite supplementing flows from Malaysia and Indonesia to Japanese refineries; and one-third was shipped as bauxite to refineries in Italy and West Germany, principally as third-party transactions (*Comalco Annual Report and Accounts 1977*). In 1974 western Europe drew 53 per cent of its bauxite imports from Australia, compared with 8 per cent from the Caribbean (principally from the nationalized Guyanese corporation at discounted prices), with most of the remainder came from the Balkans and Guinea (Cornish 1977:37). The competitiveness of Australia bauxite lay only partly in its low extraction costs (Table 2.17); the low freight rates secured on large bulk shipments also contributed substantially. When a consortium brought the Boke bauxite on-stream a decade after the Alcan nationalization in 1961, the high quality of the ore significantly reduced processing costs (IBA 1979:55–59) and secured markets at brownfield refineries in Quebec, on the Gulf Coast, and in western Europe. It was against this background of deteriorating competitiveness that the Caribbean producers adopted a policy of confrontation with the corporations in mid-1974.

TABLE 2.17 COMPARATIVE COSTS OF BAUXITE MINING[1]

Mine	Capacity (million tonnes)	Capital Cost	Operating Cost	Inland Transport	Profit	Price	Dry Grade Price
		1975 U.S.$/tonne					
Weipa (Australia)	11.0	1.57	3.77	—	0.73	6.07	6.98
Boke (Guinea)	9.0	1.25	3.21	4.88	0.62	9.96	9.91
Trombetas (Brazil)[2]	3.3	7.00	9.65	—	3.39	20.04	20.84
Onverdacht (Suriname)	3.6	3.67	5.12	1.76	1.88	14.51	16.72
McKenzie (Guyana)	3.6	4.29	4.38	0.67	2.10	11.44	11.38
Lydford (Jamaica)	3.6	1.89	1.29	2.64	0.92	6.72	8.49
Cabo Rojo (Dominican)	0.6	2.50	5.03	1.76	1.22	10.51	13.95

Notes: [1]Figures are estimated costs, with profit calculated at 10% on capital after uniform corporate taxes, but excluding levies, etc.
  [2]Capital cost for Trombetas based on June 1974 mining agreement.
Source: IBA, Bauxites of the World 1978:135.

OPEC Destabilization and the Growth of Materials-Oriented Processing

The Organization of Petroleum-Exporting Countries triggered abrupt and massive cost changes that were anathema to an energy- and capital-intensive industry like aluminium. Table 2.18 summarizes the magnitudes of cost changes within the Kaiser system between 1974 and 1979. Average energy costs for the aluminium industry throughout the world rose from 4.5 mills in 1973 to 17 mills in 1980, while the spread about these mean figures

TABLE 2.18 ANNUAL COST INCREASES IN KAISER ALUMINIUM SYSTEM,
1974–79

Cost	1974	1975	1976	1977	1978	1979	1974–79
	¢/lb						
Alumina	3.6	1.9	1.2	0.5	0.0	1.0	8.25
Power	0.6	0.3	0.5	0.9	0.2	0.0	2.44
Labour	0.7	1.4	0.5	1.4	1.0	0.3	5.21
Other	0.8	0.7	0.3	0.2	1.6	0.0	3.59
Total (U.S. and Ghana)	5.7	4.3	2.5	3.0	2.8	1.3	19.49

Source: Spector 1979.

widened from one of 2 to 7 mills to one of 3 to 60 mills. Within North
America, smelters such as those in Canada, New York state, and Washington experienced minimal rises in unit power costs; others, notably those in
the South, saw their costs more than quintuple to more than 25 mills.
Capital costs have roughly doubled over the post-OPEC period, while the
spread of these costs about the average figures has also widened appreciably.
Whereas a consortia-based mine-to-fabrication system was estimated to
cost $2,430/tonne in 1973 (Marcus 1974), the same system would cost
$4,780/tonne in 1979 (Table 2.19). The capital costs on smelters range from

TABLE 2.19 CAPITAL COSTS AND MINIMUM OPTIMUM SIZE BY ALUMINIUM
PRODUCTION STAGE, LATE 1970s

Cost	Stage				
	Mine	Refine	Smelt	Mill	System
Minimum optimum size (million tonnes)	4.0	0.8	0.2	0.025	—
Total capital cost ($ million)	160.0	560.0	500.0	40.0	1,260.0
Capital cost/tonne	40.0	700.0	2,500.0	1,600.0	4,840.0
Capital cost: total self-sufficiency ($ million)	160.0	1,120.0	2,000.0	1,400.0	4,680.0
Units for total self-sufficiency	1.0	2.0	4.0	35.0	—

Sources: Industry sources, 1980.

less than $1,000/tonne on pre-OPEC units to $2,000/tonne for current brown-field expansion to nearly $5,000 per tonne for a greenfield plant requiring substantial infrastructure in Indonesia. Costs of materials also diverged, though less dramatically, as will be shown below. The divergence of these critical cost elements paralysed investment in the world aluminium industry, eventually terminating the period of surplus capacity, and then boosting the price of aluminium ingot to levels favouring new expansion.

Expansion from about 1978 is based upon the assumption that aluminium has become a mature industry with a predicted growth rate half that of the pre-OPEC periods. It is also widely accepted that while fabrication will remain market-oriented, smelting and refining will expand where raw materials are close to cheap power, provided such sites are in regions judged to be politically stable. Even at the projected annual growth rate of 4.5 per cent for the 1980s, this implies capital investment in excess of $29 billion, in 1980 dollars, for the addition of 6.5 million tonnes of smelting capacity, 13 million tonnes of refinery capacity, and 30 million tonnes of new mines during the decade. Savings on power costs and continuing benefits of scale economies at all stages will offset the lower until capital costs of facilities established prior to the recent price escalations. In smelting, the most spectacular marginalization has occurred in Japan where one-third of the country's smelters have been idled, awaiting possible re-activation at some stage in the future price-cost spiral. In refining and mining, Caribbean governments blundered into placing their industries in an uneconomic position, at a time when more adroit policies could have expanded refining throughout the region and smelting in the southern Caribbean.

In June 1974, in desperation because of the deteriorating trade balance caused by the OPEC price rises, the Jamaican government unilaterally imposed a sixfold tax increase on the aluminium corporations and spear-headed the formation of a bauxite producers' cartel to maintain the resulting higher material prices (Auty 1980:176–78). These moves violated the aluminium corporations' critical requirements of competitiveness and stability, rendering the Caribbean an uneconomic as well as unstable area for investment. Consequently, during the first OPEC-deepened recession of 1975–76, the corporations concentrated their production cutbacks in the high-cost Caribbean, substituting materials from Australia and Guinea wherever possible to contain costs within their corporate systems and retain third-party sales. Although Jamaican officials involved claim that the size of the tax increase was based on ''scientific analysis'', they now accept the corporations' argument that they overestimated the rent their bauxite commanded. It is also clear that Caribbean political parties to the right of those ruling in Jamaica and Guyana did not favour the confrontation strategy but were unable to resist domestic economic and political pressures to follow the Jamaican lead.

As the corporations revised their cost estimates through the mid-1970s, three options for refinery expansion were being evaluated — brownfield in Australia and the Caribbean, greenfield in Latin America and West Africa and greenfield in the European markets. Fiscal policies were a vital determinant of the outcome. The unilateral imposition of massive tax increases in the Caribbean confirmed the worst suspicions raised by the Guyanese nationalization about the stability of investment in the region, while the imposition of only a nominal levy by Australia, itself an IBA member, encouraged continued rapid growth there. Incentives in the form of accelerated depreciation and tax holidays on greenfield refineries under construction in western Europe and planned for Australia, coupled with expansion for maximum economies of scale, offset the disadvantage in capital costs per tonne compared with plants established five to twenty years earlier (Table 2.20). Despite the opportunity for brownfield expansion in Jamaica, Alcoa and Alpart have halted expansion, Revere has abandoned its $83 million refinery, and Alcan has offered its two old plants to the government. Even if present moves to ameliorate taxation continue, doubts will persist about government integrity towards investment when it is most vulnerable, that is, when it has just been completed.

Towards the close of the decade, a shortage of suitable smelter locations in North America (Iron Age 1980:28–30; Thirtythree Metal Producing 1980:57–61) and the advantages of materials and power proximity (Table 2.21) made Australia and Latin America the most favoured locations for new capacity. Alcoa was invited by the Surinamese government to consider the construction of a second refinery at Apoera in the west of the country, drawing on the newly surveyed Bakhuys deposits of bauxite. The total investment required was estimated in 1977 at $355 million, of which four-fifths were the cost of the refinery and about one-tenth each for developing the mines and the associated railway. However, the reserves were evaluated as low quality, with a 2.5 per cent silica content and only 41 per cent alumina available. This meant that about 1.5 million tonnes of ore a year would be required for a "small" refinery of 575,000 tonnes, while proven reserves were sufficient for only sixteen years at double this output. Alcoa estimated that, even excluding the levy, the project would need a price of $175.22/tonne to earn a 10 per cent return on capital, while to achieve Alcoa's target of a 14 per cent return the minimum f.o.b. price would be $199.93/tonne, a figure "in the approximate order of magnitude of 35 to 50 per cent higher than some current prices of which we are aware" (Suralco 1977). These estimates compare with a cost of $133/tonne for Alcoa's

TABLE 2.20 ESTIMATED COSTS FOR ALUMINA REFINERIES BY LOCATION AND PLANT VINTAGE, 1979 ($/tonne)

Plant Vintage	New	New	New	New	New	Established		Established	
Refinery Site	EEC	U.S.	Brazil	W. Australia	Jamaica	EEC	U.S.	N. Australia	W. Australia
Ore Source	Australia	Brazil	Brazil	W. Australia	Jamaica	Australia	Caribbean	N. Australia	W. Australia
Capacity (000 tonne)	800	800	800	800	800	1,000	1,200	2,400	1,400
Bauxite	62.4	64.0	49.5	15.0	73.1	63.6	76.8	25.2	15.0
Caustic soda	13.3	8.0	7.0	3.6	14.0	13.3	9.1	17.5	4.2
Fuel	38.4	26.7	28.7	26.7	42.0	35.6	43.3	33.0	27.4
Direct labour	9.0	12.0	7.0	10.5	9.0	8.3	7.0	7.9	10.5
Maintenance and overhead	27.3	26.6	25.1	25.6	22.0	33.7	26.7	28.0	25.8
Depreciation	30.8	27.7	27.7	27.7	29.9	18.8	16.3	7.0	15.8
Sub total	181.2	165.0	145.0	109.1	190.1	173.3	179.2	118.6	98.7
Capital return	98.4	91.4	91.4	91.4	98.6	54.5	47.3	20.3	45.8
Sub total	279.6	256.4	236.4	200.5	288.6	227.8	226.5	138.9	144.5
Freight	—	—	7.0	10.0	2.0	—	—	10.0	10.0
Total	279.6	256.4	243.4	210.5	300.6	227.8	226.5	148.9	154.5
Index	188	172	163	141	195	153	152	100	104

Notes:
1. Bauxite costs include levy and freight charges where appropriate.
2. Greenfield plant depreciation calculated at 4.4%/year.
3. Finance assumed 50:50 debt: equity ratio, with interest at 10% and return on equity of 10% after corporation tax at 45%. Actual finance dependent on political risk (see Table 2.21).
4. Since capital charges (including depreciation) comprise over 40% of greenfield refinery costs, government fiscal concessions can radically modify the attraction of particular locations.

Sources: North American and Caribbean industry sources, 1980.

TABLE 2.21 ESTIMATED GREENFIELD SMELTER COSTS BY LOCATION, 1979

	U.S.	Australia	Brazil	Indonesia
	¢/lb			
Power	12.2	7.8	8.9	3.5
Alumina	14.0	11.4	13.3	17.5
Potroom materials	4.8	4.8	4.9	4.9
Plant and administrative labour	5.3	4.7	4.0	4.0
Miscellaneous plant costs	2.4	2.4	2.5	2.4
Depreciation (4.4%)	3.8	4.2	4.5	7.7
Interest cost (average on life of loan)	2.2	2.4	5.5	5.7
Transportation	1.4	4.0	4.0	4.0
Incremental sales expense	0.3	0.3	0.3	0.3
Sub total	46.4	42.0	47.9	50.0
Selling price	66.0	64.7	63.1	60.9
Pre-tax profit	19.6	22.7	15.2	10.9
Income taxes	(9.0)	(10.2)	(7.0)	(5.0)
Net income	10.6	12.5	8.2	5.9
Plus interest cost	4.3	4.8	9.2	8.6
Total income earned on capital	14.9	17.3	17.4	14.5
Return on total capital employed (%)	10.1	10.0	9.5	5.5
Return on equity capital (%)	17.1	24.1	22.4	10.1
Integrated capital cost ($/tonne)	3,287[1]	3,800[2]	4,030[2]	5,815[3]
Debt:equity ratio	58:42	70:30	80:20	78:22

Notes: [1]U.S. smelter fed by brownfield expansion of domestic refinery sources on Caribbean, Guinean or Brazilian bauxite. Price restrictions and higher power and labour costs offset market proximity and capital cost advantages. Latter advantages stem in part from superior infrastructure and quicker construction time.

[2]Smelter integrated with refinery in each case. Australia's low power and materials costs offset higher capital charges. Finance through higher debt: equity ratio than U.S. plant through loan secured on long-term contracts for supply yields highly favourable equity return. Government-backed borrowing in Brazil also gives favourable rate of return on equity.

[3]High capital costs, despite government backing, result in recourse to outside purchase of alumina, at a premium. Project capable of yielding higher return if smelter expanded by 120,000 t and bulk of output marketed in high-price Japanese market.

Source: Spector 1979.

existing Surinamese refinery (Moment 1979:58) in late 1976 and $114/tonne for Alcoa in Western Australia and $120 for the Queensland refinery of Comalco (Rivkin 1977:32). Not until measures had been implemented to reduce the Surinamese levy on bauxite in 1979 did Alcoa agree to consider a modest 25 per cent expansion of its existing plant, at a capital cost estimated at half that for greenfield expansion.

Alcan headed the counterattack of the corporations against the Jamaican government and other Caribbean producers. With its 20 to 25 year old refineries in need of modernization and carried at a book value of $56/tonne, most of its sales going to third parties in Europe and new refining capacity coming onstream in western Europe, Brazil, and Australia, Alcan was in a much stronger bargaining position than Alcoa with over $300 million invested in Jamaica, or Reynolds and Kaiser with more than 60 per cent and 70 per cent respectively, of their bauxite coming from Jamaica (*Corporate 10-K Reports 1980*). Alcan argued that Jamaican costs were more than 30 per cent above those of Australian plants, that labour productivity was half that in U.S. refineries and one-third that in Australian ones, and that the bauxite levy must be either eliminated or reduced to make cost-saving innovations worthwhile. The Jamaican negotiators, espousing the principle of indexization, nevertheless wanted to fix alumina prices with little reference to the historic 13½ to 15 per cent alumina:aluminium ratio. Thus, while ready to lock the corporations into contracts at about 13 per cent of the ingot price negotiated by the Jamaican government with the Soviet Union and Venezuela, when offered the opportunity of taking over Alcan's operations in return for agreeing to supply Alcan with nearly one million tonnes a year at 15 per cent, the Jamaicans demanded 17 per cent. Having finally convinced the Jamaican negotiators of the uncompetitiveness of the levy, neither the corporations nor the Jamaican negotiators could persuade the Jamaican government to reduce the tax owing to the government's dependence on the annual receipts of $200 million from the tax. Fortunately, the rapid increase in the price of ingot allowed a face-saving formula to be found whereby the absolute value of the levy is maintained (though not in terms of real dollars), but the tax will be reduced from the 7½ per cent of the ingot price unilaterally imposed to the 4 per cent considered tolerable by the corporations. Similar schemes were worked out by the other Caribbean producers, with the exception of the nationalized Guyanese industry which had been unable to impose a levy and, apart from occasional periods of shortage, had been unable to secure lucrative markets for its metal grade bauxite and alumina. Guyana was dependent upon its near-monopoly of refractory grade bauxite to stave off bankruptcy (though not the deterioration of its plant and operation efficiency). In mid-1978, Guyana received $138/tonne for its refractory grade bauxite compared with $17 for metal grade bauxite (Roskill 1979:57), but was forced to call in Kaiser on a management contract and pay $28 million in scarce foreign exchange to hire a U.S. company to strip away overburden in order to sustain its credibility as an effective producer.

Given past government mismanagement of the Caribbean's bauxite resource, neither Comecon nor the World Bank has been prepared to finance the region's bid for a part of the current round of smelter expansion. By mid-1977, U.S. consultant Stewart Spector estimated that the price of aluminium was close to the target price required for two hypothetical U.S. smelters, using 16.5 mill power and Australian alumina at $107/tonne or alumina based on Latin American bauxite at $142/tonne (Spector 1977). The Australian-sourced plant required a price of 54 cents/lb and the Latin American-sourced operation 56.4 cents/lb. Although the prevailing price was judged too low, the A-rated fringe company of Alumax opted for a U.S. smelter at South Holly (estimated power contract at 18 mills), allocating $400 million in 1978. Upon its completion in 1980 ahead of schedule for only $340 million, and with the aluminium price at 72 cents, the risky gamble emerges as a shrewd calculation that heralded the present wave of investment. Despite long-established plans to construct a second smelter in the Pacific Northwest, Amax has been reluctantly forced by environmentalist opposition to follow other U.S. corporations in investing abroad at cheap power sites. Political risk has now emerged as a significant criterion in locating investment with a preference, all else being equal, for Australia followed by Brazil, with the Caribbean rated high-risk along with many African countries, notably Zaire.

Table 2.21 summarizes the range of options under active consideration in 1979, as prices required to trigger investment came in line with those of major markets. An Australian smelter using 12 mill power derived from brown coal with virtually zero opportunity costs and securing alumina at $130/t has the lowest production costs. Moreover, it encourages, through political stability, a low debt-equity ratio (and consequent opportunity for a large absolute profit flow), and a return on equity above 24 per cent. The latter figure compares with a 17 per cent return in the United States where problems of price control and planning permission detract from the safe investment climate. The Brazilian plant is estimated to be capable of yielding a 22 per cent return on equity at a world price of 63 cents/lb, using hydro power at just over 13 mills, alumina at $156/tonne and drawing on government-backed loans to secure an 80:20 debt-equity ratio. The final project shown on Table 2.21 is the riskiest, involving $5,300/tonne investment including the construction of hydro facilities delivering power at only 5.4 mills. These heavy capital requirements are assumed to preclude financing a refinery, so that alumina is purchased on a third-party basis at about $200/tonne. If production is targeted for the growing protected Indonesian market, with any surplus going to the high-priced Japanese market, the chief drawback of the scheme is the commitment of $1.3 billion in high-risk Indonesia. Against such a background, the political risk introduced by the confrontation strategies of the 1970s emerges as a major disadvantage for Caribbean countries attempting to participate in current opportunities for smelter development. Significantly, the only project in the Caribbean region beyond the feasibility study stage in late 1980 was an 180,000-tonne smelter at Point Lisas in Trinidad, using natural gas and jointly financed by the American fringe company National-Southwire and the right-of-centre Trinidadian government. Plans for equity participation by Guyana and Jamaica have been postponed indefinitely.

Two other Carribbean smelter projects were being examined in 1980 by the World Bank as part of the bank's priority for energy import substitution. The two projects, involving the generation of 800 megawatts of hydro-electric power in western Suriname and 750 megawatts in western Guyana, share several important features: (1) construction is staged, and each stage reduces average power costs; (2) part of the power will replace currently high cost oil-generated power; (3) a large new power consumer is essential for full implementation of the project; (4) World Bank assistance for the total project is conditional on securing a partner to finance the smelter and guarantee ingot markets; and (5) both projects are evaluated as riskier politically than alternative Latin American schemes.

A 1978 pre-feasibility study (ASV 1978) of the proposed aluminium smelter in western Suriname, assuming a metal price of 51 cents/lb, alumina at \$180/tonne and 15 mill power, estimated the internal rate of return of a 69,000-tonne unit at 7.1 per cent and a 103,000-tonne smelter at 9.1 per cent. The Norwegian consulting firm calculated that the elimination of the bauxite levy would add one per cent to the return for each smelter scheme and considered these conservative results provided "an acceptable basis for building an aluminium plant" though they fell considerably below industry target rates and available options elsewhere. However, in mid-1980 the Surinamese Ministry of Economic Development did not expect the final stage of the project, the one involving the smelter, to be implemented. The first stage, supplying 123 megawatts at 33 mills for \$350 million, would replace the existing inefficient coastal oil-fired plant which had variable costs alone estimated at 77 mills; and the second stage would double power for \$170 million, reduce costs to 24 mills, and significantly reduce Suralco's refinery costs (Table 2.20) by replacing its existing 140-megawatts oil and gas-fired generator. Although the final phase of the scheme would bring average power costs down to 18 mills, and pricing at marginal cost to a smelter could cut this to 13 mills, the western bauxites are too limited in both quality and quantity, while shallow water would preclude economical shipment. Attempts to interest Reynolds, Pechiney, and Kaiser in the smelter project have failed, while the Surinamese subsidiaries of Alcoa (Suralco) and Shell (Billiton) favour expansion of existing installations. For example, by diverting Jai Creek north to Lake Afobaka, the full 150 mega-watt potential of the Brokopondo scheme could be realized, a 25 per cent improvement over the current position. This would permit modest brown-field expansion of the smelter, while diversion of the Bakhuys ores to the refinery would spin out available reserves and justify brownfield expansion there. Furthermore, a $40 million expansion of the existing modest refractory-grade operation would take advantage of the nationalized Guyanese company's inability to supply its customers and also prolong the viability of deep-mining at Onverdacht where the ratio of overburden to deposits is already 5:1 and deteriorating. In 1978 ore mined in the latter region cost $24/t, about three times the cost of the plateau deposits at Moengo in northeastern Suriname. It would therefore appear that Suriname will experience cautious low-cost brownfield expansion, and that the third stage of the hydro project and its associated greenfield smelter will not be executed.

Prospects are equally gloomy for the proposed Guyanese smelter, though cost and political risk rather than the availability of reserves are responsible in this case. A decade after nationalization, the Guyanese industry was close to total collapse in 1980: despite selling at prices substantially below those of other Caribbean producers, sales of metal-grade bauxite and alumina were running at one-third and one-half, respectively, of capacity. Calcined refractory bauxite, over which Guyana has had a near-monopoly, cannot be extracted in sufficient quantities to satisfy demand, while large price hikes have not compensated fully for revenue losses on other products and increasing costs (oil had reached 25 per cent of total expenses by 1975). Equipment was not satisfactorily replaced, management left, and workers experienced a sharp relative decline in remuneration. Severe strikes in 1979 were estimated to have cost $15 million in lost sales and, together with slippages, hampered production to the point where customers were actively seeking to end their dependence on the industry's erstwhile lifelines of refractory and chemical grade products. Against this inauspicious background, the Guyanese government instructed the state corporation to raise $1.5 billion to finance a 750-megawatt hydro scheme and 145,000-tonne smelter.

World Bank assistance for the scheme has been made conditional upon stabilization of the post-OPEC deterioration of the Guyanese economy, acquisition of equity as well as technical partners for the smelter scheme, and, possibly most significant of all, the inclusion of alternative smaller hydro-electric schemes which do not require a smelter to achieve scale economies, in the feasibility studies. Power cost estimates for the scheme range from 14 to 20 mills, with 16 to 17 mills being the most widely accepted figures; lower rates are available in politically less sensitive countries. Furthermore, it is not clear whether the 16 to 17 mill figure represents estimates of the average power costs, or discounted power at the expense of the coastal community. Given the Burnham government's record of unprincipled appeasement of immediate domestic political pressures and its resulting cavalier behaviour towards private investment (Premdas 1978), the corporations would require power priced at the average cost and their investment guaranteed by OECD governments. In the light of alternative sites, the Guyanese project does not appear to offer aluminium at a cost low enough to yield a return on strictly commercial funds commensurate with the political risk involved.

The scale of the Guyanese project in relation to the size of the national economy presents an additional difficulty. Of the 750 megawatts available at the Upper Mazaruni power site, 735 would be available for consumption after transmission losses over the 300 km from the interior to the Linden smelter-refinery site and the coast (UMDA 1979). The proposed smelter would require 320 megawatts and replacement of the refinery's existing oil-generated power plant a further 70, leaving 345 megawatts for absorption elsewhere in the economy. In 1979, the estimated peak demand of the Guyanese economy was 100 megawatts, and the UMDA prospectus splits the 245 megawatts surplus into a 125 megawatt "spinning reserve" for the smelter and the demands of the national grid at an assumed nine-year doubling rate. The doubts of the power consultants are reflected in the World Bank's insistence that alternative smaller schemes be included in the feasibility studies: the Upper Mazaruni smelter project is weak on technical, commercial, and political grounds. However, unlike Girvan (1971:90), even the optimistic Guyanese do not expect a domestic smelter to attract export-oriented fabrication (UNIDO 1977:41), an element of forward linkage that still eludes the more mature economies of Australia and Brazil.

Conclusions

The North American aluminium corporations functioned as a declining oligopoly during each of the three operating periods since the termination of Alcoa's U.S. monopoly in 1940. However, contrary to the assumptions of Caribbean economists, target pricing and barometric price leadership did not eliminate intense cost competition in which locational cost minimization played a central role. New entrants to the industry adopted a product strategy of vertical integration to secure both a low cost and a reliable materials flow through their systems, thereby reducing the vulnerability to supply disruption implicit in a highly capital-intensive industry. As the political risk to materials flows heightened from the end of the second period, differences in attainable debt: equity ratios between similar projects in differing regions (reflecting political risk) have come to figure prominently in the cost estimation undertaken prior to the spatial allocation of investment (compare Tables 2.10 and 2.13 with Table 2.21). The allegations by Caribbean economists (based on one erroneous set of cost estimates) of locational bias in corporate investments during the 1950s and 1960s were premature but, ironically, by fuelling demands for unilateral militant government intervention in the industry they were self-fulfilling.

The first period was characterized by rapid expansion within a largely self-sufficient hemispheric system as strategic imperatives stimulated aluminium and alumina production in North America. Falling energy costs resulted in increased market-orientation of smelters during this first period, with preference given to sites accommodating waterborne alumina supplies. The proliferating vertical chains of the U.S. corporations integrated backwards to Gulf coast refineries, drawing first on Guinanese ores and then cheap, short-hauled bauxite from the North Caribbean. Only the Canadian corporation, sensitive from the outset to long hauls, and without the benefits of rapid U.S. amortization, constructed Caribbean mine-site refineries during the first period.

The marked slow-down in U.S. demand in the late 1950s spurred the three U.S. core corporations to emulate Alcan and enter the more dynamic markets of western Europe and the West Pacific. In the face of the advantages of a market location and tax incentives in Western Europe, only Ghana and Suriname among the developing countries with potentially cheap power attracted export-oriented smelters, conceding fiscal advantages in the process. To meet incremental demand in the North American market, the core corporations restored to relatively cheap brownfield smelter expansion while the fringe opted for greenfield plants at market locations or, when cheap power together with bulk-shipped alumina became available from the mid-1960s, in the Pacific North West. The preference of the U.S. corporations for Gulf-refined bauxite gave way to greenfield Caribbean refineries which not only had the potential to supply peripheral U.S. smelters more economically than alumina railed from the Gulf, but also afforded greater flexibility in servicing the emerging global corporate smelter networks. This flexibility was further enhanced through the rapid expansion of large new competitive bauxite mines in Australia and West Africa strategically located to the dynamic Pacific and European markets. By the late 1960s the overwhelmingly market-oriented global smelter networks of the North American corporations could be supplied with alumina from brownfield or greenfield expansion in Australia, West Africa, and Western Europe; as well as from the Guianas and North Caribbean. As the hemispheric system merged into a global one, Caribbean mines and refineries yielded markets in western America to cheap Australian alumina and conceded potential markets in Western Europe to market-refined bauxite from Guinea and Australia.

Against this deteriorating background, the Caribbean policy of confrontation with the corporations in the 1970s rendered the region's bauxite both uncompetitive and unreliable, conditions incompatible with the requirements of a vertically-integrated product strategy. In the uncertain economic climate created by the OPEC price increases, the Caribbean governments overestimated the economic rent of their bauxite and underestimated corporate sensitivity to, and speed of response towards, supply marginalization. The resultant disincentives to investment in the Caribbean occurred when long-run cost trends were moving sharply in favour of the location of refineries and smelters in mining regions. Despite attempts by Caribbean governments since mid-1979 to correct past resource mismanagement, lingering mistrust compounded by the long lead-times on aluminium investment have delayed by at least a decade the substantial expansion of forward linkage, via Jamaican refining and Guianese refining and smelting, which policies based on more realistic appreciation of underlying global cost trends and corporate locational behaviour could have attracted to the Caribbean.

Acknowledgement

The financial assistance of the Social Science Research Council and Nuffield Foundation is gratefully acknowledged.

REFERENCES

Aluminium Association. 1979. *Aluminium Statistical Review* 1977. Washington, pp. 32–36.

Arthur D. Little Inc. 1961. *An Opportunity for Aluminium Refining in Peru*. Boston, p. 12.

ASV. 1978. *Prefeasibility Study for an Aluminium Smelter, Apoera, Suriname*. Paramaribo.

Auty, R.M. 1975. "Small Factories and the Measurement of Internal Economies of Scale". *Professional Geographer*, pp. 315–22.

______. 1980. "Transforming Mineral Enclaves: Caribbean Bauxite in the Nineteen-Seventies". *Tijdschrift voor Economische en Sociale Geografie* 71:161–71.

Badger Manufacturing Co. 1956. *Utilization of Domestic Ores for Production of Alumina*. Cambridge, Mass.

Bloch, I., and Moment, S. 1967. *The Aluminium Industry of the Pacific North West*. Portland.

Brokopondo Joint Venture. 1959. *Paramaribo*.

Brubaker, S. 1967. *Trends in the World Aluminium Industry*. Baltimore.

Business Week. 1978. "Alumax: Turning Aluminium Capacity Upside Down". 6 Mar. New York.

Charles River Associates, 1977. *Supply Restrictions in the World Aluminium-Bauxite Market*. Cambridge, Mass.

Cornish, J.K. 1977. "World Flow of Bauxite and Alumina, 1970–75". *IBA Quarterly Review* 2:32–44.

Davis, C.E. 1978. "Some Aspects of the World Aluminium Industry and Jamaica's Strategy for the Local Industry". *JBI Digest* 3:1–17.

Donaldson, Lufkin, and Jenrette Inc. 1966. *Major North American Aluminium Producers Offer Above-Average Earnings Growth at Reasonable Multiples*. New York, pp. 14–62.

Dunning, J.H. 1977. "Trade, Location of Economic Activity and the MNE: A Search for an Eclectic Approach". In Ohlin, B., *et al.*, eds., *The International Allocation of Economic Activity*. London, pp. 395–431.

Fulkerson, F.B., and Gray, J.J. 1965. *Economic Trends in the Pacific North West Aluminium Mill Products Industry*. Washington.

Girvan, N. 1970. "MNC and Dependent-Underdevelopment in Mineral-Export Economies". *Social and Economic Studies* 19:490–526.

_______. 1971. *Foreign Capital and Economic Underdevelopment in Jamaica*. Surrey.

Hodgson, A.A. 1965. *Aluminium Limited: A Time for Re-assessment*. Montreal, p. 30.

Huggins, H.D. 1965. *Aluminium in Changing Communities*. London.

IBA. 1977. *Influence of the Energy Factor on the Bauxite/Alumina/Aluminium Industry*. SCI/II/3. Kingston, pp. 4–35.

_______. 1978. *Bauxites of the World and Their Exploitation*. Kingston.

_______. 1979. *Processing Cost Estimates for Alternative Alumina Processing Technologies: Some Implications for a Uniform Pricing Policy*. SC-PR/VI/3. Kingston.

IBRD. 1968. *Past and Prospective Trends in the World Aluminium Industry*. Paris.

Iron Age, 1980, "Alcan Draws Bigger Bead on a Worried U.S. Market". *Iron Age*, 14 July, pp. 28–30.

Isard, W. and Whitney, V. 1952. *Atomic Power*. Philadelphia.

Jackson, C.J. 1976. *Royalties, Income Taxes and the Bauxite Levy 1950–74*. Kingston.

Karim, A. 1968. *Economics and Directional Growth in the Aluminium Industry*. Oakland.

Klagsburnn, H.A. 1945. "Wartime Aluminium and Magnesium Production". *Industrial and Engineering Chemistry* 37:608–17.

Krutilla, J.V. 1955. "Locational Factors Influencing Recent Aluminium Expansion". *Southern Economic Journal* 21:272–88.

Langton, T.G. 1978. *Investment and Capacity Characteristics of the U.S. Aluminium Industry between 1950 and 1976: Institutional and Theoretical Considerations*. NSF/RA–786152, PB 286 697. Washington.

Lanzillotti, R.F. 1961. "The Aluminium Industry". In Adams, W., ed., *The Structure of American Industry*. New York, pp. 185–232.

Lester, M.D. 1979. "The Outlook for Power in the Aluminium Industry". *IPAI Seventh AGM* (London), pp. 15–21.

Loeb, C.M., Rhoads and Co. 1951. *Aluminium*. New York.

Marcus, P.R. 1974. *Improved Environment for the North American Aluminium Companies*. New York.

Meyer, E.R. 1977. "Growth Problems in the Aluminium Industry". *Aluminium* 53:645–49.

Moment, S. 1962. *The World Aluminium Situation, 1962–63*. Portland.

_______. 1968. *World Bauxite-Alumina Review, 1967–68*. Portland.

_______. 1979. *World Bauxite-Alumina Review, 1967–68*. Portland.

_______. 1979. *The Pricing of Bauxite from the Principal Exporting Countries, 1974–78*. UNIDO IOD. 252. Vienna.

Peck, M.J. 1961. *Competition in the Aluminium Industry, 1945–58*. Cambridge, Mass.

Premdas, R.R. 1978. "Guyana: Socialist Reconstruction or Political Opportunism?" *Journal of Inter-American Studies and World Affairs*. 20:133–64.

President's Council on Wage and Price Stability. 1976. *Aluminium Prices, 1974–75*. Washington.

Radetzki, M. 1977. "Where Should Developing Countries' Minerals Be Processed?" *World Development* 5:325–34.

Reimer, J.H. 1968. *Present Status of Alumina and Aluminium Production in the World and in Developing Countries: Prospects of Developing an Aluminium Industry*. New York.

Reynolds and Co. 1947. *Aluminium Today and Tomorrow*. New York, pp. 25–30.

Rivkin and Co. 1977. *The Aluminium Industry in Australia*. Sydney.

Roach, E.H. 1957. *Bauxite, Demba, Alumina and British Guiana*. Georgetown.

Roskill Information Services Ltd. 1979. *The Economics of Aluminium*. London.
Rumelt, R.P. 1974. *Strategy, Structure and Economic Performance*. Cambridge, Mass.
Smith, Barney and Co. 1959. *Aluminium Limited*. Montreal.
Spector, S. 1977. *Spector Report* 23 May. New York, pp. 17–19.
______. 1979. *Spector Report* 3 Dec. New York.
Stamper, J.W. and Kurtz, H.F. 1978. *Aluminium*. Pittsburgh.
Stigler, G.J. 1951. "The Diversion of Labour Is Limited by the Extent of the Market". *Journal of Political Economy* 59:185–93.
Suralco. 1955. *Suriname Bauxite: A Story of Cooperation in the Development of a Resource*. New York.
______. 1977. *Prefeasibility Study for a Refinery in West Suriname (Apoera)*. Pittsburgh.
Thirtythree Metal Producinv. 1980. "Expansion-Mined Metal Producers Are Eyeing Australia". *Thirtythree Metal Producing* 18:57–61.
UMDA. 1979. *The Guyana Hydropower/Aluminium Project: Basic Facts*. Georgetown.
UNCTC. 1978. *TNCs and the Processing of Raw Materials: Impact on Developing Countries*. ID/B/209. Vienna.
UNIDO. 1977. *Aluminium Smelter Construction in Developing Countries*. ID/WG. 250/18. Vienna, p. 41.
U.S. Bureau of Mines. 1944, 1949, 1954, 1959, 1964, 1969, 1974, 1975, *Minerals' Yearbook*. Washington.
U.S. Senate Committee on Public Works. 1962. *The Market for Rampart Power, Yukon River*. Oakland.
Williams, M.L. 1975. "The Extent and Significance of the Nationalization of Foreign-owned Assets in Developing Countries". *Oxford Economic Papers* 27:260–73.
Younger, J.M. 1971. *Aluminium Industry Outlook to 1980*. Richmond.

3

Conflicts and Coincidences of Interest in the Exploitation of Natural Resources: the Ok Tedi Mining Project, Papua New Guinea

RICHARD JACKSON

The basic problem of natural resource exploitation stems from the multiple conflicts of interest associated with such activities. The basis of these conflicts lies in the fact that different societies, despite Alvin Toffler's proclamations to the contrary (Toffler 1970), still retain their different value systems, their differing levels of technology, and their own methods of economic, social, and political organization by means of which they evaluate and manage their environments. In other words, what is a resource to one society is either not a resource at all or a resource of a quite different form to another. Very large numbers of people today live in societies which depend for very few of their material requirements (and none of their minimized spiritual needs) on their own environments. Such materials are obtained from far-flung environments which are also expected to provide virtually all the material, and frequently the spiritual needs of their own inhabitants. From such variations in the assessment of the resources of any environment, interest groups inevitably emerge. This paper tries to detail the interests of groups in a particular, large-scale resource project, the Ok Tedi gold/copper mine of the Star Mountains of Papua New Guinea, and the ways in which these interests interact, whether in conflict or by coincidence.

Even before political independence was attained in 1976, the PNG government had laid down guidelines for the nation's development. These guidelines, popularly known as the Eight-Point Plan, may be generally summarized as calling for (1) rural development, (2) more equal distribution of opportunities between races, regions, and sexes, (3) greater Papua New Guinean control of the economy, (4) decentralization of much decision-making to provincial levels, and (5) greater self reliance.

In early 1980, Michael Somare's coalition government, which had ruled Papua New Guinea since 1972 and which had enunciated these development aims, was replaced after a parliamentary vote of no confidence by one led by Sir Julius Chan. With this change of government and with the onset, five years after independence, of that almost inevitable degree of disillusionment which succeeds the euphoria accompanying the end of colonial rule, it has become fashionable both inside and outside Papua New Guinea to dismiss these aims, or at least the policies used to achieve them, as having failed. It is true that regional inequalities in economic activity have not been reduced, but it is also true that such inequalities in social services have been lessened to some degree (Berry and Jackson 1981). While the economy remains dominated by foreign interests, many plantations have reverted to local ownership. Papua New Guineans now have major interests in retailing, and efficient, nationally controlled banking and financial services have been established. Political decentralization has proceeded quickly, perhaps too quickly for some, though the decentralization of staff needed to administer decentralized decision-making has not matched this, primarily because the provinces lack adequate revenue-raising powers and resources.

In all these areas, large-scale mining ventures do have some appeal. Such ventures, because of geological chance, are generally located in remote areas — providing at least the possibility for development opportunities in areas that would otherwise have none at all. The negotiation of mining agreements can certainly include provision for major roles for either nationally controlled organizations in insurance, banking, shipping, or air transport and/or local business interests. Moreover, since under the laws of Papua New Guinea the provinces receive the royalties from such mining projects, they also offer a possible solution to the financial problems of the provinces which underlie many of their administrative staffing problems.

The most conspicuous area of failure in progress towards the achievement of the eight aims has been in overall rural development, which is evidenced by the continuing rapid increase in urban populations (16 per cent a year in 1966–71, and 8 to 9 per cent per year in 1971–80), and of growing urban unemployment. While it is hardly fair to criticize a government committed to rural development for failing to provide urban jobs, the need for such jobs would probably be far less if rural conditions were themselves improving to meet the growing aspirations of rural dwellers. Mining in itself will do nothing to generate rural development, and capital-intensive mining, of the type proposed for almost all likely future projects in Papua New Guinea, will generate very few jobs, urban or otherwise, directly. In both areas, mining can only make a significant contribution if linkages are carefully planned and vigorously encouraged. On the whole, it is true to say that such linkages between mining and the rest of Third World countries' economies are neither planned nor encouraged.

Financial policy, considered in strict isolation from some of its side effects, has by world standards been very successfully pursued. The national currency, the kina, originally issued at par with the Australian dollar, has been frequently revalued and is now worth A$1.31. Annual inflation in

Papua New Guinea averaged between 6 and 8 per cent until 1979. The visible balance of payments until 1980 has been consistently and strongly positive. Import prices and imported inflation have been kept in hand. Stabilization funds have done much to smooth out fluctuations in the prices growers have received for export crops. However, the general informed opinion now is that the success of this financial policy may be chiefly attributed to good prices on world markets for most PNG products in 1975–79. Now that such prices have taken a marked turn for the worse (with the exception of gold), it seems likely that success in this area may be coming to an end. Now, more than previously, there is an urgent need for the rapid development of a new, large source of revenue. The Ok Tedi mine promises to be such a source, and in this respect taken on a vital role in PNG financial strategy.

Development of Natural Resources in Papua New Guinea

Natural Resources and their Role in National Policy

While it has not escaped well-founded criticism, particularly of the indirect effects it has had (Lam 1980), the push towards fiscal self-reliance has shown more progress than that achieved towards other aims. The architect of the strategy by which such progress was to be achieved is the now Prime Minister, Sir Julius Chan, previously the Somare government's Minister of Finance. Two major policies have been pursued. The first has been the control of consumer price indices through a "hard currency" policy. Since urban workers' demand for consumer goods is largely met by imports, the strategy of revaluing the kina against the currencies of the major suppliers of imports (most prominent of which is Australia) has tended to keep imported inflation under reasonable control (Jackson 1980). Before the introduction of this policy, urban minimum wages doubled in real terms between 1970 and 1975. The government, which employs at least one-third of the work force, could not afford to allow such a trend to continue. It successfully persuaded the public service union to accept a tie between the consumer price index and wage levels. Therefore, revaluation kept wage demands in check, theoretically allowing additional revenues to be spent on rural areas. The likelihood of the balance of payments deteriorating has already been noted; if that should occur, there would ensue serious problems for the government's wage strategy.

The second policy, of more direct relevance to this paper, is that of natural resource development. While government expenditure is held in check, revenue to replace annual Australian grants, which still account for more than a quarter of all government revenues, is to be obtained through resource projects. The policy is simple: mining is the easiest and quickest way to raise large amounts of additional revenue. Government policy has been clearly stated on several occasions:

> Mineral resources belong to the people. Mineral development is justified principally in terms of the revenue it contributes to other Government programmes. Mineral development at this stage can proceed most effectively through the involvement of overseas investors. Overseas investors are entitled to a reasonable return on their investment. A high proportion will be passed on to the people through taxation (PNG Government, Statement of Intent 1977).

In following this policy, the government clearly places itself in an intermediate position between the local interests of the areas in which resources are to be developed and those of the outside world, whether consumer countries, multinational mining companies, or the banks which finance such companies' operations. Government interests do not necessarily coincide with any others. In this self-appointed role of a far-from-disinterested middleman, the government has to deal with all other parties.

On the one hand, it has to work with companies whose technologies are sophisticated, whose investments are large, whose choice of project areas is more or less worldwide, and whose concept of a "reasonable rate" of return on investment is very likely to be different from that of its own. While the government may try to control the timing, priorities, and conditions of mining activity in its territory, in an industry where success or failure is contingent upon variations in ore content measured in one or two parts in a million, such control is severely constrained by the nature of the ore deposits as well as by events in the rest of the world.

On the other hand, most of the major mining projects and prospects of Papua New Guinea are in very remote locations (see Fig. 3.1) whose inhabitants are only but dimly aware of the existence and functions of the government in Port Moresby, let alone the logic of its development strategies. In Papua New Guinea, the overwhelming majority of the people regard land as being inalienable. Hardly a single Papua New Guinean, not even those in the highest positions of government, would admit to wholeheartedly supporting the notion that ownership of minerals beneath the land surface is separable from the ownership and use of the land itself — even though such is the law of the country. The impact of big mining projects in such remote areas where the population is barely incorporated into the structure of the state will create major conflicts of interest, with the government squarely at the centre of such conflicts.

Notwithstanding such problems, mining projects are beginning to do what was primarily expected of them — to substitute for Australian aid both directly, through the Mineral Resources Stabilization Fund, and indirectly through increased personal tax receipts, and to make a substantial

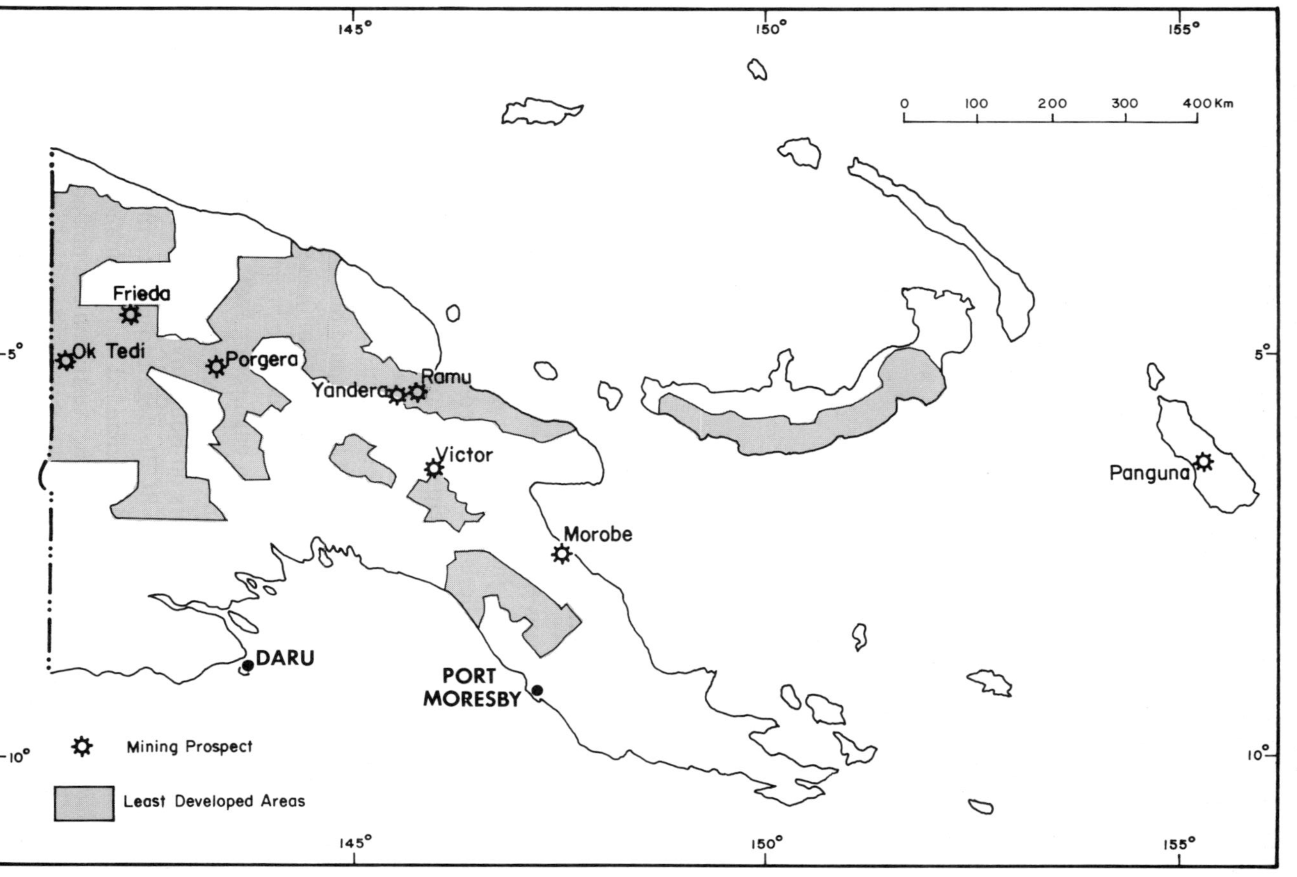

Fig. 3.1 Mining prospects and least developed areas in Papua New Guinea

contribution to government revenue (Table 3.1). Now, however, there is really only one major mining project currently operating in Papua New Guinea — that of Bougainville Copper Private Limited (BCPL), located on Bougainville Island in the North Solomons Province.

TABLE 3.1 GOVERNMENT REVENUES IN SELECTED YEARS

Revenue Source	1969 (%)	1974 (%)	1979 (%)
Australian grants	58	43	32
Foreign loans	5	14	16
Internal revenue	37	43	51
(of which)			
Direct taxation	(12)	(17)	(19)
Indirect taxation	(13)	(12)	(17)
Mineral Resources Stabili- zation Fund	(0)	(0)	(7)
Others	(12)	(14)	(8)

Source: PNG Department of Finance 1980.

The Bougainville Experience

Papua New Guineans have already experienced the problems and conflicts which arise between interest groups when large-scale mining developments occur. Bougainville Island is located about 1,000 km east-northeast of Port Moresby. In 1964, a group of geologists from Conzinc Riotinto of Australia Exploration (CRAE) discovered a large copper porphyry deposit on the island. The ore body in the mountainous interior of the island, which lies along the colliding junction of the Pacific and Solomon Sea plates, consisted of 900 million t of ore at an average grade of 0.48 per cent copper and 0.55 g per t of gold. Bougainville was a massive project, regarded by the mining fraternity as something of an engineering feat, and one which cost more than US$400 million to construct. From the start, many Papua New Guineans, particularly landowners on Bougainville, viewed the project in a somewhat different light. When BCPL began to pay its first dividends in 1973, the country's political leaders demanded that the financial agreement, negotiated between the company and the Australian colonial administration, be scrapped and a new deal negotiated. Objections to the mining project therefore came from three interest groups — the local people, the national politicians, and some international observers.

Richard West (1976) in his book *River of Tears*, a general critique of the operations of BCPL's grandparent company Rio Tinto Zinc, was particularly caustic in his analysis of both the company and the colonial administration. He described Bougainville prior to the arrival of BCPL as "a peaceful and prosperous island" and an "island paradise". However, most Bougainvilleans, whose views West tries to represent, have consistently claimed as part of their grievances that their island was shamefully neglected by the government before the discovery of copper. West himself, trying to have his cake and eat it, recounts that the period between 1880 and 1970 was one marked by the continual unrest of its inhabitants opposed to colonial rule. He argued that the benefits of the mine would be minimal: "in theory, dividends from the mine will benefit the exchequer of Papua New Guinea, although even that . . . is only hypothetical. But the mine is of little benefit to the economy of Boungainville and may even be detrimental".

A second group of objections — those of the national politicians and bureaucrats spearheaded by John Kaputin and Father John Momis — were based primarily on the inequitability of the distribution of the financial gains of the project. They believed that the agreement reached with CRAE on Papua New Guinea's behalf by the Australian administration was far from being in Papua New Guinea's best interests. In assessing the profitability of the mine before operations commenced, CRAE had based its calculations on the assumptions of: (1) copper prices in the range of US$0.30–0.40 per lb, (2) gold prices of US$38 per troy oz, and (3) a reasonable rate of return on investment of 15 per cent, discounted cash flow.

The company's judgement was that the mine would be of a low order of profitability and therefore argued that various concessions such as a tax-free investment recovery period and accelerated depreciation allowances be granted. They were. Production started in 1972, during which the average price of copper on the London Metal Exchange was US$0.465 per lb and gold prices in London averaged US$58 per oz. During 1973, however, these prices leapt to 76.6 cents per lb and US$97 per oz. The Bougainville mine was no longer marginal — it was a bonanza.

In financial year 1973, BCPL registered a profit of A$158 million, the biggest ever made by any company listed on the Sydney Stock Exchange. Of this sum, under arrangements made by the old agreement, A$80 million went to shareholders and A$68 million went towards loan repayments. Since the PNG government owned only 20 per cent equity and no company tax was to be levied during the investment recovery period, the exchequer received little more than A$16 million. The politicians were understandably furious. John Kaputin, Minister of Justice in 1974, called for the company's nationalizaton. National leaders who had previously suspected they had been duped now believed it, though, as can be seen, the large profits of the company were partly because of the steep upturn in metal prices. However, events did take a step for the better, for by October 1974 the original agreement had been renegotiated to the mutual satisfaction of both parties. The company agreed to pay normal company taxes (then 33⅓ per cent) on any profit up to A$87 million, and a 70 per cent tax on any excess, as well as agreeing to the elimination of previously agreed concessions and the additional charge of A$0.50 per t of copper mined to a special Non-Renewable Resources Fund for use in environmental reclamation projects in the mine area. Therefore, while the government still held only 20 per cent equity, on any future profit of A$100 million, it would receive: A$29 million in normal company tax, A$9 million in additional profits tax, A$12.4 million in dividends, and A$7.4 million in dividend withholding tax — a total of nearly A$58 million. This sum excluded indirect taxes. Following renegotiation, which the government handled with rather more skill than CRAE expected, the national leaders felt that a fair deal had been reached. Since then relations between BCPL and the government have remained good, if not cordial.

While this renegotiation satisfied the national government, it did not please the third interest group, the Bougainvilleans. They felt that their interests were threatened by mining over a far wider range of issues than merely finance. From the very start of the project, serious land disputes had arisen in the proposed mine area, which erupted into violence on several occasions (Bedford and Mamak 1975). Such disputes continue, though in a much-abated fashion, to this day. Demands for greater compensation to landowners affected by BCPL's activities have been made often and obtained. The occupation fees paid to landowners within the BCPL mining lease have risen steadily from A$5 per acre (A$12.34 per ha) per year at the start of the project to A$84.50 (K65.00) per ha per year today. Through pressure by its representatives in government, the landowning group received 5 per cent of an overall royalty levy of 1.25 per cent on the f.o.b. value of all exports from the mine. The remaining 95 per cent is now paid to the provincial government. In financial terms, the North Solomons is today one of the two wealthiest provinces of the nineteen in the country. It received 18 per cent of all national government allocations to provincial governments in 1980, despite having less than 3 per cent of the national population.

Although money often plays a role in other grievances held by Bougainvilleans, it is frequently not their major cause. The province had (whatever Bougainvilleans say) a flourishing smallholder cocoa sector as well as a well-established plantation sector before the development of the mine. Consequently, the island imported much of the labour for the latter activity from the PNG mainland, particularly from the less developed highlands. The Bougainvilleans, who are generally very dark-skinned, tended to look down on these less sophisticated immigrants whom they scathingly termed "redskins". As long as these outsiders remained poorly paid plantation workers, the degree of xenophobia shown towards them remained mild. But once mining construction began and the plantation-based highlanders turned to more lucrative employment with construction gangs and the mine, local grievances grew. On the one hand, Bougainvilleans protested that the mine was hiring far too few local inhabitants and far too many redskins. On the other hand, it was alleged that "large" numbers of mainlanders, unemployed at the end of the mine construction period, were a major source of petty criminal and generally anti-social activities. As a result, the company embarked upon a programme of hiring of local labour and of training of Bougainvilleans — who had previously opposed the mine's establishment. This programme, assisted by the high standard of education in the province relative to other parts of the country, has been very successful. Today, about 40 per cent of the national work force of BCPL is Bougainvillean.

The renegotiation of the BCPL agreement entailed an interesting consequence. By giving the national government the greater share of mine profits, it also ensured that the costs of any reduction in mine revenues — through payment of bigger royalties, higher compensation, or loss through stoppages — hurt the government as much, if not more, than any other single interest. This tended to place national and provincial interests very much in opposition to each other, while placing the company in a far less vulnerable relationship with either group. Therefore, early violent opposition to BCPL was later turned on the newly independent state and the secession of Bougainville from Papua New Guinea became a real possibility. Richard West (1976) likened Bougainville to Katanga, and forecast a similar, grisly fate for this erstwhile tropical paradise. In his attempt to castigate the machinations of multinational companies, West — like the company — underestimated the ability of local politicians not only to look after national financial interests (which they did rather well) but also to retain national unity. The Bougainvilleans led by Father Momis, Leo Hannett, and Dr Alexis Sarei were no latter-day undisciplined followers of Tshombe, and Michael Somare was both far more flexible than Patrice Lumumba and far less volatile than any Zairean general. Between them, provincial and national leaders turned the potential disaster of secession into a veritable triumph for national development by agreeing to a general decentralization of certain national decision making and financial functions, not only to the North Solomons but to all nineteen provinces of the country.

Today, opposition to the mine on Bougainville is mild and muted, while in the eyes of the national government BCPL has taken on something of the stature of a model company which other foreign-owned companies should attempt to emulate. Bougainville Island retains a strong and prosperous rural economy based on smallholder cocoa and copra production, suffers little from the disruption of village life by urban drift, loses few of its educated sons and daughters to other provinces through migration, has a fast-growing manufacturing sector, has considerable political influence in national affairs, and receives more finance from central revenues per capita than any other province. Though there is reason to believe that a national government could have done even better out of BCPL, the company's contribution to national revenues and to the country's healthy (up to 1980) balance of payments, its excellent training facilities for nationals, and its accommodating attitude to local landowners, to union activity, and to government requests — have all given BCPL a good name in Papua New Guinea, if not with those international observers who would oppose the whole concept of multinational investment in such countries as Papua New Guinea.

Ok Tedi

Potential Mining Developments in Papua New Guinea

Since 1972, BCPL has provided about 8 per cent of national government revenues, 45 per cent by value of all PNG exports, 3,700 jobs directly in the mining industry, and another 2,000 indirectly on Bougainville. It remains the only major mining project in operation. However, in recent years, theoretical advances in the investigation of the mineralogical characteristics of plate collision and subduction zones have been paralleled by an intensification of mining exploration and the discovery of commercially viable deposits in such areas. Papua New Guinea, except for its southwest corner, is almost entirely the result of plate collision processes. The vastness of its mineral resources is only now becoming apparent. A list of potential projects currently under study is shown in Table 3.2 (the list is not exhaustive).

In considering the probability of whether these projects will eventually proceed, one must take into account changing circumstances both internal to Papua New Guinea and those taking place in the outside world. Within Papua New Guinea, times have changed in the past five years. In 1974, assured of continuing Australian grants in aid to the extent of A$200 million each year, the PNG government could afford to take a tough line with foreign companies. Its position was made even stronger in October 1974 when the renegotiation of the BCPL agreement assured it of larger revenues from that source. It was able, in March 1975, to reject the proposals made to it by Kennecott for the mining of the Ok Tedi deposits which Kennecott had first discovered in 1968. However, in the next five years economic growth had not been rapid, and the rural cash crop sector had been fairly

TABLE 3.2 POSSIBLE FUTURE MINERAL DEVELOPMENTS IN
PAPUA NEW GUINEA

Project Location	Mineral	Earliest Date of Construction	Capital Cost (US$ Million)	Company
Ok Tedi, Western Province	Au/Cu/Mo	1981	1,300	BHP Australia Amoco U.S.A. KEG W. Germany
Porgera, Enga Province	Au	1983	300	Placer Aust. MIM Australia Cons. Gold U.K.
Bundi, Madang Province	Cr/Co/Ni	1984	600	MIM Nord U.S.A.
Frieda, West Sepik Province	Au/Cu	1988	1,000	MIM and others
Yandera, Madang Province	Au/Cu	1990	1,000	BHP and others
Morobe, Morobe Province	Cr	?	150	Amax U.S.A.
Mt. Victor, Eastern Highlands	Au	1985	?	Kerea Rubberland, Aust.
Panguna, Bougainville	Cu/Ag/Au	1983	375	BCPL expansion

Sources: Various announcements at various dates in the Australian financial press.

stagnant, buoyed up only by very good world prices for coffee and good prices for cocoa and copra. In 1980, however, almost all prices for agricultural commodities entered a period of low prices more or less simultaneously. At the same time, Australian grants were being cut, by mutual agreement, in real terms, while lower ore grades provided lower revenues at BCPL. By 1980, therefore, Papua New Guinea was much more anxious than it was in 1975 to ensure that more mining ventures should take place.

Events in other parts of the world, especially in southern Africa, the Middle East, and parts of Central America, are all resulting in a redirection of mining effort and investment towards the countries of the Pacific rim, particularly Australia, and also Papua New Guinea. Practical politics and geological advances are both pointing to the likelihood that extensive mining development will occur in the southwest Pacific area over the next two decades. In Papua New Guinea's case, the question is not really whether mining investment can be attracted — although some companies have expressed doubt on Papua New Guinea's internal stability — but whether the PNG government can negotiate agreements which both raise the necessary revenues for national development purposes and safeguard the social fabric and environment of the mining areas as much as possible. For the moment, however, given PNG's short-term financial problems, there is little doubt that at least one more major mining project will proceed quite quickly.

The project nearest to construction and also one of the biggest projected is that in the headwaters of the Ok Tedi ("ok" means river) in the Western Province of Papua New Guinea, 15 km from the country's border with the Indonesian province of Irian Jaya. The project will exploit a monzonite/diorite porphyry intrusion at Mount Fubilan. The location of the mine alone creates several problems (see Fig. 3.2).

The first problem is its extreme remoteness, 1,000 km by river from the sea and in an area without roads, which will give rise to severe transport and supply difficulties. Concentrates will be piped 30 km to a concentrator at a new mining town (Tabubil), trucked 150 km to a small existing port on the Fly River (Kiunga), and then barged 1,500 km down the Fly and across the Gulf of Papua to an ocean terminal at the capital, Port Moresby. Transport will account for about one-fifth of all capital costs and a quarter of all recurrent costs.

Second, several streams in the mine area drain across the border into Indonesia, while the barge route to Port Moresby uses the Fly which forms the border with Indonesia for more than 100 km. Although relations with Indonesia at government level are superficially cordial, there is an underlying tension stemming from widespread popular feeling in Papua New Guinea for the cause of fellow Melanesians under Indonesian rule in the western half of the island, feelings which were additionally fuelled by Indonesia's takeover of partly Melanesian East Timor. Such feelings would not accord well with any action by the PNG government requiring favours of Indonesia in respect of the Ok Tedi mine's environmental impact or transport route.

Third, Mount Fubilan, the site of the orebody, and the people (the Min) of the Star Mountains were first contacted and tenuously brought under colonial rule in 1963, only five years before the ores were discovered by Kennecott geologists. There had been virtually no "development" prior to mine exploration work in the area. Moreover, the Western Province as a whole, of which the area is a part, has played the very humble role throughout the past century of being a labour reserve for the remainder of the country.

Fourth, the area is environmentally very difficult. It is criss-crossed by limestone precipices up to 1,300 m high, and it receives, in places, more than 10,000 mm of rain a year. Rain falls in measurable quantities on more than 330 days a year, in what is one of the wettest places anywhere in the world.

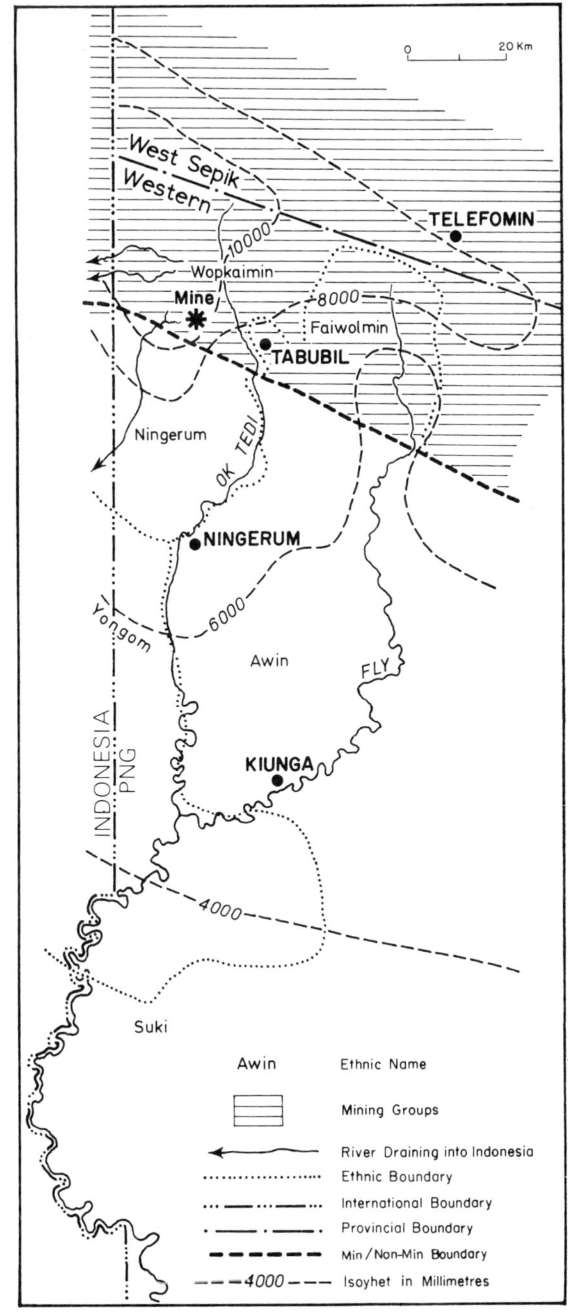

Fig. 3.2 Some locational problems associated with the Ok Tedi Project

Fifth, the Min people of the Star Mountains who are the landowners of the mine site form only a small segment of the Min as a whole. Some live across the border in Indonesia. The majority live in West Sepik Province across the old internal colonial border between Papua New Guinea. Given the degree of decentralization of decision-making powers that has occurred recently, ironically brought about as a result of strains caused by mining on Bougainville, the coordination of any planning for the project, especially those aspects which might positively tie mining to the propulsion of the local economy, is extraordinarily difficult. A further aspect of this same question is the strength of the Papua Besena movement in the 1970s, towards the separation of Papua from the rest of the country. Although this party is now a member of the Chan coalition government, it is likely that any attempt to spread deliberately the benefits associated with mining to the Min as a whole would raise problems and objections at the national political level.

In comparison with Bougainville, Ok Tedi's ore reserves are somewhat smaller but have higher grades.[1] Depending on whether one accepts the mining consortium's estimates or those of the government's own engineering consultants, metal content is between 70 and 105 per cent of those on Bougainville (Table 3.3). The deposit was first discovered in 1968 by Kennecott geologists during an exploration programme which also uncovered several smaller but similar deposits in the region. In March 1975, Kennecott abandoned the project, partly as a result of internal company problems, partly because of a certain degree of ignorance on its part of the extent of

TABLE 3.3 COMPARATIVE SIZES OF OREBODIES, BOUGAINVILLE, AND OK TEDI

Mine	Ore Reserves (million t)	Av. Copper Grade (%)	Cu Content (million t)	Av. Gold Grade (t. oz/t)	Gold Content (million troy oz)
Bougainville	900	0.48	4.3	0.018	16.2
Ok Tedi (company)	410	0.67	2.8	0.027	11.1
Ok Tedi (govt)	780	0.61	4.7	0.023	17.5

Sources: Bougainville — *BCPL Annual Reports* (various dates); Ok Tedi — *Ok Tedi Consortium Feasibility Study* (1979), and unofficial estimates by engineering consultants to the PNG government (1980).

---

[1]Average grade depends upon the assumed lower cut-off rate for ores mineable. Reduction of the lower limit would increase the ore reserves but reduce the grade.

Ok Tedi's ores, particularly gold ores, and partly because of the tough negotiating stand taken by the PNG government team (Davies 1978) fresh from its triumphs with BCPL.

In many ways, it was fortunate for everyone — except Kennecott — that this abandonment occurred, not only because since 1975 ore reserves have been proved up by a minimum of 250 million t of ore, but also because the rocketing price of gold has now made what many people thought to be a marginal project into an extremely attractive one, one of the most attractive, indeed, in the world today. What was a hundred million t of waste rock overburden in 1974 is now a gold ore bearing metal worth at least US$2 billion at current prices.

The structure of the interest groups involved with the Ok Tedi project resembles a Russian doll — they lie inside one another like so many onion skins. One may perhaps isolate four major groups in descending order of scale and add on one external, and very big, interest group — the mining consortium. Unlike the Bougainville case in which a single subsidiary was established (BCPL) by a parent company (CRAE), itself an offshoot of a worldwide company (RTZ), Ok Tedi will be mined by a very mixed consortium of interests made up of four parties.

The first party is the Broken Hill Proprietary Limited (BHP), the largest registered Australian company. Before Ok Tedi, it had been involved in very few ventures outside Australia and very few involving copper mining. It would be difficult to label BHP as a multinational. It will hold 30 per cent of the Ok Tedi equity.

The second party is the Mount Fubilan Development Company, a wholly owned subsidiary of Amoco Minerals of the United States, which in turn is a branch of the Standard Oil group. It too will hold 30 per cent of equity.

Third, Kupferexplorationsgesellschaft, a mini-consortium of three West German metal smelting, cable making, and electrical companies, will hold 20 per cent of the equity.

Fourth, the PNG government will take up the remaining 20 per cent of the equity.

The nature of the consortium members' interests, while complementary, is certainly not uniform. Apart from all members sharing the wish that the mine project be profitable immediately, their other interests vary. The German group is certainly also interested in ensuring for itself a regular supply of copper concentrates. BHP, from the start of the project, has appeared to want to use Ok Tedi as a springboard to show off its capabilities as a project manager to future possible customers for such services. Amoco, on the other hand, has almost certainly had some reservations about BHP in this role, given the latter's limited experience in overseas mining ventures. The varying nature of these interests (and of the different countries which they represent) is important because they have been reflected on several occasions in some degree of potential disagreement and weakness between the partners. Such differences of opinion have given the government some leverage it would not otherwise have had. At the least, the image of the monolithic multinational so commonly conjured up by the literature is erroneously misleading in the Ok Tedi case. As the cost of developing mining projects increases throughout the world, one might suggest that it will become more and more common for such development to be undertaken by consortia of interests rather than by single companies. As can be seen from Table 3.2, most of the likely future mining projects will be operated by such consortia. In extracting the best deal for itself while negotiating the conditions for the development of such projects, this fact alone will be of great advantage to the government negotiating teams.

The PNG Government.   All governments are a species of administrative consortia, and like the company consortia with which they deal, they necessarily have their structural weaknesses and internal disagreements. When one talks of the ''government'' negotiating a deal with mining companies, one is actually talking about representatives from several different departments who do most of the real negotiating work — in Papua New Guinea's case, representatives from the Department of Minerals and Energy, the Department of Finance, the Department of Justice, and the National Planning Office. Routine negotiations are undertaken by senior bureaucrats in these departments, and even major sessions find the government represented by the heads of each of these departments. Politicians play a very irregular role, coming in on only one or two very major occasions, even though all decisions negotiated by the bureaucrats have to be ratified by the National Parliament or the Cabinet.

From such a structure certain inevitable weaknesses might be expected to arise from some areas.

First, there are conflicting interests between departments. Obviously, the Department of Environment and Conservation will hold a very different view of mining to that held by the Finance Department. Such potential for conflict within government might be expected to be even greater when it is remembered that the PNG government has, since 1972, always been a coalition in which different leaders of the parties involved have held different ministries. This is a serious weakness, but it has been reduced by the overriding importance of maximization of revenues to the government.

Second, there are the rivalries between departments for project control. These stem directly from the nature of the process whereby minerals are discovered, proved, exploited, and marketed. At first, this process falls directly within the ambit of the Department of Minerals and Energy and only later do most other departments develop an interest. At the negotiating stage, Finance and Justice clearly take precedence, and at the revenue raising stage, when the mine is in production, National Planning considerations become more prominent. If organization by government fails to follow this shifting order of things, then one might expect frustrations and rivalries to arise.

Third, there is potential conflict between politicians and bureaucrats. In the Ok Tedi case, this has been minimal mainly because the politicians, on the whole, have been very anxious for the project to proceed. If anything, conflicts have arisen because the public servants have believed that some proposals from the companies were not in Papua New Guinea's best interests and have asked for improved proposals, thus causing short delays in Ok Tedi's progress towards production.

Fourth, there is lack of government expertise. Underlying much of the debate on the role of multinational companies operating in the territory of Third World nations seems to be the assumption (a rather unflattering one to Third World governments) that the match is totally unequal; that governments must rely on the company for its supply of basic, particularly technical, data; that the company is staffed by smooth-talking financial wizards and the most cunning of negotiators backed by both modern communication and computational equipment and the most devious of skilled legal minds, while governments are staffed by well-meaning but generally inadequately trained and poorly briefed officials. It is up to others who know the situation in other countries to challenge these assumptions. In Papua New Guinea's case they are wrong. First, because in general the PNG government personnel were certainly as well-qualified and experienced in almost all areas of importance to the project as their company counterparts. In some areas, for example, knowledge of local conditions, they were far better informed. In fact, the government has occasionally had to lend personnel to the BHP consortium to assist in such matters as the computation of financial returns. Second, because in such areas as international finance, mine engineering, and environmental/social impact, the government was able to hire consultants who were often better and always as good as the consortium's own staff. In all these three areas, the government generally had better and usually more information than did the consortium. In the case of the Bougainville renegotiations and the withdrawal of Kennecott from Ok Tedi, the government was clearly far more astute than its multinational opponents and came off better in those negotiations.

To summarize, the most serious weakness of government lies in its own internal management, in ensuring that all its constituent departments work together towards an agreed common aim. This is difficult since some departments cannot, by their very nature, agree with a goal of, say, revenue maximization. Papua New Guinea has not found multinationals to be impregnable. Indeed, not only has the government been able to out-manoeuvre them on several occasions, but it has found them far more prepared to enter into agreements on training of nationals and the encouragement of local business development than other, smaller expatriate controlled petty trading and service companies.

The Provincial Government.   As noted earlier, one of the most important results of the establishment of the Bougainville copper project was the resultant inauguration of provincial governments throughout Papua New Guinea. To this new tier of government has been given powers of decision making in primary and secondary education, most land matters, health care, agricultural extension work, and general administration. The Fly River Provincial Government (FRPG) of the Western Province has a direct interest in the Ok Tedi project for two major reasons.

First, it will bear responsibility for the provision of many of the services in the mining area, although these costs will be borne initially by the national government. But more important, it will also bear the responsibility for ensuring that the mine is well integrated into the general development of the area under its control.

Second, the province will receive 95 per cent of the 1.25 per cent of f.o.b. value royalties generated by mining. Since such values are likely to be in excess of US$120 million in terms of provincial receipts over the next twenty-five years in comparison with the current annual receipts of US$1.6 million, this is of particular concern to the FRPG. It is therefore very much in the interests of the FRPG that the mine should be developed since there is very little prospect of the FRPG obtaining even much smaller revenues from any other source. The more the mine exports, the more money the FRPG will receive.

In addition to the above, the importance of the Ok Tedi mine to the national government should give the FRPG considerable political clout in its future financial dealings with Port Moresby — something the poorly developed province has never had before.

As noted above, one result of mining on Bougainville was the strong — but eventually aborted — move of the province to secede from Papua New Guinea. This move has a parallel in the Ok Tedi case, but here the question is not so much of the province seceding from the state but of a part of the province, the Kiunga District, wishing to establish itself as a new province. Kiunga, the administrative centre of the district in which the mine will be located, is more distant from the provincial headquarters in Daru than Daru is from Port Moresby. The Western Province, although having a total population of fewer than 75,000, is larger in area than Belgium and Holland put together. Communication between Daru and Kiunga has always been poor, and many people in the district blame Daru officialdom for the lack of development in their area in the past. Since 1975, the move to establish a new province has grown. Until the beginning of 1980 there seemed little chance that such a move could succeed, since the brother of the Premier of FRPG was also the Deputy Prime Minister in the national government, while the Member of Parliament for the area wanting secession was an opposition backbencher. With the overthrow of the Somare government, the latter is now a minister and the former has little influence. Therefore, the prospects for secession have now become real.

Local People's Interests.   The prime focus of local interest in the Ok Tedi project is participation, either through employment or through associated business development, in both construction and operations of the mine. The project is located in the outermost periphery of the country, itself located on the world's economic periphery, at least until the most recent past. Average annual income through money-generating activities does not exceed US$25 a head, mainly from a few rubber smallholdings. The area's health and educational facilities have always been left in the hands of the missions, supported rather reluctantly and certainly inadequately by the national government. The only reliable form of transport has been the air service provided by the Ok Tedi Development Company and that operated by the Catholic Mission. Paid employment for anyone from Kiunga generally meant simultaneous migration out of the area. Therefore, ever since Kennecott first moved into the area in 1968, the mine has been looked upon by the local people as an economic saviour and local opposition to its development has been minimal. The disputes which have arisen largely reflect frustration that mining has been slow to start.

The mine site is in the mountains, in the land of the Min. Most of the rest of Kiunga District is composed of low-lying ridges and swampy valleys rarely reaching more than 100 m above sea level. These lowlands are occupied by groups of mobile sago gatherers, the Aekyom (or Awin), and by the Yonggom whose culture and economy are quite different from those of the highland Min. A third group, the Ningerum, occupies a foothill zone between the Min and the lowlanders. The Min are themselves divided by traditional enmities, and more recently by the mining proposals. Those Min occupying the mine site (the Wopkaimin) seem to have been on the losing end of a simmering and longstanding feud with their Faiwolmin neighbours to the east. However, almost all the major payments of compensation, occupation fees, and royalties which should total more than US$10 million over the next twenty-five years will go to the Wopkeimin landowners, of whom there are not many more than 300. A few Ningerum landowners will also receive direct cash payments for the land to be used for a tailings dam for the mine. Virtually nothing of this form of cash payment will go to any other group; to obtain mine benefits these others will have to participate either as workers for the mine or in businesses supplying the mine. As a result, disputes are beginning to arise between these different groups; old rivalries are taking on a new form. Even so, the general situation is that the long-neglected and unsophisticated people of the areas are, unlike their Bougainvillean predecessors, strongly in favour of the project.

They have had no impact on the negotiations over the mine, except at an informal level, and have received very little information as to the exact nature of the project. What information they have had has come from field staff of the mining company who have also played a major role in establishing local business groups in each village. These groups, with the assistance of company field staff, now own 97 per cent of the shares in a small company owning property in Port Moresby, wholesaling facilities in Kiunga, a partnership in a small air transport company, and various small retailing outlets. This company is negotiating for joint-venture contracts with outside companies for activities as varied as the supply of explosives and arms to the PNG Defence Force, the establishment of a stevedoring company at Kiunga, and supermarkets for the proposed mine. In addition to these efforts, and despite their own lack of involvement in the negotiating process, the agreement between the PNG government and the consortium specifies that the company shall, wherever possible, give preference in the hiring of labour and in any assistance given to the development of PNG businesses to people from the Kiunga area and the Min.

International Banking Interests. Mining in developing countries is carried out mainly by multinational companies. But as Radetski has pointed out (1980), mining in the past two decades has been increasingly financed by international banks rather than by the companies themselves. Capital for mining projects is increasingly raised through loans rather than through internal company finance or shareholder equity. For example, the debt to equity ratio in the Ok Tedi case will be 70:30. As a result, initial control of such projects is primarily in the hands of banking institutions and not in those of the mining companies. Though some analysts may feel that such a distinction is so fine as to be meaningless in the face of the exploitation of the Third World's resources by metropolitan capitalism, banking interests are far from being identical with those of the mining companies, and the distinction does give able Third World governments some potential to lever themselves into a somewhat better bargaining position. When this factor is taken along with that mentioned earlier — the internal differences of interest between members of mining consortia — there is clearly greater scope for Third World manoeuvre. For example, in the Ok Tedi project, the banks would certainly not be keen to lend US$600 million to a project for which the PNG government was less than eager to give its seal of approval. Such a seal would be given by the state taking its 20 per cent share of equity. If there was any hint that the state would not do this (and that, therefore, the banks were faced with the possibility, however remote, of losing their loan through expropriation), then no mining company, no matter how prestigious, could prise a loan of such proportions from lending institutions. The PNG government has already demonstrated that it was prepared to reject Kennecott's proposals for Ok Tedi. Therefore, if it wished to ensure that any mining company with which it was negotiating provided adequate environmental plans or labour training plans, it could, by withholding its uptake of equity, let its doubts on such a project be transmitted to lending institutions. Any widening of interests on the consumers' side of resource development is to the advantage of a government like Papua New Guinea's.

Conflicts and Coincidences of Interest

The basic interests of the above parties can be summarized as follows:

First, there is consortium interest in the overall profit, with the variations that the German part of the consortium is also very much interested in an assured supply of copper, while BHP is interested in gaining a reputation as a manager of projects — a very sensible interest in view of the declining significance of mining companies in mining projects.

Second, there is national government interest in maximizing revenues to the exchequer for use on "real" development.

Third, the provincial government is interested in maximizing exports to gain maximum royalties, so as to enable it to improve the staffing situation, and generally to use these royalties to develop the province.

Fourth, the local people are interested in the mine as the only means available to them to develop the area.

Fifth, there is international banking interest in debt-servicing and loan repayments.[2]

Areas of conflict that can arise are therefore multiple. Not only has the national government to ensure that a good agreement is reached with the consortium; it also has to try and keep provincial and local demands to a minimum since any expenditure which does not lead to increased mining revenues affects it more than any other party. The consortium, while it has the incentive to maximize its profits, has to pay 70 per cent taxes on all profits over and above a reasonable rate of return, which in this case is about 20 per cent discounted cash flow on investment. If it calculates that it

---

[2] The geology of the Mt. Fubilan deposit is extremely favourable in this regard. The topmost ores are those richest in gold, thus allowing a rapid return of project loans to the lending institutions.

is more in its own interest to match production with metal prices to such a degree that profits are spread out over a longer period, or that profits just fall short of the threshold at which additional profits taxation applies, it could easily come into conflict with both the national and provincial governments over the question of timing over profits and revenues. The provincial government is in potential conflict with both the national government and the local people. With the former, it will almost certainly argue for a bigger share of the profits from mining, as did the North Solomons, particularly since it is one of the three least developed provinces in the country to date. It faces conflict with the local people who turn exactly the same argument against the province, in which the southern areas are far better provided with services than are the northern areas.

However, there are also counterbalancing areas where interests of at least two of the parties coincide. It is in the company's interests to reduce the costs of the project and even more so to assure itself of a loyal, stable work force. Before the BHP-led consortium took over responsibility for the Ok Tedi project, Kennecott had already established the Star Mountains' first school, first aid post for health services, and the first job opportunities. In particular, it established, at a very early stage, training facilities for the local people, and in so doing won for itself enormous respect from the people of an area which almost totally lacked any form of government-provided services. There is clearly a coincidence of sorts between local and consortium interests here, one which is not shared by the provincial government which would like the whole province to benefit in this way, and not shared even by the national government which will be faced with the difficult task of trying to prevent migrants from other parts of the country obtaining work at the Ok Tedi when preference is being given to Star Mountains people.

Once agreement is reached between the national government and the consortium, then a firm coincidence of interests arises between them — to obtain from lending agencies loans on the best possible terms and to keep non-productive expenditure to a minimum, the latter bringing them in conflict with both provincial and local interests. It is fortunate that in the Ok Tedi case gold prices had tripled since the consortium took a firm interest in the project. By late 1978, most points of negotiation between Papua New Guinea and the consortium had been more or less sorted out — taxation, arbitration, revenue sharing, and so on. What remained were the questions relating to such things as how many concessions could both parties afford to give to local interests. The sharp rise in gold prices in 1979 meant that both the national government and the consortium were given the opportunity to be generous in this regard, since total expected revenue was not much higher than had been estimated previously, far higher than either side's definition of a reasonable rate of return. For its part, the consortium could now afford to pay for the advantages of a loyal work force mentioned above through training. At the same time, the national government could now afford more generous financing of provincial development projects, whether or not such projects were directly associated with mining development. In short, now that the Ok Tedi cake had an icing of gold, all interests could be more easily accommodated.

Conclusions

Many people would regard Papua New Guinea as one of the luckier Third World nations. It is rich in mineral resources and it has received massive financial support since self-government (in December 1974) in the form of Australian annual grants of about US$240 million a year for its three million people. Some would criticize such assistance as illusory, since it continues to tie Papua New Guinea to high-level, expensive expatriate labour, and self-reliance is not fostered. Others would criticize it as forcing Papua New Guinea to live beyond its means and to follow a developmental path unsuited to its peoples' traditional ways. The PNG budget for 1981 envisages expenditure by government of more than US$900 million. Mining projects, like Ok Tedi or Bougainville, it could be argued, would not be necessary if such aid had not been so easily available. But what are the alternatives for Papua New Guinea? Should it immediately dismiss half its bureaucrats and do without both aid and services? Most industrialized countries are criticized for their totally inadequate transfer of funds. Can Australia be rightly criticized for giving such amounts and Papua New Guinea for accepting them?

Perhaps Papua New Guinea should replace these grants from Australia with the commercial loans that have driven so many of its Southeast Asian neighbours into spectacular debt. Perhaps it should expand its stagnant cash crop sector or its miniscule manufacturing sector. But the first would make Papua New Guinea no less dependent on a world market situation which is even more difficult than that which applies to minerals and would reduce the rural populations' self-reliance. Increased manufacturing entails increased urbanization (already a problem for PNG), vastly increased import bills, and very little foreseeable government income. Both extended cash cropping and more manufacturing would destroy traditional ways of life far more effectively for very much larger numbers of people than does mining. Indeed, on this point of social impact, one of the most commonly made criticisms of mining — that it all too often remains an enclave — can be made into one of the strongest points in its favour.

Morcover, Papua New Guinea certainly does have a very rich mineral resource base, while its agricultural resources are not outstanding and its human resources are still poorly developed. Mining can raise capital for development in other areas more quickly and more reliably than almost any other activity. The two vital issues are, first, can Papua New Guinea protect its own financial interests adequately in its dealings with the multi-national companies interested in mining its resources? Second, can Papua New Guinea spend any capital raised by government through mining on real development, development as defined in its own eight aims and constitution? On the first question, Papua New Guinea has quite clearly shown, so far, that it can deal with multinationals on their own terms and do well. Its ability to do so has been assisted by the fact that there is no desperate urgency to develop the mining projects, its use of outside specialist skills where necessary, the fact that the mining companies' own powers have been reduced by the increased capital costs of mining, and the lack of total identity of interest between the mining companies themselves and between such companies and the international banking community.

It is with regard to the second question that most real doubts seem to remain. The Ok Tedi project may, by virtue of its profitability, bring very positive economic benefits to one of the remotest areas in the world. Such remote areas otherwise offer almost no prospects for the advancement of their inhabitants other than that well-beaten path of education immediately followed by permanent migration. But will Papua New Guinea be able to use the revenues it obtains from mining to bring improved services and opportunities to the rest of the country? Its record at least since 1974 has not been a particularly good one in this area. Whatever the riches mining can bring to the government's exchequer, the real problem still remains the same — how does one establish the means to achieve rural development?

REFERENCES

Bedford, R.D., and Mamak, A. 1977. *Compensating for Development: The Bougainville Case*. Canterbury, New Zealand.

Berry, R., and Jackson, R.T. 1981. "Provincial Inequalities in Papua New Guinea". *Third World Planning Review* 3, no. 1:57–76.

Jackson, R.T. 1979. "Papua New Guinea Trade Patterns since Independence". *Australian Geographer* 14:251–53.

______. Forthcoming. *Ok Tedi*. Port Moresby.

Lam, N.V. 1980. "Unemployment and Poverty: Some Neglected Issues". In Jackson, R.T.; Batho, P.; and Odongo, J., eds., *Urbanization and Its Problems in Papua New Guinea*. Port Moresby.

Papua New Guinea Government, Department of Minerals and Energy. 1977. *Statement of Intent on Mining Policy*. Port Moresby.

Toffler, A. 1970. *Future Shock*. London.

West, R. 1976. *River of Tears*. London.

4

Mining and Land Dereliction in Chotanagpur, India

MAHESHWARI PRASAD

There are many differing views on the physical state of an area of land which identifies it as being derelict. In legal terms, derelict land is that which has been abandoned by its owners. In common parlance, the state of being abandoned is a prerequisite of dereliction. Therefore, theoretically derelict land is such land "which has been abandoned as useless or as too badly damaged to repay a private person to improve it" (Goh and Morgan 1975). This definition reduces the scale of the problem. For example, it excludes all land which does not fall into these categories — land being used for tipping of mining wastes, subsided land, underground fire-damaged area, contaminated land, and many active mines and quarries having the physical characteristics of dereliction.

In practice, "derelict land has proved to be more difficult to define than might be supposed by the casual observer of the landscape" (Wallwork 1974). The problem of defining dereliction becomes more complex if both the economic and aesthetic aspects are included. Land can be spoiled or degraded without being derelict in the economic sense, as for example, the waste tip forming a part of an industrial plant. On the other hand, a piece of land may be economically derelict without becoming an eye-sore, such as a derelict woodland. It may be argued, however, that in general derelict land resulting from mining is both economically useless and aesthetically dead.

Derelict land was first defined in Britain by S.H. Beaver in his pioneer survey of the Black Country in 1945 as "land which has been so damaged by extractive or other industrial processes or by any form of urban development that in default of special action it is unlikely to be effectively used again within a reasonable time, and may well be a public nuisance in the meanwhile" (Beaver 1946). In 1964, a much shorter definition was produced, when it was decided to collect annual returns of derelict land from local authorities, as "land so damaged by industrial and other development that it is incapable of beneficial use without further treatment" (Explanatory Memorandum 1966). However, this definition was inadequate as the statistics based on it were inaccurate to the extent of excluding at least half the derelict land in England and Wales.

Various attempts have been made to embody the aesthetic component in the definition of derelict land with the intention of producing a more balanced definition (Oxenham 1966; Bush 1969). The visual impact of dereliction, no doubt, is important in evaluating the landscape, but its inclusion creates ambiguity.

There is no official definition of dereliction in India. The problem here is that the resources for investment are very limited and expansion of the mining industry has priority over the restoration of derelict land. There is no public concern for or consciousness of damage to land that can result from mining.

In the absence of an official definition, the following is proposed for the purpose of this study. Land which has been already rendered useless, or is being made so, by the effects of mining and other forms of activities connected with mining, and which is incapable of becoming productive within a reasonably short time without artificial improvement.

Data Availability

There is a complete absence of statistics concerning mining dereliction in Chotanagpur. There are several agencies concerned with mineral exploitation but none concerned with land restoration. No agency, private or public, is involved in the compilation of data on dereliction. It reflects the emphasis on exploitation as well as the profit-oriented nature of modern mining in this region. No one is concerned over how much land is mutilated, mined, abandoned, and made useless.

The mining operators in Chotanagpur are private, captive, government, or illegal. However, none of the published figures pertain to either mined-over areas or the active mining sites. Even the leasehold areas of different mines are buried deep in the official records, and it is difficult to compile figures from divergent and scattered sources. It is just as difficult to gather statistics on the numbers and areas of abandoned mines in the region. Moreover, the prevalence of illegal mining has complicated the matter.

Official returns of the areas under mineral extraction are non-existent. These areas are not shown in the land use statistics. There is no government department concerned even indirectly with estimating the extent of damage of land caused by extraction of minerals, subsidence, underground and surface fires, and related matters. The limitations placed by the absence of data on dereliction in Chotanagpur has been partially reduced by obtaining some facts and figures from discussions and personal communications with officials of the mining companies, and by direct personal observation of the nature and extent of the problem. It has, however, made this study essentially interpretative in nature.

Organization of Major Mining Industries

The coal mining industry has various central agencies involved in the exploration and exploitation of this mineral in Chotanagpur. The National Coal Development Corporation (NCDC), established in 1956, was the first public company to undertake the mining of coal. Subsequently, coking coal mines were nationalized in 1972, and the non-coking coal mines in January 1973. The Bharat Coking Coal Limited (BCCL) was incorporated as a public company to own and operate the nationalized coking coal mines, while the Coal Mining Authority Limited took over the non-coking coal mines. The Authority has three divisions: Eastern Coalfields Limited, Central Coalfields Limited, and Western Coalfields Limited. Eastern Coalfields Limited covers only 49 km² in Chotanagpur and Santhal Paraganas, and Central Coalfields Limited controls all the other non-coking coal fields of the regions. The reorganization of the nationalized coal industry from November 1975 brought the entire industry under a holding company — Coal India Limited. To initiate the planned development of coal deposits in the country, the Central Mines Planning and Design Institute Limited was created with three important wings — Geology and Exploration, Coal Preparation, and Research and Development.

There are many other minerals in Chotanagpur (Fig. 4.1). The development and exploitation of mica, bauxite, and limestone are private sector undertakings. Iron ore deposits are mined by the Tata Iron and Steel Company, Indian Iron and Steel Company, and the National Mineral Development Corporation. The Hindustan Copper Limited and the Uranium Corporation of India are concerned with the development of copper and uranium deposits respectively.

History of Dereliction in Chotanagpur

The extent, pattern, composition, and characteristics of dereliction depended upon various conditions, but they result chiefly "from the interaction of a complex set of influences, notably geological structure and physical location of mineral outcrops, techniques of extraction and refining, and methods of waste disposal related both to technical competence and the economics of alternative modes of operation" (Wallwork 1964:42).

The genesis of the problem of dereliction in Chotanagpur is linked with the development of mineral resources in this region. The scale of exploitation

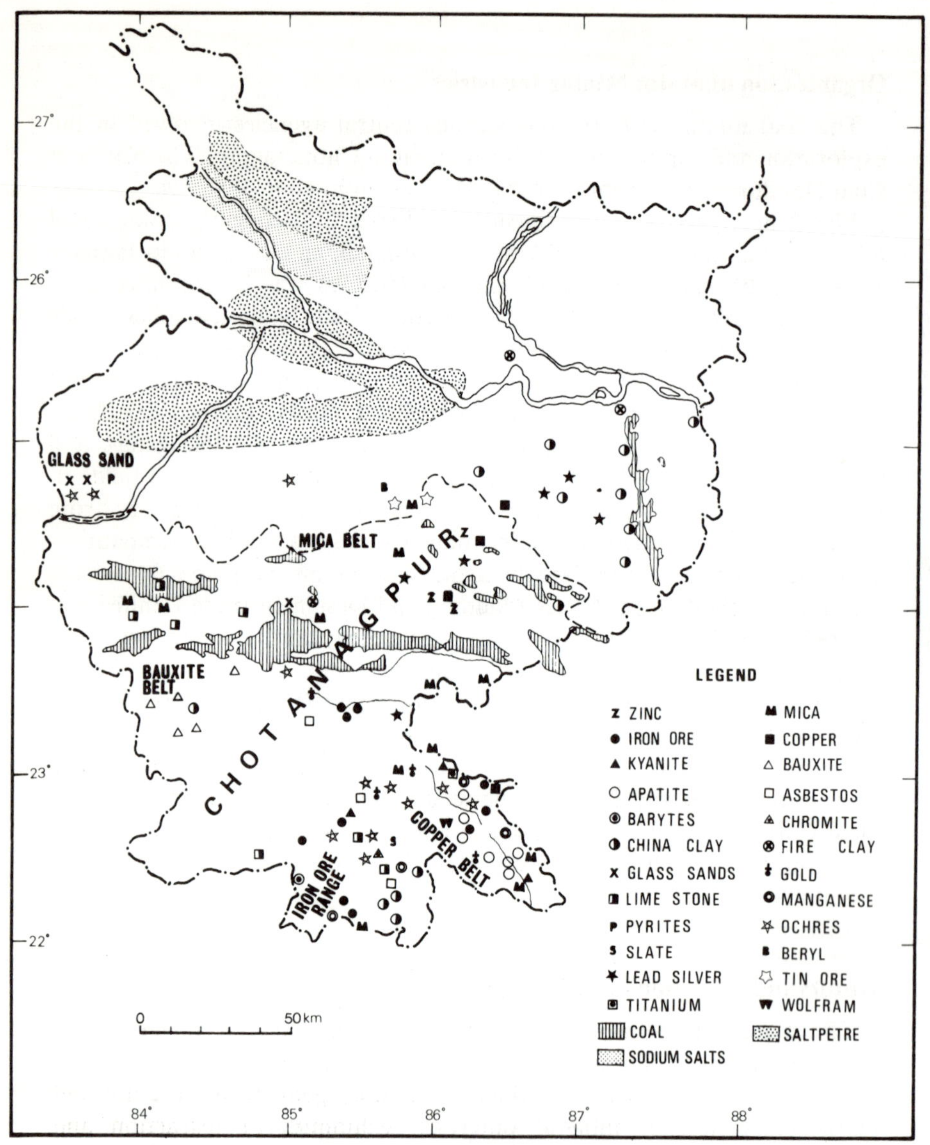

Fig. 4.1 Major minerals and mineral belts, Bihar, India
Source: Techno-Economic Survey of Bihar, Vol. 1–1959

of the different minerals has been an important factor of dereliction. But no precise data exist on the areas affected by extraction of the minerals, some of whose scales of operation are quite small. However, six major minerals — coal, iron, bauxite, limestone, copper, and mica — exploited for a considerable time on a fairly large scale, are primarily responsible for causing extensive damage on the landscape.

Coal-Mining Dereliction.   Chotanagpur has the richest deposits of coal in India. The contribution of this region to national requirements by the year 2000 would be 2,000 million t of coking coal and 2,250 million t of non-coking coal. The reserves of prime coking coal are estimated at 5,651 million t. The entire reserve is in the Jharia coal field of Chotanagpur. Out of this reserve, only 3,500 to 4,000 million t are considered workable. It is, therefore, the premier coal field of the country, in respect of both potentiality and development.

The coal field had nearly 500 collieries in the pre-nationalization period, out of the total of 900 for the whole of India. Nearly 60 per cent of the coal output of Chotanagpur (48 per cent of the total output of India) is obtained from this field. On account of the high quality of the coal and the thickness of seams, it has become the most developed coal field in India (Records, GSI 1973).

The exact date of discovery of the Jharia coal field is uncertain, but coal was known to occur here before 1858. Coal mining, however, did not begin until 1894 when the field became the supplier to the East Indian Railway system. Further extension of the railways was very soon followed by a "coal rush". Private speculators and business men swarmed the coal field to obtain mining leases. The subsequent history of this coal field has been one of the gradual development of a network of roads and railways and the beginning of what is referred to as "slaughter mining". "More holes, more coal" became the slogan of the profit-seeking private operators. The number of collieries increased from 6 in 1891 to 222 in 1908. As early as 1919, this coal field produced nearly 53 per cent of the total output of coal in India. This pattern of exploitation of coal persisted until nationalization in 1973 and produced the worst type of mining dereliction in the country.

The other coal fields shared a similar exploitation history, although on account of various limitations associated with the inaccessibility of the coal fields and the small and low-quality deposits the scales of operation have been modest and the resulting damage to land less severe.

The East Bokaro coal field, covering an area of 208 km^2, was developed first to supply the railways. The main source of coal, even at the present time, is the 21–23 m thick Kargali seam (Memoirs of Coal Explorations 1963). The Bokaro colliery was opened in 1917 and the Kargali seam was exposed by manual excavations. It was in 1948 that mechanized methods of open-cast mining were introduced to extract coal from the good quality Kargali and other seams. The reserves of Kargali coal are estimated to be 628.28 million t (Brown and Dey 1955). Mining at Kargali has resulted in very serious land dereliction.

The major part of the West Bokaro coal field of more than 259 km^2 was practically undeveloped in the pre-nationalization period. Illegal mining in this area has caused severe damage to land in recent years. Mining activities, principally open-cast, have intensified in this field, leading to further dereliction.

West of the Bokaro coal field lie the North and South Karanpura fields. The Indian Bureau of Mines has established the proved reserves at 148 million t. Some of the seams, particularly the Argada (15.36 m) and Sirka, (5.15 m) are thick, and the ratio of over-burden to coal is small, thereby favouring their extraction by open-cast methods. But because of poor accessibility and inferior deposits, only about 31 km^2 were developed in the North Karanpura field in the pre-nationalization period. The rest of the field is underdeveloped, but is exploited by illegal miners extracting coal from shallow pits.

The South Karanpura coal field, with its thick Sirka and Argada seams, was developed in 1924 when the railway was extended between Gomoh and Daltonganj. Subsequently, several open-cast and underground workings were developed in the areas. The National Coal Development Corporation introduced mechanized open-cast and underground methods in the major collieries of this field in 1958. It is a very actively worked coal field in Chotanagpur, with an estimated reserve of nearly 2,034 million of non-coking coal.

The other coal fields lying in the North Koel valley and the Barakar basin are smaller and less developed. The coal is of poor quality. Here the workings are small and manually operated, and damage to the land is localized.

Among the smaller coal fields, dereliction is most intractable in the Giridih coal field, which has an area of 28.47 km^2 and is located near the thickly populated town of Giridih itself. Coal was mined in 1871 by the East Indian Railway. It is the oldest colliery in Chotanagpur. Underground fires, subsidence, abandonment of quarries, and continued illegal mining have combined to devastate the landscape. Except for a very small area, the entire coal field is derelict. It is a good example of extant dereliction in Chotanagpur.

Very little damage to land is noticeable in the Daltonganj coal field, on account of both the small size of the deposits and the scale of workings. Mining began in 1920 in the Ramgarh coal field which has a reserve of 1,855 million t. The area has a large number of old manually-worked pits. Recently, the Central Coalfields Limited launched a massive open-cast mining project which would eventually lead to extensive dereliction in this area.

Non-Colliery Dereliction. Dereliction resulting from exploitation of minerals other than coal is different in several respects to that resulting from coal mining. The regional setting, the mode of exploitation, and the age of mining are the areas of differentiation. Colliery dereliction is more serious and widespread, is spatially well-defined, and has a longer history. It is a product of a combination of factors — deep mining, open-cast strip mining, subsidence, and underground and surface fires. Dereliction stemming from the extraction of the major non-coal minerals is of smaller magnitude and extent, younger in age, and a product of open-cast mining (iron ore, bauxite, limestone), localized shaft mining (copper), or surfacial shallow pit mining (mica).

The extraction of iron ore is responsible for extensive dereliction in Chotanagpur, second only to coal-mining dereliction in its seriousness. The Gua-Noamundi region of Singhbhum district has by far the richest and the largest reserves of iron ore in India. It is a part of the Iron Ore Range, stretching for about 48 km from Gua in the north (Singhbhum district) to Bonai in the south (Keonjhar district of Orissa). The range is between 120 and 300 m wide, and rises about 450 m from the level of the plain. It is capped for the most part by massive hematite ore, averaging more than 60 per cent of iron throughout its entire length. Percival estimated the total reserve of high grade ores of this range at 3,180 million t (Percival 1947).

There is large-scale open-cast mining in different localities of the Iron Ore Range. Most of the mines are still active. Dereliction from iron mining is visually less objectionable than colliery dereliction. The iron ore mines are sited on high tops and ridges. They are smaller in area and located in the midst of dense forest, unlike the collieries in large deforested areas. Consequently, unless one is very close to the actual sites of operation, dereliction is not visually obvious.

The Tata Iron and Steel Company (TISCO) discovered iron ore at Joda in 1909 and at Noamundi in 1917. Mining leases were obtained for Joda in 1916 and for Noamundi in 1923. Besides iron ore, the Noamundi-Joda complex also supplies ferro-manganese and manganese ores. Mining was started by TISCO at Noamundi in 1923. The reserves of iron ore at Noamundi have been estimated at 242 million t (Percival 1931). Mining in the earlier days was highly labour intensive. The extent of dereliction was small. Mechanization was gradually introduced, and a major mechanization programme was started in 1953 and completed in 1957. Modern methods of open-cast mining was introduced at Noamundi and Joda, leading to the stripping-over of extensive areas and the removal of great quantities of overburden. A series of large benches were formed for extracting iron-bearing banded hematite quartzite.

The Iron Ore Range was proved, through vigorous geologic prospecting in the latter part of the first decade and beginning of the second decade of the present century, to have one of the largest and richest deposits of iron ore in the world (Jones 1934). The Indian Iron and Steel Company began extracting iron ore at Gua. This mine has an estimated reserve of 100 million t. The waste from open-cast mechanized mining is a threat to the dense Nuia Protected Forest. The Kiriburu iron ore mine was developed by the National Mineral Development Corporation under the terms of a 1958 agreement between the governments of India and Japan. It is one of the most recent and modern iron ore mines in the country. Its lease area covers 1,937 ha of land in the Singhbhum district of Chotanagpur. The ore deposits, as elsewhere in the range, occur on the tops of hills. Production began in 1963. Further development of open-cast mining at Meghahatu Buru of the Kiriburu Complex and many other potential sites in the Iron Ore Range would lead to very extensive dereliction in the region. In view of the importance of iron ore as a foreign exchange earner (Report, Central Wage Board 1967) it is unlikely that the threat of dereliction would impede such development.

The open-cast mining of bauxite (alumina content 49.68 per cent) on the high "Pat" (flat-topped hills) surfaces of Ranchi district and that of limestone in the hilly terrain of Palamau district has created derelict areas similar to but smaller than those resulting from iron-ore mining in Chotanagpur. The bauxite deposits occur over an area of more than 3,870 km^2; the estimated reserves are 12.7 million t (Kumar 1970). Bauxite mining in Ranchi district dates back to 1933. Mining is by open-cast methods, but several lease-holders also operate bauxite quarries. Although operations are manual in some localities, in others they are mechanized, with extensive mining benches being formed by stripping out lateritized mesas. The high ratio of overburden to bauxite (Cameron 1955) contributes to the amount of dereliction in the region.

Small limestone workings occur throughout Chotanagpur (including Rohtas district) as the deposits of this mineral amounting to 130 million t occur in all the districts of the region. No appraisal of the extent of dereliction associated with these workings is possible. But extensive extraction of the Vindhyan limestone by open-cast mining (presently by manual labour) near Bhawanathpur which has a reserve of about 18 million t, is bound to cause dereliction in the area which would be more difficult to restore than many other sites in Chotanagpur.

There is a well-defined mica belt of Chotanagpur, part of which extends northward beyond this region (Fig. 4.1). It is about 144 km long and 32 km wide. The most productive segment of this belt spreads over 147 km^2 in the Kodarma Reserve Forest. The whole area is divided into 900 squares, each measuring about 14.2 ha. It has been intensively searched for valuable ruby mica, and the area has more than five hundred productive mines. In the pioneer stage of development, numerous holes were made, many pits abandoned, and the entire area recklessly mutilated.

The manufacture of splittings began at the end of the nineteenth century. Mica was exploited by private operators using primitive methods of mining. Such methods persist today. This may be attributed to the mode of occurrence of its deposits. Mica mining is a hazardous enterprise. "Muscovite occurs generally very sparingly in a rock known as pegmatite, which forms small veins, lenses or irregular masses in schists. Because of the small size and irregular nature of the pagmatites and of the sparse and irregular occurrence of the mica within them, mica mining is somewhat hazardous undertaking as compared with certain other minerals" (Adarkar 1945). Limited mechanization was introduced as early as 1918, but large-scale mechanized operation is impractical because of the peculiar mode of occurrence of mica.

Mica mining is very sensitive to fluctuations in demand in the international market. Although it is still a highly disorganized industry, it was even more disorganized in the past. Considerable damage to land was done during the initial stage of its development, particularly during the First World War when there was a great demand for this strategic mineral. The number of holders of mining licenses in 1918 was nearly 200, and the number of operative mines 650. Most of the mica was obtained from shallow pits known locally as *uparchala* workings. Many mines were abandoned after the war. A similar pattern of rise and fall was associated with the Second World War.

The number of mica mining leases in Chotanagpur is 421 and that of mines nearly 700. The region produces nearly 80 per cent of India's mica and 70 per cent of world's sheet mica. Some of the mines have been abandoned in recent years, owing to the depressed state of the industry (The Indian Nation 1975).

Dereliction in the mica belt of Chotanagpur is related not only to shallow (*uparchala*) workings and the abandonment of the pits and mines but also to huge deposits of muscovite scrap and waste. "India has abundant sources of scrap. It is estimated that it produces more than 50,000 t of scrap every year" (Mica and Mica Wastes 1957). In the Kodarma region of Chotanagpur, there are large quantities of accumulated scrap near the pits of the mines. Mica mining is thus responsible for extensive, almost intractable, dereliction in Chotanagpur.

Copper occurs in a well-defined belt extending over 160 km in Singhbhum district of Chotanagpur (Fig. 4.1). Copper mining is 2,000 years old. But modern mining was started only in 1857. The Hindustan Copper Limited holds two leases covering an area of 80 km². The three productive mines of Mosabani, Surda, and Pathargora are covered by one mining lease of 70

km². The actual workings are spread over an area of 392 ha². Some of the
mines such as Dhobani and Kendadih have become unproductive and consequently abandoned. Dereliction is partly related to the dumping of wastes
by concentration plants in and around Mosabani. The Mosabani shaft is
more than 600 m deep. The ore contains 2.51 per cent of metal content, but
this is being progressively exhausted. The proved reserves of ore with 1.5
per cent metal content in the copper belt are 98.5 million t. As this ore
contains a lower percentage of metal, working it would produce a great
amount of waste material. Subsidence has not yet complicated the problem
of dereliction and is rare in the area. Of all the major minerals associated
with modern mining, copper has produced the least amount of dereliction in
Chotanagpur.

Causes of Dereliction

Problem of Mining

Dereliction in Chotanagpur owes its origin to a number of factors, of
which the most important are the following.

Shifting Mining. Shifting mining is associated with two important
minerals of Chotanagpur — coal and mica. Here it is being explained with
reference to coal. In the initial stage of coal mining, the mines were in
private hands. To maximize profits, the owners dug in all places where coal
could be easily obtained. When mining became difficult without the use of
machinery, the sites were abandoned and diggings were shifted to another
location. This practice appears similar to that of shifting cultivation in
the tropics. The discarded quarries abound in the Jharia field as well as in
the coal fields of Hazaribagh district, particularly in the Kedla-Jharkhand,
Kuju, and Bermo-Kargali areas.

Primitive Methods of Mining. The private owners used to work minerals
for quick profits. Little machinery was installed so as to minimize capital
costs. The mines were not very deep, and extraction was limited to the easily
worked outcrops. When these were exhausted, the mines were abandoned
and new outcrop sites selected for mining. This very destructive way of
mining coal is known as ''slaughter mining''.

Mines were closed during the monsoon as they became flooded. It was
also difficult to work underground under hot and humid conditions. Many
miners had to carry out agricultural tasks in the villages. Mining was therefore seasonal in character, and only those sites were re-worked after the
monsoon which were not damaged by floods and subsidence. The incidence
of abandonment of mines was thus very high in the past.

Open-cast mining was carried out at selected places where the ratio between the coal and overburden was most favourable. Because of the lack of transport for bulk removal, the overburden was not taken away but was heaped over the underlying seams close to the mines. The waste heaps were usually flat and spread over large areas, making mining extremely difficult.

Small Scale of Mining.   The Chotanagpur region, particularly the coal fields and mica belt, passed through a period of "mineral rush" in the first quarter of this century, when mining prospectors and exploiters swarmed into it. It led to the development of numerous small mining concerns. Before the nationalization of the coal industry, for example, there were about 800 collieries in Chotanagpur, some producing less than 10 to 15 t of coal a day. The situation in the mica belt was not much different. The numerous small mines resulted in the wasteful exploitation of coal and mica in Chotanagpur and in large-scale dereliction in the region.

Illegal Mining.   Although it is very difficult to say illegal mining was absent in the past, it is now widespread in the coal fields of Chotanagpur. Illegal mining also prevails in the mica belt of the region because of the high price of this mineral. The price of coal has also increased after the nationalization of coal mining industry, so that the margin of profit for the illegal operators has widened considerably.

In illegal mining the operators establish themselves on outcrops of coal and start extracting it manually with a small work force. Because of the high incidence of unemployment in the region, labour is cheap and the inputs are minimal. The pirate miners are therefore able to offer coal at lower prices. The sites of active illegal mining are Bermo, East Ghutway, Meghatanr, Sirka, Jagesar, Bundu, Hesalong and Lehrilonri in Hazaribagh, and Damuda in Dhanbad district. In Giridih district, illegal mining is confirmed to the abandoned collieries which are estimated to have a reserve of 12 million t of coking coal. In these two districts, nearly 40,000 labourers work in 15 unauthorized mines. Occasionally the pirate miners even take out coal from the areas leased to the government agencies. As these operators are very poorly equipped, they cause considerable damage to land.

Illegal mining also contributes substantially to the problem of dereliction by increasing fire hazards in the coal fields of Chotanagpur.

Subsidence and Fire.   The problem of dereliction is complicated by subsidence and fire in the coal fields of Chotanagpur. For example, the underground fire in the Giridih coal field, first noticed about five years ago, has reduced more than 100,000 t of "D" grade coking coal to ashes. Similarly, underground and surface fires have damaged very extensive areas in the

Jharia and Bokaro coal fields. In the Jharia coal field, about 11 km² have been damaged by surface fires. The scars remain as there has been no effort at land reclamation (Srimany 12 Dec. 1980). In the Dhori colliery, the fire at the outcrop of coal is creeping along the coal seam towards a developed area. There are also underground fires in Bhurkunda Mine no. 1, Sounda D, and the West Tunang collieries of the Central Coalfields area of Chotanagpur. But these fires are kept isolated and are under control.

Subsidence and fire are interrelated problems. As early as 1929, an official committee appointed by the Mining, Metallurgical, and Geological Institute of India had drawn the government's attention to the dangers of subsidence caused by primitive methods of mining. The Mines Act of 1901 was amended in 1933, spelling out the procedures to take in case of a premature collapse of mine workings. The Act was further amended in 1938 to safeguard against premature collapse, fires, and other hazards. In 1939, the Coal Mines Safety (Stowing) Act came into force, under which a stowing fund was created to provide for schemes of stowing to prevent and control fire collapses and surface subsidences. A Coal Mines Fires committee was appointed by the Planning Commission of India in 1964.

The practice of sand stowing is no longer common in the Central Coalfields Limited coal fields of Chotanagpur. Out of 63 collieries in the Central Coalfields only nine have sand stowing, namely, Bhurkunda, Swang, Saunda, Associated Karanpura, Sayal D, Jarandih, Kargali, Chalkari, and Giddi A (Safety and Conservation Dept., CCL). Sand stowing is done only where it is necessary. A limiting factor is the shortage of sand. Furthermore, the collieries are quite distant from densely populated areas, and sporadic subsidences do not seem to affect the economy or social well-being of the people. It is also expected that by the time the mining areas become thickly populated the subsided areas would have been colonized by natural vegetation.

Recent Expansion in Coal Production.   Mining dereliction is likely to increase in extent in Chotanagpur, owing to the various measures taken to expand coal production in the region. The development of coal mining was handicapped in the past because of the large deposits of coal which have high ash content. Mining was confined only to those deposits which could be used without washing. But the National Coal Development Corporation is now extracting large reserves of washable coal by highly mechanized methods, from both open-cast and underground mines.

In the past decade, thermal power plants set up in the coal fields of Chotanagpur have also contributed to an increased exploitation of non-coking coal, and thus have led to more extensive dereliction in the region. The super as well as captive thermal power plants presently proposed would be located in the coal fields themselves.

Coal mining is a rapidly expanding industry and, consequently, a major cause of increasing dereliction in Chotanagpur. The Fuel Policy Committee have fixed a target of 113.5 million t of coal output for 1980. Coal India Limited has an output target of 225 million t for 1983/84 and 340 million t for 1989. The 53 coal mines of the Central Coalfields Limited in Chotanagpur are expected to make a major contribution to the targetted production of Coal India Limited. The CCL has, therefore, launched a programme to open 24 mines by 1985, but the total number would be only 66 as some of the existing mines would be exhausted by then. However the average size of mines has increased, and the production capacity of each colliery would not be less than 500,000 or more than 3 million t per annum.

The rehabilitation of old mines and the launching of new large mines have compeltely changed the coal-mining industry of Chotanagpur. Mechanization is the key feature of the modern phase of this industry. Central Coalfields Limited has launched a massive rehabilitation programme involving mines such as those of the Kuju and Kedla-Jharkhand group in Hazaribagh district. A daily output target of 400 t of coal has been fixed for the Kuju sub-area. The Kedla-Jharkhand area is also endowed with abundant medium grade coking goal. Central Coalfields Limited has launched a multi-million rupee programme of mechanization of nationalized coal mines to achieve the target of production of 300 million t of coal by the end of 1986. In the first phase of mechanization, it would introduce 277 scrappers, the number of which would increase to 477 by 1986.

Several new major open-cast projects have also been launched by Central Coalfields Limited in Chotanagpur (Programme Report, CCL 1980). The Kedla-Tapin open-cast project of Hazaribagh was approved by the Indian government in September 1979. The production target for 1980/81 was 3.36 million t and that of overburden removal 0.99 million m^3. In the advanced stage of the project, the overburden removal would amount to 2.75 million m^3 a year. The life of the project is 44 years, and the mine would have an area of 526 ha in the final stage.

The Sirka Open-Cast Project of Argada was approved by the government in October 1978. It will produce one million t of grade B coal a year for the railways and power plants. The project will cover 485 ha of land. The target of coal production in 1980/81 was 0.36 million t, and that of overburden removal one million m^3.

The Religara Open-Cast Project of Argada now spread over 37 ha of land was approved by the government in August 1978. It will produce 0.37 million t of selected grade B coal a year. The targetted production of coal for 1980/81 was 0.115 million t, and overburden removal 0.50 million m^3.

The Dakra-Bukbaka Project of North Karanapura was approved by the government in October 1978. It will have an annual production of 1.3 million t of grade E coal a year. The project covers 138 ha of land. Targets of one million t of coal and one million m³ of overburden were fixed for 1980/81.

The K.D. Hesalong is the second major open-cast project in North Karanpura. It covers an area of 113 ha. It was approved by the government in January 1979, and the target of coal production for 1980/81 was 0.53 million t and that of overburden 0.75 million m³. The Kedla underground, Karo special (Main), and Sawang Reorganization Projects are important mine rehabitation schemes in Chotanagpur. The Dhori open-cast project was approved in August 1979. It has a rated capacity of 2.25 million t of coal and peak removal of 1.6 million m³ of overburden a year. It would be spread over 38 ha of land in the first phase of development.

The Ramgarh open-cast project, approved by the government in August 1977, will have a production capacity of 3 million t of grade B coal a year by 1982/83. It is the most ambitious of the open-cast projects in Chotanagpur. The coal will feed the steel plants. The project will last for about 41 years and will cover an area of 8 km². Its area contains proved reserves of about 128 million t of coal. The project will need a capital outlay of Rs 41.61 crores. In terms of capital outlay, area, and targetted production it is the biggest open-cast mining project of Central Coalfields Limited in Chotanagpur. The target production for 1980/81 was 0.47 million t, and that of overburden 1.50 million m³. In the advanced stage, the overburden removed would be 8.5 million m³ a year. The project is being worked by a shovel-dumper combination. It would be the first open-cast mine in the country with modern communication and signalling systems. A washing plant will be located close to it to wash the coking coal for the steel plants. It is designed to produce 1.8 million t of clean coal, 0.66 million t of midlings, and 0.54 million t of rejects.

Mechanization and Open-cast Mining.   The problem of dereliction in Chotanagpur has been greatly accelerated in recent years, owing to the increased tempo for mechanization of mines and the development of large-scale open-cast mines. In fact, both are interrelated — mechanization is necessary to open-cast exploitation. The extent of overburden removal depends on the degree of mechanization. Mechanization has led to a 15 per cent increase in overburden removal in the collieries of Central Coalfields Limited between June 1979 and June 1980 (Performance Report, CCL 1980).

Coal production from open-cast mines has more than doubled in Central Coalfields Limited during 1971–79 (Table 4.1). It is evident that in the last decade the open-cast method has been the mainstay of coal mining in Chotanagpur. This trend would be further accelerated in the coming years as more and more mechanized open-cast mines become operative in the region. Open-cast mining is now limited to those deposits where the ratio of ore to overburden averages 1:1. Later, with new equipment, this mode of mining may be extended to those areas which have a ratio of up to 1:3. Table 4.2 shows the extent of mechanization in coal mining in Chotanagpur.

Open-cast mines account for nearly a quarter of the total number of mines of Central Coalfields Limited. Mechanized open-cast mines make up about 20 per cent of the total number of mines. Coal production from open-cast mines considerably exceeds that from underground mines. The trend of production of coal from both open-cast and underground mines is shown in Table 4.3.

Open-cast mining leads to the removal of large quantities of overburden. During 1973–80, the overburden removal increased from 8.61 to 19.55 million m³. During this period, open-cast coal production increased from 7.33 million t to 10.99 million t. The present phase of coal mining has therefore led to more widespread dereliction in Chotanagpur, especially in the Argada, Hazaribagh, Kathara, and Kargali areas.

The present phase of intensive mechanized mining has greatly added to the volume of waste as it does not reject non-combustible material. The ratio of waste to saleable coal has also gone up. Before nationalization, waste disposal was not a major problem in the coal-mining areas as the shafts were shallow and were worked manually. Very little waste required tipping at the surface. The waste heaps were both small and low. Mechanization has resulted in the progressive deepening of shafts, enlargement of individual pits, and amalgamation of small units into large ones. As the volume of waste has increased, large waste heaps are now common in the coal fields in Chotanagpur. These heaps with ridge tips up to 40 to 50 m high are more prone to spontaneous combustion and are also unstable because of lack of compaction in the tipping process.

TABLE 4.1 PRODUCTION OF COAL BY OPEN-CAST
MINING, CENTRAL COALFIELDS, LTD.
(million tonnes)

Year	Production	Year	Production
1971/72	8.23	1975/76	14.11
1972/73	9.24	1976/77	13.91
1973/74	9.80	1977/78	14.25
1974/75	12.34	1978/79	16.35
		1979/80	17.19

Note: The table also includes production from mines outside the
Chotanagpur region.

TABLE 4.2 MINE MECHANIZATION, CENTRAL COALFIELDS LTD., CHOTANAGPUR

Areas	Mixed Mines		Open-Cast Mines		Underground Mines	Total
	Mechanized	Unmechanized	Mechanized	Unmechanized		
Barkakana	1	1	—	—	6	8
Argada	3	—	—	1	1	5
Ramgarh	—	—	1	—	1	2
Kuju	2	3	—	—	—	5
Hazaribagh	—	1	5	1	1	8
N. Karanpura	1	1	1	—	7	10
Kargali	1	—	2	—	2	5
Dhori	2	2	—	1	1	6
Kathara	1	1	1	—	1	4
Total	11	9	10	3	20	53

Source: Statistics Section, CCL, Darbhanga House, Ranchi, 1980.

TABLE 4.3 PRODUCTION OF COAL IN CENTRAL COALFIELDS LTD., CHOTANAGPUR, JUNE 1979 AND JUNE 1980
(million tonnes)

Areas	Underground Production		Open-Cast Production		Total	
	June 1979	June 1980	June 1979	June 1980	June 1979	June 1980
Barkakana	1.89	2.12	0.27	0.49	2.15	2.65
Argada	0.57	0.64	1.29	1.00	1.87	1.61
Ramgarh	0.05	0.04	0.21	0.38	0.26	0.42
Kuju	0.65	0.70	0.80	0.28	0.81	0.98
Hazaribagh		0.12		1.23	0.64	1.33
N. Karanpura	0.75	0.83	0.90	1.44	1.65	2.27
Kargali	0.48	0.45	2.52	1.57	2.09	2.02
Dhori		0.06		1.52	0.91	1.18
Kathara	0.29	0.30	0.81	1.89	1.10	2.19
Total	4.68	5.23	6.80	9.36	11.49	14.59

Source: Statistics Section, CCL, Darbhanga House, Ranchi.

Open-cast mining is being envisaged as the principal mode of coal extraction in India. A national survey reveals that open-cast mining accounts for about 30 per cent of the total production of Indian coal. By 1990, the share of open-cast production would have increased to 50 per cent of the total coal production in the country. Coal production by open-cast mining from the fields of the Central Coalfields Limited in Chotanagpur is expected to increase from 12.8 million t in 1981/82 to 19.9 million t in 1984/85 (Production Department, CCL).

Open-cast mining would also become more important in the mines of Bharat Coking Coal Limited in the next decade. For example, the Jharia Coalfield Reconstruction Scheme of 1978 envisaged an annual production of about 20 million t by open-cast mining by 1990. It would further increase to 25 million t out of the total production of 56 million t from this field by the year 2000. The surface area that would be finally covered by open-cast mining in the Jharia Coalfield alone is likely to be nearly 10 km² out of its total area of about 450 km². In other words, more than 20 per cent of the total area of the Jharia Coalfield would eventually be stripped of top soil and surface layers by open-cast mining (Feasibility Report 1980:4).

Absence of Statutory Control on Dereliction. The long history of mining in Chotanagpur and the lack of any statutory check on dereliction have resulted in its becoming a major problem in the region. Legislation exists on different aspects of mining but none of it deals specifically with dereliction. In the pre-independence period, mining laws were concerned only with labour and safety problems in the coal-mining industry. Scant attention was given to enacting laws for regulating the coal industry. It was only during the Second World War that the government realized, for the first time, that the coal industries should be controlled and regulated. As a result, the Colliery Control Order was passed under the Defence of India Act in 1945. The Mines and Minerals (Regulation and Development) Act, 1948 was passed by the government for the rational, scientific, and proper exploitation of mineral resources. The Mineral Concession Rules, 1949 were framed under this Act. The Act was amended in 1957, and is now known as the Mines and Minerals (Regulation and Development) Act, 1957 and the rules framed therein as the Mineral Concession Rules, 1960.

The Coal Mines (Conservation and Safety) Act of 1952 was brought into force for the conservation of the resources of metallurgical coal in the country and to rationalize the administrative machinery. All the subsequent Acts and Regulations such as the Coal Mines (Conservation and Safety) Rules, 1954, the Coal Mines Regulations, 1957, the Coal Bearing Areas (Acquisition and Development) Act, 1957, the Coking Coal Mines (Nationalization) Act, 1972, the Coal Mines (Taking Over of Management) Act, January 1973, and the Coal Mines (Nationalization) Amendment Act, May 1976, are concerned only with aspects of acquisition, development, safety, and conservation, and not at all with the reclamation of derelict land. The emphasis, as reflected in the statement of objects and reasons incorporated in the Act of 1976, is on bringing coal mining under the full control of the government, on stopping wasteful methods of extraction, and on halting illegal mining in the country. The former Minister of Energy has outlined the problems of the coal industry in the following words:

> After the nationalization of coal mines a number of persons holding coal mining leases unauthorizedly started mining of coal in the most reckless and unscientific manner without regard to considerations of conservation, safety and welfare of workers. Not only were they resorting to slaughter mining by superficial working of outcrops and thereby destroying a valuable national asset and creating problems of water-logging, fires etc. for the future development of the deeper deposits, their unsafe working also caused serious and fatal accidents. . . . Thefts of coal from adjacent nationalized mines were also reported after the commencement of these unauthorized operations. The number of these unauthorized operations had shown an increasing trend of late (Pant 1976).

Jharia Coalfield: A Case Study

The problem of mining dereliction in Chotanagpur is exemplified in the Jharia coal field, located in the oldest mining area in the region. This coal field is in the Damodar valley and covers an area of 450 km², of which about 250 km² fall under the operational jurisdiction of Bharat Coking Coal Limited. The coal-bearing sedimentaries of Permo-Carboniferous age, covering 387 km², are in a sub-basinal structure with rock formations exposed in an arcuate fasion (Fig. 4.2A).

Development of Coal Mining

Coal has been mined in this field on a large scale since 1894. The area between the small rivers Bans Jor and Khudia was the first to be developed. It includes the collieries of Ekra, Lowabaid, Kankani, Sendra, Sijua, Angarpathar, Katras, and Kailudih. "It was probably the first tract of the Jharia coal field to be opened up when railway communication between Katras and Dhanbad, through Kusunda was established in 1893. There were probably few areas of equal size anywhere in which so many coal seams occur and crop out at the surface" (Fox 1930:220).

Mining in the Jharia coal field is a recent activity. But because of its high quality coal, it has become the most extensively worked coal field in India. "On account of extensive mining activity in Jharia coal field and progres-

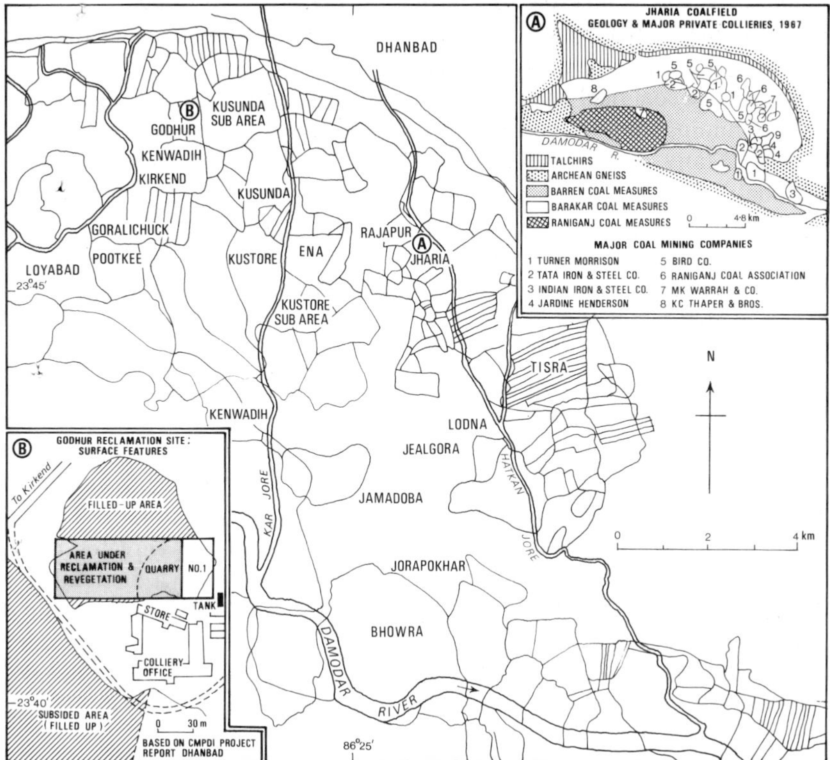

Fig. 4.2 Pattern of mining leasehold in a part of the Jharia Coalfield before 1973

sively increasing industrialization, effective change in the surface configuration of this area has taken place very fast in the last four decades. It is all
the more accentuated by sporadic subsidences in the depillard mining
sectors coupled with dumping of overburden at places and opening up of
new inclines and open-cast quarries'' (Pande and Kar Ray 1980).

Genesis of Dereliction

Dereliction in the Jharia coal field owes its origin to many factors, principal among which are the primitive methods of mining, subsidence, underground and surface fires, and other related phenomena.

Coal has been worked in a primitive fashion for a long time in the Jharia
coal field. Mining was confined to outcrops which could be mined with
minimum skill and investment. As soon as it became difficult and uneconomic to work at a particular site because of increasing depth, the mine was
abandoned and a new site opened and developed. Such mining not only

made the seams vulnerable to fire but has made the introduction of large-scale mining difficult. Accidents have occurred, claiming many lives and damaging land and the coal deposits.

The Jharia coal field was worked by numerous private operators interested only in mining and not concerned about the effects of mining on the land surface. Many of the nearly 400 different mining leaseholds, in the pre-nationalization period, were so small that they may be appropriately termed "handkerchief" leaseholds (Fig. 4.2). Coal was mined from shallow bell pits or adits driven into the outcrops. A large number of collapsed shafts and spoil mounds bear testimony to the damage caused by such workings. Many of the abandoned mines are 150 years old, and some of them have become mining pools.

Fox drew attention to the problem of dereliction in his study of the Jharia coal field published in 1930. Subsidence due to underground coal mining as well as fire has caused widespread dereliction not only in the Jharia coal field but also elsewhere in Chotanagpur. Furthermore, the depth and size of workings have increased because of improved mining technology. This has accentuated the problem of subsidence dereliction. The Director of the West Bengal Fire Service has predicted that "Jharia town within a radius of 20 miles would be wiped out within five years. . . . The entire township of Jharia and its adjoining areas with a population of one million will perish in one swift and devastating subsidence, if immediate measures are not taken to put down the subterranean fire that has been raging since 1940" (*Indian Nation*, June 1979).

Subsidence owes it genesis, to a large extent, to underground fires. The caving-in of the land surface over the fire-affected areas is a common feature of the mining landscape of the Jharia coal field. It has given rise to the worst type of mining dereliction in Chotanagpur, as the soil is completely baked and burnt. Furthermore, it is extremely hazardous to deploy heavy equipment on the fire-affected surface because of the danger of subsidence.

The first coal seam fire around Jharia town broke out in 1911 in the Bagdigi colliery. Its subsequent spread was caused by subsidence which made control difficult. It was finally brought under control by the Coal Mines Stowing Board at a cost of Rs 0.314 million. The recorded major fires, some of which are still alive, are shown in Table 4.4.

Fox (1930:254) drew attention to the destruction of the valuable coal deposits of the Jharia region as early as 1930: "This subject has been brought to the notice of those engaged in the Jharia field. Nevertheless, little has been done and a retrograde movement has in fact taken place. In this respect it appears hopeless at present to expect any serious action to be taken by firms in the coal industry."

TABLE 4.4 RECORDED MAJOR FIRES IN JHARIA COALFIELD

Fire Area	Year Fire Started	Fire Area	Year Fire Started
Bhowrah Quarry	1916	Kusanda (Karijhore)	1938
Bararee (Chiratand)	1919	Mudidih	1933
Ekra Khas	1923	Gaslitand	1934
W. Gopalichak and Central Kirkend	1926	Angarpathar	1936
Jharia Fire Area	1930	Jogta	1940
Loyabad	1930	Kustore	1942
Bhulan Bararee	1931		

Source: Bhatt 1976.

Since then underground fires have penetrated to a much greater depth because the water-table has been disturbed and lowered on account of extensive quarrying and deep mining in this coal field. Fires have also spread laterally and have remained unchecked until very recent times. These indicate the persistence of the use of the very wasteful, primitive methods of coal mining in Jharia. It is relevant to quote Fox (1930:254) here: "Making allowances for losses from fires and subsidence and including the coal left in pillars (as depillaring operations are not actively engaged in) it may be broadly stated that for every ton of coal raised from the seams in question an equal quantity has been rendered unavailable by existing methods of workings".

Fire cannot be altogether eliminated as the process of mining can lead to the spontaneous combustion of coal. There have been about 110 fires in the Jharia coal field, of which 50 were surface fires. Of these 110 fires, 70 are active today. They cover about 10 km^2 of this field (Bhatt 1976:2).

Fire in the Jharia coal field may be caused by spontaneous heating, by negligent acts, or by accident. Spontaneous heating may be caused by the collapse of pillars, or the caving-in of thick seams underground. Spontaneous heating may also occur in quarry overburden or in washery rejects. Negligent acts include the dumping of hot ashes from coke ovens or boilers and/or the making of soft cokes at quarry edges or along outcrops of coal or over surface subsidences. Accidental fires have been caused by Bantulsi (*Osiumum basilicum*) catching fire and by illicit alcohol distillation either in abandoned quarries or in underground workings. Belt fires, explosions, crossing over of fires from neighbouring mines, and the breakdown of the insulation of electrical cables, are among the other causes of fires in the coal fields.

"Once such a fire has come up, it will go on intensifying with the aid of sufficient amount of coal and oxygen available to it. When the coal is burnt, the void created inside the earth causes further subsidence of the surface, resulting into more cracks, which will further increase the area of fire. Thus, it becomes a cyclic operation and goes on increasing its area of action by itself" (Kumar and Singh 1969). The land thus affected is rendered useless.

A contributory factor to the fire hazard is pit-head accumulation of coal in Jharia as well as in the other coal fields of Chotanagpur. Millions of tonnes of coal lying as pit-head stocks in recent years have increased the danger of surface fires in the coal fields. Poor ventilation and coal dust treatment also lead to the accumulation of gas. Some of the mines are gassy and more prone to fire.

Fires can be controlled by systematic sand carpeting. The cracks are closed and oxidation of coal is checked by this method, but it is not possible to spread sand in active fire areas. The abolition of the Coal Board, which was responsible for sand carpeting, has been a setback to the fire control programme.

The extent of damage caused by fire in the Jharia coal field is an outcome of the fact that surface rights are not vested in the Central Government. Surface fires can be put out in the initial stage by heavy earth-moving machinery. But in some cases, delay in acquiring surface rights has led to the fires getting out of control and damaging extensive areas of land.

It is a difficult task to make any comprehensive evaluation of the damage caused to coal deposits, human settlements, topography, drainage, soil, vegetation, and the social well-being of the people of the Jharia coal field by more than four decades of underground and surface fires. But the fact that the region is almost entirely ruined, both physically and socially, cannot be denied. As much as 80 million t of coking coal have been either lost in the coal seam fires or locked in them. About 34 million t of coking coal have been burnt to ashes. The Coal Mines Stowing Board and Coal Board have spent about 16 million rupees on protective works against such fires between 1939 and 1970 (Kumar and Singh 1972).

Jharia town — the oldest and main town of the coal field — inhabited by 150,000 people, looks like an inferno. There is a constant glow in the air caused by heat escaping from fires raging in abandoned, underground mines. The surface above is completely burnt. The burning mines are only a few hundred metres from the outskirts of the town, and the fire may close in on it in sixty to seventy years. A series of stone-filled trenches temporarily separate the fire from the town (*Indian Nation*, July 1979).

In the Jharia coal field, many buildings have crumbled and many others have developed cracks. The guest-house of Bharat Coking Coal Limited has tilted because of subsidence resulting from underground fires. Many roads passing through the fire areas bear the warning signboards "Drive at your own risk". A major underground fire near Lodna has endangered the Dhanbad-Patherdih railway line and the Lodna Washery and Coke Oven Complex. The whole town of Jharia would have to be shifted because of the fire danger and also because nearly 300 million t of prime coking and 58 million t of non-coking coal lie underneath the town.

A small part of the deposit beneath the Jharia town was mined prior to nationalization and, consequently, a part of the town stands on coal pillars — even a mild tremor can cause this area to cave in. Despite this fact, multi-storey buildings continue to rise and business to flourish in the township. A sum of nearly 350 million rupees is being spent by Bharat Coking Coal Limited to contain the major underground fires surrounding it. A major obstacle to the shifting of the town is the huge private and public investment tied up in it and its growth as the hub of the coal industry in the Dhanbad belt of Chotanagpur.

Bharat Coking Coal Limited has now become more conscious of the damage done by fire by the Jharia coal field. Nearly 46 million t of coal lie locked in the fire areas. In some cases where it is not possible to deal with the fires, the locked-out coal might be lost for ever. Some experts believe that no project for dealing with the fires and releasing the coal will be economically feasible (Bhatt 1976:6). The aim would be to control the fires; the coal recovered should be considered only as a bonus. In most cases it would not be possible to extract coal immediately for lack entry points in the area and because of working losses.

Several measures are being taken to contain the fire areas in the Jharia coal field. Vigorous stowing is being done in some of the badly affected areas. Another method is to seal the area with doors. If sealing off does not work, then a mine may be flooded with water from the Damodar river. But this may be in turn lead to another problem, as pointed out by the Director of the Central Mining Research Station, Dhanbad: "If water is removed fire spreads. On the other hand, if the area is submerged, fire is extinguished but danger of inundation is created. . . . The affected area should be quenched with water from a distance and then the area should be dewatered gradually. This water can again be used for quenching purposes and ultimately the fire would die out" (*Indian Nation*, Dec. 1979).

According to the latest survey conducted by Bharat Coking Coal Limited, about 700 million t of coking coal still remain locked in standing pillars because of poor mining practices in the past. The company has further studied, in depth, the background of the Jharia fires and launched preventive measures to conserve the prime coking coal reserves of the country. It has divided the Jharia field into four major fire areas — Kusunda, Jogta, Lodna, and Rajapur. The fire of Kusunda area is widespread, and a series of quenching-cum-salvaging measures have been launched. The Jogta fire is also very extensive, as it was fed by oxygen through numerous subsidence cracks and fissures. A scheme has been launched to quarry out the nearby coal patch with the help of heavy earth-moving machinery and to use non-carbonaceous overburden for heavy blanketing and compacting to prevent air ingress into the fire zone. The locked-up coal on the dip side and in the lower seams would be available for extraction when the fire is eventually extinguished. This would improve the physical environment as well as allow coal to be quarried in the area.

In the case of the Rajapur fire area, the measures undertaken include the excavation of a large trench to save 1.5 million t of coking coal in the Basta-cola colliery, in the vicinity of Jharia town, from the approaching fire. More than 100,000 t of coal would be excavated when the trenching operation is completed in two years time.

The Lodna fire area covers 2.2 km² of the Jharia coal field. Experts have recommended a series of measures similar to those in Rajapur to save the pits and the industrial and office complexes. Surface sealing of nearly 4.4 km² by bulldozers and blanketing has also been recommended.

To control the extensive underground fires, massive sand stowing would be required in the Jharia coal field. However, the supply of sand from the Damodar river has been seriously reduced owing to the construction of a series of dams on its upper course. Bharat Coking Coal Limited has drawn up a Rs 50 million plan to transport stowing sand from the Maithon Dam to Jharia. A special rail system on the south bank of the Damodar river will be constructed to transport sand from the Durgapur barrage in West Bengal to the coal field. An alternative source of sand for stowing is the Sone Canal. In addition, crushed stone from quarries would also be used for stowing to reduce the cost of using distant stowing-sand. It is estimated that more than 10 million t of stowing material would be required by 1984.

Reconstruction of the Jharia Coal Field

The recovery of large coking coal deposits now locked under fire and sub-sidence-damaged areas of the coal field is one of the prime objectives of Bharat Coking Coal Limited.

The Central Mine Planning and Design Institute Limited has prepared a ninety-year programme of construction of the Jharia coal field. Starting with a small outlay of Rs 300 million in 1978/79, the project would involve an expenditure of Rs 20,000 million by the end of this century. This pro-gramme is phased to reach virgin coal by 2070.

An ambitious production target of about 60 million t of coal (40 million coking and 20 million non-coking) per year is envisaged in the feasibility report prepared by Polish and Indian experts (CMPDI) on the modernization of the Jharia coal field. The project when completed would be the biggest of its kind in the world.

The amalgamation of the 400 or so collieries into viable units and the merger of colliery boundaries to minimize loss of coal are part of the modernization programme. Immediately after nationalization, these units were reorganized into 85 collieries in the Bharat Coking Coal leasehold area. The project envisages further integration of these into 21 underground and 9 open-cast blocks (Fig. 4.3). The open-cast quarries are to be worked to a depth of about 250 m, whereas the underground blocks would operate at a depth of 360 m and below. The thirty integrated mines would be capable of producing 133,000 t of coal a day by the year 2000.

The project is to be completed in three stages. The first stage consists of reorganization and modernization to increase output from the present mines. The second stage would cover individual units for developing deeper horizons. Development of new mines and horizons for increasing production is proposed for the third stage. The feasibility report envisages the production of coal from both open-cast and underground mines in the Jharia field to increase to 50 million t per year by 1990/91.

The final feasibility report indicates that up to a depth of 600 m there are about 12,800 million t of coal, including 5,600 million t of prime coking coal, in the Jharia field. About 700 million t of prime coking coal are reported to be available in standing pillars up to a depth of 200 m in the old workings. For utilizing low-grade coking coal of the field, the reconstruction project envisages the setting up of twelve new washeries. Modernization in this field involves the substitution of pick mining by mechanical devices. Only 6 per cent of the coal is being produced by pick mining and the remainder by mechanized operations.

The project also calls for the evacuation of built-up areas so that the coal beneath the settlements could be extracted. This has been postponed for socio-political reasons. It is believed that within the next ten years the major urban centres of the coal field such as Jharia, Katras, Kerkend, and others would be so much affected by dereliction and related problems that they would no longer remain fit for human habitation. The population would perhaps move out, making it possible to mine the coal presently inaccessible. The displaced population would be accommodated in seven satellite townships to be established on the periphery of the coal field. The overburden would be used in making a long embankment along the northern border of the coal field. The upper courses of the streams now draining the area would be dammed, and the water stored in a large reservoir for multi-

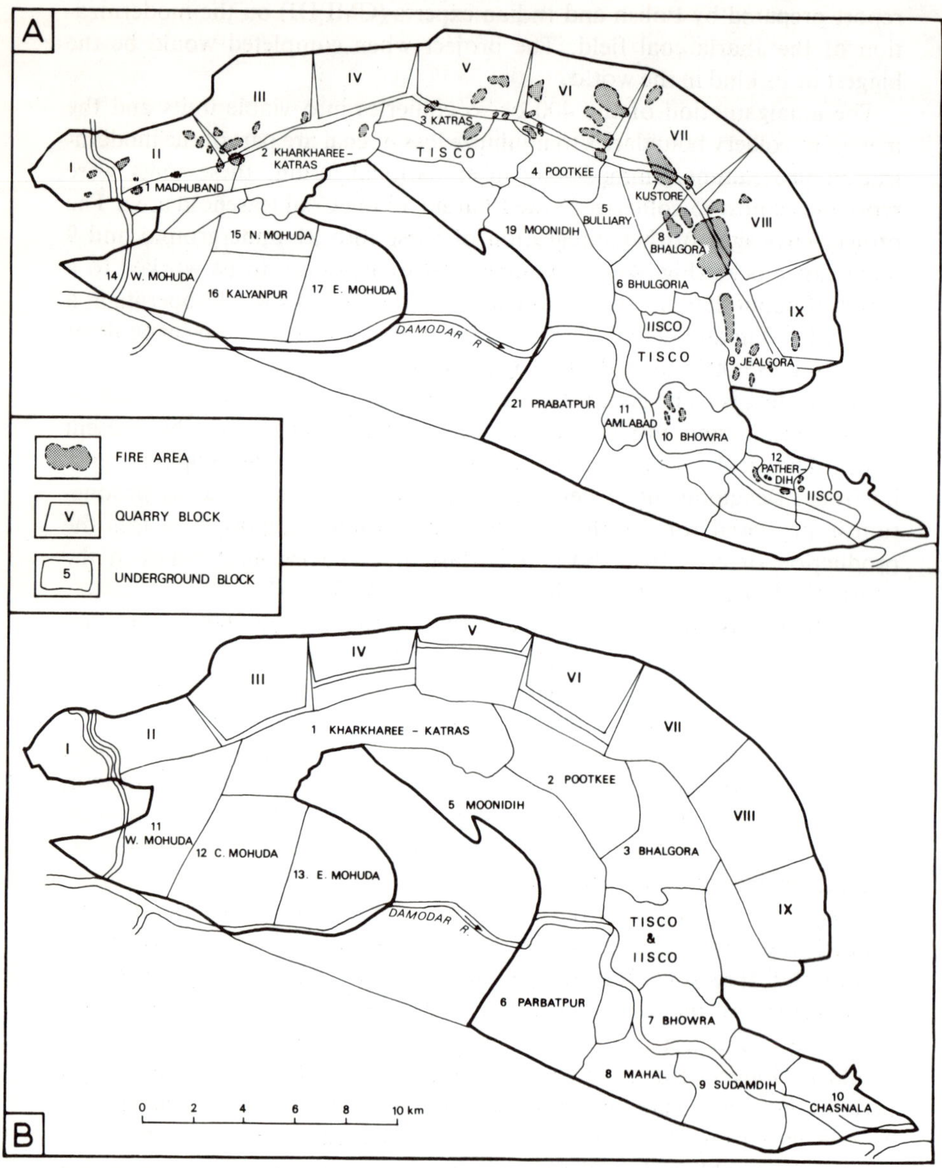

Fig. 4.3 A. Distribution of quarry and underground blocks in Jharia Coalfield, 1980
B. Proposed integration of Bharat Coking Coal Limited Mines, Jharia Coalfield

purpose use. If the reconstruction scheme is fully implemented, it would be a novel example of reclamation of an area extensively damaged by unscientific and uncontrolled mining.

Physical and Social Impact of Dereliction in Chotanagpur

A full appraisal of the physical and social impact of dereliction prevailing in the Chotanagpur region is beyond the scope of this paper. However, any discussion of the problem of dereliction cannot ignore these aspects. In the following section the Jharia coal field is examined as a representative area because the physical and social impacts of dereliction are more widespread and obvious in this mining region than anywhere else in Chotanagpur.

The topography of the Jharia coal field has been badly damaged and transformed by careless coal mining. Deep abandoned quarries, decade-old heaps of overburden, subsided tracts, and blazing grounds have devastated the area. There is practically no trace of the original surface. Excavations and diggings are so widespread that the top soil can no longer support grass or plants without artificial treatment. Moreover, the nutrients of the top soil have been either washed down to great depths through numerous cracks in the surface or burnt to ashes by underground fires. Wild bushes and thorny plants have replaced the tree vegetation in the Jharia coal field. Both the surface and underground drainage systems have been seriously affected by mining.

Everywhere the landscape presents a picture of physical chaos and destruction. It would be difficult to point out an area in any other mining region of the world that has suffered more from mining dereliction.

The social impact of dereliction is as severe as the physical. Water is in short supply, and whatever of it is available is badly contaminated and polluted. The population suffers from all types of diseases and ailments accompanying the consumption of polluted water. Road transportation is chaotic. All the mining settlements are connected with a single, metalled but dusty road. Driving on this road is perilous because of heavy mixed traffic, subsidence, and underground fires. The land around the settlements is so badly damaged by careless mining, subsidence, and fires that no other road can be built in the area. The environment is enveloped by a permanent cloud of suffocating dust and smoke. Dereliction in the Jharia coal field has motivated industrialists to set up their establishments and factories outside the coal-bearing areas, notably along the Grand Trunk Road north of the town of Dhanbad.

All kinds of violent crimes are committed in the derelict localities surrounding the settlements because the abandoned pits and quarries, subsided areas, thick bushes, and blazing flames provide hideouts for criminals. The

heavy addiction of workers to wine and country liquors and the high incidence of indebtedness and social exploitation are closely related to the
"derelict mentality" of the people.

Constraints to the Reclamation of Derelict Land

There are several constraints operating against the reclamation of old
derelict sites. Reclamation of derelict land becomes feasible only when a
particular country has reached a stage of sound economic development and
is equipped with adequate financial and technological resources. In a
developing country such as India, there is a strong lobby working against
diverting limited resources for the reclamation of derelict land because
important development projects of irrigation, power, industry, and other
sectors must have priority in investment.

It has been argued that there is a greater need to conserve the forest
resources of Chotanagpur which are being rapidly depleted than to start
expensive re-vegetation projects in derelict areas. Afforestation would be
more successful in areas where the top soil is in good condition. Reclamation requires costly mechanical equipment which must be imported from
outside. This would put a further strain on the slender foreign exchange
resources of the country.

The reclamation of the derelict land in Chotanagpur is a difficult task.
Spreading a layer of the recommended thickness of 0.6 m of fresh soil on
reclamation sites in the Jharia coal field alone would be a gigantic task.
Reclamation in a tropical region such as Chotanagpur is also difficult
because the top soil layers on reclaimed sites are easily washed away unless
protected during the heavy downpours of the monsoon.

The torrential rains have severe gullying effects in the tropics. The pattern
of surface drainage would have to be designed very carefully on reclaimed
sites. The drainage channels would further be required to be protected by
artificial covers in such a tropical situation. The development of grass
covers and tree plantations on the reclaimed sites is difficult, arduous, and
expensive as the summer is very hot and even mature trees and plants may
wilt during a hot spell. Although plantations may be established during the
monsoon, and are likely to survive through the winter, the watering of
plants during the hot summer would be very costly. Moreover, water is in
short supply, particularly in the Jharia coal field, so that even the household
requirements cannot be met. The temperate areas have a distinct climatic
advantage, in this respect, over their tropical counterparts. Almost all the
scientific research and experiments pertaining to the reclamation of mining
dereliction land have been conducted in temperate countries such as the
United States, Great Britain, and West Germany. The experience of these

countries may not be of much relevance to the tropics. The absence of a proper research base is an impediment to land reclamation in Chotanagpur.

The reclamation of derelict land should be an integral part of the development process of a country. In this study, four stages in the hypothetical evolution of dereliction, synchronizing with four stages of development, are suggested (Fig. 4.4). The characteristics of each of the stages of economic development and mining dereliction are as follows.

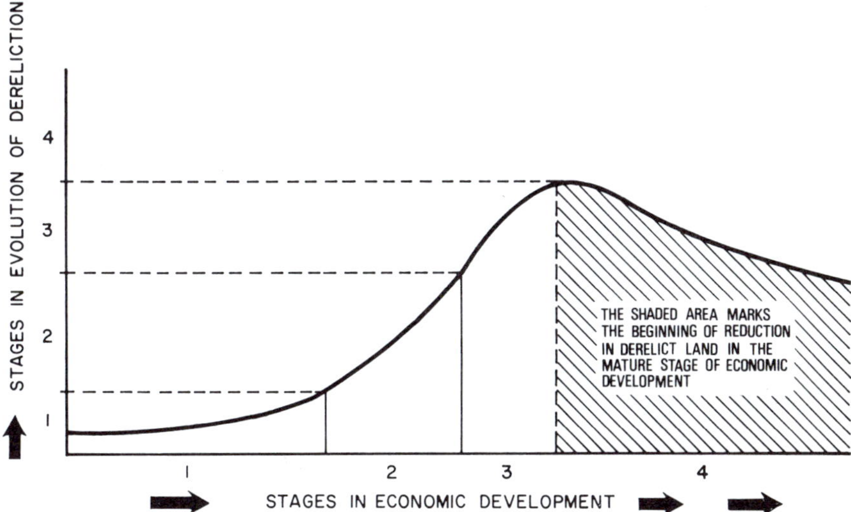

Fig. 4.4 Hypothetical evolution of mining-dereliction and economic development

Characteristics of Evolutionary Stages of Dereliction

1. Dereliction is limited in extent, type, and amplitude. It is characterized by a complete absence of public consciousness of this problem. Parts of the mining areas are devastated because of careless "slaughter" mining.

2. An increasing tempo of dereliction accompanies a high rate of mineral exploitation. The extent, amplitude, and types of dereliction are increased and diversified; the problem becomes complex. Public concern over this problem begins to manifest itself, but no action is taken by the government because of the high cost of reclamation and priority to other development projects.

3. Dereliction proceeds at an accelerated pace. Public concern becomes stronger. Reclamation begins to be considered necessary by the government and preliminary policy decisions are taken. Scientific, planned development of mineral resources is initiated.

4. Dereliction continues to grow in extent and diversity, but reclamation becomes an integral part of the exploitation of mineral resources. Derelict areas of the past are reclaimed. The growth of new dereliction is controlled by statutory regulations and restoration of land damaged by mining proceeds concurrently with exploitation of minerals. This state is presumed to be a protracted one, striking a near balance between dereliction growth and land reclamation. Dereliction would be much reduced but it would not altogether disappear. The large-scale exploitation of different minerals would result in more extensive dereliction than in the first two stages, but such dereliction would only be temporary as restoration and exploitation would proceed *pari passu*.

Characteristics of Stages of Economic Development

1. The economy in the first stage of evolution of dereliction would be undeveloped and primarily agro-based. The demand for minerals would be low and the method of mining primitive.

2. The demand for minerals would increase sharply as the industrial base of the economy expands. The exploitation of increasing quantities of different minerals is the sole objective of the mining concerns. The application of improved mining technology, indigeneous or borrowed, continues to accelerate the growth of dereliction.

3. Minerals play the key role in the process of economic development. The economic infrastructure is dominated by such vital sectors as industry, power, and transport.

4. Economic development maturity, accompanied by sound financial resources, a developed industrial base, and sophisticated technology, would permit quick, large-scale reclamation of derelict areas. A part of the mining spoils would be re-used for improving the infrastructure. The use of rock-shale as filling material in road construction in Great Britain is an example in this case. Large quantities of urban refuse and industrial effluent would be used in restoration schemes. Some of the derelict areas may be preserved for their floral and faunal significance.

Prospects of Reclamation

Chotanagpur may be placed at the end of the second and the beginning of the third stage of economic development and dereliction. Dereliction continues unabated in this region, but public concern over this problem is now growing stronger. The reluctance of the government to take restorative measures is now less marked.

At a meeting of the Energy Research Committee held in August 1976, a decision was taken to set up a study group to investigate the surface ecological imbalance caused by mining activities and to suggest measures for reclamation. The committee further proposed, in 1977, to submit a project report on the reclamation of quarried land.

The study group identified and selected a small quarried-out area, measuring 2.73 acres (1.1 ha) and marked as Quarry no. 1 on the map (Fig. 4.2B). The site is near the Godhur colliery of Bharat Coking Coal in Area VI of the Jharia coal field. A project report on it was formulated through the combined efforts of the Department of Science and Technology through its Environmental Planning Wing, Central Mine Planning and Design Institute Limited, Bharat Coking Coal Limited, and the Indian School of Mines. As no work has yet been done to make the land re-usable once mining ceases, the project proposes that such a pilot study be conducted. The primary aim of the project is to study, over five years, the measures "to reclaim a quarried-out land by complete filling of the excavation with mine-spoils and to vegetate the area thus filled and partly spread with top soil" (Feasibility Report 1980). The growth of different species of plant life would be studied under two different conditions, (1) mine spoils covered with top soil and mixed with different types of manures, and (2) mine spoils without top soil covering and manure application. Simultaneously, changes in soil and rock characteristics resulting from development of plant life would be closely observed. Nearly three-quarters of the quarry has been filled by overburden obtained from a nearby active quarry. Because of the small scale of the project, the possibility of agricultural use or the creation of a wild life area is precluded, but ultimately the small woodland may be converted into a community park.

This pilot project aims at identifying plants which would grow best on the reclaimed collieries of Chotanagpur. Vegetation is quite sparse in the Jharia coal field. The natural vegetation is primarily composed of Chor Kata (bottle brush), Ban Tulsi (*osimum bacilicum*), Putush (*lantana camara*), and Thethar (*ipomea cornea*). There is evidence to indicate the possibility of other vegetation growth on the reclaimed sites. Generally, trees such as Arjun (*terminalia arjun*), Gulmohar (*caesalpinia*), Nim (*azadirachta indica*), and a few flowering plants grow well in the area. The species selected for planting are Radhachura (*peltaforum innermi*), Sisoo (*dalberzia sisoo*), Neem, Arjun, Gulmohar (*caesalpinia*), Ashok (*saraca indica*), Laburnum (*casia fistula*), Australian acacia (*acacia auriculoformis*), and Jhau (*thuja compacta*). These species are known to grow fast under semi-arid conditions. Mulch would be provided for creating a grass bed because the nutrition and moisture-retention capacities are poor in minespoils.

The total cost of the project has been estimated at 37.7 million rupees. It has been further calculated that about 0.163 million t of coal had been mined from the quarry site, requiring removal of 0.232 million m^3 of over-burden (i.e., overburden coal ratio of 1.4 m^3/t). Therefore, the cost of reclamation works out to about 12.5 million rupees per acre or 2.31 rupees per t of coal mined from this particular project site (Feasibility Report 1980:25). The cost of reclamation, however, would vary from site to site as it would depend on many factors such as coal overburden ratio, size of the project area, thickness of coal seams mined, the nature of dereliction (subsidence/underground fire or combination of many other causes), and other factors.

The above reclamation project, the first of its kind in Chotanagpur, would generate data for further investigation. It would also pave the way for formulating suitable schemes for the reclamation of land damaged by large-scale open-cast mining, which would be the dominant form of coal mining in this region. The success of this project would encourage further investigation to develop other techniques of reclamation and revegetation for different kinds of minespoils.

The management, control, and improvement of the environment depend on the political will of a nation. The Parliamentary Consultative Committee has taken a policy decision to fill the open-cast mines as soon as possible and spend a large amount of money on the re-vegetation of such sites of the Central Coalfields Limited areas (*Indian Nation*, 2 Nov. 1980:4). It is a positive indication of the recent change in political attitude to the problem of dereliction in Chotanagpur. The creation of the Department of Environment under the Union Ministry of Science and Technology is the latest step forward in the direction of controlling damages to land resulting from the exploitation of mineral resources. The problem of dereliction is gradually becoming so acute in Chotanagpur that very soon, it is hoped, reclamation would be made a statutory responsibility of the mining agencies.

Conclusions

The dereliction of land can result from mining and mineral-connected activities. This is only one of the many problems of an industrial civilization based on the exploitation of non-renewable resources. The problem of dereliction of land is encountered in every country in some form. It is a major problem in some parts of India. It is particularly serious in Chotanagpur because it is the most richly endowed region in the country with respect to the variety and size of mineral deposits.

The modern phase of mining in Chotanagpur, associated with coal, began in the 1774 (Barraclough 1955). Other important minerals were exploited more than a century ago. Much of the dereliction is therefore a legacy of the past.

The problem of land dereliction exists in all the major mineral belts of Chotanagpur (Fig. 4.1), although the worst type of dereliction, caused by a complex set of physio-cultural factors, is in the Damodar valley coal fields. Mining is the major industry here, and the land has been damaged. Likewise, the entire area bearing mica deposits is pitted with holes dug by small-scale operators — rendering it unfit both for cultivation and large-scale exploitation of minerals and forest resources.

What has happened in Chotanagpur as a consequence of careless, uncontrolled exploitation of minerals is not unique to this particular region. Almost every country which has attained a high level of technological development has had to pass through this phase of the evolution of the mining industry. Dereliction arises because operators in the initial stage of this industry are unwilling to spend money on rehabilitation which will give them no direct financial return. The responsibility of restoring land damaged by private mine owners is generally not shouldered by the government. It appears that no matter whether the minerals are extracted by private or government operators, the prime objective is greater output coupled with high profit, with no concern for environmental degradation, in the initial stage of the development of the mining industry.

The problem of dereliction is accepted as inevitable at a particular cultural level of human development. Even in advanced countries, dereliction was accepted as a natural consequence of mining in the past. Man has often been profligate in his use of resources; until very recently there was little attempt to clean up the mess left behind after mining. The change in the attitude towards dereliction comes slowly with the overall development of society. In the developed countries, for example, it is accepted that while a mineral is being worked there are bound to be pits and mines and piles of waste materials. It has become usual to enforce the rehabilitation of minedover land. This had led to a situation in which pits must be filled in wherever possible, tip-heaps flattened or contoured into more natural shapes, and vegetation replanted. Suitable regulations for restoring mined-over land are now commonly written into the operating conditions laid down by governments. It is happening not only in the developed countries but also in some others where population pressure on land is not very great, for example, Australia, New Guinea, Nigeria, and Malaysia. In Malaysia large expanses of agricultural land were damaged by tin mining, so that mining enactments had to be passed requiring proper control measures to be taken by the miners. Experiments on the rehabilitation of mined-over land were concluded in the early 1960s; these laid the foundation for land reclamation in subsequent years (Ooi 1963).

If the picture of mining dereliction in most of the developing tropical countries is depressing, it is even more depressing in India. The problem of mining dereliction assumes greater significance in these countries because of the almost total absence of public concern and mining regulations. Financial considerations override concern for the restoration of damaged land. In the tropics dereliction is accepted as inevitable — a natural consequence of the exploitation of mineral resources.

The developing countries, however, are better placed to control dereliction by taking advantage of the experience, research, and technology of the developed countries. The magnitude of the problem of restoring derelict land is far more serious in developed countries because they have inherited dereliction as a legacy of the past. The developing countries, on the other hand, have only recently begun to extract minerals on a large scale. They can therefore prevent widespread dereliction within their territories by strictly enforcing appropriate legislation. It is a challenge for the developing countries but an opportunity for the developed countries to contribute towards the betterment of the tropical regions of the world.

Acknowledgements

The author wishes to thank officials of the Central Mines Planning and Design Institute Limited, the Central Coalfields Limited, and the Bharat Coking Coal Limited for their kind assistance in providing material and holding discussions with him. He is also grateful to Prof. J.I. Clarke of the Department of Geography, University of Durham, under whose supervision aspects of British mining dereliction and reclamation were studied when the author was a Visiting Research Fellow of the British Council during September–November 1979. The inspiration and encouragement received from both Prof. E. Ahmad and Prof. P. Pandey of the Department of Geography, Ranchi University, Ranchi are also gratefully acknowledged.

REFERENCES

Adarkar, B.P. 1945. *Report on Labour Conditions in the Mica Mining and Mica Manufacturing Industry*, p. 2.

Barraclough, L.J. 1955. *Early Development of Coal Mining*. Progress of the Mineral Industry of India, 1906–1955. Golden Jubilee Commemoration Volume. The Mining, Geological and Metallurgical Institute of India, 27, Chowringhee, Calcutta, 13, p. 141.

Beaver, S.H. 1946. *Derelict Land in the Black Country*. Ministry of Town and Country Planning.

Bhatt, K.B. 1976. *Dealing with Coal Field Fires: Jharia Coalfield Area*. Proceedings of Seminar on Underground Fires and Coalfield Fires, 15–16 Nov. 1976. Central Mine Planning and Design Institute Ltd. Ranchi, pp. 1–7.

Brown, J. Coggin and Dey, A.K. 1955. *India's Mineral Wealth*. London, p. 19.

Bush, P.W. 1969. *Spoiled Lands to the South-East of Leeds*. Proceedings of the Derelict Land Symposium, pp. 3–9.

Cameron, J.W. 1955. *Aluminium*. Progress of the Mineral Industry of India, 1906–1955. Golden Jubilee Commemoration Volume. The Mining, Geological and Metallurgical Institute of India, Calcutta, p. 326.

Central Coalfields Ltd. 1980. *Performance Report* (*June, 1980*). Ranchi, p. 2.

Dunn, J.A. 1943. *Presidential Address: Geology and Geography Section*. Proceedings of the Thirtieth Indian Science Congress. Lucknow, p. 91.

______. 1973. *Explanatory Memorandum on Grants for the Reclamation or Improvement of Derelict Land Under the Local Employment Act 1972 and the Local Government Act, 1966.*

______. 1980. *Feasibility Report for Reclamation and Revegetation of Quarried Land.* Central Mine Planning and Design Institute Ltd. Regional Institute, Dhanbad.

Fox, C.S. 1930. "The Jharia Coalfields". *Memoirs, Geological Survey of India* 56:220, 254.

______. 1924. *Mineral Resources of India for a Domestic Steel Industry.* Report of the Indian Tariff Board Regarding Production to the Steel Industry (Annex), p. 91.

______. 1960. *Indian Mineral Year Book.* Indian Bureau of Mines, summary, p. xi.

Goh Cheng Leong and Morgan, Gillian C. 1975. *Human and Economic Geography.* New Delhi, p. 421.

Imperial Institute. 1937. *The Mineral Position of the British Empire.* London, p. 78.

Jones, H. Cecil. 1934. "The Iron Ore Deposits of Bihar and Orissa". *Memoirs, Geological Survey of India* 63, 1:183–207.

Kumar, N. 1970. Ranchi District Gazetteer. Patna, p. 162.

Kumar, R., and Singh, B. 1969. "Surface Subsidence: Its Impact on Economy". *Coal Field Tribune*, Mines Safety Week Supplement, 7, no. 8, p. 12.

______. 1972. "Protection of Surface Structures Against Damages Caused by Underground Mining in Jharia Coalfield". *The New Sketch.* Seminar on Reconstruction of Jharia Coalfield, 12–13 Nov., p. 1.

______. 1957. "Mica and Mica Wastes". *Research and Industry*, 2 (March), pp. 60–62.

______. 1963. *Memoirs of Coal Exploration.* Second Five Year Plan, Government of India, Ministry of Steel and Mines, Indian Bureau of Mines, p. 94.

Ooi Jin-bee. 1963. *Land, People and Economy in Malaya.* London, pp. 306–07.

Oxenham, J.R. 1966. *Reclaiming Derelict Land.*

Pandey, S.K. and Kar Ray, M.K. 1980. "Potentiality of Jharia Coalfield, the Store House of Prime Coking Coal in India". Unpublished paper presented on 21 March 1980 at Jamshedpur, Bihar, in the Seminar on Energy Management, p. 4.

Pant, K.C. 1976. "Statements of Objects and Reasons". *The Coal Mines (Nationalisation) Amendment Act, 7 May 1976.*

Percivil, F.G. 1931. "The Iron Ores of Noamundi (Singbhum)". *Trans. Min. Geol. Inst. of India.* 26, no. 3, pp. 169–271.

______. 1947. "Estimated Reserves of Iron Ore in Singbhum-Orissa Field". *Transactions of the National Institute of Sciences, India* 11:373–75.

Raja, S.T. 1965. "Coal and Its Part in Our Fuel Economy". *Industrial India*, p. 85.

______. 1973. *Records of the Geological Survey of India*, 1, p. 64.

______. 1967. *Report of the Central Wage Board for Iron Ore Mining Industry.* New Delhi, p. 8.

Srimany, S. 1980, Superintendent to Mines and Technical Secretary to the Regional Director, CMPDI Ltd., Dhanbad. *Personal Communication*, 12 Dec., p. 1.

The Indian Nation (a daily newspaper of the State of Bihar, published from Patna). 7 Dec. 1975, p. 5; 15 Nov. 1975, p. 5; 19 June 1979, p. 4; 28 July 1979, p. 5; and 27 Dec. 1979, p. 5.

Wallwork, Kenneth L. 1974. *Derelict Land: Origins and Prospects of a Land Use Problem.* London, pp. 17, 43.

5
Rural Energy Resources in Tropical Countries, with Reference to Tanzania

H.M. MUSHALA

The African nations like many other countries, have had to re-examine their energy resources more carefully when the OPEC price rises triggered off a global energy crisis in the early 1970s. As most of Africa's population live in rural areas, the energy crisis has taken a different form in this continent, and is in effect a rural energy crisis. Its effect in Africa has been a general deterioration, due to intensified use, of arboreal resources, specifically fuelwood for cooking, heating, and agricultural tendering of crops. In 1978 about 1,220 million m^3 or 47 per cent of the total wood consumed in the world was used as fuelwood, and developing countries used about 1,075 million m^3, which is about 80 per cent (Mnzava 1980). A century back, tropical forests were the main sources of supply of energy in the form of fuelwood for the poor of the world. Today, the forests are disappearing at the rate of 15 ha a minute, a fact which has grave implications for the developing world (Matheson 1976).

This paper seeks to examine the nature of the rural energy crisis in some African countries, particularly Tanzania (Fig. 5.1), and to evaluate its impact on development. Attempts at harnessing alternative sources of energy and their implications are also treated briefly.

Fuelwood and Charcoal

The African continent covers an area of about 30.4 million km^2, and had a population of about 450 million in 1977 with an annual increase averaging 2.6 per cent. Most of the people (about 85 per cent) live in rural areas and depend upon agriculture (including animal husbandry) for their livelihood. For most of these people, including urban dwellers, fuelwood remains important for domestic cooking and heating; at the same time, it plays a significant role in small-scale industries such as baking, brick making, tobacco curing, pottery firing, fish smoking (drying), and iron smelting (for

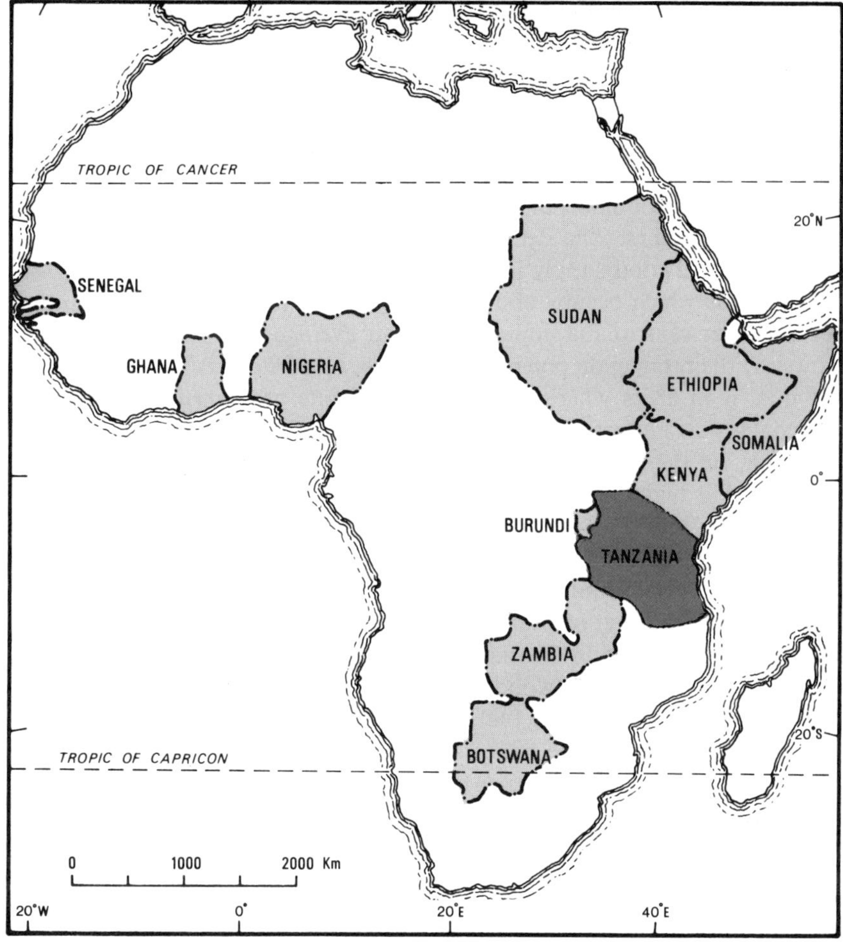

Fig. 5.1 Location of countries of study

the manufacture and repair of farm tools). The energy situation in Africa is well expressed by Robert Lamb (1978) who observes that:

> Everyday, the peasant must go a little further away from the village to find firewood. Everyday the charcoal burner charges a little more, forests are felled and soil erosion grows worse. This is the poor man's energy crisis.

The effect of the crisis on countries in Tropical Africa varies according to their geographical location, size, physical resource endowment, and the degree of colonization experienced by individual countries as colonization has greatly influenced land use and resource exploitation in many areas. Fuel-wood remains a significant domestic energy resource in all these countries.

Nigeria, with a land area of about 927,420 km² had a forest land area of about 214,600 km² in 1965. Wood is the chief fuel for domestic cooking and sometimes for heating. Other fuels are charcoal, animal waste, and crop refuse. It was estimated that the local demand for fuelwood in the country in 1955 was about 3 million m³ a year. Ten years later, the consumption had increased five times. The demand is likely to escalate with time (Adeyoju 1965). The fuelwood supply in Nigeria has for most of the time been from the savannas which occupy about 85 per cent of the land area and support about 64 per cent of the population at an average density of 45 per km². However, the production potential has decreased northwards, and the poorest potential is in areas where transport is an acute problem. In some areas, charcoal is now imported from overseas (Morgan 1978).

Ghana, on the other hand, is generally highly wooded and fuelwood is plentiful. The high (closed) forest extends from the sea northwards as far as latitude 7°45 'N and covers an area of about 82,258,400 km². Almost all the domestic requirements of energy in rural areas have to be met by fuelwood or charcoal. The amount of fuelwood consumed annually is estimated to be 5,547,200 m³, and that of charcoal is about 1,992,400 solid m³ (Kesse n.d.). Most of the fuelwood in Ghana is collected from dead wood derived from farm clearings. Some fuelwood is commercially produced along the coast, and used for fish curing, as fuel in main urban centres such as Accra and Kumasi, and also in some mines. There is a shortage of fuelwood in the heavily populated northeast corner and in the southeastern coastal plains, which are almost treeless. Charcoal is consumed mainly in Accra, although its production is primarily in the savanna woodland east of the Volta river. Charcoal is made from woodland trees, as well as from the slabs and offcuts from upcountry saw mills in the forest zone. Commercially produced char-coal has recently been used for smelting iron ore in the southwestern part of the country.

Compared with Ghana, the situation in the other African countries is grim. Owing to the intensified use of the arboreal resources, energy experts predict, Senegal will be without trees in about thirty years, Ethiopia in about twenty years, and Burundi in seven years if reafforestation campaigns are not launched and implemented soon. About fifty years ago, 60 per cent of Ethiopia was covered with forests. Today these forests have been cleared, mainly for fuelwood, and only 5 per cent of the land remains under forest (Darkoh 1979).

Most of Somalia is arid. There is no production of primary energy except for fuelwood and charcoal used for domestic purposes. It is estimated that 51,000 to 61,000 t of charcoal are consumed annually, and of these about

36,720 t are consumed in Mogadishu. Although the production of charcoal is managed by the government in the form of cooperatives controlled by the Forest Department, the main problem is scarcity of wood in the savanna area. The general deforestation of the country necessitates moving the charcoal camps progressively further from consumer centres at a rate of about 15 to 30 km every year.

In Sudan wood remains the main fuel source since imported fuel such as kerosene is too expensive for the majority of the population. Fuelwood makes up 71.5 per cent of the total energy consumption. Although productive natural forests cover 18 per cent of the country, and 5 per cent of the country is covered by protective forests, on dry and semi-arid lands clearing of the thorn-scrub savanna in western Sudan has left vast areas denuded of tree cover. The current consumption of fuelwood is estimated at about 14.28 million t out of an estimated net allowable annual cut of 23.46 million t. Fuelwood is essentially a rural energy resource, and it has been estimated that an average weekly consumption per family was between 50 and 70 kg in 1977 (Ibrahim 1978). In addition to these domestic uses, wood has to provide for building huts and enclosures. Consequently, the consumption of building timber and fuelwood per family can be as high as 200 trees a year (see Table 5.1). This rather gloomy situation is aggravated

TABLE 5.1 ANNUAL CONSUMPTION OF WOOD PER FAMILY IN NORTHERN DARFUR, SUDAN

Usage	Number of Needed Trees or Bushes
Hut building	2.5
Enclosure for the huts and court	40.0
Enclosure for the field (600 m³)	100.0
Fuelwood (1 tree weekly)	52.0
Total no. of trees needed	194.5

Notes: For hut building a minimum of 2 huts are required per family. This requires 16 trees. The approximate durability for these huts is about 6 years.

    Huts and enclosures required about 80 m of wood, half of which are renewed annually.

    For nearly one-third of the families half of the wood lot required is renewed annually.

Source: Ibrahim 1978.

by the nomads who, it is estimated, uproot a minimum of 548 million acacia shrubs a year for cooking purposes only. Despite the high consumption pattern of wood for various purposes, there are very limited efforts at producing wood for these needs. Generally, trees are planted as "green belts" around towns for protection against sand accumulation but not for these other basic uses.

Ninety per cent of Kenya's estimated population of 15 million depend on fuelwood and charcoal for domestic cooking and heating energy. Fuelwood is also used in agriculture for tobacco curing and in industry for firing steam boilers. Fuelwood consumption is estimated at 10.2 million t of wood and 1.44 million t of charcoal (excluding commercial consumption). Of this, 83 per cent of the fuelwood is used for cooking and 16 per cent for heating. Ninety-one per cent of the charcoal is used for cooking and 8 per cent for heating (Nyoike 1972). The situation threatens the country's forest cover, estimated at 1.8 million ha or about 3 per cent of the total land area of Kenya. Desertification, which already threatens the northern half of the country, may become a more widespread problem with over-rapid deforestation.

Zambia depends on fuelwood and charcoal as the main sources of domestic energy. All the rural population (which is about 60 per cent of the country's population) depend on fuelwood, and the low income sector of the urban population depends on charcoal and fuelwood. During 1978 it was estimated that at least 250,000 t of charcoal were produced and consumed. This has caused an alarming rate of woodland devastation, and constant price increase of these fuels. The problem is compounded by the irregularity of supply in some areas and during certain seasons.

This energy situation has affected the level of production in different sectors of the economy. The Five Year Development Plan has identified the principal issues involved in the fuelwood and charcoal industry, especially their supply, marketing, and use. Efforts are being made to establish rural and peri-urban forest plantations of fast growing trees to provide fuelwood and charcoal. To curtail production costs and wood wastage, ways are being sought to improve the technology for charcoal production. Simultaneously, methods of improving charcoal stoves to achieve maximum utilization of charcoal energy are being tried out.

In neighbouring Botswana, as in many other African countries, fuelwood — the traditional source of energy for cooking and space heating — is getting scarce. It has been estimated that annual consumption of fuelwood in rural areas is between 762 and 1,016 kg per capita. Seventy per cent of that is used for domestic purposes. When Gaborone, the capital, was built, fuelwood could be collected a few kilometres from town; today it is collected 30 to 40 km away (Thipe *et al.* 1979).

Rural Energy Resources

The rural energy crisis in developing areas originated with the incorporation of these countries into the global mode of commodity production for external markets. With the advent of colonialism, there was a change in the production process of colonized areas in that some of the areas covered by diverse natural forests and woodland had to be opened up for cash crops such as coffee, sisal, cotton, cocoa, sugar, rubber, etc. The affected rural people had to forgo their previous subsistence production and opt for either working in big estates, which claimed much of the fertile vegetated areas, or establishing small cash crop holdings and using the proceeds from their crop sales to pay the taxes imposed by the colonial authorities. The establishment of large plantations not only increased the distances rural people had to travel to look for fuelwood for domestic purposes but also changed the nature of the vegetation in those areas. In some areas it prompted the opening up of subsistence farm plots next to the large estates, thereby encroaching further on the natural vegetation.

The ensuing deterioration of the vegetation led to gradual desert encroachment. In the West African Sahel, for example, land in the pre-colonial period could be left fallow for up to twenty years between seasons of planting. The system allowed a variety of crops to be grown and did not result in the lowering of soil quality. Nomads could then drive their mixed herds over vast grazing grounds. With the arrival of the French in the nineteenth century, all the best land was usurped for export crops (mainly cotton and groundnuts) and the local population evicted to marginal areas. Intensive cotton cultivation depleted the soil, eventually leaving it too poor for production and paving the way for desert encroachment. In general, desertification in the West African Sahel is a heritage of the colonial past when the disruption of socio-economic management led to the breakdown of social structures, and the overexploitation in time and space of the limited natural resources (Darkoh 1979).

In some African countries, the fuelwood crisis has been exacerbated by socio-political structures that persisted in the regions for a very long time. In Ethiopia, for example, the agricultural system did not provide any incentive for the peasants to adopt land conservation measures because the feudal system required the tenants to pay their landlords heavily for the privilege of farming. Trees were also felled indiscriminately for fuel, with the result that the country lost its forest cover within fifty years. In other countries, whatever the mode of production, wood has remained the sole source of energy for domestic and other purposes. In such cases, the felling of trees for these purposes has led to the irreversible destruction of the natural vegetation. Around Khartoum, the acacia tree has been pushed about 90 km south within a span of about twenty years (*The Courier*, Jan./Feb. 1978).

The depletion of arboreal resources outlined above is due to the intensification of agriculture prompted by the imposed market-oriented mode of production on subsistence systems, by overgrazing, and by the collection of wood for domestic and other purposes. Levels of consumption of fuelwood to meet rural domestic energy needs are determined by the physical availability of the arboreal resources. Fuelwood that could be collected in the immediate vicinity of most households a few years ago now has to be gathered and carried from a distance a half-day's walk away (Arnold and Jongma 1978).

Implications

The progressive depletion of the vegetation resources, which are the basic fuel sources in rural areas, leads not only to a shortage of fuelwood but also to a reduction in the availability of wood for other essential needs such as house construction. In some countries, the fuelwood shortage has forced people to use animal dung and crop residues, which are otherwise essential inputs for increased crop yields, as domestic fuel. This has created additional pressure in that more land has to be cleared for agricultural production to meet an equal amount of crop demand satisfied earlier by a smaller portion of land.

The scarcity of wood for household domestic needs imposes an arduous burden on the wood collectors, who are mostly women. As a result, more rural labour has to be expended for the collection of fuelwood. In Tanzania, for example, numerous man-days of work are "lost" in this manner. Rural labour productivity, which is already low, is lowered further as energy which could be used for other productive activities is spent instead on collecting fuelwood.

The rural poor are seriously affected by the present high fuel prices due to the fuelwood shortage. In Tanzania, commercial fuels have become so expensive that the rural people cannot maintain enough light during the evenings for important social activities. In parts of West Africa which are acutely short of fuelwood, the people now have reduced their cooked meals from two to one a day (Arnold 1979:238).

Some Alternatives

Fuelwood, despite its low calorific value, is the universal domestic energy source in developing countries because of its physical availability and because it is easy to use and obtain. Commercial sources of energy such as electricity, coal, kerosene, and gas are not easily available in rural areas as there are no established sources, or transport, or other supply facilities for them in rural areas. In many cases, the facilities are non-existent. In relatively developed areas where the basic facilities for the commercial energy sources are available, the change from fuelwood to other fuels is inhibited by the high initial cost of appliances for, say, kerosene, gas, or electricity compared to open wood fire devices. Even where individual families are able to install the basic appliances, the ready supply of these energy sources can never be guaranteed.

One would argue for solar energy as being equally available and probably more easily procured when compared to fuelwood in tropical countries. However, solar energy devices such as cookers and heaters for domestic purposes are usually out of the reach of the majority of the rural population. Moreover, many rural families are not used to cooking in the open. They would also have to adjust their cooking times to the availability of solar radiation. The uses of solar energy on a massive scale in rural areas therefore remains more experimental than practical.

Besides solar energy, the other viable alternative is biogas (gobar gas). Compared to solar energy, the appliance costs of biogas are lower. In Tanzania, a small 2 m³ unit for single household purposes costs about T.Shs. 5,400 (1978 prices). Most of the rural people cannot readily afford to buy such a unit. Biogas provides a viable option only for wealthier households and is also suitable for use at a community level (e.g., rural community welfare centre, rural health centre, rural industrial organization).

In all these cases, the factors working against the use of alternative energy sources in rural areas in place of fuelwood are the high initial installation costs, the obvious intermittent supply to the rural areas under current global constraints, and probably the resistance of the peasantry to change. As Arnold (1979:243) has put it,

> The main scope for improving the rural energy situation rests in using existing fuel resources more efficiently and creating additional fuel resources.

Fuelwood can be used more efficiently with properly designed cooking appliances and facilities. Traditional means for cooking and heating are very inefficient in that most of the heat is wasted. Several attempts at improving cooking stoves have been made in tropical countries such as Kenya, Zambia, Senegal, and Tanzania. The improvement of cookers has to be done hand in hand with new pot designs to cope with the variations in temperature due to the efficiency achieved. It is expected that by having efficient appliances for cooking and heating, the quantity of fuelwood used could be reduced.

Insofar as charcoal is concerned, its use in place of wood will not reduce the fuelwood shortage as much of the energy in the wood is utilized in the manufacturing process or is lost as processing waste. Uncontrolled charcoal production poses as great a danger as uncontrolled fuelwood collection.

Apart from increasing the efficiency of fuelwood use, rural afforestation programmes should be initiated to supplement existing fuelwood resources. In many countries, the programmes are constrained by the competition for land for agricultural development and other activities. Another problem is the long gestation periods for afforestation. In some cases, there are management problems due to lack of trained manpower. Nevertheless, these programmes are given priority in many tropical countries. Arnold (1979:246) has suggested that in addition to growing suitable tree species for fuelwood, available forests should be protected against destructive practices. This can be institutionalized within the local rural government systems.

All these methods have been experimented with success in Korea, including the intercropping of trees for agricultural crops and the growing of multi-purpose trees. Many tropical countries can learn from Korea's experience. In all cases, carefully made and implemented land-use plans can play a significant role in the endeavours to cope with the energy crisis.

Rural Energy Resources in Tanzania

The Fuelwood Situation

Tanzania has a population of about 17 million (1978 Census) and covers an area of about 887,000 km^2. More than 95 per cent of the population live in rural areas and depend on agriculture as a means of subsistence. Like many other tropical countries, the rural people depend on muscle power to perform the basic tasks for survival. Such tasks include the clearing and preparation of land for cultivation, the hoeing of land, planting, weeding, irrigation, harvesting and threshing or husking of crops, the transport of goods, and the grinding and preparation of various crop products.

As in many developing countries in the tropics, Tanzania's rural energy is derived from an arboreal base. By 1977 the wood harvest in the country was 40 to 50 million m^3, and more than 95 per cent of this amount was used as an energy source (Mnzava 1980). It has been estimated that fuel for domestic consumption (mainly cooking) accounts for about 80 per cent of all extra-muscular energy that is used among Tanzania's rural population (TNSRC and U.S. National Academy of Science 1977). The per capita fuelwood consumption is almost 0.7 m^3 a year. However, Tanzania is only marginally forested in absolute terms in that only one per cent of the total land area is dense tropical forest. Man-made forests cover only 0.1 per cent of the land area. Consequently, most of the fuelwood (more than 98 per cent of the supply) is harvested from poorly stocked natural forest which is dwindling at a very fast rate. The result is that many households are obliged to use animal remains and agricultural residues such as cotton and maize stems or sisal stalks for cooking and heating. It is estimated that about 6.12 million t of cow dung are produced annually; of this amount about 20 to 30 per cent is used as an energy source (Mnzava 1979:13). Whereas in 1970 household fuelwood consumption was 30.48 million m^3 and represented 96.3 per cent of total wood consumption, in 1980 wood consumption was about 33.44 million m^3 and would reach about 44.79 million m^3 by the year 2000. Dunkerley (1979) estimated that in 1968, Tanzania's rural energy consumption was 21 giga-joule per capita, and that for Gambia (1973) was 13, Kenya (1960) 9.5, Sudan (1962) 16, Uganda (1959) 14, and Nigeria (1975) 16. This indicates that Tanzania is among the highest consumers of fuelwood and charcoal in Africa, to the extent that it now faces an annual fuelwood shortage of about 18 million m^3, which is expected to reach 20.7 million m^3 by 1985 (Table 5.2).

There is uncontrolled felling of trees to meet the increasing demand of fuelwood, poles for house construction, and fuelwood for commercial activities such as bread baking and tobacco curing. Although Tanzania was estimated to have 44 million ha of forest in 1978 (Quorro 1978), the forested area is rapidly shrinking because of this indiscriminate tree felling, inadequate forest management systems, inadequate afforestation measures, and uncontrolled wild forest fires. It is forecasted that Tanzania would be a desert in sixty years' time if urgent afforestation measures are not implemented immediately (Mashalla 1978).

Fuelwood is already in critically short supply in the semi-arid districts of Maswa, Bariadi, Magu, Dodoma, Kyela, Mpwapwa, Singida, Manyoni, Kiomboi, and Shinyanga (Fig. 5.2). Tanzania's rural population living in nearly 8,000 planned (*ujaama*) villages require a forest reserve of 75 ha of consolidated forest per village if the annual demand for fuelwood is to be met. At the moment, none of these villages has even half of the estimated area of forest reserve. Besides domestic energy demands, fuelwood is required for non-household purposes such as tobacco curing, brick making, and bread baking. Tobacco cultivation exerts intensive pressure on the forest because tobacco plantations require the clearing of large forested areas and, at the same time, need large fuelwood supplies for the curing factories (Table 5.3).[1] There are few villages in Tanzania, if any, which have afforestation programmes implemented simultaneously with tobacco growing programmes.

---

[1] It is estimated that one tree is needed to supply the fuelwood for every 300 cigarettes made in Africa (*Daily News*, 19 July 1979).

TABLE 5.2 DEMAND FOR FUELWOOD/POLES IN MAINLAND
              TANZANIA BY 1985

Region	Demand (m³ × 1000)	Potential Supply (m³ × 1000)	Deficit (m³ × 1000)	Required Planting (ha)
Arusha	1,882	753	1,129	7,000
Coast and Dar-es-Salaam	2,134	573	1,561	9,760
Dodoma	2,116	840	1,276	6,990
Iringa	2,532	1,413	1,119	2,580
Kagera	1,897	227	n.a.	n.a.
Kigoma	1,271	859	412	11,880
Kilimanjaro	2,010	109	1,910	n.a.
Lindi	1,169	1,169	n.a.	10,260
Mara	1,736	94	1,642	4,720
Mbeya	2,373	1,618	755	3,630
Morogoro	1,929	1,348	518	10,080
Mtwara	1,873	360	1,613	18,980
Mwanza	3,146	109	3,037	n.a.
Rukwa	990	990	189	1,180
Shinyanga	1,770	1,770	2,069	12,930
Singida	1,272	1,272	176	1,100
Tabora	2,361	2,185	1,649	10,310
Tanga	2,208	559	1,620	10,130
Total	37,404	16,625	20,779	129,570

Source: Forest Division, Ministry of Natural Resources and Tourism, Dar-es-Salaam, 1978.

Fuelwood is also needed to fire brick and tile kilns. With the escalating
prices of building materials, people are encouraged to use baked bricks. The
demand for bricks has in turn increased the demand for fuelwood to fire
the kilns. In districts such as Kyela and Mbozi, this has caused a rise in the
price of the fuelwood and the felling of immature trees to meet the demand.
In a nutshell, the rural energy crisis in Tanzania manifests itself in the
wanton felling of trees and bush for various household and non-household
purposes, with the demand for the fuelwood exceeding the supply.

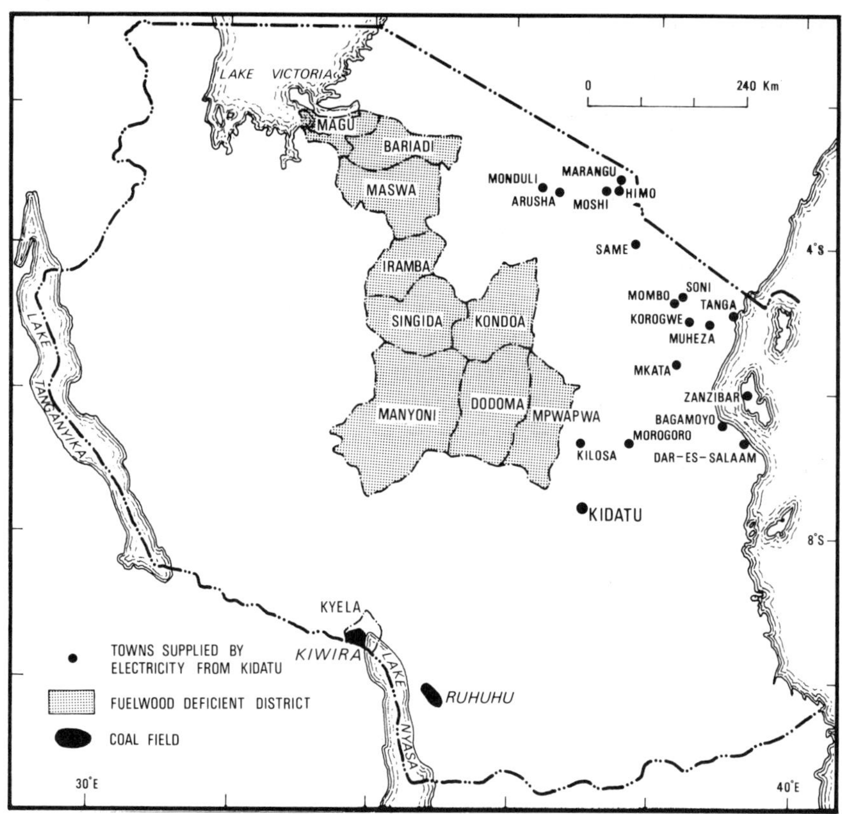

Fig. 5.2 Location of coal fields, fuel wood deficient areas and towns with electric supplies, Tanzania

TABLE 5.3 FUELWOOD DEMAND FOR TOBACCO
CURING, 1985, FOR MAIN GROWING REGIONS

Region	Area of Tobacco (ha)	Fuelwood Required (m³)
Iringa	8,000	400,000
Mbeya	2,000	100,000
Rukwa	2,000	100,000
Ruvuma	11,000	500,000
Tabora	19,000	950,000
Total	42,000	2,100,000

Source: Mnzava 1979.

Other Energy Resources

The vegetation resources have been overexploited to meet rural energy requirements, although the country is endowed with other energy resources. These, however, remain wholly or partly untapped for socio-technological reasons. With the oil import bill rising drastically, all possible alternatives are being carefully explored.

Coal Resources

Tanzania has coal fields in the southern and southwestern parts of the country, with Ketewaka-Mchuchuma, Songwe-Kiwira, and Ngaka being the major ones (Fig. 5.2). The development of the coal industry in the country is at a very embryonic stage. But there are efforts to develop it at a faster rate. The geological and proved reserves in the three major coal fields are shown in Table 5.4.

TABLE 5.4 MAJOR COAL RESOURCES OF TANZANIA

Coal Field	Proved (million t)	Geological Reserves (million t)
Songwe Kiwira	20.4	606.9
Ngaka	99.654	155.04
Ketewaka Mchuchuma	190.332	504.9
Total	310.386	1,266.84

Source: Nkonoki *et al.* 1978.

Tanzanian coal in workable seams are of average quality. A small colliery operating in Ilima in the Songwe-Kiwira coal field can supply coal at a rate three times higher than the present demand. The demand is expected to rise sharply in Mbeya and Iringa regions once the Mbeya Cement Factory and the Mufindi Pulp and Paper factory are in operation. In future, coal may also be used to meet the fuel requirements in brick making and in the tea industry. Coal is replacing fuelwood for tobacco curing in the southern regions. At present about half of the fuel requirements at the Katumba tea factory in Mbeya region is met from the Ilima colliery.

It is apparent that coal can only reduce the current demand of fuelwood for non-household purposes. Coal in Tanzania is not presently used for individual household needs because it is of high calorific value. Before coal can be used for domestic purposes, the present domestic appliances have to be redesigned to cope with the extra heat generated from coal. For rural communities, the solution will have to be found in other directions.

Electricity

Tanzania has an estimated hydro-electric (installed) potential of 1,400 MW, but so far less than 300 MW (including about 14.5 MW generated by parastatals, individuals, and other organizations) has been harnessed (Njau 1976). Most of this power is generated at Kidatu station which supplies power to Dar-es-Salaam, Kilosa, Bagamoyo, Muheza, Mkata, Morogoro, Korogwe, Mombo, Soni, Tanga, Same, Monduli, Himo, Marangu, Moshi, Arusha, and recently Zanzibar (Fig. 5.2). Other hydro-electric stations feed into the Kidatu system to form the coastal grid, which accounts for 85 per cent of all Tanzania's current electric generating capacity. In 1977, about 16 per cent of this power was used for domestic purposes, 33 per cent for light industrial purposes, 30 per cent for industrial work, 20 per cent for Industrial KVA, and 1 per cent for public lighting.

At present electricity is used by the few urban dwellers and an insignificant number of rural centres close to urban areas. The rural electrification programme launched during the Third Five Year Development Plan aimed at extending electric power to some small urban centres and neighbouring villages. The programme encountered various economic constraints, thereby slowing down the implementation pace. In areas where it has succeeded in reaching the rural population, many people have not used it because of the initial costs of installing equipment such as fuse boxes, bulbs, wires, and switches. For example, it costs about T.Shs. 8,000 (US$1,000) to install the initial equipment in a modern family house. In these rural areas, electricity is mainly used for lighting, and only a few rich peasants use it for other purposes like laundry work, refrigeration, and some for cooking. Some villages are using electricity to run small grinding and saw mills. Except for a few areas in Tanzania where people have modern houses, the prospects of utilizing electricity in the typical rural areas remain small. With the current escalating prices of materials and appliances, the prospective number of users is unlikely to increase.

Biogas (gobar gas)

Biogas is produced when waste matter (of human, animal, or agricultural product origin) ferments in the absence of air. During the process, a slurry residue is produced and this is suitable for use as a fertilizer. Tanzania has learnt from various countries, especially Asia, to use biogas to generate energy cheaply and make use of the fertilizer byproduct. There are nearly 50 biogas plants in the country, distributed in the main cattle-rearing regions of Mwanza, Mara, and Arusha. There are plans to install more community-size gas plants in Shinyanga and Dodoma regions.

Most of these plants are community-size plants serving village communities, especially rural health centres. At this stage of the country's development, it is preferable to install and operate these larger plants (10.5 m^3) which cost about US$5,000 each, rather than family size plants (2 m^3) which cost about US$770 in 1978.

There are several advantages in the use of biogas plants in rural areas. Except for a few materials such as metal barrels and pipes which have to be bought, other equipment and materials can be obtained locally. This reduces the foreign dependence constraint. Second, installation is simple and easily taught. Third, the energy is produced cheaply but efficiently and its consumption can be controlled. Last, the sludge that is produced can be used as a fertilizer to increase crop productivity.

The biogas plants are operated with the assistance of SIDO (Small Scale Industries Development Organization) and AATP (Arusha Appropriate Technology Project). OXFAM has planned a rural biogasification project in the Shinyanga region which will include one family biogas digester in each of the official homes of village managers.

Among the problems experienced at the rural family level are the initial installation costs of the plants. Many rural families cannot afford the costs. Furthermore, it has not yet been established whether the plants are socially accepted by the majority of the rural people. The plants have been installed initially on an experimental basis, with assistance from outside the rural communities themselves.

Solar Energy

Solar energy is widely used in Tanzania for drying cash crops, for the preservation of foodstuffs in general, and to dry clothes. Such traditional uses are likely to persist, though modern methods of harnessing and controlling solar power are being explored. Research into the fabrication and use of solar cookers has been initiated at the University of Dar-es-Salaam. Preliminary results have indicated the possibility of preparing some local dishes under specified environmental conditions. It is certain that even when prices and other related problems are resolved, it will take some time before solar cookers are disseminated for use in rural areas.

Experiments are being conducted on the use of solar stills to convert otherwise dirty water into drinking water. Many villages in Tanzania do not have adequate fresh water supplies. It is expected that when the experiments are completed, solar stills will find ready markets in the semi-arid regions which are already faced with the energy crisis.

As noted earlier, large quantities of fuelwood are used in the production of bricks, in fish smoking and tobacco curing. Much of this fuelwood can be saved if solar driers are used instead. Besides solar drying there are community services whose costs of energy supply can be minimized if solar energy were used. In Tanzania solar water-heaters have been installed in the staff houses of the Arusha-based Technical School. These are working efficiently. Many more could be installed in rural community centres.

Solar energy, though in plentiful supply in Tanzania, has not been effectively harnessed because of the technology involved and the costs of installing the appropriate appliances by individuals. Much of what is being done is only experimental. Attempts by the government and private organizations to diffuse the technology need to be boosted through a proper national energy policy.

Impact of the Rural Energy Crisis in Tanzania

Fuelwood remains the main energy source for rural communities in Tanzania. The depletion of these resources to meet domestic and other commercial demands has had a great impact on the country. The indiscriminate felling of forests and other arboreal resources has led to desert encroachment, especially in the semi-arid districts of central Tanzania. The distances walked to gather fuelwood are increasing and so are the distances for collecting water. In these areas, distances travelled to collect fuelwood are normally more than 5 km for each trip. If fuelwood is collected three times a week, this would involve about 30 km of travel. In Dodoma region, it has been established that 200 to 300 man-days (equivalent to T.Shs. 3,000–4,500) a year are spent in fuelwood collection (Mnzava 1979:14). In a study conducted in the Usambaras in northeast Tanzania, it was found that fuelwood collection takes about one-sixth of the time women spend in productive activity. In all these cases, if food preparation and fuelwood collection could be simplified, labour input into the critical area of subsistence production could be larger (Fleuret and Fleuret 1978).

The fuelwood shortage has not only affected productive activities but also caused forest encroachment on established forest reserves and hindered afforestation programmes. In areas where fuelwood was once available freely, it has now acquired a commercial value. In Tabora, for example, a bundle of firewood sells for about T.Shs. 60. It has been estimated that because of the current energy crisis a peasant in parts of central Tanzania may spend up to 20 per cent of his annual income on fuelwood resources alone. In time the burden may assume crisis proportions.

Developments in Research

Research on the energy crisis in Tanzania is not coordinated. The National Scientific Research Council has some research projects on the energy question. Most of these studies are preliminary and their findings have not been published. Other studies have been attempted by institutions such as the University of Dar-es-Salaam or by individual scholars.

Preliminary findings of the Rural Energy Consumption Survey indicate that trends in Tanzania are similar to those of many developing countries. For example, fuelwood is used for cooking almost exclusively in rural areas. Kerosene is the sole energy source for lighting in the rural areas and a few peri-urban areas. However, kerosene is expensive and not available in rural areas for most of the year. There is a tendency in parts of central Tanzania for people to turn to cow dung as an alternative to fuelwood, a situation which threatens agricultural production in these areas.

What this points to is that the fuelwood question remains an important element in the socio-economic organization of rural communities. As there is much more to be covered in the area of energy resources, more studies are being encouraged and a Rural Energy Centre has been established in Dodoma to coordinate these activities.

Discussion and Conclusions

The rural energy consumption pattern based on fuelwood is based on vegetation resources. These resources are fast disappearing and cannot be replenished at a rate equal to their rate of depletion. The best solution lies in the careful management and use of the available resources. This can be achieved in the first instance by using the available resources more efficiently. To attain greater efficiency in fuelwood utilization, cooking devices must be improved. Cooking in open fires wastes much of the fuelwood, as the efficiency attained is only about 6 per cent. In countries such as Indonesia, India, Paraguay, Haiti, Senegal, Kenya, and Zambia, improved wood cookers have been introduced and have worked more efficiently. Studies and experiments on improved wood cookers in Tanzania are being undertaken in Shinyanga region with the assistance of OXFAM staff. Studies on clay cookers instead of metal ones have indicated that clay cookers use up to 50 per cent less charcoal and therefore save time and money. They are also cheaper. If the improved cookers are adopted, fuelwood consumption would be cut down. Further savings on fuelwood can be achieved if fuelwood is obtained from specific tree species with good burning properties. The only constraint here is that such species are usually also in demand for their timber.

More control must be exercised on forests and forest reserves. Many forests in rural areas are felled indiscriminately because there are no regulations prohibiting their clearance. If such legislation is introduced, it should be in collaboration with village governments. Evidence from Nepal and other areas shows that legislation alone in a situation where there are no cheaper and easily available substitutes cannot work, but has to be accompanied by other conservation measures and extensive mass mobilization.

It has been noted that fuelwood competes with other wood requirements such as for house construction. Wood requirements for house construction can be reduced if the timber used is treated to give it longer life. Timber life can be increased four- or fivefold upon treatment. This would increase the life span of the rural houses and reduce the demand for wood. Forests would have more time to regenerate.

The supply can be boosted by afforestation measures. In Tanzania afforestation programmes go back as far as 1920 when the local government launched a programme for rural communities. Later in 1967/68, the central government launched a Village Afforestation Programme whereby the government raised tree seedlings specifically for fuelwood and poles and distributed them to villagers. One problem is that most of the seedlings are raised in regional or district headquarters and are expensive to transport to villages far away.

Another problem is that the villages lack the expertise and the discipline to tender the nurseries as required. In some cases, the village wood lots are gutted down by bush fires caused by hunters, livestock keepers, and even crooks. In Tabora region most of the village wood lots have been reduced to stubs by these bushfires (*Sunday News*, 21 Sept. 1980). In Kondoa district, the problem of soil erosion has exacerbated the fuelwood question. The seedlings that are raised and distributed to villages, schools, and other institutions do not as yet meet the demand for fuelwood. Marauding livestock are a menace to the tree-planting projects in the area.

The afforestation programmes often require sustained administrative efforts to protect the new plants. At present the country-wide afforestation programmes are accompanied by campaigns through the mass media. Recently, a campaign was launched by the Institute of Adult Education in cooperation with the Ministry of Natural Resources to raise the awareness of the people on the harm, damage, and consequences of deforestation and other land-use malpractices. It also seeks to educate the people on the ways of arresting the deterioration of arboreal resources especially, and on a more rational use and conservation of natural resources. The campaign is still in its embryonic stage, but there are bound to be serious administrative problems at the village level as no officials have been posted to villages.

As the fuelwood crisis continues, villages are being advised to plant tree species which can be used for a variety of purposes. These include *acacia albida*, *acacia mearnsii* (wattle), *anacardium occidentale* (cashew nut), and eucalyptus. *Leucaena lecucocephala* (the Hawaiian variety) is mainly used in central Tanzania for land reclamation.

The rural energy crisis is in fact a manifestation of mismanagement of available resources. The main concern here is the extent to which fuelwood collection for use in rural Africa has contributed towards the depletion of these resources. Many tropical countries including Tanzania are looking for solutions within their borders, but multinational corporations are also ready to tap this market. For example, charcoal cookers are designed and fabricated in developed countries and imported by developing countries. This tendency will certainly increase dependence on foreign countries. Even some of the alternatives that seem to work have to depend on foreign funding and the import inputs may be very great.

Since the impact of the energy crisis was felt first by the industrialized nations, solutions that are being sought are biased towards the industrial sector, hence isolating rural man even further. What it means is that the rural population may become even more impoverished as the natural resources are depleted or become unavailable to them.

REFERENCES

Adeyoju, S.K. 1965. "The Forest Resources of Nigeria". *Nigerian Geographical Journal* 8: 115–26.

Arnold, J.E.M. 1979. "Wood Energy and Rural Communities". *Natural Resources Forum* 3, no. 3:229–52.

———, and Jongma, J. 1978. "Fuelwood and Charcoal in Developing Countries". *Unasylva* 29, no. 118:2–9.

Darkoh, M.B.K. 1979. *Man and Desertification in Tropical Africa*. University of Dar-es-Salaam Inaugural Lecture Series.

Dunkerley, J. 1979. "Patterns of Energy Consumption by the Rural and Urban Poor in Developing Countries". *Natural Resources Forum* 3, no. 4:343–63.

Earl, D. 1975. "A Renewable Source of Fuel". *Unasylva* 27, no. 110:21–26.

Fleuret, P.C., and Fleuret, A.K. 1978. "Fuelwood Use in a Peasant Community: A Tanzanian Case Study". *Journal of Developing Areas* 12, no. 3:315–22.

Ibrahim, F.N. 1978. "Anthropogenic Causes of Desertification in Western Sudan". *GeoJournal* 2, no. 3:243–54.

Kesse, G.O. n.d. "Energy Resources of Ghana and the Programme for Their Development". Paper presented at the UNITAR Conference on Long-Term Energy Resources.

Lamb, R. 1978. "One Answer to African Energy Crisis". *Daily News*, 11 Oct. Tanganyika Standard Newspapers.

Miti, R.; Siamwiza, M.N.; Yamba, F.D.; and Chiyabwe, B.O.M. n.d. "A Review on Conventional and Alternative Energy Resources in Zambia". Mimeo.

Mnzava, E.M. 1979. *Village Afforestation: Lessons of Experience in Tanzania.* Dar-es-Salaam.

______. 1980. "Firewood Collection: Formidable Task for Women". *Daily News,* 10 July. Tangayika Standard Newspapers.

Morgan, W.B. 1978. "Development and the Fuelwood Situation in Nigeria". *Geo-Journal* 2, no. 5:437–42.

Myers, N. 1978. "Whose Hand on the Axe?" *Mazingira,* no. 6:66–73.

______. 1979. "Disappearing Forests: Draining the Earth's Genetic Resources". *Daily News,* 31 Oct. Tanganyika Standard Newspapers.

Njau, E.C. 1976. "Hydropower and Rural Electrification in East Africa". In Khamala, C.P.M., and Castellino, J.B. eds. *Energy Resources in East Africa.* Proceedings of the 12th Annual Symposium of the East African Academy.

Nkonoki, S.R., *et al.* 1978. "Energy Systems and Development in Tanzania". Paper tabled at the Symposium on African Goals and Aspirations in the United Nations Conference on Science and Technology for Development. Arusha, Tanzania.

______. 1980. "Prospects and Problems of Research and Planning Activities in the Energy Sector for Rural Transformation in Tanzania". Paper presented at the Conference on Energy for Rural Communities. Arusha, Tanzania.

Nyoike, P.M. 1972. "The Energy Situation in Kenya". Paper presented at the African Energy Group Meeting. Arusha, Tanzania.

Openshaw, K. 1976. "Woodfuel: A Time for Reassessment". Energy Resources in East Africa.

Quorro, J. 1978. "Disappearing World Forests, Tanzania's share". *Daily News,* 20 Mar. Tanganyika Standard Newspapers.

Tanzania National Scientific Research Council and United States National Academy of Sciences. 1977. *Workshop Report on Solar Energy Utilization in Tanzania.* Dar-es-Salaam.

Thipe, G.N., and Mokobi, F.K. 1979. "Current Status of Energy Needs and Field Experiences with Solar Technologies in Botswana". Paper presented at the African Energy Group Meeting. Arusha, Tanzania.

II

RENEWABLE RESOURCES

6
The Marine Fish Resources of Southeast Asia

A. HAMID ABDULLAH

The importance of marine fisheries in Southeast Asia varies from one country to another. The contribution of marine fisheries to the gross domestic product ranges from one per cent in Indonesia to about 18 per cent in the Philippines. Except for Indonesia, the level of national fish consumption is two or three times the world average of 11 kg per capita (Table 6.1). Thus a substantial proportion of marine fish landed is consumed by the local and/ or national population. The proportion exported ranges from 4 per cent in the Philippines to 21 per cent in Malaysia (SEAFDEC 1980). In terms of employment, fishermen make up about 2 to 3 per cent of the total national work force, and about 4 to 5 per cent of the total population depend directly on fisheries for their livelihood.

Fisheries in Southeast Asia vary widely in terms of specialization of employment and specialization of techniques. At one end of the scale, fishing forms part of a whole system of food-getting, which also includes hunting and collecting commonly practised by certain tribes. At the other end of the scale, fishing is modern in technology and highly specialized, employing large capital inputs. Similarly, at one end of the spectrum fishing is purely subsistence, using rudimentary gear such as single rod, line, spear, throwing net or traps to catch fish in waters very close to shores. At the other end, fishing is highly commercialized, using power-driven vessels, equipped with sophisticated and expensive gear such as radar, echo-sounders, and electronic "fish-finders" to capture fish in deeper and more distant waters (Hill 1979:102–3). Fisheries development programmes in Southeast Asia are changing fisheries from the purely subsistence activities towards the more commercialized, capital-intensive, and technologically advanced activities. As a result, fisheries have become more competitive, productive, and likely to become more damaging to the common property fishery resource.

In recent years, fishery authorities have become increasingly concerned about the size and future of the resources, and have been actively promoting fisheries development. Fishery development and management, which are two sides of the same coin, are beset by problems of overcapitalization,

TABLE 6.1 STATUS OF MARINE FISHERIES IN SELECTED COUNTRIES OF SOUTHEAST ASIA

Country	Total Population (millions)	Number of Fishermen	Landings (tonnes)	Consumption		Kg/Capita
				Total ('000 tonnes)		
				1974	1980	1974
Indonesia	141.6	831,965	1,227,386	1,324	1,706	10.1
Malaysia						
Peninsular	13.3	110,871	629,912	341	457	26.0
Sabah						38.6
Sarawak						39.2
Philippines	46.3	517,386	1,281,772	1,330	1,727	33.0
Thailand	45.1	76,000	1,837,807	869	1,219	22.0

Source: *Far Eastern Economic Review*, 1980; SEAFDEC 1978; and UNFAO 1977, Tables 5 and 6.

biological overfishing, and resource allocation (Marr 1976:39). These are common problems faced by fishery authorities throughout Southeast Asia. This paper examines the magnitude of the available marine fish resource, the recent development of the fishing industry, and the problems facing marine fishery resource management in the region. The availability of data for analysis will limit the study to four countries of Indonesia, Malaysia, the Philippines, and Thailand. The case of Peninsular Malaysia will be analysed in greater detail.

Marine Fish Resources

Types

Marine fishery resources are broadly categorized into pelagic and demersal, though some authorities say that there is no clear distinction between the two. A group of semi-pelagic species is also recognized. Pelagic fisheries are those based on species living near the surface or in mid-water, but generally not associated with the layer near the bottom. In general, pelagic species tend to migrate widely in the sea, and therefore their abundance, caused mainly by natural factors, is unstable. They are the free swimming types of marine fish. In this sense, it is quite a problem to assess the stock of pelagic resource at any one time.

The important pelagic species available in the Southeast Asian waters are the roundscads (*Decapterus* spp.), sardines, chub and Spanish mackerels (*Rastrelliger* spp., *Scomberomorus commersonii*), and anchovies. Other species include squid, cuttlefish, octopuses, tunas, sharks, and rays. Table 6.2 shows the proportions of pelagic and demersal fish landings in various countries of the region.

TABLE 6.2 THE LANDINGS OF MAJOR PELAGIC AND DEMERSAL FISHERIES IN SELECTED COUNTRIES OF SOUTHEAST ASIA, 1977

| Countries | Major Pelagic Fisheries | | Major Demersal Fisheries | | Total |
	(000 tonnes)	%	(000 tonnes)	%	(000 tonnes)
Indonesia	666	58	107	10	1,158
Malaysia	176	31	37	7	570
Philippines	918	75	179	15	1,231
Thailand	463	25	54	3	1,916

Note: Total includes many other marine, mainly non-fish products.
Source: UNFAO 1978, SEAFDEC 1980.

Pelagic fishes are caught by a wide variety of fishing gear including trawl net, while the demersal species are usually caught by trawl nets, bottom gill nets, long lines, and handlines. The main habitat of demersal fishes is the layer near the bottom. Their other major ecological characteristics include wide adaptability to environment, limited migratory range, smaller schools, slow rates of growth, and late maturity. Their stable habitat and rates of reproduction, unlike pelagic fishes, renders stock taking less problematic because their stock size tends to be influenced by the size of the territory they occupy in groups rather than in schools.

Some of the major demersal fishes available in Southeast Asian waters include the threadfin (*Nemipterus* spp. and *Polynemus* spp.), red snappers (*Lutianus* spp.), croakers and jew fish (*Sciaena* spp.), fusilier (*Cavesio erythrogaster*), pony fishes (*Leiognathus* spp.), and marine cat fishes (*Tachysurus* spp.). Horse mackerel (*Caranx* spp.), squids, shrimps, sharks, and rays are also included in the demersal catches of Southeast Asian marine fisheries.

Magnitude

It is extremely difficult to give an accurate assessment of the magnitude of both pelagic and demersal fish resources in Southeast Asian waters. However, a rough estimate can be obtained from information and data gathered from studies done by marine biologists and technologists in the various parts of Southeast Asia.

Elaborate stock assessment procedures need not be of concern here, but it should be pointed out that estimations of the magnitude of the marine fishery resources are based on the catch, effort, and biological data (the yield model). Such data are obtained from the actual fish landings and fishing effort of commercial fisheries as well as from research vessels invariably using trawl nets to catch both pelagic and demersal fish. Results of research into the magnitude of fish resources are usually expressed in terms of standing stock and potential yield/catch.

Information and data on the magnitude of demersal fish resources in Southeast Asia are more comprehensive and recent. Table 6.3 shows the size of standing stocks and the potential and present catches in the various areas of the South China Sea including the Straits of Malacca and the Java Sea.

Table 6.3 shows three significant features of the demersal fish resources of Southeast Asia: (1) variations in the resource and catch size between sub-areas, (2) similar variations between fishing zones within each sub-area, and (3) the sizeable differences between the present and potential catches of demersal fish in the various sub-areas of the South China Sea.

In terms of stock density and size of demersal fish, the richer areas appear to be the Straits of Malacca, the Gulf of Tonkin, the east coast of the south

TABLE 6.3 STOCK DENSITY AND SIZE, STANDING STOCKS AND POTENTIAL HARVEST OF DEMERSAL FISH IN THE SOUTH CHINA SEA AND ADJACENT WATERS

		Indices of Stock		Standing Stock			Harvest				Area (km²)
		Density kg/h	Size ('000 tonnes)	Present ('000 tonnes)	Maximum ('000 tonnes)	Optimum ('000 tonnes)	Potential ('000 tonnes)	Present ('000 tonnes)		Percentage Exploited	
Gulf of Tonkin	A	185	27,962	250	542	271					108,300
	B	175		99	91	45	424	260.2		61.4	45,300
Southern Mainland Shelf	A	120	8,661	19	50	25					12,600
	B	85		89	167	83					83,400
Northern Sunda Shelf	A	100	9,648	71	227	113		375.9			56,700
	B	95		42	71	35					35,300
Gulf of Thailand	A	35	21,826	78	895	447		604.9			179,000
	B	95		194	328	164		49.0			163,800
Central Sunda Shelf	A	160	25,894	99	198	99	1,618	30.6			49,500
	B	95		225	378	189			1,192.6	73.7	189,200
Southern Sunda Shelf	A	115	35,304	154	214	107		72.0			107,200
	B	80		287	287	143					287,200
Eastern Sunda Shelf	A	180	33,796	155	276	138		60.2			69,000
	B	115		256	366	183					178,400
Luzon Palawan Region[1]	A	180[2]	6,187	35	46	23	52				15,400
	B	115[2]		43	59	29					29,700

(con't. overleaf)

TABLE 6.3 (*con't.*)

		Indices of Stock		Standing Stock			Harvest				Area (km²)
		Density kg/h	Size ('000 tonnes)	Present ('000 tonnes)	Maximum ('000 tonnes)	Optimum ('000 tonnes)	Potential ('000 tonnes)	Present ('000 tonnes)		Percentage Exploited	
South China Sea[1]	A	100	9,440	64	64	32	59	0			51,200
	B	100		54	54	27					43,200
Sulu Seas & around Philippines[1]	A	180[2]		286	509	254	368	335.0[3]		79.8[3]	127,300
	B	115[2]		165	229	114					
Straits of Malacca[1]	A	250	48,830	298	478	239	400	428.8			95,500
	B	155		312	322	161					161,000
Java Sea	A	190	94,155	491			275	99.0		36.0	166,930
	B	209		938			469				298,750

Notes: Waters in various sub-areas are divided into zones A (up to 40 to 50 m deep), B (between 50 and 500 m) and C (deeper than 500 m).

[1] These sub-areas have more extensive waters deeper than 500 m.

[2] Rough estimates.

[3] For both Luzon Palawan region and Sulu Seas and around Philippines.

Source: Aoyama 1973; Buzeta *et al.* 1979.

Malay Peninsula, the coastal zones of Sarawak and Sabah, and the Sulu Seas. The poorer areas include the Luzon-Palawan region, the South China Sea Basin, and the southern mainland and the northern Sunda shelves (both off the coast of Vietnam). The Gulf of Thailand and the central Sunda Shelf (off the northeast coast of the Malay Peninsula) have considerable stocks of demersal fish.

There are substantial differences in the relative abundance of demersal fish between the inshore or coastal zones shallower than 40 to 50 m (zone A) and the offshore zones shallower than 500 m (zone B). In all sub-areas except the Gulf of Tonkin, the Gulf of Thailand, the northern Sunda Shelf, and the Luzon-Palawan region, the offshore zones are richer in demersal resources than the coastal zones. In the Tonkin Gulf, the northern Sunda Shelf, and the Luzon-Palawan region, the offshore resources have been underestimated owing to the lack of information. The case of Thailand is interesting in that the intensive coastal fisheries have also caught fish which are common to the offshore zone. The overall picture for the whole of the South China Sea region is that the offshore zones are richer than the coastal zones. The opportunity for future expansion of marine fisheries in the region certainly lies in the offshore zones deeper than 50 m because resources in those areas have been less exploited.

In the region as a whole, the present catch of about 2.5 million t is about 73 per cent of the total potential catch estimated to be about 3.5 million t. Fisheries in this region can thus support more production than at present. However, how much more production that can be supported varies from area to area. In most areas, a bigger catch of demersal fish can no longer be expected from the coastal zones, except those in the central, southern, and eastern parts of the Sunda Shelf. It also appears that fishing intensity in the South Kalimantan and Eastern Sumatra waters and the offshore areas of the Java Sea can be considerably increased.

A comparable study has also been carried out on the pelagic fishery resources of South China Sea. The findings are summarized in Table 6.4 in terms of the estimated sizes of the potential yields of pelagic fish. The relative abundance of pelagic fish varies from area to area with the richest being the waters off the Mekong Delta and the Sarawak Bay. Some of the other richer areas which have an abundance of zoo-plankton include the Gulf of Tonkin, the Gulf of Thailand, the east coast of Peninsular Malaysia, the Karimata Straits, and the west coast of Peninsular Malaysia.

Based on the estimated potential yield and the 1971 pelagic catches, several areas in the region show considerable potential for further development in pelagic fisheries. Notable among them are the less exploited fisheries of the east coast of Peninsular Malaysia, the Karimata Straits, the waters off the Mekong delta, and to a lesser extent the west coast of Peninsular

TABLE 6.4 PELAGIC FISH RESOURCES IN THE SOUTH CHINA SEA AND ADJACENT WATERS: ESTIMATED SIZES OF THE POTENTIAL YIELDS

Sub-Area	Area (000 km²)	Potential Yield		Landings (000 tonnes)	
		(000 tonnes)	tonne/km²	1971	1978
Gulf of Tonkin	131	164	1.25	100	837
East coast of Vietnam	113	45	0.4		
Off Mekong delta	236	566	2.4	188	
Gulf of Thailand	304	380	1.25	262	
East coast of Peninsular Malaysia	100	130	1.3	32	154
West coast of Peninsular Malaysia	335	300	0.9	124	411
Sarawak Bay	288	533	1.85	295	119
Karimata Straits	258	336	1.3		
East Sarawak & Sabah	98	64	0.65		
Sulu Sea	308	154	0.5	598	1,282
Philippine waters	693	450	0.65		
Central Basin	1,418	284	0.2	—	—

Note: 1978 landings taken from various sources.
Source: Menasveta, D. *et al*. 1973.

Malaysia and the Gulf of Tonkin. A preliminary assessment of the pelagic resource in the Java Sea also reveals a higher potential than is indicated by the present catches of pelagic fish in the area.

Structure and Development of the Marine Fishing Industry

In Southeast Asia the marine fishing industry engages a sizeable proportion of the labour force. The largely coastal population in the various countries have easy and ready access to the calm waters of the South China Sea and its adjacent seas (Fig. 6.1). Moreover, these seas are rich in fish resources, which have become the base for both the small-scale artisanal (traditional) and the larger-scale commercial (modern) fisheries of the region. The artisanal fisheries tap the inshore or coastal resources while the commercial fisheries have been extending their operations into deeper offshore waters as a result of recent development in fishing technologies, declining coastal resources, and the increasing local demand for fresh fish.

Fig. 6.1 Southeast Asia and the South China Sea

Nevertheless, small-scale artisanal or traditional fisheries still dominate the region. A very high proportion of fishermen engaged in coastal artisanal fisheries operate small vessels propelled by sail or outboard or small inboard engines. In Indonesia, for example, only about 9,000 (3 per cent) out of 295,000 fishing boats are power-driven. In Malaysia the proportion of power-driven boats is much higher, but most of the boats are still small. The small-scale fishermen use a variety of fishing gear such as hook and line, bag net, cast net, and beach seine to exploit the inshore fisheries. Although their productivity is generally low, their collective catches are substantial and make a significant contribution to the total catch in the various countries. For example, in the early 1970s the municipal (artisanal) fisheries of the Philippines contributed more than 50 per cent of the total production of fish in the country (Tiews 1976).

In Southeast Asia, the fishing industry is rapidly developing into a larger-scale, more modern, and commercial operation. Various government-sponsored programmes and private sector profit-motivated projects have led to an increase in the number of powered and larger boats, a wider use of large and modern gear, and an improvement and modernization of shore and service facilities. Indonesia, for example, witnessed an increase of 74 per cent in the number of powered boats between 1969 and 1973. In the Philippines, the gross tonnage of boats with high powered engines increased from 81,950 in 1968 to 99,554 in 1972. In the same period, the number of other trawlers and purse seiners increased from 653 to 690 and 202 to 319 respectively, while the number of bag nets (a traditional type of gear) decreased from 883 to 650.

The development of the Southeast Asian fishing fleets has resulted in the considerable increases in the nominal catches of marine products (Table 6.5). The development of the fishing fleet in Malaysia is almost similar to that of its neighbours. To handle the operation of larger and more powerful fishing vessels, and the increasing volume of catch they land, government programmes also include the development of fishing harbours, marketing complexes, and cold rooms.

Development of the Fishing Industry: Peninsular Malaysia

Since 1970 the objectives of the Malaysian government have been: (1) to promote an expansion of job opportunities in the fishing industry, (2) to increase the productivity and income of fishermen, (3) to manage and control the exploitation of fish resources with the view to optimizing the utilization of the resources, and (4) to restructure and consolidate the fishing community to assure the bona fide fishermen a fair share of the wealth of the sea (Second Malaysia Plan 1971–75).

TABLE 6.5 NOMINAL CATCHES OF FISH, CRUSTACEANS, MOLLUSCS, ETC. IN SELECTED
COUNTRIES OF SOUTHEAST ASIA, 1971–75

Country	1971	1972	1973	1974	1975	Percentage Increases 1971–75
			tonnes			
Indonesia	1,244,500	1,269,800	1,265,100	1,336,208	1,389,861	12
Malaysia						
Peninsular	326,100	314,500	374,400	432,652	376,690	16
Sabah	27,300	27,700	31,700	32,000	33,000	21
Sarawak	14,400	16,500	38,600	51,920	63,923	344
Philippines	1,043,700	1,125,700	1,244,700	1,291,396	1,341,636	29
Thailand	1,586,900	1,678,800	1,678,800	1,515,500	1,369,900	14

Source: UNFAO 1975.

To achieve those objectives the government through its Fisheries Division provides supporting services in various forms, namely: (1) research in the production and management of the fishery resources, (2) training of fishermen in modern fishing methods, (3) provision of basic physical infrastructure such as cold rooms, fishing harbours, and other ancillary services, (4) provision of subsidies and grants to fishermen, (5) control and regulation of fishing through the licensing of fishing boats and equipment, and (6) collection, compilation, and dissemination of statistics and other information related to fishing and fisheries development.

Active government participation in the fishing industry was made possible by the establishment of the Fisheries Development Authority (FDA) in 1971. The FDA has been directly involved in the production, processing, and marketing of fish with the cooperation and assistance of the Fisheries Division. The private sector has also contributed its share towards the development of the fishing industry.

With the active participation of the public and private sectors, the industry has recorded considerable progress in the modernization, mechanization, and commercialization of the hitherto small-scale artisanal fisheries during the period between the First Malaysia Plan (1971–75) and the first half of the Second Malaysia Plan (1976–80) (Tables 6.6 and 6.7).

Table 6.6 shows the significant features of the development of fishing fleet in Peninsular Malaysia. First, there was a 25 per cent increase in boats with inboard engines. There are indications that the inboard powered boats have been getting larger, heavier, more powerful, and more efficient in terms of the gear types used. The most notable is the number of increase in the inboard powered boats measuring more than 25 feet (7.6 m) in length, weighing between 15 and 45 tons (15.2 and 45.7 t), and fitted with engines in the 20–39 HP range and those over 100 HP. The increase is even more marked on the east coast where fishing has long been an artisanal and traditional occupation (Firth 1941; 1966). The larger, heavier, and more powerful fishing vessels are more seaworthy in the open and more extensive waters of South China Sea facing the east coast of Peninsular Malaysia. The fishing grounds on the west coast are limited to the calm and narrow waters of Malacca Straits. Outboard powered and non-powered boats are therefore more numerous on the west coast than on the east coast. In fact, in almost all aspects, fishing fleet development has been more marked on the east coast. Greater resource potential in the east coast might well be one of the major reasons for such a development in the 1970s.

One significant feature of fisheries development, especially on the east coast of Peninsular Malaysia, is the improved design of the fishing vessels. Before 1968 all fishing boats constructed in the country were of traditional design. The improved design takes the form of a 38 ton (38.6 t) vessel with a

TABLE 6.6 PENINSULAR MALAYSIA: DEVELOPMENT OF FISHING FLEET, 1971–78

| Boats and Gear | West Coast | | | East Coast | | | Peninsular Malaysia |
| | Number | | % +/− | Number | | % +/− | % +/− |
	1971	1978	1971–78	1971	1978	1971–78	1971–78
Licensed boat	14,216	19,700	+39	6,925	7,797	+13	+30
Powered boat	11,549	15,354	+33	4,771	6,105	+28	+32
Inboard powered boat	9,972	12,310	+24	4,312	5,465	+27	+25
Dimension: <25 ′	618	407	−35	593	574	−4	−19
25–54 ′	8,754	11,293	+29	3,667	4,651	+27	+29
55 ′ & over	600	610	+2	52	240	+362	+31
Tonnage: <5	3,584	4,380	+23	1,292	1,312	+2	+17
5–14.9	4,692	5,842	+25	2,623	3,106	+19	+23
15–44.9	1,523	1,997	+32	385	1,012	+163	+58
45 & over	173	91	−48	12	35	+192	−32
Engine: <9 HP	3,885	2,960	−24	2,226	2,023	−9	−18
10–19	1,428	1,834	+29	1,168	1,279	+10	+20
20–39	2,620	5,220	+100	501	1,325	+165	+110
40–99	883	513	−42	314	325	+4	−30
100 & over	1,156	1,783	+55	103	513	+398	+82
Gear types: Trawl nets	2,142	3,254	+52	509	884	+74	+56
Seine nets	1,661	1,473	−12	419	601	+44	−0.2
Gill nets	4,472	6,098	+37	906	1,019	+13	+32
Lines	491	515	+5	1,630	2,147	+32	+26
Traps	232	245	+6	299	377	+29	+17
Lift nets	248	21	−92	405	342	−16	−80
Outboard powered boats	1,577	3,044	+93	459	640	+40	+81
Non-powered boats	2,667	4,346	+63	2,154	1,692	−22	+25
Fishermen	43,397	54,091	+25	24,564	29,603	+21	+23

Source: Calculated from *The Annual Fisheries Statistics*, Fisheries Division, Ministry of Agriculture, Malaysia, 1971 and 1978.

TABLE 6.7 PENINSULAR MALAYSIA: LANDINGS OF MARINE FISH BY GEAR TYPES 1971–78

Gear Types	West Coast				East Coast				Peninsular Malaysia		
	1971	1978	% +/– 1971–78	% of Total 1978	1971	1978	% +/– 1971–78	% of Total 1978	1971 (000 tonnes)	1978 (000 tonnes)	% +/– 1971–78
	tonnes				tonnes						
Trawl nets	93,241	231,564	+149	57	18,930	48,874	+159	32	112	280	+150
Seine nets	77,403	67,195	–14	17	11,081	57,961	+423	38	88	125	+41
Gill/drift nets	12,952	19,662	+52	5	5,856	11,340	+94	7	19	31	+65
Lift nets	147	151	+3	0	15,523	10,092	–35	7	16	10	–35
Bag nets	20,630	19,544	–6	5	5,355	6,940	+30	5	26	26	+2
Other nets	2,350	2,242	–5	0	1,304	1,403	+8	1	4	4	0
Lines	6,379	4,336	–32	1	5,049	10,606	+110	7	11	15	+31
Portable traps (bubu)	2,936	871	–71	0	4,356	3,551	–19	2	7	4	–39
Stakes	4,210	3,410	–19	1	1,376	884	–36	1	6	4	–23
Shellfish	28,899	55,197	+91	14							
Total landings	249,147	404,172	+63	100	68,826	151,648	+121	100	318	556	+75

Source: Calculated from *The Annual Fisheries Statistics*, Fishery Division, Ministry of Agriculture, Malaysia, 1971 and 1978.

hold capacity of 5 tons (5.1 t), powered with a 125 HP unit and equipped with trawl winches. Such development is evident from Table 6.6, which shows the large increase in the number of vessels measuring longer than 54 feet (16 m), weighing more than 15 tons (15.2 t), and powered with over 100 HP engines. The number of such vessels using trawl nets increased by about 56 per cent between 1971 and 1978. The inboard powered boats using other gear types did not increase as much. The number of trawl nets in operation increased by 735 per cent, from 734 in 1969 to 5,392 in 1978 but that of seine and gill/drift nets remained fairly constant during the same period.

Technology in a primary industry such as fisheries plays an important part in that it serves to reduce the influence of physical conditions inhibiting the production process. Technology makes the exploitation of an otherwise unexploitable resource possible. A boat — its power, size, and gear — constitutes the basic fishing technology that may largely determine the productive capacity of the industry. A mechanized boat is more powerful, faster, larger, and able to operate larger gear and more load than a non-mechanized one. While the productive capacity of a mechanized boat is greater than that of a non-mechanized one, power, size and gear capacity of the former are the major determinants of its productive capacity. Analysis of the Malaysian fisheries statistics has shown that the development of fishing technology has undoubtedly led to an increase in fish landings (production). Table 6.7 shows that total landings of marine fish in Peninsular Malaysia increased by 75 per cent between 1971 and 1978. The east coast has, however, recorded a much higher rate of increase than the national (Peninsular Malaysia) rate, owing to the more recent modernization of fishing in what was previously a traditional fishing region.

It is also evident from Table 6.7 that trawlers have become the greatest single contributor to marine fish landings in the country. The proportion of fish landed by trawlers increased from 35 per cent in 1971 to about 50 per cent in 1978. During the same period, the amount of fish landed by trawlers in Peninsular Malaysia increased by about 150 per cent, with the east coast recording a slightly higher percentage than the west coast. Trawling grew rapidly in the peninsula after it was introduced in the mid-1960s. The trawling gear, designed explicitly for the capture of demersal or bottom fish (Munro and Chee 1978:18), is dragged over the sea bed by mechanically powered boats, collecting all fish in its path. The catch is greatly increased largely because the gear makes fishermen independent of the habits and morphological characteristics of the fish (Ackerman 1941:79). As a consequence although the number of inboard powered boats using trawl nets increased by 56 per cent, the total catch by trawl nets increased by 150 per cent between 1971 and 1978. Trawl nets are supposed to be an offshore gear, but trawling in Malaysian waters has come into direct competition with other major inshore gears such as the seine nets and the gill/drift nets.

A major fishing gear used in coastal fisheries is the seine net of which the purse seine is very important, especially on the east coast. In fact, because of the wide use of trawl nets on the west coast, the seine net catch dropped by as much as 14 per cent between 1971 and 1978. On the east coast, however, the contribution of seine nets to total landings is the highest, with their share of the catch increasing by 423 per cent since 1971.

It is necessary to go back briefly into the history of the use of seine nets to understand the reasons for the high increase in landings on the east coast. The most common seine net in Malaysia today is the non-indigenous purse seine locally known as *pukat jerut* (*pukat* means net and *jerut* means to tighten a slip-cord). This type of net was first brought to the west coast (Kuala Kedah and Pangkor) by a group of Chinese fishermen who migrated from Pakhoi in South China to Thailand and later to Malaya (now Peninsular Malaysia) at the end of the nineteenth century (Gopinath 1959: 75). The purse seine is an effective method of catching schooling fish such as chub mackerel and anchovies, usually found very near the surface of the sea. Owing to its effectiveness in catching small pelagic fish and the progress in mechanization, the use of *pukat jerut* has spread throughout the peninsula. Kuala Kedah, Perlis, Pangkor island off Perak, and Mersing on the east coast of Johore are the major purse seine fishery centres in the country. This fishing gear was introduced to the east coast states of Kelantan and Trengganu in the late 1950s or early 1960s.

The Malay fishermen of the east coast already had their own versions of seine nets. The boat seine (*pukat payang*) closely resembles the modern purse seine, while the beach seine (*pukat tarek*; *taret* means to pull or haul) is used to fish close to the beach. After surrounding the fish, *pukat tarek* is hauled onto the beach, while the *pukat payang* operated from a special boat known as *perahu payang* some distance away from the shore is hauled onto the boat. These nets were important to the Malay fishing community because of their large capacities and also the number of men required to operate them. For example, about 18 men were required to operate a large *pukat payang* of about 200 fathoms long and 4 fathoms deep. *Pukat tarek*, the same size as a large *pukat payang*, used to be hauled onto the beach by the beach party, which could include an unlimited number of men and even women and children.

However, the superiority and effectiveness of *pukat jerut* soon put many of the traditional *pukat payang* nets out of use, especially when more boats were being motorized. *Pukat payang* could not operate effectively with a motor boat because the motor noise frightens off some of the ground-feeding types of fish usually taken by this net (Firth 1966:307). The modern purse seine is used to catch pelagic fishes, which are less frightened by motor noise. Likewise, the beach-hauled seine, a very important Malay fishing gear until the late 1950s, has now declined in numbers and importance. The shallow inshore waters in which this gear is operated has probably been overfished, leaving the density of reasonably sized fish too low to make such fishing worthwhile. However, it was the lack of capital required to meet the greater costs associated with purse seine fishing that limited the use of the gear during the 1960s in spite of its superiority and effectiveness (Elliston 1971:2). Therefore, when more government subsidies, loans, and grants as well as private capital were made available to fishermen in the 1970s the purse seine became a dominant fishing gear on the east coast. However, if trawling continues to increase in popularity, it may one day reduce the important purse seine's status as a "leader in mechanization and in innovation in the fishing industry" (Elliston 1971:1).

Another inshore fishing gear of some national importance is gill or drift netting. Although the gear contributed about 6 per cent to the total catch in 1978, landings from this gear increased by 65 per cent between 1971 and 1978. It consists of nets set out to drift in the currents and, hence, unlike the trawl and purse seine which are encirclement types of gear, it is an entanglement gear. The nets can either be set to float near the surface or be weighted down to the sea bed. They can thus be used to catch both pelagic and demersal fish, although Fisheries Division statistics suggest that this form of fishing is more a pelagic than demersal undertaking (see Table 6.8). An important advantage of this gear to small-scale fishermen is that it requires only a small capital investment, and hence a considerable proportion of smaller motorized boats operate this gear in the Malaysian coastal waters. However, like the purse seine, drift netting is also facing direct competition from trawling.

While other lesser gears are losing importance, there is one that merits discussion. This is the line gear whose catch not only contributes very little to the fishery in the west coast but has also been decreasing. On this east coast, however, this gear contributes about 7 per cent of the total catch. Landings increased from about 5,000 t in 1971 to about 11,000 t in 1978. Long lines with baited hooks operated from single boats are very popular among east coast fishermen. These boats can fish in shallow waters or in deep waters, and may remain at sea for several days. Although it is not necessary to operate lines from powered boats, about 40 per cent of the powered boats in the east coast are used to catch larger pelagic fish such as banito, little tuna, Spanish mackerel, and the demersal variety of red snapper (Hamid 1979:43). Because this gear is more mobile and largely used to fish in deep waters around many offshore islands, it can probably better resist the competition from trawling.

Effects of Development

The preceding analysis of the development of the fishing fleet and catch data has amply demonstrated that fishing efforts have increased to a con-

TABLE 6.8 PENINSULAR MALAYSIA: PERCENTAGE OF TOTAL LANDINGS OF MARINE FISH BY GEAR GROUPS AND TYPES OF FISH, 1978

Gear Groups	Types of Fish (ISSCAAP Classification)									
	24	33	34	35	36	37	38	39	45	57
Trawl nets	0.3	12.8	4.3	1.7	—	5.3	2.2	47.9	17.4	5.5
Seine nets	—	1.1	36.8	23.6	4.3	19.5	—	7.9	6.5	—
Gill/drift nets	5.2	5.4	16.9	11.0	9.0	20.3	2.9	13.3	13.5	—
Lift nets	—	0.6	75.5	14.4	0.3	5.7	—	2.9	—	—
Bag nets	—	0.5	—	—	—	—	0.8	29.5	69.1	—
Hooks and lines	—	32.9	17.5	0.4	26.1	12.1	7.5	2.5	—	1.0
Traps	0.3	41.5	11.9	1.5	—	2.1	0.7	20.1	18.1	3.1
All gear groups	0.5	15.5	14.9	8.0	2.4	9.6	1.7	31.7	16.6	3.3

Note:
24. Shad, milkfish, etc.
33. Redfish, bass, conger, etc.
34. Jack, mullet, saury, etc.
35. Herring, sardine, anchovy, etc.
36. Tuna, banito, billfish, etc.
37. Mackerel, snoek, cutlassfish, etc.
38. Shark, ray, chimaena etc.
39. Miscellaneous marine fishes (consisting largely of trash fish)
45. Shrimp, prawn, etc.
57. Squid, cuttlefish, octopus, etc.

Source: Calculated from *Annual Fisheries Statistics, 1978*, Ministry of Agriculture, Malaysia.

siderable extent. It is quite premature, however, to state categorically that they have reached a level which will lead to the overexploitation of the available resource. While some of the traditional gears have fallen into disuse, other newly introduced gear has resulted in greater catches so that the total landings of marine fish in Peninsular Malaysia have increased considerably over the years. However, on the west coast the landings of certain groups of fish and the yield per unit effort of certain gear types seem to have declined (Table 6.9). Serious declines in catches are recorded in the chub mackerel (*kembong/Rastrelliger* spp.) and the anchovy (*bilis/Anchoviella* spp.) fisheries. This is an indication of biological over-exploitation (Khoo 1976: 41). Catches of other species such as those belonging to Groups 33 (mostly demersal) and 34 have also decreased. Such decreases may be partly related to the expansion of trawling because trawl nets are more appropriate to capture bottom living fish such as those belonging to Group 33. But if reference is made to Table 6.7 and Table 6.9, it is clear that the expansion of trawling has led to a significant increase in the capture of miscellaneous marine fish, of which the largest proportion is made up of trash fish. Trash fish not only fetches a very low price but also contributes little to the food

TABLE 6.9 PENINSULAR MALAYSIA: LANDINGS OF MARINE FISH BY GROUPS OF FISH, 1971 AND 1978

Fish Groups	West Coast		East Coast	
	1971	1978	1971	1978
	tonnes		tonnes	
24. Shad, milkfish, etc.	1,180	2,449	35	234
31. Flounder, halibut, sole	524	2,979	11	295
33. Redfish, bass, conger, etc.	13,036	31,754	8,510	16,828
34. Jack, mullet, saury, etc.	13,494	30,909	16,348	44,352
35. Herring, sardine, anchovy, etc.	28,717	24,687	7,900	15,558
36. Tuna, bonito, billfish, etc.	1,765	3,139	2,555	9,148
37. Mackerel, snoek, cutlassfish, etc.	38,082	30,327	7,648	18,127
38. Shark, ray, chimaera, etc.	2,632	5,241	697	3,277
39. Miscellaneous marine fishes	66,824	129,303	20,658	29,985
45. Shrimp, prawn, etc.	50,760	72,506	2,559	10,630
57. Squid, cuttlefish, octopus, etc.	1,719	11,589	1,894	4,856

Source: *Annual Fisheries Statistics, 1971 and 1978*, Ministry of Agriculture, Malaysia.

economy of the country. The increased catch of trash (manure) fish is a consequence of overfishing. It is very wasteful and may contribute towards the extinction of some valuable edible fish such as white pomfret. Although there has been a similar increase in the catch of trash fish on the east coast, overfishing does not seem to have occurred there yet because the catches of other fish types have also increased.

Another indication of overfishing is the decline in catch per unit effort as fishing intensity increases (Khoo 1976:40). Table 6.10 shows no overall decline in catch per unit effort in Peninsular Malaysia. There are indications, however, that the catch per unit effort for most gears on the west coast is beginning to decline. The peak catch per unit effort has been reached, and therefore further increases are highly unlikely. The trend on the east coast is somewhat different in that catches per unit effort for major gears such as trawlers, seine, and bag nets have increased since 1971 despite some minor fluctuations in certain years. Catches per unit effort of traditional gears such as drift nets and hooks and lines have decreased steadily since 1974, indicating that their fisheries are being exploited by other more effective and larger gears. Under these circumstances, therefore, resource and fishery management is necessary to perpetuate the fish stock on the east coast.

Resource and Fishery Management

According to J.C. Marr (1976), there is a clear distinction between resource management and fishery management. While the former is purely a conservation effort, the latter is an attempt to maximize economic yield from fisheries. He writes:

> (Resource management) is management by any means either to ensure the perpetuity of the stock . . . or to take advantage of some biological property of the stock, that is, attempting to maximize the productivity of a stock by limiting the catch to the level of the maximum sustainable yield. Resource management must be accomplished perforce through regulating or managing a fishery, but the orientation is still toward the properties of the resource. . . . Fishery management, on the other hand, is oriented toward users of the fishery and is management by any means to achieve some economic, social, or political objectives, for example, limitation of entry in order to maximize economic yield, allocation between competing or conflicting user groups, etc.

In the past, most management has been resource management which took little account of the national, economic, social, and political objectives. Recently, however, with the increase in fishing effort the need for fishery management has been generally realized. Fishery policies limiting entry and stipulating fishing zones for the inshore and offshore gear have become

TABLE 6.10 PENINSULAR MALAYSIA: ESTIMATES OF CATCH PER UNIT EFFORT OF MARINE FISHERIES BY MAJOR GEAR TYPES (tonnes)

Year	West Coast					East Coast				
	Trawl	Seine	Drift	Bag	Lines	Trawl	Seine	Drift	Bag	Lines
1971	28.7	53.0	3.5	30.7	5.4	18.6	54.1	4.5	15.4	3.4
1972	23.0	28.5	3.1	47.6	4.3	12.6	90.1	6.6	12.3	4.7
1973	37.9	43.8	2.7	48.5	5.2	37.6	87.8	9.4	11.1	7.8
1974	38.5	37.9	3.0	39.1	5.4	52.1	136.7	16.3	14.0	8.4
1975	37.4	24.4	2.7	33.2	4.0	38.4	109.9	13.1	13.0	9.9
1976	44.0	31.3	2.7	27.3	5.1	42.0	114.8	11.7	19.9	9.3
1977	53.1	43.3	2.7	35.3	5.1	34.6	135.0	11.6	13.9	11.1
1978	51.9	48.2	2.8	29.5	5.1	52.6	185.8	12.6	23.2	7.2

Source: *Annual Fisheries Statistics, 1971–78*, Ministry of Agriculture, Malaysia.

more common in some Southeast Asian countries. Nonetheless, policy implementation faces many problems.

The maximum sustained yield (MSY) is the annual catch that is equal to the natural rate of increase in the fish population (bio-mass). According to the logistic model of population dynamics, a fishery in a given aquatic environment would grow to a certain maximum size if there was no fishing. At this level, the gross addition to the stock of fish is balanced off by loss due to predation and mortality. If for some reason the bio-mass was reduced below the maximum size, the natural rate of increase would become positive so that the fishery would always rebuild itself to its natural equilibrium level, that is, its maximum size. The rate of natural increase depends on the population size so that for every population level there exists a maximum sustainable yield equal to the natural rate of increase. If the yield is greater than the natural rate of increase owing to the intensification of fishing effort (thereby reducing the bio-mass), biological overfishing is said to occur. If, on the other hand, the yield is smaller than the natural rate of increase owing to lower level of fishing effort, thereby allowing the bio-mass to gradually grow to its maximum size, the fishery is being under-exploited. To many marine biologists the MSY is the goal of fishery management, in which case the level of fishing effort should be limited to attain the MSY. Such an effort constitutes resource management rather than fishery management, because the former does not take into account the economic, social, and political objectives with regard to the fishing industry.

From a purely economic point of view, fishing effort should be at a level where rent from the fishery is maximized, that is, the positive difference between total revenue and total cost is the largest. The level of effort is not necessarily equal to the level of effort which brings in maximum sustainable yield from the view of a marine biologist. Any extra cost to raise that level of fishing effort without bringing the correspondence extra revenue would reduce the benefits to society from fishery. That which constitutes economic overfishing does not necessarily mean biological overfishing. In this sense, economic overfishing occurs when there has been an overallocation of resources such as capital and labour to the fishery. Since fish is a common property resource, any rent generated from the fishery will always appear as additional returns to labour and capital. If fishery policy also takes into consideration the social and political objectives, economic and biological overfishing is bound to occur because of the competitive nature of the industry. Regulation of entry into fishery is always difficult to enforce.

Public Fishery Policies

Since fish is an important source of protein and acceptable to the major religions of Southeast Asia, the underlying objective in public fishery policy in the region is to increase fish production (Robinson 1976:33). Invariably, this policy objective is to be achieved by developing the national fishery resources. Such development is commonly designed not only to contribute to the growth of the national economy but also to improve the socio-economic conditions of fishermen, particularly those engaged in the small-scale traditional inshore fishing. In Peninsular Malaysia, for example, the public policy regarding fishery since 1950 has been to make more and better quality fish available to the people and to raise the level of fishermen's income (Malaya 1956:1). The promotion of offshore (or deep-sea) fishing has recently been added to the list of policy objectives, probably because of the general awareness that much of the inshore resources have been exploited. Coupled with this new objective is the felt need to manage and control the exploitation of fishery resources so that optimum utilization of the resources may be achieved.

The main concern here is with the reasons for such policies rather than with their success or failure. Besides being great fish consumers, Southeast Asian countries have easy access to the rich fishing grounds in the seas bordering them. These tropical seas are the habitat for about 2,500 fish species (Marr 1976:1) which grow fast and are short-lived. Such species can stand higher fishing intensities than those in the temperate and sub-Arctic regions (Kasahara 1976:27). It is only natural that given such conditions the coastal population of these countries continue to fish for their livelihood. A variety of traditional fishing gear has been developed to catch fish of different species and sizes. The development of the present wide range of fishing techniques and equipment reflects not only a wide range of habitats and a wide range of species exploited whose habitats are well known to fishermen, but also a long history of exploitation (Hill 1979:102; Firth 1941). However, the large-scale commercial exploitation of fishery resources by local fishermen in this region was limited until very recently.

In many Southeast Asian countries, the initial expansion of fisheries was partly due to the lack of a rigid institutional framework to control or regulate access to fishing. As domestic demand for fish increased (partly because of population increase and partly because of general improvements in living standards), the investment of private capital in fisheries increased. Such investment has made possible the mechanization of fishing boats and the introduction of more sophisticated and expensive fishing gears such as trawl nets and purse seines. As a result, fish landings in most countries of Southeast Asia have increased substantially. However, this rapid expansion of fisheries has at least two serious consequences relevant to fishery policies in this region. One is socio-economic in nature. In Peninsular Malaysia, for example, the technological revolution in fisheries (Firth 1966:305) that has led to the rapid expansion of fisheries has resulted in the displacement of traditional fishermen who lack the capital to finance technological progress.

It has also led to the division of fishermen into modern commercial and traditional subsistence or owner operators and non-owner operators (Hamid 1979:49). The other consequence is the constant threat of fishery resource depletion, especially in the coastal waters. It is already evident that certain fisheries such as sardine and mackerel have been overexploited in several parts of Southeast Asia.

Invariably, these two issues have led governments to intervene in fishery operations by public policies that tend, on the one hand, to favour small-scale inshore fishermen and, on the other, to control and regulate the fishing industry. The programme to improve the socio-economic conditions of small-scale fishermen through grants, loans, and subsidies has resulted in the proliferation of the number of larger mechanized fishing vessels and more effective fishing gear such as trawl nets and purse seines. The development of landing jetties, fishing ports, and the improvement of fish marketing and processing as part of the fishery development package have intensified fishing activities and made the industry more competitive. This rise in the level of fishing effort may have resulted in biological overfishing, depending on the size and the natural rate of growth of the fishery resource. For certain fisheries in the Gulf of Thailand, the Straits of Malacca, and the Manila Bay, the current level of fishing effort has already resulted in both biological and economic overfishing because the catch rate of certain gears has declined and certain groups of fishermen have found it no longer worth their while to go fishing because of poor returns relative to their capital investment.

While fisheries development is actively being promoted, governments have taken certain measures to control and regulate the traditional free access to fisheries. The licensing of fishing boats and gears is an example of such measures. Although governments collect revenues from these licences, licensing is not an effective control measure because there are always many more boats and gears in operation than those licensed. In Peninsular Malaysia, for example, there were 4,504 licensed trawlers in 1978, but those in operation were estimated to be about 5,392. In the same year, the Malaysian government collected about half a million dollars in licence fees.

Another control measure is the stipulation of territorial limits to trawlers. But trawlers, legal or illegal, have constantly violated the three or five-mile (5 to 8 km) limit from the coast, thereby disturbing or threatening the inshore fisheries. Trawlers may deplete fish stock if they are not effectively controlled and regulated because, in Malaysia for example, about half of their catch is made up of trash (small inedible) fish. In spite of the fact that trawlers constitute a real threat to the fish resources of Southeast Asia, trawling is dominant in many parts of the region.

To explain why trawlers were allowed to operate in this region in the first place, it is necessary to turn again to the Malaysian case. At the time of their (trawlers') inception, it was officially argued that marine fisheries in Malaysia were directed primarily towards the exploitation of pelagic stocks (UNFAO 1968:2ff). It was believed that a large unexploited, or at least underexploited, demersal fish resource existed then, and that trawlers were capable of harvesting demersal fish (Munro and Chee 1978:27). To ensure that they would not encroach on inshore fisheries, trawlers were subjected to various regulations. For example, trawlers were restricted to fishing in waters 12 miles (19 km) from the coast and in waters of not less than 15 fathoms deep.[1] There has, however, been a gradual relaxation of the original regulations because of their ineffectiveness. In the early 1970s, trawlers of 25 to 99 tons (25.4 to 100 t) having engines of 60 HP or more (large trawlers) were not allowed to fish in waters less than 7 miles (11 km) from shore, and small trawlers (< 25 tons with < 60 HP engines) were forbidden to fish in waters less than 3 miles (5 km) from shore.[2] The latest regulation with regard to territorial limits, which came into effect on 1 January 1980, forbids trawlers weighing more than 40 tons (41 t) to fish within 12 miles (19 km) from shore, and the smaller trawlers are not allowed within 5 miles (8 km) from shore.[3] These new trawling limits are designed to protect inshore fishermen and to preserve the resources of the sea.[4]

Problems of Resource and Fishery Management

Issuing fishing licences and limiting trawlers to waters some distance from shore constitute a common effort to manage both fisheries and their resources. It involves the allocation or distribution of the catch between groups of fishermen or gear groups at the national and local levels. At the international level, the southern Sunda Shelf fishery resources which lie primarily in the extended coastal zones of Indonesia and Malaysia are also being exploited quite freely by fishermen from Taiwan, Thailand, and Singapore who have historically been fishing in the area (Marr 1976:42). While allocating fishery resources to their nationals is already a problem, Indonesia and Malaysia also face difficulties in exercising jurisdiction over their resources, which are being actively tapped by foreign fishermen, usually operating in large trawlers.

Previous accounts have indicated that most national governments in the region have greatly expanded their fisheries through either encouragement

---

[1]The basic legislation under which regulations are established is the Fisheries Act of 1963. Section 2 of this Act gives the Minister of Agriculture wide powers pertaining to the establishment of regulations with respect to the administration of marine fisheries. These initial regulations were formally codified in the Fisheries (Maritime) Regulations, 1967, P.U. 49/1967, Fifth Schedule.

[2]Fisheries (Maritime) (Amendment) Regulations, 1971, P.U. A8/1971, Fifth Schedule.

[3]Fisheries (Maritime) Amendment Act, 1980.

[4]Press statement made by the Minister of Agriculture (*New Straits Times*), 18 Feb. 1981.

of private capital investment or/and their own development programmes. The consequence of fishery expansion is succinctly described by J.C. Marr (1976:39–40):

> When a fishery first starts, it is usually very profitable. This causes the entry of additional boats into the fishery, either by new construction or by movement from another fishery. The yield of any fishery resource is finite; as the number of boats entering the fishery increases, the total catch is divided among more and more boats and individual profits decrease. For a while it will continue to be profitable for the new entrants, even though their profits cause reduced earning by those boats already in the fishery. Eventually the point is reached at which income just covers operating costs; beyond this number of boats there will be a net loss, and presumably the less efficient boats will be forced out of the fishery.

In the case of Peninsular Malaysia, such an expansion has forced out some of the traditional gears whose catch rates have been declining. This over-capitalization has primarily been due to the entry of too many trawlers (some of them illegal) into the coastal fisheries.

Overcapitalization is one of several problems of fishery development and management. It is a direct result of unlimited entry into fisheries. Once allowed it is almost impossible to control (or cut back) unless and until the yield is too low to make fishing a profitable enterprise. But declining yields will seriously threaten the supply of a vital source of protein. However, fish production is still rising and, in the case of Malaysia, the increase has mainly been due to trawling. Instead of de-licensing trawlers, the Malaysian authorities have tried (without much success) to limit trawling to the stipulated distances from the shore. Trawlers, however, have kept coming closer to shore because the fishermen claim that the inshore waters are rich in resources. This encroachment has led to conflicts between the inshore fishermen using traditional gears and the trawlers which tend to destroy the potential catch of the former. Conflicts between the inshore and the off-shore fishermen, and between the present and future fishermen, lie at the heart of the fisheries management problem, and possibilities of such conflicts are as numerous in Southeast Asia as elsewhere in the world (Robinson 1976:293).

Another problem facing fishery management is the lack of accurate and up-to-date knowledge of the size and nature of the fishery resource. To establish and enforce fishery regulations so that the level of fishing effort permitted does not result in greater than the maximum sustainable yield, it is necessary to determine whether a resource is capable of producing the MSY. Although it is possible to determine the MSY based on the catch and effort data alone (Marr 1976:41), knowledge of the size and nature of the fish stock in a given area is necessary to manage fisheries on a species-by-species basis. Surveys have been carried out to assess the stocks of both pelagic and demersal resources in the South China Sea and the adjacent waters, but the multiplicity of species in the region poses difficulties in fishery management. For example, trawling was initially permitted to exploit demersal species, but it has also captured substantial quantities of pelagic species.

The fact that the same fishery resources may be the basis for the development plans of more than one country adds another dimension to the problems of fishery management in South China Sea Area. For example, the development of pelagic resources has been one of the needs of almost every South China Sea country. National and local fishing zones might have been delineated for the purpose, but pelagic fish is known to migrate widely. Technologically more advanced fishermen are able to pursue a pelagic fishery at a substantial distance from their own ports. These fishermen may be fishing in their neighbours' or foreign waters. For example, Thai trawlers are often sighted (and sometimes confiscated) by Malaysian authorities in Malaysian waters, especially on the northeastern coast. Such a situation not only causes international conflict or at least embarrassment but also makes it difficult for national governments to exercise controls to limit over-capitalization and overfishing in their territorial waters. Related to this problem of controlling entry into fisheries is the desire of each South China Sea country to extend its economic zone to 200 miles (322 km). For example, Malaysia will soon put into effect a licensing policy for the 200-mile exclusive economic zone for modern, commercial deep-sea fishing.[5] If Indonesia takes the same action, the fishing grounds for Singapore and Thailand fishermen will be substantially reduced. Such an allocation of fishery resources may not auger well for international relations among these countries which have claims on this common property resource.

Oil and industrial pollution in the coastal waters of Southeast Asia constitutes another set of problems facing fishery resource management in the region.[6] The Straits of Malacca, which is not only one of the busiest waterways in the world but also one of the main fishing grounds for Peninsular Malaysia, has been the scene of several major oil spills since 1975. It is estimated that about 21,000 tons of oil were spilled in Malaysian waters between 1975 and 1979 (Tan 1979). Besides, marine pollution surveys carried out in various parts of the Straits of Malacca recorded very high coliform (E. Coli.) counts ranging from 1800/100 ml in the coastal waters of Penang island to 18,000/100 ml in the Straits of Johore. In the east coast of Peninsular Malaysia, about 57 per cent of the beaches stretching from Kota Bahru to Mersing have been contaminated by oil residue in the form

---

[5]Press statement by the Minister of Agriculture, Malaysia, *New Straits Times*, 18 Feb. 1981.

[6]Pesticides, suspended solids, oil, detergents, sewage, and heavy metals are some of the major pollutants affecting the marine and inshore aquatic environments.

of tar. Such contamination is largely due to the increase in activities related to oil exploration, rigging, and production off the east coast of the Peninsula (Maheswaran 1979).

Crude and refined oils affect marine organisms, although their concentrations do not result immediately in deaths (Pham 1979). These effects include chronic toxicity interference with fish feeding and reproduction, abnormal growth and behaviour, and susceptibility to predation. These effects would normally lead not only to changes in abundance and distribution of individual species but also to shifts in species composition within the oil-affected area. Nevertheless, many factors either individually or in combination govern the effects that an oil spill may have on marine life. Generally, the biological damage is more severe if the spill occurs in a coastal or estuarine environment, especially if the intertidal zone is affected, than if it occurs in the open ocean. In the case of Malaysia, there has been no study on the possible and actual impact of marine pollution, especially oil, on the income of coastal communities in the form of reduced catches and contaminated fish.

Conclusion

The seas in Southeast Asia are undoubtedly rich in fish. A variety of fisheries has developed in the region. Recent expansion of these fisheries has led to overcapitalization and wasteful exploitation of certain resources or species. Trawling in the inshore waters is a case in point. It has rendered traditional inshore gears which are species-specific obsolete. Trawling is productive, but unselective and wasteful because it captures a substantial amount of small trash fish unfit for direct human consumption. Such fish should be left to replenish stocks for future exploitation. Regulations to control or limit wasteful fishing abound, but enforcement is difficult and ineffective. Fishery management must aim at making regulations work effectively so that the level of fishing effort allowed would only produce the maximum sustainable yield from all fisheries. Catch and effort data should constantly be made available for analysis to determine the optimum effort level. Surveys are necessary to indicate the size of fish stocks. This is vital for the management of fishery resources. Regulations to prevent and control marine pollution are also required to protect and conserve the marine resources.

REFERENCES

Ackerman, E.A. 1941. *New England's Fishing Industry*. Chicago.
Aoyama, T. 1973. *The Demersal Fish Stocks and Fisheries of South China Sea*. UNFAO, UNDP, Rome.

Buzeta, R.B.; Dwiponggo, A.; and Sujastani, T. 1979. *Report of the Workshop on Demersal and Pelagic Resources of the Java Sea*. Marine Fisheries Research Institute and SCSFDC, Manila.

Elliston, G.R. 1971. "A Survey of the Economics of the West Malaysian Pukat Jerut Industry in 1966". Mimeo. University of Hull.

Firth, R. 1941. *The Malay Fishermen: Their Peasant Economy*. London.

Gibbons, D.S. 1976. "Public Policy towards Fisheries Development in Peninsular Malaysia: A Critical Review Emphasizing Penang and Kedah". *Kajian Ekonomi Malaysia* 13, no. 182:89–120.

Gopinath, K. 1959. "The Malayan Purse Seine (Pukat Jerut) Fishery". *Journal of the Royal Asiatic Society Malayan Branch* 23, no. 3:75–96.

Hamid, A. 1979. "Physical and Technological Factors in the Development of Marine Fishing Industry in Peninsular Malaysia". *Ilmu Alam*, no. 8, pp. 27–51. Department of Geography, National University of Malaysia, Bangi.

Hill, R.D. 1979. "Tribal and Peasant Agriculture, Fisheries". In Hill (ed.), *Southeast Asia: A Systematic Geography*. Kuala Lumpur: Oxford University Press.

Kasahara, H. 1976. "Fishery Policies". In K. Tiews (ed.), *Fisheries Resources and Their Management in Southeast Asia*, pp. 17–22.

Khoo, K.H. 1976. "Optimal Utilization and Management of Fishery Resources". *Kajian Ekonomi Malaysia* 13, nos. 1 & 2:40–50. Faculty of Economics and Administration, University of Malaya, Kuala Lumpur.

Kurogane, K. 1977. "Some Considerations of Research and Study on Pelagic Fishery Resources". *Proceedings of the Technical Seminar on South China Sea Fisheries Resources*, Bangkok, May 1973, Japan International Cooperative Agency.

Latiff, M.S. 1976. "Demersal Fish Resource Surveys and Problems of Fisheries Resource Management". In K. Tiews (ed.), *Fisheries Resources and their Management in Southeast Asia*, pp. 224–47.

Liu Hsi-Chiang. 1977. "The Demersal Resources of the South China Sea". *Proceedings of the Technical Seminar on South China Sea Fisheries Resources*.

Maheswaran, A. 1979. "The Steps Being Taken by the Government to Safeguard the Living Resources of the Sea". Paper presented at the Third Annual Seminar on The Seas Must Live, Penang, Mar. 1979.

Malaya. 1956. *Report of the Committee to Investigate the Fishing Industry*. Govt. Printers, Kuala Lumpur.

Malaysia. *Annual Fisheries Statistics, 1971–1978*. Fisheries Division, Ministry of Agriculture, Kuala Lumpur.

Malaysia. 1971. *Second Malaysian Plan, 1971–1975*. Govt. Printers, Kuala Lumpur.

Marr, J.C. 1976. *Fishery and Resource Management in Southeast Asia*. Resources for the Future, Washington, D.C.

Menasveta, D., *et al.* 1973. *Pelagic Fishery Resources of the South China Sea and Prospects for Their Development*. Rome

Munro, G.R., and Chee, K.L. 1978. *The Economics of Fishing and the Developing World: A Malaysian Case Study*. Universiti Sains Malaysia, Penang.

Pathansali, D., *et al.* 1974. *Demersal Fish Resources in Malaysian Waters Fisheries*. Bulletin, No. 1. Ministry of Agriculture and Fisheries Malaysia, Kuala Lumpur.

Pham, N.K. 1979. "Oil Discharges: Chemical and Physical Effects". Paper presented at the National Seminar on the Protection of the Marine Environment and Related Ecosystems, Kuala Lumpur, June 1979.

Robinson, M.A. 1976. "Reconciling Conflicts among Economic Interest Groups in Southeast Asian Fisheries". In K. Tiews (ed.), *Fisheries Resources and their Management in Southeast Asia*.

______ . 1976. "Report on the IPFC/IOFC Symposium on the Economic and Social Aspects of National Fisheries Planning and Development". In K. Tiews (ed.), *Fisheries Resources and Their Management in Southeast Asia*.

SEAFDEC. 1980. *Fishery Statistical Bulletin for South China Sea Area, 1978*. Bangkok.

Tan, P. 1979. "Marine Pollution and Its Control: The Malaysian Experience". Paper presented at the National Seminar on the Protection of the Marine Environment and Related Ecosystems, Kuala Lumpur, June 1979.

Tiews, K. 1976. *Fisheries Resources and Their Management in Southeast Asia*. GFID, FRBF and UNFAO, Bonn.

UNFAO. 1968. *Report of the FAO/IBRD Fisheries Identification to Malaysia*. Rome

______. 1977. *The South China Sea Fisheries Report on Asian Fisheries Products: Development Potentials*. Manila.

______. 1978. *Yearbook of Fishery Statistics, 1977*, vol. 44. Rome.

7

The Fisheries Resources of Northern Zambia

J. DENNIS HUCKABAY

Zambia's fisheries provide Zambians with more animal protein than all other sources combined. The varied fisheries of this central African country encompass deep open waters, swamps, flood plains, and *dambos*.[1] They range from the second deepest lake in the world, Lake Tanganyika, in the north, to the world's third largest man-made lake, Kariba, in the south. They also include the open waters of Lake Mweru and the associated swamps and backwaters upstream along the Luapula River. The Mweru/Luapula fishery has for decades supported a large population and a busy trade in fish to Zambia's industrial heartland, the Copperbelt. The vast Bangweulu Swamp has also been an important traditional and commercial source of fish. Remote and shallow Mweru Wantipa, whose name means "lake of mud", is Zambia's primary fishery.

Flood plain fisheries, such as those of the Upper Zambezi River and on the Kafue Flats, hold great promise for increased production, especially from the Kafue River. In addition to these major fisheries, many small rivers, streams, and *dambos* provide fish to Zambia's widely scattered rural population. These subsistence fisheries may never reach commercial importance, but they are the only source of animal protein for many people.

Zambia's most important fisheries are in the north. Tanganyika, Mweru/Luapula, Mweru Wantipa, and Bangweulu supplied two-thirds of Zambia's estimated commercial catch in 1978 (Muyanga 1980b). These fisheries, like all those in Zambia, are now suffering increasing pressure on their fish resources. Zambia's population growth rate of more than 3 per cent a year is one of the highest in the world. In less than twenty-five years its population will be doubled. Many more people, plus rising standards of expectation and consumption, have already begun to strain the ability of the fisheries to

---

[1] *Dambo* is the name given in Zambia to the seasonally water-logged, grassy, and often treeless depressions which frequently occur in the upper basins and headwaters of streams on the gently undulating plateau surfaces which characterize such a large part of the country's landscape.

supply the fish demanded. In 1964, at the time of Zambia's independence from colonial rule, fish production from the main fisheries was estimated at 30,723 t. The catch has increased substantially since then (Fig. 7.1). For the past several years, Zambia has produced more than 50,000 t annually. In 1978, the latest year for which fisheries statistics are readily available, unusually heavy rains and rising waters inundated many fishing camps and limited production to an estimated 43,260 t.

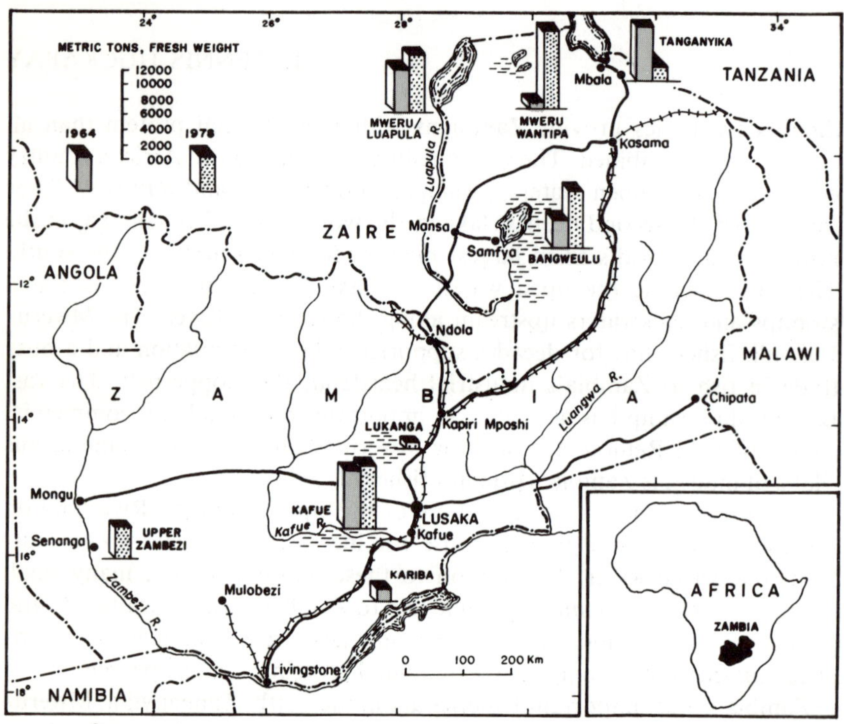

Fig. 7.1 *Zambian fisheries production 1964 and 1978*
Source: Department of Fisheries Annual Reports

The increased catch, however, has not kept pace with demand. In 1972 it was estimated that 90,000 t FWE[2] of fish would be required by 1985 just to maintain annual per capita fish consumption, which was then reckoned to be 14 kg per person per year, with consumption very unevenly distributed throughout Zambia (Christensen 1972). Writing in 1978, Muyanga and Chipungu estimated the country's annual demand for fish and fish products

---

[2]FWE stands for fresh weight equivalent, for most of the fish eaten in Zambia is sun-dried prior to sale and shipment.

in 1976 to have been 79,000 t. They estimated an annual increase in demand of 2,616 t/year, and foresaw that demand would reach 94,700 t by 1982. In 1978 estimated production was little more than half that, as Table 7.1 shows.

Muncy estimated in 1972 that

> if the country is to produce sufficient fish without relying on importation, then production would have to be increased by a minimum of about 20,000 t per annum or more than 50 per cent immediately. In order to meet the needs of the growing population the increase must be close to 5 per cent per annum to maintain this level. If such a scheme could be implemented over a ten-year period it would need an increase in the annual production of about 8 per cent; this compares with the growth rate of 3 per cent over the last decade. An enormous effort in fishing as well as in marketing would be necessary but this certainly would be worthwhile.

TABLE 7.1 FISH PRODUCTION, 1978

Fishery	Production (£)	
Mweru Wantipa	10,351	
Bangweulu	8,941	
Kafue	8,634	
Mweru/Luapula	7,628	
Upper Zambezi	4,474	
Tanganyika	2,171	(poor weather and an outbreak of cholera drove 1978 fish catches down to about one-fourth of its recent production)
Kariba	n.a.	(no data collected because of the security situation along Rhodesian frontier)
Lukanga	785	
Lusiwasi	276	
Total	43,260	

This has not occurred. Demand has far outstripped supply despite the increase in production from Zambia's fisheries. By 1979, the Director of Fisheries in a newspaper interview estimated production to be about 40 per cent short of national requirements (*Sunday Times*, 25 Mar. 1979).

The potential production of Zambia's fisheries should be much higher than present levels of exploitation. In 1968, the United Nations Food and

Agricultural Organization (FAO) estimated maximum sustained production to be 59,000 t. Estimates of actual production in recent years (except for the low 1978 fish catch) have been around 55,000 t, indicating that the FAO seriously underestimated the potential production from Zambian fisheries. Welcomme, in a general survey of the inland waters of Africa, estimated potential yields from Zambian waters to be 108,920 t, but warned that the "present state of knowledge . . . (is) imprecise and may therefore be subject to revision after further studies" (1972:1). Zambian fisheries production statistics, both actual and potential, are not likely to be reliable. Perhaps the most that can be said is that yields could be substantially increased if the fishing industry were completely reorganized. This would require more intensive management to protect fish stocks, more efficient fishing gear, better storage, transport and marketing facilities, and a fairer price structure. Christensen, writing in 1972 when production from Zambia's major fisheries was about 38,000 t, provides a useful summary:

> the yield from the present fisheries can be substantially increased, (but) it is quite unrealistic to assume that their combined yield could be more than doubled without intensive research and management. Such intensive control must, in practice, be considered prohibitively expensive in relation to expected returns in fish production. In order to meet the long-range goals of fish production in Zambia, a large expansion of aquaculture appears to be necessary.

It is beyond the scope of this essay to discuss the development of Zambian fish farming. The Fisheries Department has devoted considerable research and effort to this; but so far the result in terms of the quantities of fish produced has been disappointing. The potential for doubling production from Zambian fisheries may well be there, but under present levels of management many of the fisheries are already in trouble. The Mweru/Luapula and Bangweulu fisheries have generally suffered declining catches over the years, especially of favoured species. On Lake Tanganyika, commercial purse-seining for sardines has led to the virtual disappearance of fish which prey on them. Mweru Wantipa, the most productive fishery, is inherently unstable and susceptible to sudden collapse and, it should be added, rapid recovery. Bangweulu is not productive in relation to its great expanse because of the lack of oxygen in its swamp waters.

The situation in each of the northern fisheries of Zambia is reviewed below.

Lake Mweru and the Lower Luapula

Lake Mweru and the lower Luapula River below Mambalima Falls have been one of the most important fisheries for many years. They have also been a valuable source of food for Zaire, for the river forms part of the boundary between Zambia and Zaire and the two countries share the waters of Lake Mweru.

It used to be the most important fishery in central Africa. It attracted a constant stream of migrants in the pre-colonial past, when fish was an important item of trade and tribute (Musambacime 1975). When the railway to the Copperbelt and Katanga was completed in 1910, copper mining began to develop rapidly, requiring large inputs of labour and foodstuffs. Rinderpest and tsetse restricted the supply of meat, so that the mining companies began to explore the possibility of substituting fish for meat in miners' rations. A fishing industry was established in the 1920s, which provided fish for the mine workers of Katanga in the Belgian Congo, as it then was. An ice-plant was built at Kasenga, which permitted fish to be transported without spoilage to Elisabethville and other mining centres in the Congo. Greeks were encouraged to settle along the Belgian side of the Luapula, and they not only fished but also bought and transported fish to the Congo mines. "By 1924, a line of large villages had been built along the banks of the Luapula river to utilize the . . . marketing facilities offered by the Greeks" (Musambacime 1975).

The northern Rhodesian authorities were content to let the fish trade flow to Katanga, for their own growing mining centres were largely supplied with fish from Bangweulu until about 1948 (Musambacime 1975). On the northern Rhodesian side of the Mweru/Luapula fishery, fishing was reserved for Africans, who sold their catches to Greek trading boats. Africans became increasingly involved in the trading and transporting of fish. The establishment of an ice-plant at Kashikishi in 1954 greatly encouraged trade on the northern Rhodesian side by freeing traders from the need to carry ice up from the Copperbelt. Increasing demand, the availability of ice, and improved roads and transport led to the opening and growth of many large fishing camps alongside Lake Mweru (Soulsby 1959). This trade with the Copperbelt has continued until today.

In 1978, Mweru/Luapula contributed 18 per cent (7,628 t) of Zambia's total estimated catch. In Luapula, fish is by far the most important source of animal protein. Per capita consumption of fish there is four times the national average; more than half the catch each year is consumed locally rather than exported to the Copperbelt (Muncy 1973).

Below Mambalima Falls, the Luapula River forms an extensive swamp about 160 km in length and up to 18 km wide when flooded during the rains (Fig. 7.2). This swamp system with its many oxbows and lagoons is interlinked with the open waters of Lake Mweru, for many lake species spawn in the shallow, quiet swamp waters of the river.

Fishing in the Mweru/Luapula fishery is carried out in the tributary rivers, in swamps, in and around the shoreline vegetation, and in the open

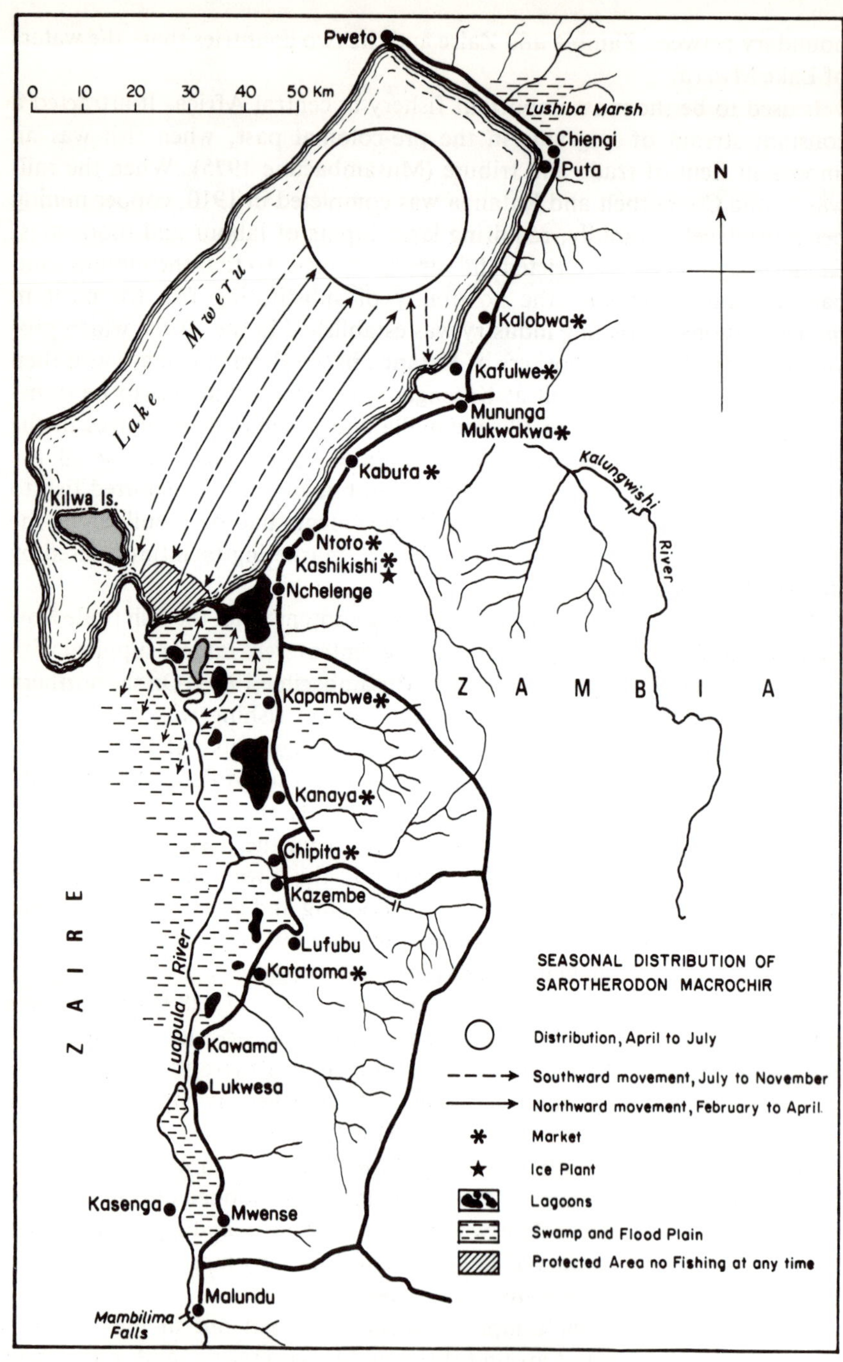

Fig. 7.2 *The Mweru-Luapula fishery*
Source: based on Mortimer, 1965, p. 68

water of the lake. Gill nets are the most common method of fishing, being set overnight on the bottom. This is the main type of commercial fishing, and is carried out in the open lake. A variant of gill netting, called *kutumpula*, is commonly used in the swamps at the mouths of the Luapula and Kalungwishi rivers (Carey 1965). Nets of small mesh-size (50 and 75 mm) are set around vegetation in shallow water. The fish are then driven out of the vegetation into the nets by thumping the water with poles (Fig. 7.3). Nets are moved at short intervals and the process repeated (Mortimer 1965b). As will be seen below, this method of fishing has had some very harmful effects.

Fig. 7.3 Kutumpula, a variant of gill-netting in swamps

Another type of gill netting practised on the rivers which flow into Lake Mweru is known as *kasenswa*. Nets are set across the river when certain species are migrating upstream to spawn. The nets are allowed to drift for short distances downstream, and then retrieved. The main species caught in this manner are catfish such as *Synodontis spp.* (*bongwe* or squaker), *Chrysichthys mabusi* (*monde*), and *Auchenoglanis occidentalis* (*mbwa*), mormyrids or snoutfishes such as *Cyphomyrus discorhynchus* (*chipumamabwe*), and *Gnathonemus spp.* (*kalimulimo, lusa,* and *ndomodomo*), and the tiger fish *Hydrocynus vittatus* (*manda* or *mcheni*). *Kasenswa* fishing, too, can be a destructive practice if, as is all too often the case, migrating fish are taken before they have had a chance to breed (Carey 1965; Jackson 1961). A particular form of *kasenswa* fishing is used in the Kalungwishi River to catch *Alestes macropthalmus* (*misebele*) when the first rains appear:

> The *Alestes* of Lake Mweru run up the Kalungwishi river in November, travelling as far as the Muaunga rock barrier where they spawn as close to the rocks as possible. As each wave returns downstream local fishermen using staked nets across the river take tremendous toll. *Kusowelesha* fishing, as it is called, is carried out amidst great excitement and is a magnificent spectacle. Fishermen who stand in the river supporting stakes of the net are often knocked over by the flying *Alestes* as they attempt to leap the barrier. (Soulsby 1960)

In addition to gill netting, traps and weirs are used in the tributary rivers at times of low water and in swamps when the Luapula floods. The highly prized *Sarotherodon* (formerly *Tilapia*) *macrochir* (*pale*) is caught this way, as well as in nets, as are *Barbus spp.*, *Petrocephalus simus* (*chise*), *Ctenopoma spp.* (*akome*), and *Clarias spp.* (*barbels: mita, milonge,* and *akabukula*). Long lines are also used, mainly in the two main rivers and in the south end of the lake, to catch *Clarias spp.*; *Synodontis spp.*; *Chrysichthys mabusi*; *Auchenoglanis occidentalis*; and *Hydrocynus vittatus* (Carey 1965).

As the lengthy list of commercially valuable species indicates, Mweru/ Luapula is a well-balanced fishery, not overly dependent on one or a few species. Out of 94 species present (Welcomme 1972), 27 are of commercial importance. There is also reasonably good fishing throughout the year, with different parts of the fishery — the Luapula River, the shallow southern part of the lake, and the northern deep end — having peak or maximum catches at different times of the year. Two commercially important species, *Serrahochromis stappersii* (*makobo*) and *Tylochromis mylodon* (*tembwa*), breed during the dry season and another, *Alestes macroptalmus* (*misebele*), breeds late in the dry season (October) as well as during the rains. The breeding season of most other species is concentrated in the rain and flood season from November to March. However, the *makobo* is not a popular fish with traders and customers, who maintain that "the body is too thin and that the flesh lacks natural oil and therefore lacks flavour" (Jackson 1961). Similarly, the abundant *misebele* is unpopular in fresh form, but can be marketed after drying (Clarke 1972). *Tembwa*, on the other hand, is an important as well as tasty species, particularly in September and October in the northern part of the lake.

The lake is also hydrologically quite stable, since the Luapula marshes have a dampening effect on the major source of flood waters by extending the period over which they are released into the lake, and the outlet, the Luvua River, is fairly large (Williams 1972).

Therefore, the lake is stable and the fishery less seasonal than others in Zambia, with fishing pressure spread over a number of species. None the less, it is important to emphasize that desirable species have greatly declined or even disappeared from the waters of Lake Mweru. The first and most serious loss (because it has never recovered) is that of *Labeo altivelis*, the famous and beautiful *mpumbu* or Luapula salmon. The *mpumbu* has a very short breeding season, lasting only a few days each year, preceded by a mass migration upstream when the rivers are first swollen by the rainy season floods:

> The habit of mass migration . . . makes the adult fish very vulnerable to attack at a critical phase of their life history, and the story of the Pumbu in the lower Luapula is a sad one, of insufficient management leading in a few short years to the ruin of a valuable natural resource. From time immemorial the mass migration here was enormous, and justly famous, known as the *kapata*, when densely packed shoals of the species used to move shoulder to shoulder, slowly up the river. In the immediate post-war years, however, local fishermen on both the Congo and Rhodesian sides of the river developed a large export fishery to the copper mines for the first time, and the brunt of the attack fell first upon the Pumbu. Gravid fish were killed in thousands during the spawning run and caviare made from the ripe eggs in the ovaries was a popular delicacy in the Belgian Congo. Nearly half of the total catch of Mweru/Luapula fishery was made up of this species. Stocks were decimated and the conservation measures that were at last applied came too late to protect them. The last *kapata* of any size occurred in 1949 or 1950 and the fish no longer is of much commercial significance. (Jackson 1961:53)

In 1948, *mpumbu* constituted 55.6 per cent of the fish catch purchased by Union Miniere for the Congo mines (Matagne 1950). Prior to 1950 it was always well over 40 per cent of the catch, but by 1960 it had dropped to little over 2 per cent (Soulsby 1960). By 1970 the recorded catch was less than 0.5 per cent (Central Statistical Office 1971), and in 1971 fewer than ten *mpumbu* were observed in total catches (Muncy 1973).

More recently, *Sarotherodon macrochir*, the highly desired bream known in Lake Mweru as *pale*, has suffered a similar drastic decline. As recently as 1965, *pale* had formed as much as 75 per cent of the catch from Mweru/Luapula: four years later, this had dropped to 10 per cent (Clarke 1972). By 1975, the commercial landings of *pale* had "paled into insignificance" (Mabaye 1977). Jackson (1961) has described the fish as one "which is a national asset and of the greatest importance to our country". Although it is immensely prolific, spawning five or six times a year, and under pond conditions producing 4,000 to 5,000 developed fry per female per year (Jackson 1961), *pale* catches have declined considerably in Lake Mweru, largely because of uncontrolled and illegal fishing. *Pale* is caught legally in deep-water lake using bottom set gill-nets. However, there is a pronounced migration of *pale* from the north end of the lake to the south, near the mouth of the Luapula River, from July to October. By September, much of the breeding population has moved up the Luapula, and as soon as the river floods its banks, *pale* move out with the flood waters into surrounding swamps and spawn in shallow water. At this time of year, *kutumpula* fishing, where the fish are driven out of the swamp vegetation into nets, becomes very harmful for large numbers of young *pale* and *Tylochromis mylodon* (*tembwa*) female carrying eggs and fry in their mouths are caught (Carey 1965).

In the colonial era, a closed season of two months was enforced from late January to late March, during which industrial fishing was prohibited and no export of fish to the Copperbelt was allowed (Soulsby 1959). In addition, the swamps at the mouth of the Luapula, known as *mifimbo*, were closed to fishing throughout the year. Therefore, the main breeding grounds were protected and fishing pressure during the breeding season was controlled. Mesh size throughout the Mweru/Luapula fishery was controlled at a minimum of 10 cm (Soulsby 1959). These measures "saved the fishery from what appeared to be a certain collapse at the time" (Mabaye 1973). Recently, however,

> enforcement of the fishing regulations has been minimal. . . . Fishermen were increasingly using 50 mm nets and moving into permanent fishing camps in the Mifimbo restricted fishing area. In the southern end of the lake (which probably harbours the bulk of the juvenile *Tilapia* (*Sarotherodon*) unrestricted use of the *kutumpula* method of driving fish into gill nets resulted in very high catches of young *T. macrochir*. This technique was used in the Mifimbo area, which came under heavy fishing pressure from Zambian and Zairean fishermen (Clarke 1974).

With the decline of *pale*, an effort has been made to find suitable alternatives. *Alestes macropthalmus* (*misebele*) is probably the most abundant single species in the lake, but it is underexploited, except during its spawning run up the Kalungwishi, because it is chiefly found in the deeper, offshore waters. Fishermen rarely venture more than 10 km from shore in their dug-out canoes and plank boats; yet the lake is as much as 51 km wide. Since, as noted earlier, this fish is unpopular in its fresh form, an FAO fish-processing consultant was called in 1970 to test its acceptability to the consumer in dried form. As it is more effectively caught in top set gill nets of 50 mm mesh size, it was suggested that the ban on the use of such nets on Mweru be lifted to enable fishermen to exploit *A. macropthalmus* (Clarke 1972). Since the soft-bodied fish quickly deteriorates if left in nets for any length of time (overnight, for example), short netting periods along with faster open-water transport (and/or ice) would be necessary; in other words, an industrial-type fishery (Muncy 1973), such as that which was practised by the Greeks from Kasenga.

Another species traditionally underexploited is *Tilapia rendalli* (*chituku*). Not ordinarily caught in gill nets, experimental seine netting has caught large quantities of *T. rendalli* (Muncy 1973). Seine netting, however, is illegal in the Zambian portion of Lake Mweru because it destroys the nests and breeding females of the important cichlid species.

The disappearance of desirable species such as *mpumbu* and *pale* and the steady decline in catch per effort (from about 10 kg/100 m of net in 1959 to less than 5 kg/100 m by 1973) indicate that the Mweru/Luapula fishery is reaching the point of maximum sustained yield (Clarke 1974). Enough is known about the fishery to protect and sustain it (and bring back the population of *pale*), but that knowledge has not been forcefully applied.

In 1977 fifteen fishermen fishing illegally in the Mifimbo breeding area were arrested and fined. Two hundred nets were confiscated. Catches of *pale* were reported to have increased in 1977 and 1978 (Muyanga 1980a; 1980b). This indicates how quickly this highly-sought after species can breed and replenish an area if existing regulations are enforced. What is needed now is the elimination of poaching in the Mifimbo prohibited fishing area, and the creation and enforcement of new regulations concerning the breeding seasons of the other important fish in Lake Mweru (Clarke 1974).

Lake Tanganyika

Enormous in size and depth, surrounded by precipitous cliffs and mountains, with an interesting geological history and an unbelievably rich endemic fauna, Lake Tanganyika has, as Beadle (1974) suggests, an immense aesthetic and scientific attraction (Fig. 7.4). Of 214 fish species in the lake, 176 or 82 per cent are endemic — they are found there and nowhere else.

Lake Tanganyika supports an important commercial fishery. In normal years, Zambia derives almost 15 per cent of its total fish catch from its toehold on the foot of Lake Tanganyika.[3] In 1976, 7,866 t were estimated to have been taken from the Zambian waters of the lake. Perhaps one-eighth of the annual catch is consumed locally; the rest is sun-dried or frozen and shipped to the distant urban centres on the Copperbelt and along the line-of-rail, 1,000 km or more away.

Emphasizing Lake Tanganyika's unique character, the major product of its fishery is not a member of the cichlid family (including *Sarotherodon* and *Tilapia*) which forms such an important part of most African lake, swamp, flood plain fisheries, including those of Zambia; rather, the major components of the Lake Tanganyika fishery are the two small (up to about 9 cm), endemic genera and species of herring — *Stolothrissa tanganicae* and *Limnothrissa miodon*, known in Zambia as *kapenta*, *dagaa*, *nshembe*, or sardines. The *kapenta* fishery is worthy of attention not only for its com-

---

[3] Fishing activity in the years since 1977 has been disrupted by adverse weather and the imposition of a quarantine against cholera.

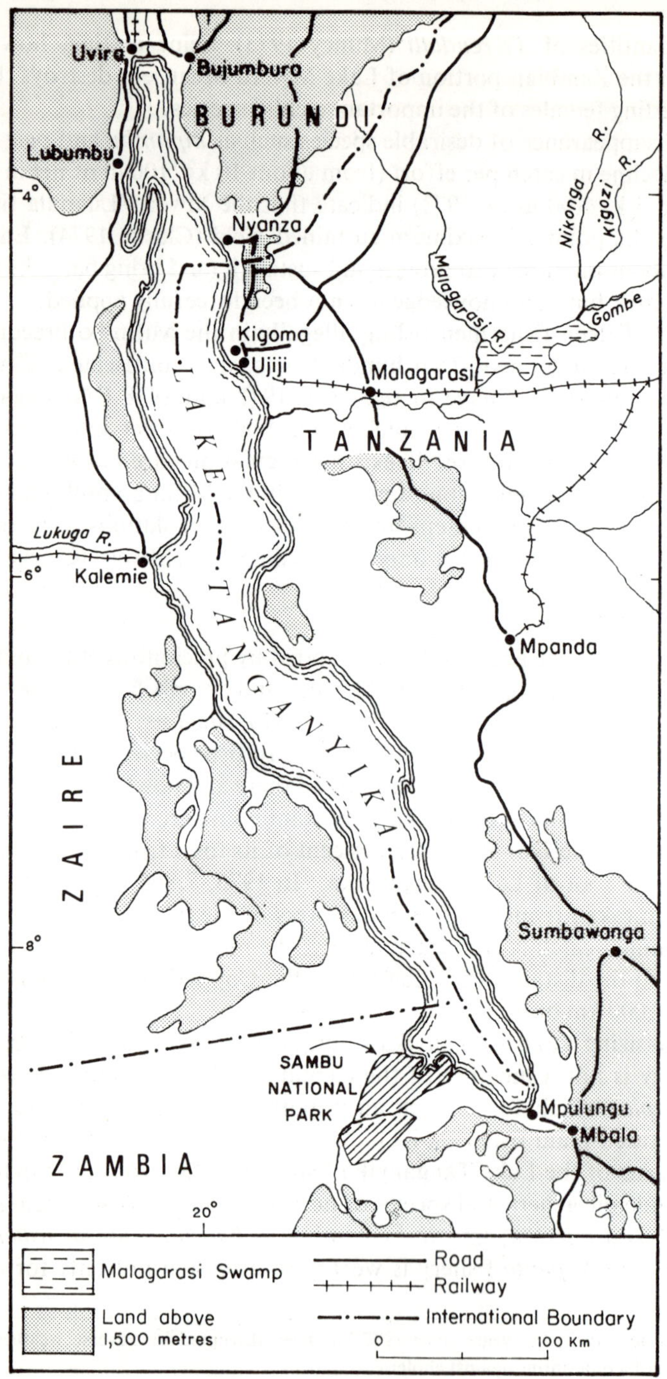

Fig. 7.4 Lake Tanganyika

mercial importance but also for its biogeographical features. The *kapenta* are members of the Clupeidae, or herring family, one of the premier commercial fish families of the world. They are distributed throughout the world and are everywhere valued for their oily, tasty flesh. They are almost exclusively marine, but they have invaded some of the larger tropical river systems, including the Zaire River, to which Lake Tanganyika has an intermittent connection via the Lukuga overflow.

The waters of Lake Tanganyika, 1,470 m at their greatest depth, are permanently stratified, with an epilimnion (the frequently windstirred water layer at the surface) 100 to 200 m deep, and a hypolimnion (the unstirred water below about 200 m) without oxygen and with little life apart from bacteria. Thus about three-quarters of the total volume of the lake is permanently anoxic. It has sometimes been assumed, as Beadle (1974) points out, that the pelagic region, or deep open waters, of such lakes is unproductive or oligotrophic, since so many nutrients must be lost to the anoxic abyssal levels of the lake and never recirculated to the epilimnion containing aquatic life. The abundance of Lake Tanganyika's sardines suggests, however, that the epilimnion of Lake Tanganyika is unusually productive. Strong southeast trade winds which blow from April to September up the long axis of the lake cause a flow of surface water to the north, so that considerable upwelling of nutrient-rich waters occurs, especially in the southern (Zambian) end of the lake. The same winds may also bring nutrients in the form of organic aerosols from the vast savanna woodlands of south central Africa which burn yearly during the dry season when these winds are strongest.

When the wind-stirred nutrients reach the sunlit surface zones, there is a rich blooming of phytoplankton, which in turn supports zooplankton in abundance. The zooplankton is the primary food source of the *kapenta* which have so successfully exploited this source of food that they have multiplied to prodigious numbers, and which in turn support an abundant flock of predators. Most important among these predators are the large centropomids (perch family), three endemic *Lates* species, and two *Luciolates* species, which feed almost entirely on the sardines.

All of the members of this pelagic community migrate diurnally. The zooplankton descends to a depth of 60 to 120 m where it spends the day, then migrates through about 40 m towards the surface at night. It is followed by the sardines and in turn by their predators (Lowe-McConnell 1975). The surface water, "which in mid-lake is wonderfully clear and seems to be devoid of life during the day, may towards evening become almost turbid and seething with massive shoals of sardine. The large predator fish are seen darting around them" (Beadle 1974).

Sardines are most abundant between June and November, when the surface levels are rich with stirred-up nutrients. During this time, small canoes with two or three men attract sardines on moonless nights by means of a small paraffin pressure lamp in the front of the boat. "Attracted by this light the sardines swarm around the canoe in such numbers that they appear to fill all the available space" (Beadle 1974). The canoe fishermen use large dip nets made of mosquito netting mounted on long poles, called *lusenga* nets (Fig. 7.5). With this method, a fisherman can ladle out an average of 200 kg of *dagaa* each night; "there is no other small-scale fishing method in Zambia that gives such a high catch per night of fishing for so little capital outlay" (Bowmaker and Coulter 1965). With the approach of dawn, the catch progressively declines as the zooplankton, the sardines, and the perch migrate downwards.

Fig. 7.5 Night fishing with lusenga nets

In addition to the small-scale *lusenga* net fishing there have been, since 1962, large-scale commercial ring net or purse seining operations. A string of small boats with powerful lights are towed into position by a tug in the evening to attract schools of fish. A net barge then sets a large (400 m long and 150 m deep), small-mesh ring net around each pair of lamp boats, and the shoals of sardine and the attendant perch are hauled in. In 1972, for the first time, such commercial fishing accounted for more than half of the recorded *kapenta* harvest (Clarke 1973). Catches are large, averaging 33 t per night fished. With such large catches, refrigeration is required, and both *dagaa* and perch are frozen and exported to the distant line-of-rail markets (Bowmaker and Coulter 1965).

Unlike the *lusenga* net fishing, this commercial purse seining is conducted throughout the year. It also catches perch and allows immature sardines to escape; *lusenga* net fishing does neither, resulting in the destruction of tens of millions of tiny fry (Jackson 1971). Against these advantages of purse seining is the high capital cost of boats and equipment, so that for the most part it has been undertaken by expatriates, especially Greeks. It requires considerable skill to set the nets correctly; and it requires a processing and refrigeration plant to cope with the large catches. The refrigeration facilities are run by commercial concerns at Mpulungu which, in addition to storing their own catch, buy *kapenta* from small-scale canoe fishermen at wholesale prices. This is helpful when they are unable to dry their fish and retail it. However, the long distance to markets and the high cost of refrigeration are such that fish must be obtained cheaply from fishermen if they are to compete in price on the markets with fish from other fisheries (Bowmaker and Coulter 1965). At times the price offered to fishermen has been so low that they have preferred to throw away their catch rather than sell it wholesale. In 1975, Lakes Fisheries offered only 5 ngwee per kilo to fishermen, a price they refused to accept (Mabaye 1977). A kilo of dried *kapenta* retailed in Lusaka then for about twenty times that price.

A consequence of commercial purse-seining has been the virtual disappearance of fish (*Lates* species, especially *L. microlepis* and *Luciolates stappersii*) which prey on the sardines. As mentioned above, purse seining catches perch as well as sardines. By 1966, four years after the commencement of purse seining, the large predatory centropomids, once abundant in the catches, had diminished markedly in numbers. This is not surprising since the large predators have a life cycle which is many times longer than the small, short-lived sardines, almost none of which lives more than a year. Obviously, the replacement rate is very different, so that if predator and prey are subjected to the same rate of exploitation, the numbers of the predators will quickly fall. This seems to be just what has happened in the southern end of the lake, repeating what had occurred earlier in the northern end, where the *Lates* fishery collapsed after about five years of intensive clupeid fishing (Coulter 1968).

At the same time, there has been a steady increase in the annual sardine catch. Regression studies suggest that "changes in fishing effort are not significantly correlated with clupeid abundance" (Muncy 1973). Rather, the increased *kapenta* catch suggests that the sardines have proliferated in those heavily fished areas of the lake where the perch which prey on them have been fished out.

Stages in the food chain of most fresh-water fish communities are characteristically few (Coulter 1965), and in Lake Tanganyika the transformation from plankton to predator is particularly short. The predator-prey interrelationship, density-dependent and naturally balanced, has now been disrupted. *Kapenta*, the mainstay of the Lake Tanganyika fishery, is prodigiously abundant. With the removal of control by its predators, its numbers are increasing, and neither *lusenga* fishing nor purse seining seem to have an effect on the harvest from one year to the next (Clarke 1973; Mabaye 1976). Nevertheless, the sardine fishery in Lake Tanganyika can experience year-to-year variations of considerable magnitude. The short life-span of the sardines, together with the variability of plankton production, makes the fishery inherently unstable. Certainly the experience of California and Peru indicates how problematic and unpredictable sardine fisheries are even when subject to the best available scientific scrutiny and assessment (Allsopp 1975).

The *kapenta* fishery is still underexploited, and it is quite possible that production from this fishery could be trebled if the fishermen were well trained and provided with good fishing equipment and marketing facilities (Muyanga 1980a). Nonetheless, the *kapenta* fishery will have to be carefully monitored and regulated to avoid a sudden collapse like that which occurred in the Peruvian anchovy fishery in the early 1970s.

In addition to the *kapenta* and perch fishery, Lake Tanganyika had a substantial bottom gill net fishery involving mostly cichlids in the benthic or sublittoral areas of the lake. These catches had reached such a level of intensity that by 1972 stocks at the southern end were reported to be fully exploited down to a level where oxygen is totally absent and life for aerobic organisms impossible (Muncy 1973). These benthic or bottom living stocks have proved to be very susceptible to even modest increases in fishing effort. By 1974, the "existing technology and fishing efforts (had) been sufficient to drive gill-net catch rates significantly below their levels of 10 years ago in all Zambian waters that have been exploited" (Mabaye 1976). The benthic areas near Mpulungu and Nsumbu have been fished out; experimental gill netting carried out there in 1975 by the Department of Fisheries yielded a highest average catch per unit effort of only 2 kg/100 m of net. In contrast, gill netting in Mweru/Luapula and Mweru Wantipa fisheries yielded 23 kg and 24 kg/100 m respectively (Mabaye 1977). Thus it is no surprise that many of the artisanal fishermen who cannot afford the ever-rising cost of nets, engines, and fuel to go offshore after *kapenta* have left Lake Tanganyika in recent years and gone to Mweru Wantipa. In areas of Lake Tanganyika remote from centres of population, presently lucrative benthic fish concentrations could be driven to sub-economic levels in a few years if gill netting intensifies (Mabaye 1976).

Another important type of fishing on Lake Tanganyika is beach seining. Unlike gill nets, where the fish swim into or are driven into stationary nets and become entangled, seine nets are carried out in a boat or canoe away from the shore and then pulled slowly and steadily back onto the beach. Here, too, there are signs of overfishing, at least near human settlements; catch per effort has declined steadily in recent years, as has the catch of desirable species, such as the large cichlids *Tilapia tanganicae*, *Boulengerochromis microlepis*, and *Tylochromis polylepis*, and the perch, *Lates* species. These have been replaced by catches of small cichlid species known collectively as *mutununu*, and taken only by seines. This heterogeneous component of the seine net harvest has increased from 4 to 60 per cent during 1967–73 (Clarke 1973). There is some fear as well that beach seining may also have reduced the supply of young *Lates* to the offshore fishery (Mabaye 1970).

There is also a very specialized type of fishing in Lake Tanganyika — the collection of live fish for export to tropical fish fanciers in Europe and North America. Since 1976, two small companies have flown live fish in insulated, oxygenated containers on the Zambian Airways tourist flights from Kasaba Bay to the Copperbelt and Lusaka, where they connect with international flights to Europe. Prompt flight connections are essential, and the risk is high that the fish will be left on the airport landing apron in the hot sun or freezing cold too long for them to survive.

Tropical fish enthusiasts will, however, pay a very high price for many of the more colourful, endemic (and therefore rare) species. The fish involved in this trade are mostly cichlids. Lake Tanganyika has more genera of cichlids than any other lake, with 33 genera and all 126 species endemic (Lowe-McConnell 1975). This rich diversity of cichlids is also of great biological and evolutionary interest. The Zambian Department of Fisheries charges a small levy on each live fish exported, but lacks the manpower and means to check the numbers of fish actually exported. The small scale of the trade at present may not justify stricter controls, but as Jackson (1973) has warned with regard to the tropical fish trade from Lake Malawi: "the value of the trade and the laws of supply and demand may, in future, lead to evasion of controls, so a potential threat looms for these beautiful creatures".

These, then, are the patterns which have developed in the fishery at the southern end of Lake Tanganyika. Around population centres, such as Mpulungu and Nsumbu, the artisanal fisheries — beach seining and gill netting — have already greatly reduced the numbers of the larger and more desirable species. Offshore, the perch have disappeared as the *kapenta* catches have steadily increased. *Kapenta*, with their brief life-span and rapid turnover of populations, are a fishery subject to rapid collapse if over-exploited, as has happened in the case of heavily fished marine sardine fisheries. The *kapenta* of Lake Tanganyika are not yet overfished, and present fishing intensities seem to have little effect on the sardines. Indeed they offer the greatest potential for fisheries development on the lake. There is room for many more *lusenga* fishermen in the pelagic waters closer to shore; they should be encouraged because "the requirements of a *lusenga* net unit operating in shallow water — a plank canoe, a pressure lamp, a few metres of mosquito gauze, and a few bush poles — are within the existing financial resources of a large number of fishermen (Mabaye 1976).

Bangweulu

Bangweulu is one of the great swamps of Africa (Fig. 7.6). Five shallow lakes up to about 7 m in depth are connected to a vast area of swamps to the east, the whole system covering about 9,850 km^2 during the rains (Welcomme 1972). A number of rivers drain into the basin from the north and east, chief of which is the Chambeshi, which forms the headwaters of the great Zaire river system.

Interspersed among the vast swamps are canals, channels, and lagoons that are used as waterways for canoes and boats. Some of the channels have been enlarged to allow for water transport; for example, the Chisale and Goodall channels were opened in the First World War to transport war material north towards German East Africa. Again, in the 1940s, considerable efforts were expended on opening navigable channels for motor launches, so that maize and other "dry land" requisites could be taken into the swamps in exchange for fish right through the low water season, September to December. During the main fishing season, from June to November, fishermen — whom Inoue estimated in 1971 to number 11,780, using 4,348 boats, 97 per cent of which were dugout canoes — move off the lake shores and dry islands and set up fishing camps in the swamps. These camps are made of papyrus and set on floating mats of swamp vegetation. Here the fish are caught and, because the swamp area is so large and access and communications to it are so poor, the fish are dried before being transported via Kapalala or Katanshya to the Copperbelt. There is no fresh fish trade in the Bangweulu area, because the distances are too great and access too difficult to transport and ice fresh fish or bring in the wood required to smoke them. Therefore, all the fish exported from Bangweulu is sun-dried.

As late as 1964, 40 per cent of all the dried fish from the swamps was transported to Mwamfuli (near Samfya) by the Bangweulu Water Transport Service (Tait 1965). Though this service still exists, it has been curtailed by lack of spares and inadequate servicing of boats and engines, and by the lack of maintenance of the cuts through the swamps. Its service now is largely confined to the open water of the lake to and from Samfya and the large sand bar islands Mbalala, Chisi, and Chilubi. Today fish traders

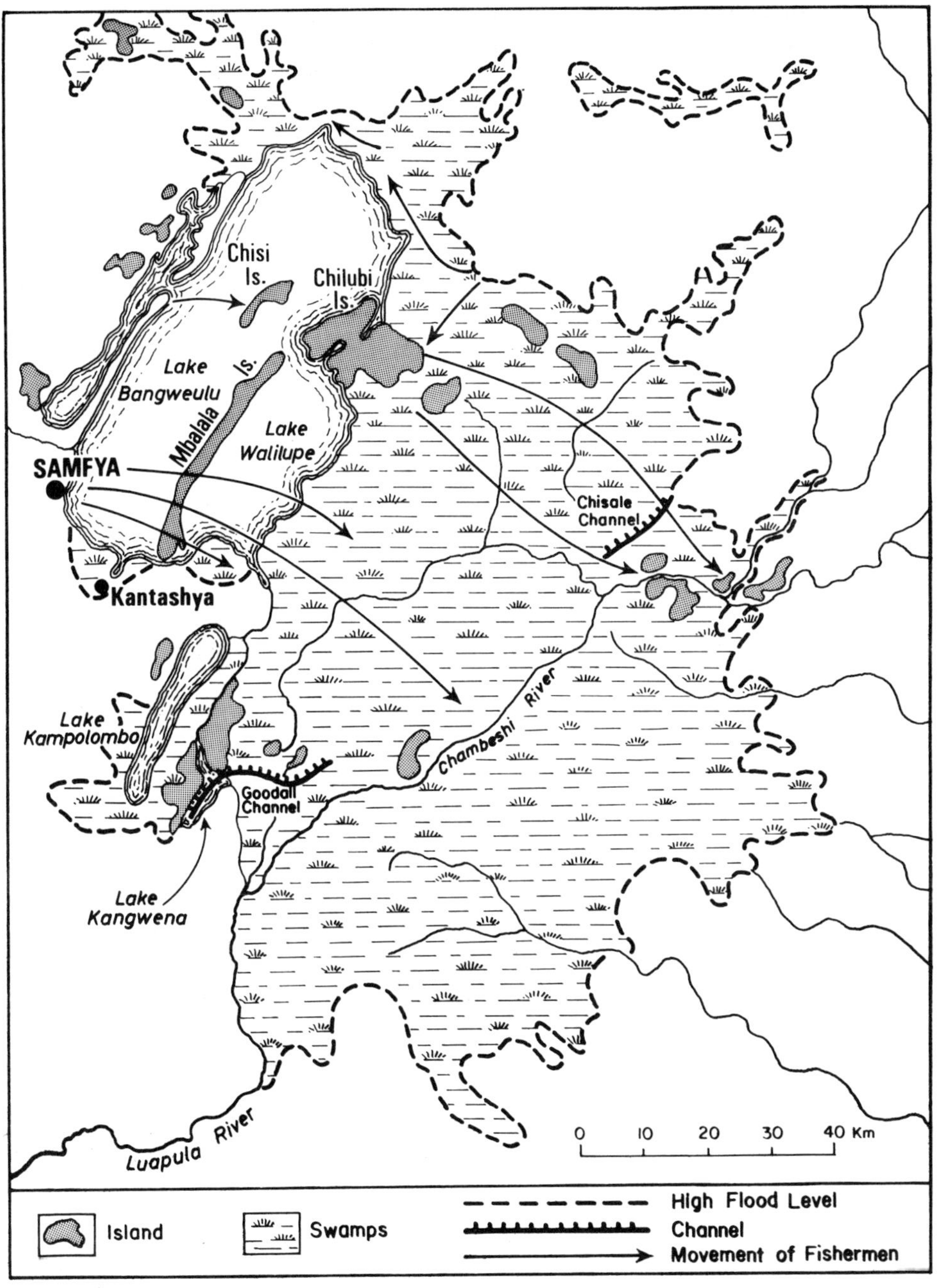

Fig. 7.6 The Bangweulu swamps

often hire canoes for their transportation into the swamps where they buy fresh fish and dry it within the swamps, come out again by hired canoes thence by lorry or bus to the line of rail markets. Other influential dry fish traders purchase fishermen's catches in advance by means of money advances or stores and simply wait with their vehicles at landing sites such as Kapalala and Katanshya for the goods to be delivered (Mabaye 1977).

There is great potential for the reintroduction of a ferry-boat scheme to collect fish and bring it out of the swamps, thus enabling fishermen to spend more time in actual fishing rather than ferrying their catches to distant markets (Clarke 1974).

Since the Bangweulu basin is so flat and shallow, the waters of the swamp expand and contract greatly with fluctuations in rainfall. The catchment area is more than 100,000 km² (Welcomme 1972), and has an average total rainfall of more than 1,300 mm, which may vary by 20 per cent in any year. Thus the area of the swamp varies greatly from one year to another. Of course there is also a regular seasonal expansion and contraction, with the water level rising 1.5 m from December after the rains have begun in earnest, to May or June, a month or two after the rains have ended but when the rivers continue to pour their waters into the Bangweulu basin. The highest water level is reached in June, after which there is a gradual fall which becomes pronounced during the dry, hot period from September to November.

The first rise in water level occurs in the river channels; it soon floods out into the swamps, where the oxygenated river water mixes with and replaces the stagnant swamp water, and finally reaches the lakes to the west in June to July (Tait 1965).

Ecologically, the swamp areas are very different from the open water areas. In the swamps, the plants (mainly *Vossia cuspidata*, *Cyperus papyrus*, and *Phragmites* species) grow from a floating mat of entangled roots and rhizomes embedded in decomposing plant remains. These dense stands of emergent vegetation grow very rapidly. These ultimately die and fall onto the surface of the mat, sink and become waterlogged, so that the mass of vegetable matter undergoing anaerobic decomposition is being continuously and copiously augmented (Beadle 1974). The emergent swamp vegetation forms a closed canopy which Beadle likens to a tropical rainforest, so that little light reaches the shaded water below. As a result

> there is little or no photosynthetic production of oxygen in the water, which is also well protected from wind stirring. The only source of oxygen is therefore by diffusion from rather still air. On the other hand, the consumption of oxygen by decomposing organic matter is so rapid that except during flood periods, exposed water under dense papyrus may be devoid of oxygen to within a cm of the surface. The straight unbranched leaves of grasses and reeds provide less shade from the vertical sun, but under very dense stands of . . .

> tall swamp grass . . . deoxygenation can be almost as extreme as under papyrus
> (Beadle 1974:247).

In the open, unshaded waters of the lagoons, lakes, and channels, ecological
conditions are very different:

> Here the water is well exposed to the sun and dissolved oxygen is abundant,
> and during the day it is often supersaturated by intense underwater photo-
> synthesis. This . . . well-aerated and nutrient-rich water is of great ecological
> importance as it supports a large invertebrate fauna and is the feeding ground
> of many fish, especially the young and growing stages which also find protec-
> tion from predators among the submerged vegetation (Beadle 1974:244–45).

Without the oxygenated open waters of the deeper channels and numerous
lagoons, aquatic life in the Bangweulu swamps would be poor. Scarcity of
oxygen, highly reducing conditions, and a high level of carbon dioxide — all
the result of the rapid decomposition of vegetable matter — combine to
create anoxic, stagnant water conditions which fish avoid, especially during
the dry season when such conditions are most severe. Even in the open
channels, measurements have shown that oxygen is extracted from waters as
it moves through the swamps, so that oxygen saturation in the channels
during the dry season is as low as 5 per cent (Beadle 1974). Even the bottom
mud is "completely unproductive", numerous samples "nearly all yielding
indications of an almost desert-like . . . bottom". Such poor conditions
indicate that there is something essentially wrong with the decomposition of
organic material in the mud (Joint Fisheries Research Organization 1964).

Such a substrate offers little opportunity for nutrient cycling, and results
in an extremely poor bottom fauna. Phytoplankton provide food for
bottom fauna; they also oxygenate the water, convert inorganic materials to
organic matter and serve as food for zooplankton (Mabaye 1973). Com-
paring the Bangweulu and Mweru fisheries, Soulsby (1959) wrote that in
Mweru,

> it is significant that of the three top species two (*T.* = *S. macrochir* and
> *Tylochromis mylodon*) are planktonic feeders to a large extent and herein lies
> the difference between the productivity of lakes Mweru and Bangweulu. In the
> latter system, the predatory *Hydrocyon vittatus* is one of the top two. If pro-
> ductivity be guaged in terms of catch/100 yds of net, the Mweru figures average
> out at 25.1 lbs and the Bangweulu area approximates to 10.8 lbs. Water
> samplings by Research Officers indicated that there are not sufficient nutrient
> salts in the waters of Lake Bangweulu to support a good plankton population,
> while . . . in Lake Mweru . . . the water is very rich by comparison and thus
> able to support heavy populations of fish which are close to the foot of the
> food chain.

Since bottom organisms and phytoplankton represent important links in
the aquatic food chain, they are, together with the lack of oxygen, important
limiting factors in Lake Bangweulu. Fish avoid the swamps and move in

only when the water is oxygenated by the floods (Tait 1965). The breeding season is concentrated during times of highest temperature at the end of the dry season, so that when the rains and rising floods rejuvenate the swamp waters and flood the surrounding flats, the young are able to disperse into the swamps and newly flooded land for protection and feeding (Tait 1965).

The open waters of Bangweulu are also the home of the tiger-fish, *Hydrocyon vittatus*. This voracious predator has an enormous influence on the lives of other fishes. Jackson (1961a) postulates that wherever this predator is found, both the absolute number of fish and the number of fish species small enough when adult to be preyed upon by tiger-fish are governed by the amount of protective vegetal cover available throughout the year. Thus in the open waters where there is sufficient oxygen for fish to thrive, prowling tiger-fish limit fish populations by the pressure of their predation. Tiger-fish, one of the mainstays of the Bangweulu fishery, are themselves limited by their place in the Bangweulu ecosystem. They are exclusively carnivorous, so that much of the ecosystem's net available energy has already been dissipated at lower trophic levels. Therefore, through the laws of energy transfer and their predation pressure, tiger-fish along with a dearth of oxygen and an unproductive bottom fauna restrict man's offtake from the Bangweulu ecosystem.

Fishing in Bangweulu is carried out in a variety of ways. Brelsford (1946) has described seventeen methods of catching fish in the swamps. Fishing is largely seasonal and reaches its peak during the period of low water when the fish are more accessible and concentrated in the river channels and lagoons (Joint Fisheries Research Organization 1959). Weirs are commonly used in the flooded margins of the swamp when the water level is dropping. Enormous numbers of *Gnathonemus macrolepidotus* (*nchesu* or *mintesa*), a small mormyrid (snout-fish family) are caught. It is a popular fish with fat which can be extracted by merely frying the fish in a pot (Brelsford 1946: 61–62). Long lines are used for subsistence fishing, and fish are speared when seen in the clear water of channels and lagoons. Baskets are used by women and children to catch small fish in very shallow water, but this is entirely for subsistence. Beach seining is carried out on the sandy beaches of the lakes and along the Chambeshi and Luapula rivers (Tait 1965). Seine nets catch mainly cichlids, but the lack of effective control over seine net sizes has done great harm to this fishery.

The main commercial fishing in Bangweulu is done with gill nets usually set on the bottom in quiet lagoons and backwaters (Tait 1965). It is such open water areas in the swamps that produce the best fishing; the open lakes, where fish find little protective cover from predators, produce very little fish.

The majority of fishermen concentrate in the southwestern part of the swamp, especially around Lake Chali, a body of open water close to the major exit points of the fishery. "Considerable concern is being felt regarding the present status of this fishery . . . the general impression is that production is falling despite a fairly even fishing pressure" (Mabaye 1977). Admittedly, total recorded catch has increased from 6,854 t FWE in 1964 to 8,941 t in 1978. However, catch per unit effort and catch of desirable species have declined alarmingly. Tait, writing in 1965, listed 33 out of the 86 recorded species of fish in Bangweulu as being of commercial importance. He went on to list 13 species caught commercially in large numbers. As seen in the passage below, the situation has since deteriorated drastically. Catch data gathered by the Department of Fisheries in 1972 were compared with data taken in 1955 and 1958 by the Joint Fisheries Research Organization:

> A sizeable decline in catch had occurred during the past seventeen years. Although records were incomplete, the average catch for 90 m of mounted net in 1955 (21.8 kg) was more than twice that of 1958 and more than ten times that of 1972. . . . In addition to the drastic decline in quantity of catch, changes in its quality had taken place since 1958. The most serious trend was the shift towards a fishery that almost entirely depended upon two species, *Hydrocyon vittatus* and *Alestes macropthalmus*, which together composed about 75 per cent of the research catch in the Samfya area. This indicated a potentially dangerous situation where poor survival of one or both of these species in any one year could seriously affect the fishery. In contrast, these two species comprised only 44 per cent of the total catch in 1958, whilst the seven most important species together formed 78 per cent of the catch. Two species of *Labeo*, *L. altivelis* and *L. simpsoni*, taken in 1958, were not found in research catches during 1972. Also significant is the drop in the proportion of *Tilapia*. *Tilapia* (*Sarotherodon*) *macrochir*, the most important cichlid in 1958, accounted for less than one per cent of the 1972 catch and its status as the most common cichlid was lost to other species which are less valuable because of their smaller size (Clarke 1973).

Muncy (1973) points out that the proportional decrease in bottom species (cichlids and *Labeo* species) and increase in the much less desirable surface feeding species, the tiger-fish or *nsanga* and *Alestes macropthalmus* or *manse* suggests that netting may be disproportionately changing populations. Results of experimental gill netting carried out by the Fisheries Department's Bangweulu Research Unit in 1977 showed that "while cichlid catches remain a diminishing asset, the less desirable species such as *Alestes* species and *Clarias* species are increasing" (Muyanga 1980a). Muncy concludes by saying that

> there may be under-fished stocks in isolated areas; however, all data indicate that fish stocks in accessible areas of the system are far from underutilized. All the classical warnings of overfishing are indicated by
>
> 1. decreased catch per effort of experimental gillnets and per fisherman,
> 2. decreased size of fish within catch as well as within species,
> 3. actual drop in total catch even with increasingly efficient gear.

Greater exploitation of the presently fished species would not appear to be in the best interest of the fishery. At present there are no indications of greatly under-exploited stocks which could be developed into extensive yield since catch per 90 m for 51 mm gillnets averaged only 4.39 kg (highest 7.61 kg) at all stations in Lake Bangweulu although *Alestes macropthalmus* was taken at all test netting stations. In Lake Mweru . . . 51 mm top set gillnets took up to 73.4 kg/100 m², mostly *A. macropthalmus* . . . (Muncy 1973).[4]

The Bangweulu fishery, then, is showing signs of overexploitation. Swamp conditions leading to oxygen scarcity and intense predation pressure by tiger-fish limit the production potential of the fishery for man. Improper fishing has exacerbated these ecological limitations. This has certainly been the cause of declining catches of desirable cichlid species. *Kutumpula* fishing — driving bream into gill nets as they congregate for spawning and rearing fry — has greatly reduced the populations of *Tilapia rendalli* (*mpende*), *Tylochromis bangwelensis* (*nsungula*), and *Sarotherodon macrochir* (*nkamba*). Increasingly, fishermen have resorted to smaller meshed nets as the size of fish caught has decreased. The result has been that more and more immature cichlids are caught, which have not yet had a chance to breed and raise fry. Data from Lake Mweru indicate that nets of less than 76 mm catch mostly immature *S. macrochir*; the same should apply to the similarly-sized and shaped *T. rendalli*. Research data from Bangweulu show that about half of the *Tylochromis* taken in nets of 50 mm mesh are immature. In 1973, a Fisheries Department survey of draw seines around Samfya found that half the nets used were of 50 mm mesh or less, and none were larger than 75 mm (Clarke 1974).

In Bangweulu, as at Mweru/Luapula, effective enforcement of fishing regulations will be required if the catch of immature and breeding cichlids is to be curtailed. Since Bangweulu cichlids, unlike those of Lake Mweru, do not have a distinct and restricted breeding ground, a closed area is not feasible. Instead, small mesh sizes must be banned, and driving of fish into gill nets must not be allowed between the first of September and the end of December, the major spawning and breeding period for Bangweulu cichlids (Clarke 1974). Without such controls on fishing, there will be no increase in the biomass available for human exploitation from Bangweulu.

Mweru Wantipa

The fishery of Mweru Wantipa, the "lake of mud" (Fig. 7.7), has developed spectacularly in the past decade. Throughout most of the colonial period, it was neglected because of its inaccessibility. Little or no research

---

[4]Since Murcy's investigations in 1972, total catch since 1972 has generally increased, but catch per fisherman has decreased, as has fish size.

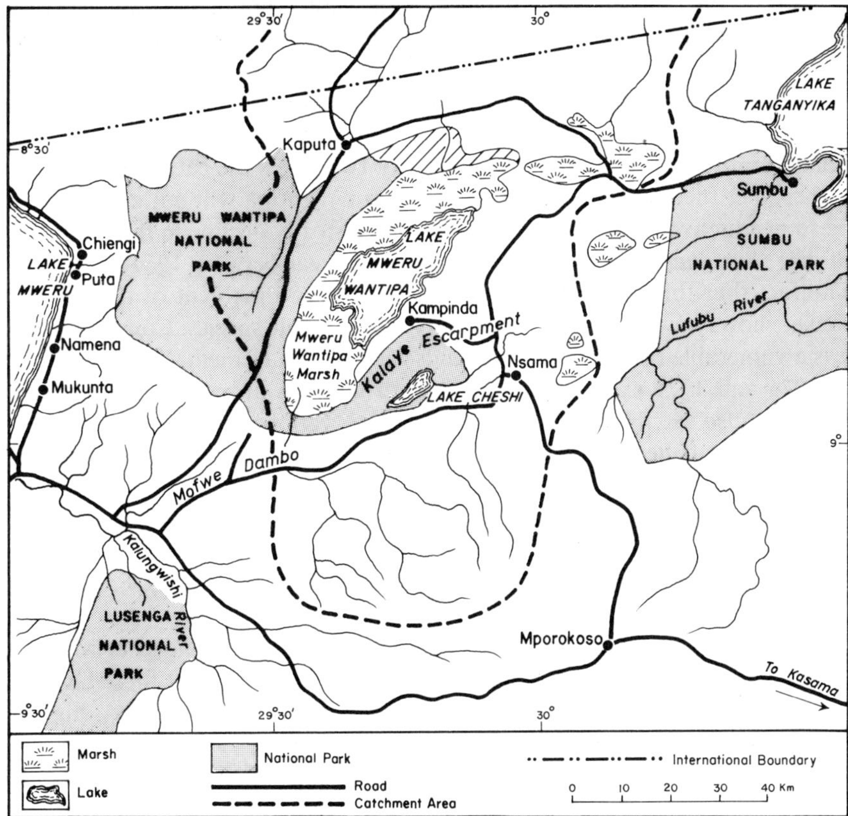

Fig. 7.7 The Mweru-Wantipa fishery

was carried out, and the fact that it was located within a game reserve meant that access could be controlled, so that management consisted largely of regulating the number of fishing permits granted each year. Soulsby (1960a) points out both the prevailing (and mistaken) assumption of the times and also, then as now, the main obstacle to the fishery's development:

> The fishery lies far from main markets where heavy populations are concentrated and exports are largely local. With no ice facilities the fresh fish export does not extent over 10 miles from the camps, except in the case of Lake Chishi, where lorries and vanettes, supplied with ice, carry the popular bream as far as the Copperbelt. Most trading is carried out on foot or cycle and only recently has the main Mporokoso-Sumbu road, which runs alongside the marsh, been upgraded. Feeder routes are still bad or non-existent and the crux of fishery development here lies in the improvement of these roads, though with increased production at other fisheries in closer proximity to the main markets, it is difficult to visualize this area in any role other than local supplier.

Prior to 1965, about 500 t of fish were harvested annually from Mweru/ Wantipa. By 1970 the harvest had multiplied eight times, to more than 4,000 t per year. Mweru Wantipa had become a major supplier of dried fish to the Copperbelt. In 1975, the fishery was estimated to have yielded 16,765 t of fish, double the harvest of any other Zambian fishery, 29 per cent of the Zambian total catch, and more than thirty-two times the catch of a decade earlier. Yet there are indications that Mweru Wantipa will not continue to be so productive. Williams (1972) recorded a steady decline in the numbers of pale, *Sarotherodon macrochir*, caught each year between 1964 and 1969. The fact that this species still accounts for over 85 per cent of catches by weight shows how dependent this fishery is on a single species. Consequently, it is a vulnerable fishery which demands careful management. A comparison of 1974 and 1975 experimental catches "showed a drop not only in total catch but also catch per unit effort. These observations suggest unfavourable trends in the fishery" (Mabaye 1977). By 1977, the fishermen were using smaller mesh size gill nets, an indication that fish catches were declining. They were also resorting to illegal and destructive *kutumpula* fishing to augment their catches (Mabaye 1977; 1980). Production has indeed declined repeatedly since 1975, with an estimated harvest of 13,330 t in 1976, 12,512 t in 1977, and 10,351 t in 1978.

However, the phenomenal rise in the amount of fish landed at Mweru Wantipa over the past fifteen years might seem to contradict these indications of a fishery in the early stages of decline. Mweru Wantipa is a highly productive yet highly variable and vulnerable fishery, easily susceptible to disruption. It is also subject to a variety of conflicting interests, for the fishery is largely confined within the boundaries of a national park, which is also a recognized outbreak area for the red locust, *Nomadacris septemfasciata*. All these variables complicate its management.

Mweru Wantipa's high productivity is mainly due to its morphology. The lake and the extensive swamps surrounding the open water occupy a shallow depression between the north end of Lake Mweru and the south end of Lake Tanganyika. This depression is a *graben*, a trough between parallel faults, and forms a part of the larger Africa rift system. The depression is filled by rain water draining off the surrounding country. There is no outlet although in times of very heavy rainfall there can be a connection via Mofwe *dambo* to the Kalungwishi River, which flows into Lake Mweru. Because there is seldom any outlet to Mweru Wantipa, the nutrient salts washed in from the surrounding country stay in the drainage basin. Thus the lake and swamp waters are very rich in nutrients.

In addition, Mweru Wantipa is a shallow body of water; its maximum depth in December 1971 was 9 m (Lee 1972). Thus the open waters of the lake are shallow enough to be stirred right to the bottom by wind action.

The nutrient salts which have been taken up and cycled through the several trophic levels sink to the bottom in the form of decomposing bodies of dead plants and animals. A rich organic sediment accumulates on the bottom, and wind stirring brings these products of decomposition up from the mud and lower water levels to the upper illuminated regions of the lake where photosynthesis occurs and where they can be recycled again through the food web of the aquatic ecosystem.

The productivity of Mweru Wantipa is not only due to its shallowness and lack of outflow. It is also very much a result of the great fluctuation of the water level in the lake basin. On at least four occasions since 1867, the lake virtually dried out completely (Lee 1972); when this last occurred, in 1949–50, thousands of fish, hippopotami, and crocodiles died. "The periodical drying out of the area assists in maintaining a high water fertility by the action of the air and sun in oxydizing the bottom muds, which when flooded again, release salts to the water" (Bowmaker 1965). During drought periods, however, the salts become too concentrated to be tolerated by most fish, and as the water recedes fish concentrate in pools in which the water becomes deoxygenated and thousands die. The *pale* suffer most, while the silurid fishes such as *Clarias mossambicus* (*mita*) which can breath oxygen from the air, are better able to withstand dessication (Bowmaker 1965). After the last drought,

> with the rise in water level, fish repopulated the marsh so quickly that it was possible to reopen the fishing camps early in 1952, when extremely good catches were made. It is considered in this respect that vast numbers of fish "made their escape" to tributaries as conditions became untenable, but before they could be isolated in pools on the mud flats. (Soulsby 1960a)

Because Mweru Wantipa varies so greatly in the volume and salinity of its waters, it is not surprising that its fish fauna is limited. Two — *Sarotherodon macrochir* and *Clarias mossambicus* — are extremely abundant and form the basis of the commercial fishery. A few other — *Synodontis* species (squeakers or *mbongwe*) and *Alestes* species (*chitololo* or *ntuntu*) — are caught in considerable quantities, and are very popular table fish (Mabaye 1976). In 1974, twenty-three fish species were found in Mweru Wantipa; ten more than were recorded in 1965, when *Alestes* was unrecorded (Mabaye 1976; Bowmaker 1965). This suggests that it is only the high water conditions and diluted salinity that have existed more or less continuously since the 1961–62 rainy season that have allowed most species to reach the basin and survive. In times of high rainfall, water floods back into Mweru Wantipa from the Kalungwishi River, and salt concentrations are not too high for most species. The abnormally high rainfall in 1961–62 initiated a dramatic rise in the level of the lake. Between September 1961 and September 1962, Mweru Wantipa rose 5.5 m, and increased its area by 41 per cent. Since that time, the lake elevation has remained at a higher level than for any period in the last century (Lee 1972). The rise in lake level has reduced the area suitable for highly successful breeding of *Sarothrodon macrochir*. The depression has been flooded so that the lake is presently well confined by high ground, and only a narrow margin along the western shore is less than 6 m deep (Lee 1972). *Pale* make their nests in water from 0.5 m to about 7 m deep (Coulter and Mortimer 1965) and thus "highly suitable conditions for fish breeding only occur soon after the water breaks out of the deeper area, and . . . no 'elastic' marsh area exists when the water level is at its present height" (Williams 1972).

Management of Mweru Wantipa will always be a problem. Its fluctuating water levels will always give a "boom or bust" production of fish, so that management should try, as Bowmaker (1965) suggests, "to make the most of good periods". Perhaps a steadier state could be attained by hastening the recovery of *S. macrochir* populations during the recovery phases of the lake. This could be achieved either by limiting fishing in marginal freshwater lagoons and swamps during recessions or by restocking after recessions, as Morgan (1971) has suggested for Malawi's ecologically and morphologically similar Lake Chilwa. Certainly *S. macrochir* is immensely prolific, as Soulsby noted in the rapid recovery of the fishery after the 1949–50 drought, so that artificial rearing and restocking might not be necessary.

Another possibility for reducing the fluctuations in the fishery would be to consider the introduction of a more alkaline resistant species of *Tilapia* or *Sarotherodon*. Several small species are found in the highly alkaline lakes Magadi, Natron, and Manyara on the floor of the Eastern Rift Valley in Kenya and Tanzania when the pH is as high as 10.5. *Tilapia nilotica*, distributed over the whole soudanian region of Africa and up into the Jordan valley,

> is often found, and can be experimentally cultivated, in water too saline for most purely freshwater fish. The most adaptable of the large commercially valuable species seems to be *T. mossambica*. It is widespread in the warmer parts of southeastern Africa south of the Zambezi. It inhabits coastal brackish waters and inland waters both fresh and saline due to evaporation during the dry season. It has been experimentally acclimatised to seawater (Beadle 1974: 270).

As Morgan (1971) points out, however, it is uncertain whether a more alkaline resistant fish would be more successful than the indigenous fish fauna, for it is asphyxiation as much as alkali-stress which leads to fish kills at times of low lake level.

There is also the problem of conflicting land use in the basin. Almost the entire swamp is enclosed in Mweru Wantipa National Park, where access is supposed to be temporary. Cutting firewood for cooking and smoking fish and the clearing and planting of gardens are all prohibited, at least in theory. The establishment first of a game reserve and in 1972 of a national park was intended primarily to protect the elephant, buffalo, puku (*Kobus vardoni*), zebra, and hippo which roam the grassy plain of the exposed lake bed. These species were formerly found in large numbers on the margins of Mweru Marsh, where the seasonal drying and flooding and periodical water level fluctuations created a habitat for terrestrial herbivores just as fertile as that of the complementary aquatic ecosystem. The rise in water level which has greatly reduced the area of exposed floodplain since 1961/62 and the more or less concurrent rise in poaching, much of it originating across the nearby Zairean border, have greatly reduced the populations of large mammals. A wild life count made in April 1973 recorded small numbers of elephant, buffalo, roan and sable antelope, hartebeest, and puku (Clarke 1974). Peripheral and inaccessible, the park suffers from a low priority compared with other parks in the system. It also suffers from an acute shortage of staff of all grades (Zyambo 1978), "manned by only a token force which offers little resistance to the challenge made by fire and illegal hunting" (Clarke 1974). Crocodile, a protected species, have multiplied greatly. It has been estimated that there are between 12 and 20 crocodiles per ha. Most sand beaches are laden with crocodile eggs (Muyanga 1980b). Crocodiles feast on fish caught in gill nets, and tear the nets badly. The result has been that fishermen are now reluctant to use gill nets and are concentrating on seine netting in an attempt to avoid damage by crocodiles (Muyanga 1980b). However, seine nets capture females of mouth-breeding cichlid species (e.g., *S. macrochir*) while carrying eggs or fry in their mouth. The result is heavy losses of fry and future generations of adult fish. Seine nets also scrape the bottom as they are dragged in, thus destroying the nests of other valuable cichlid species, such as *Tilapia rendalli*.

Unfortunately, the best and most accessible fishing is found within the national park, so that the conflict between fishermen and crocodiles is exacerbated. Crocodiles are a hazard not only to the fishermen's nets but also to the fishermen themselves. In turn fishermen, by their presence and poaching, disturb and destroy the wild life. Efforts have been made to relocate fishing camps which lie within the park, but these will not succeed so long as the park boundary circumscribes the swamp.

Aggravating the problem is that fact that the park's boundaries have never been properly defined nor adequately surveyed. The Kaputa road bisects the park along the western margin of the swamp and the Sumbu road lies an undetermined distance south of Lake Cheshi and the southern edge of the swamp. On a flood plain with a fluctuating water level and few obvious topographical features for references, where large numbers of fishermen live and work in proximity to wild life inside a gazetted but unsurveyed and largely unadministered national park established to protect wild life, the opportunities for abusing, confusing, or simply ignoring park regulations are legion.

The roads leading to the east shore markets are among the poorest in the country, so that traders prefer to keep to the Kaputa and Sumbu roads, which themselves are poor. Canadian assistance has been obtained to improve the access roads to the fishing camps at Mweru Wantipa, and this should ease the movement of fish from this remote fishery. Improved access will lead to lower transport costs, and should put the fishermen in a stronger position to insist on a fair price for their fish from the traders. A higher producer price would bring more fishermen to the lake, which might well precipitate and hasten a further decline in the fishery. Increased access and fishing activity will certainly increase conflicts with the interests of the national park.

In addition to the conflicting interests of fishermen and the National Parks and Wildlife Service, there is also the fact that Mweru Wantipa is one of the breeding grounds of the Red Locust and was the source, together with Lake Rukwa in southwestern Tanzania, of the last great locust plague which spread over central Africa between 1929 and 1944. Here, too, there is a conflict of interests. As outlined above, the productivity of the lake is largely a result of its fluctuating water level, and high water levels are inimical to successful breeding of *pale*. However, it is high water which most effectively prevents an outbreak of locusts; low water conditions create extensive plains of damp but not water-logged bare soil which are ideal for the incubation of locust eggs. Under especially favourable conditions, numbers of locusts increase until overcrowding occurs. Overcrowding triggers behavioural and morphological changes in the insect in which solitary grasshoppers are transformed into the swarming locust phase.

A bare patch of moist soil near standing grass provides an ideal locust egg-laying site. Burning the flood plain grasslands increases the area of bare soil and thus favours successful oviposition; burning also makes more nutrients available to fish and game. In unburnt grasslands the extent of bare soil is limited.

After 1949, when the lake bed was almost completely dry, a rather costly and largely unsuccessful effort was made to divert part of the Kalungwishi River into the Mweru Wantipa basin by constructing canals (Brelsford 1955; Lee 1972). Throughout the 1950s, the International Red Locust Control Service (IRLCS), an organization established in the early 1940s by the British and Belgian Colonial administrations, was proposing expensive, fanciful (in view of present knowledge of the topography) schemes to divert the Kalungwishi River or even the Lufubu River, which flows into Lake Tanganyika in order to raise the level of Mweru Wantipa sufficiently to flood most of the potential locust breeding grounds exposed at low water (Gunn 1957). It was mistakenly thought that this would benefit not only the locust control operation but also the local fishing industry (Brelsford 1955). By the end of the decade, it was realized that such diversion schemes would entail considerable conflict as well as cost. Diversion of the waters of the Kalungwishi would not be acceptable to the riparian interests of the then Belgian Congo. The Department of Game and Tsetse Control pointed out that permanent flooding of the plains around the lake would destroy the habitat of the grazing animals of the game reserve. Moreover, it was suggested by fisheries personnel that periodic exposure of the Mweru Wantipa flood plain benefited fish productivity, provided that complete lake extinction was avoided.

Therefore, attempts at ecological control of locusts foundered in a morass of conflicting resource interests and, more immediately, because of prohibitive costs. Both flood control and fire control cost far more than the aerial spraying of persistent pesticides which the IRLCS began to practise in the mid-fifties. Ground operations were soon abandoned in favour of aircraft for all purposes of locust control in outbreak areas and, by 1970, recurrent expenditure had decreased by more than half and the effectiveness of the service had greatly increased (DuPlessis and Kuhne 1970). However, the International Red Locust Control Service is now in danger of collapsing altogether, for its successful prevention of locust outbreaks since 1944 has made the financially pressed African governments which support it reluctant to provide the funds needed for control operations.

Therefore, projects with serious environmental consequences — controlled burning and interbasin canalization — were discarded in favour of a cheaper technology — aerial spraying of persistent insecticides — whose long-lasting, non-selective, widespread and unintended ecological and regional effects on the red locust outbreak areas of central Africa have never been investigated.

The valuable fishery at Lake Wantipa then is but one of several conflicting resource interests. Its inherent instability, due to fluctuating water levels and its overwhelming reliance on a single species, makes management difficult enough. Complicating management still more is the fishery's incompatible location within a national park. Crocodiles and fishermen compete directly for the fish resource, and both are present in large numbers. More fishermen and better access roads will increase fish production but also increase the pressure on this vulnerable fishery to the point where it will require much more management to avoid collapse. Locust control is yet another factor to consider in the multiple use of the Mweru Wantipa basin. Fisheries production will have to be balanced with locust control, wild life protection, and in the case of crocodiles, possible cropping and utilization. In the coming years, the concept of optimum sustained yield ["a deliberate melding of biological, economic, social, and political values designed to produce the maximum benefit to society from stocks that are sought for human use, taking into account the effect of harvesting on dependent or associated species" (Roedel 1975)] — will be put to the test.

Conclusion

Foin (1976), in his book dealing with the management of ecological and environmental systems, lists eight management practices for renewable resources: (1) exploit if you can, (2) obtain yield at minimum cost, (3) do whatever you can get away with, (4) run the resource down, and . . . (5) scale consumer demand to supply, (6) increase supply by ecological engineering, (7) try to operate at sustained yield, and (8) reach for sophistication in resource management with new scientific tools. These may provide a useful framework for analysing trends in Zambian fisheries.

Economic principles govern the use of Zambian fisheries as they do for the use of all natural resources. When economic considerations run counter to the best long-term interests of the resource, then management becomes necessary.

Exploitation alone, without attempts at management, tends to occur mostly in the pioneering stages of the development of the resource. However, as Foin points out, demand tends to outstrip supply, so that the longer exploitation continues, the more need there is to develop management strategies. When exploitation persists beyond the pioneering stages, resources are run down for lack of management. This is exactly what caused the disappearance of *Labeo altivelis*, the *mpumbu* from Mweru/Luapula. The short breeding season, concentrated in a few days at the beginning of the floods and preceded by a mass migration upstream, made these fish very vulnerable to attack at a critical phase of their life history. Lack of management quickly led to the ruin of this valuable natural resource, and conservation measures that were at last applied came too late to protect the fish (Jackson 1961). In cases like this, it seems that successful management requires the effective intervention of an outside authority, such as fish wardens, willing and able to stop harmful overuse of the resource.

Foin's second management practice emphasizes short-term efficiency and economic growth above other considerations — do what needs to be done to maintain an income from the resource. This has characterized all Zambian fisheries. Management of Zambian fisheries has mainly been concerned "with monitoring and where necessary, controlling the exploitation of commercial fishing areas to avoid depletion of resources through the use of harmful fishing methods, over-intensive application of acceptable fishing methods or fishing in recognized breeding areas" (Williams 1977). Fortunately, the pressure on most Zambian fisheries has not yet led to their destruction, although intensive fishing has led to declining catches (as in the case of *Sarotherodon macrochir* at Mweru Wantipa), or even severe depletion of desirable species (*S. macrochir* and *L. altivelis* from Mweru/Luapula, *S. macrochir* and other cichlids from Bangweulu, and *Lates* species from Lake Tanganyika).

Zambian fishermen have generally responded to the efforts by the Fisheries Department to conserve fish and maintain fisheries by doing whatever they can get away with. Regulations against fishing in the Mifimbo swamps and lagoons at the mouth of the Luapula River do provide protection, when enforced, for Mweru's breeding stocks of *S. macrochir*. Similarly, regulations prohibiting certain fishing techniques — seining in Lake Mweru, or *kutumpula* fishing in all Zambian waters — are intended to allow fish to breed successfully before being taken, just as regulations setting minimum net mesh sizes are intended to do. The problem is that these regulations lack any means of effective enforcement, and the situation has deteriorated over the past decade.

Williams (1977) suggests that what is required is the vitalization of district fisheries development committees which exist on paper in all Zambian administrative districts where fishing is an important economic activity. Local involvement is needed to create a sense of responsibility within the community, so that local fishermen feel obligated to conserve the resources of their own fishing area. "This is essential to the effectiveness of any regulations brought out to control fishing intensity and fishing methods" (Williams 1977). Fisheries Department research and catch assessment data must be taken into account by these local fishing development committees, so that they can then propose reasonable and enforceable regulations for their own fishery. In this way, fisheries regulations will become an aspect of fishery development rather than simply a conservation measure intended to curb fishery development (Williams 1977). The training and extension programmes carried out by the Fisheries Department should also emphasize to fishermen the common benefit to be gained from protective regulations.

In the absence of effective enforcement and management, the trend has been to run down Zambian fisheries, on the assumption that fishing pressure has not yet been sufficient to destroy them. Where species have been seriously depleted, efforts have been made to find substitutes. Hence, at Lake Mweru, where present yield data have shown significant reductions in population abundance, species composition, and catch per effort (Muncy 1973) the effort has been made to switch from the popular *S. macrochir* to the less desirable but unexploited species such as *Alestes macrophthalmus*.

In a recent economic appraisal of Zambia's fishing industry, Williams (1977) states that the number of fishermen in almost every fishery "has the potential to attain a level of production well above the present estimated sustainable yield''. The existing fishing population is operating well below its capacity. For example, the capacity of existing fibreglass boats is roughly 25 nets, yet most fishermen set fewer nets — perhaps 10. The present capacity is therefore underutilized. Similarly, fishing effort is low; ''as few as 20 per cent of resident fishermen go out fishing on any given day'' (Williams 1977). This is partly a result of overfishing; low effort is more prevalent on fisheries that also show low catch rates and low utilization rates. ''A second factor in low motivation relates to the marketing system, the uncertainty of being able to sell all his fish at a predictable price which in turn encourages a reluctance by fishermen to go on the water frequently and achieve high catches which may be wasted'' (Williams 1977).

Fisheries development in Zambia — credit facilities to purchase improved craft and fishing gear; training and extension programmes designed to teach fishermen improved methods of fishing, handling, processing; a better understanding of the financial aspects of fishing; a revamping of fish marketing, with a price system that allows for differences in quality, species, state of processing, seasonality of supply, and cost of production and distribution over time and between fisheries and markets; and development of infrastructure, such as improved access roads, more and better road and water transport, and ice-plants — will increase the pressure on Zambia's fisheries. It is short-sighted to simply run down the resource and hope that harmful uncertainties will be kept to a minimum. Zambian fisheries management policies must be drawn up immediately to anticipate problems and allow fishermen to achieve the potential of the fishery without destroying the resource. Reasonable regulations, enforced with local involvement, are part of the solution. An adequately staffed and financed Department of Fisheries, able to monitor and assess fish stocks and control their harvests, is also required. Williams (1977) suggests that increasing the numbers of fishermen and large-scale industrial fishing should be discouraged on all Zambian fisheries except Lake Tanganyika.

Fish production in Zambia is low partly because development funds have been diverted to mining, agriculture, transport, and social services, and partly because, as Christensen (1972) points out, the fish marketing system is inadequate. The present marketing system is inefficient because of (1) a bad price system, in which the official controlled price is far too low to encourage production and the uncontrolled actual selling price is too high for many low-income people to afford; (2) a lack of transport facilities, so that fishermen have to spend time drying and smoking their fish instead of fishing, for there is no way for most fishermen to market fresh fish; (3) a lack of a collection system from the scattered production places, with traders coming irregularly to the various landing places — the result is that fishermen are left with no regular outlet for their catches, and often have to retail their fish themselves, again at the expense of fishing time; and (4) bad handling of fish with few or no facilities for treating or storing fish — most of the fresh fish sold in urban areas are not fresh, and dried fish are sold which are broken, infested with beetles or maggots, and covered with sand.

Foin's sixth practice, to increase supply by ecological engineering, has been attempted in Zambian fisheries management with the introduction of *kapenta* from Lake Tanganyika into the Kariba reservoir and with the development of fish farming. The former has been very successful, but fish farming has not. A lower percentage of ponds are stocked now than in the mid-1960s, and production has been negligible, despite hopes expressed in the current Third National Development Plan that fish culture would produce 20 per cent of the projected 94,500 t of fish produced in Zambia in 1982 (Williams 1977; National Commission for Development Planning 1979).

Seventh in Foin's list of management strategies is to try to operate at sustained yield. "It should be possible at any point to stop and assess the resource, determine how fast it is reproduced, and from this information define the permissible take in terms of future population stability. This is routine management for fisheries, which must monitor catch statistics and other population indicators to determine population trends" (Foin 1976). Operating at sustained yield has not been practised in Zambia. Certainly information is available concerning the state of fish resources. At the very least, scarcer fish and declining catches suggest overexploitation in some fisheries already. Yet emphasis has and continues to be on increasing yields, and studies have not yet been done which would allow reliable estimates to be made of the sustained yield to be taken from each of Zambia's fisheries. The problem is more than simply one of fisheries science and more than just applying the principles of population ecology and mathematics to determine sustained yield. If calculated sustained yields conflict with economic goals and national or local political interests, then "the concept is likely to be dismissed by decision-makers as overly theoretical and inapplicable to their particular situation" (Foin 1976). This has happened repeatedly, which is why so many fisheries suffer from overexploitation today. With better craft and fishing gear, better transport and marketing facilities, and better access to credit, there will be a growing vested interest in the exploitation of the resource, with little attention paid to the consequences.

Foin's eighth practice — the utilization of sophisticated resource management tools — is something to which Zambia may aspire. Elsewhere, computer simulation has provided a powerful tool with which to estimate sustained yield from fisheries under various conditions while avoiding irreversible trial-and-error procedures that had been used in the past. A start in this direction was made in the early 1970s with the introduction of improved statistical data collection, which allows for a better analysis and assessment of fish catches. But in general, such tools as computer modelling must remain a distant goal for Zambian fisheries management.

In closing, a statement of general principles may be in order. Canada's Fisheries and Marine Service proposed that the goals for that nation's commercial fisheries should be to "maximize food production, preserve ecological balance, allocate access optimally, provide for economic viability and growth, optimize distribution and minimize instability in returns, ensure prior recognition of economic and social impact of technological change, minimize dependence on paternalistic industry and government, and protect national security and sovereignty" (Larkin 1977). This statement of principles can also be applied to Zambia's fisheries but it does not provide a management strategy. There is no priority implied in the goals listed; trade-offs and compromises will be required between goals of protection and production, and of course, one cannot optimize for two things at once, let alone a dozen. Perhaps the best that can be said is that the goal of the Zambian Fisheries Department should be to protect fish stocks by setting permissable catches at moderately safe levels of biological risk and then let the economic and social problems resolve themselves within the biological restraints (Larkin 1977). This approach requires little in the way of sophisticated management techniques from a department already desperately short of trained personnel and funds. It does require adequate means of enforcing safe levels of harvesting. Such an approach also keeps future options open. If the Fisheries Department is stern in its efforts to ensure that the resource is not abused, and if the Zambian government is equally firm in seeing that the nation's fisheries are exploited economically, then future generations of Zambians will continue to enjoy this valuable source of protein.

Acknowledgements

I wish to acknowledge the financial support provided by the United States Fulbright-Hays Doctoral Dissertation Research Abroad Programme, the Commonwealth Geographical Bureau, and the University of Zambia.

I am also indebted to the Director of the Department of Fisheries, Zambia, for answering questions, helping to obtain hard-to-find reports, and allowing me to use the departmental library.

Members of the Geography Department of the University of Zambia have patiently withstood presentations on aspects of Zambian fisheries on several occasions, and one, Dr. Adrian Wood, kindly read an earlier draft and took the time and effort to make many helpful suggestions. I am especially grateful for his assistance. Neither he nor anyone else, however, is responsible for any mistakes, omissions, or opinions. The Geography Department's cartographic staff gave valuable advice while preparing the maps.

REFERENCES

Allsopp, W.H.L. 1975. "Management Strategies in Some Problematic Tropical Fisheries". In Van Dobben, W.H., and Lowe-McConnell, R.H. (eds.), *Unifying Concepts in Ecology*. The Hague.

Beadle, L.C. 1974. *The Inland Waters of Tropical Africa*. London.

Bowmaker, A.P. 1965. "Mweru-wa-ntipa". In Mortimer, M.A.E. (ed.), *Natural Resources Handbook: The Fish and Fisheries of Zambia*. Ndola.

______, and Coulter, G.W. 1965. "Lake Tanganyika". In Mortimer, M.A.E. (ed.), *Natural Resources Handbook: The Fish and Fisheries of Zambia*. Ndola.

Brelsford, W.V. 1946. "Fishermen of the Bangweulu Swamps". *Rhodes-Livingstone Papers*, no. 12. Lusaka.

______. 1955. "The Problem of Mweru-Wantipa". *Northern Rhodesia Journal* 2, no. 5:3–15.

Carey, T.G. 1965. "Lake Mweru-Luapula". In Mortimer, M.A.E. (ed.), *Natural Resources Handbook: The Fish and Fisheries of Zambia*. Ndola.

Central Statistics Office. 1971. *Fisheries Statistics (Natural Waters), 1970*. Lusaka.

Christensen, G. 1972. *Development Plan for the Fishing Industry in Zambia*. Central Fisheries Research Institute. Mimeo.

Clarke, J.E. 1972, 1973, 1974. *Annual Reports*. Department of Wildlife, Fisheries and National Parks. Lusaka.

Coulter, G.W. 1965. "Research on Lake Tanganyika". *Fisheries Research Bulletin Zambia, 1962–1963*, pp. 4–7.

______. 1968. "Changes in the Pelagic Fish Population in the South-East Arm of Lake Tanganyika between 1962 and 1966". *Fisheries Research Bulletin Zambia* 4:29–32.

______, and Mortimer, M.A.E. 1965. "The Biology of Fish". In Mortimer, M.A.E. (ed.), *Natural Resources Handbook: The Fish and Fisheries of Zambia*. Ndola.

DuPlessis, C., and Kuhne, K.W. 1970. "Recent Developments in Preventive Control in Outbreak Areas of the Red Locust, *Nomadacris Septemfasciata* (Serv.)". *Journal Entomological Society of Southern Africa* 33, no. 1:95–117.

FAO. 1968. *Wildlife, Fisheries and Livestock Production*. Rome. Multipurpose survey of the Kafue River Basin, Zambia FAO/UN Final Report, vol. 5, FAO/SF 35/ZAM.

Foin, T.C., Jr. 1976. *Ecological Systems and the Environment*. Boston.

Gunn, D.L. 1957. "The Story of the International Red Locust Control Service". *Rhodesian Agricultural Journal* 54:8–24.

Inoue, K. 1971. *Frame Survey on Lake Bangweulu*. Central Fisheries Research Institute. Mimeo.

Jackson, P.B.N. 1961. *The Fisheries of Northern Rhodesia*. Lusaka.

______. 1961a. "The Impact of Predation, Especially by the Tiger-fish (*Hydrocyon vittatus* Cast.) on African Freshwater Fishes". *Proceedings of the Zoological Society of London* 136, no. 4:603–22.

______. 1971. "The African Great Lakes Fisheries: Past, Present and Future". *African Journal of Tropical Hydrobiology and Fisheries* 1, no. 1:35–49.

______. 1973. "The African Great Lakes: Food Source and World Treasure". *Biological Conservation* vol. 5, no. 4:302–4.

Joint Fisheries Research Organization. 1959, 1964. *Annual Reports*, no. 8, 1958 and no. 11, 1961. Lusaka.

Larkin, P.A. 1977. "An Epitaph for the Concept of Maximum Sustained Yield". *Transactions of the American Fisheries Society* 106, no. 1:1–11.

Lee, P.S. 1972. *Hydrology of Mweru Wantipa: A Study of the Water Balance of an Enclosed Lake in Northern Zambia*. Lusaka.

Lowe-McConnell, R.H. 1975. *Fish Communities in Tropical Freshwaters*. London.

Mabaye, A.B.E. 1973. "The Role of Ecological Studies in the Rational Management of Fish Stocks". *African Journal of Tropical Hydrobiology and Fisheries*, Special Issue 2, pp. 143–60.

————. 1976, 1977. *Annual Reports 1974, 1975*. Department of Fisheries, Zambia. Lusaka.

Matagne, F. 1950. "Premiers notes au sujet de la migration des Pumbu (*Labeo* sp.), bief du Luapula-Moero". *Bulletin de l'Agriculture du Congo Belge* 41:793–833.

Morgan, P.R. 1971. "The Lake Chilwa *Tilapia* and Its Fishery". *African Journal of Tropical Hydrobiology and Fisheries* 1, no. 1:51–58.

Mortimer, M.A.E., ed. 1965. *Natural Resources Handbook: The Fish and Fisheries of Zambia*. Ndola.

————. 1965a. "Fishing Gear, Methods and Craft". In Mortimer, M.A.E. (ed.), *Natural Resources Handbook: The Fish and Fisheries of Zambia*. Ndola.

Muncy, R.J. 1973. *A Survery of the Major Fisheries of the Republic of Zambia*. FAO, FI: DP9/10 ZAM 511/3. Mimeo.

Musambacime, M.C. 1975. *The Development and Growth of the Fishing Industry in Mweru-Luapula Area: Its Social and Economic Effects*. University of Zambia, History Department Seminar Paper. Mimeo.

Muyanga, E.D. 1980a, 1980b. *Annual Reports*. Department of Fisheries, Zambia. Lusaka.

————, and Chipungu, P.M. 1978. "A Short Review of the Kafue Fishery since 1968". Paper presented at the National Seminar on Environment and Change: the consequences of hydro-electric power development on the utilization of the Kafue Flats, April 1978. Lusaka.

National Commission for Development Planning. 1979. *Third National Development Plan, 1979–83*. Lusaka.

Roedel, P.M. 1975. "A Summary and Critique of the Symposium on Optimum Yield". *American Fisheries Society*, Special Publication no. 9, pp. 79–89.

Soulsby, J.J. 1959. "Status of the Lake Mweru Fishery". *Joint Fisheries Research Organization Annual Report*, no. 8. Lusaka.

————. 1960. "Some Congo Basin Fishes of Northern Rhodesia". *Northern Rhodesia Journal* 4:231–46.

————. 1960a. "The Mweru-Wantipa Fishing Record". *Rhodesian Agricultural Journal* 57, no. 4:331–37.

Tait, C.C. 1965. "Bangweulu". In Mortimer, M.A.E. (ed.), *Natural Resources Handbook: The Fish and Fisheries of Zambia*. Ndola.

Welcomme, R.L. 1972. *The Inland Waters of Africa*. FAO/Committee on Inland Fisheries Africa. Technical Paper no. 1.

Williams, C. 1977. *Economic Appraisal of the Fishing Industry in Zambia*. Department of Fisheries, Mimeo. Zambia.

Williams, R. 1972. "Relationship between the Water Levels and the Fish Catches in Lakes Mweru and Mweru Wantipa, Zambia". *African Journal of Tropical Hydrobiology and Fisheries*, 1, no. 2:21–32.

Zyambo, G.C.N. 1978. *Annual Report, 1977*. National Parks and Wildlife Service of Zambia. Lusaka.

8
The Use of Wild Life in the Development of Zambia

ROBERT A. PULLAN

Zambia, with a large game estate, large numbers of animals, and a diversity of species and of endemic sub-species, is well endowed with wild life. But, like all developing countries, it has had and will have many conflicts of interests stemming from alternative land uses. These conflicts will intensify as the human population increases and the government is faced with problems of urban growth, resettlement, and allocation of land for cultivation, ranching, wild life, forestry, industrial development, and the supply of water for domestic uses and power generation.

The government therefore needs to have clearly defined specific policies and attitudes to wild life. But these policies cannot be rigid and doctrinaire. They should be flexible and must be based on sound information and ideas.

This study examines the roles of wild life in the economy and development of that area of land which was Northern Rhodesia and is now the Republic of Zambia. It looks briefly at the pre-colonial and colonial periods and looks in greater depth at the post-Independence period. It presents some of the findings of a major study, recently started, and attempts to look at one very important aspect of wild life which all governments must consider. Is wild life a valuable resource? From this follow the questions how best it can be used to contribute to the development of the country and what the costs will be. The study also looks at the equally important questions of whether different forms of wild life use are compatible and whether a developing country can afford conservation measures. There are various approaches to these questions. This study looks at what has happened in the past, for this can provide insights for the future. The experience of other countries may be of help but may not always be relevant, for there are both environmental differences and economic and cultural differences between nations.

Role of Wild Life

Wild animals have been used in various ways by different human societies in the past, and their importance as a resource has also changed as societies have changed. Eight uses have been identified in Africa:

1. The use of wild life in the economies of subsistence in which the hunting of wild animals may be associated either with gathering of vegetable produce as the sole occupation of the people, or with various forms of farming. Only the second type exists now in Zambia, but it is very widespread.

2. The use of wild life as an exploitable resource providing goods for local or international trade. Such products are usually meat, skins, furs, teeth (hippo and elephant ivory), and horns (especially rhino horns).

3. The use of wild life for sport (recreation) hunting. This may take the form of hunting by local residents, non-resident nationals, or by aliens. It may take the form of organized safari hunting. In all cases, the hunting is for trophies — usually heads, horns, or ivory — but in some cases for skins. Exceptional size is desirable.

4. The use of wild life for cropping, whereby an operator will cull many animals of either one or several species for meat and other products, but with the objective of maintaining a sustained yield from the area where cropping is undertaken. Cropping may also be a management process in national parks.

5. The use of wild life for the provision of meat by ranching. There will usually be some provision of water and confinement by fences. The species are usually selected and may be kept with cattle in tsetse fly-free areas.

6. The use of wild life as semi-domesticated animals on farms, where they are either controlled by herders or kept in paddocks, possibly with improved grazing. They may be given supplementary food, provided with water and probably given some veterinary care. Breeding stock will be maintained and meat or feathers produced. In the past, a few animals (zebra and ostrich) were used for transportation.

7. The use of wild life, within the general context of tourism, for viewing and photographing, usually associated with environments which are frequently beautiful or interesting, and designated as national parks. They may also be used in the general context of education.

8. The protection and conservation of wild life species for scientific and aesthetic reasons, often within the context of preventing local, national, or world extermination. This use will usually be associated with game reserves or government sanctuaries.

The Environment

Zambia is situated in the heart of south-central Africa. It consists largely of extensive plateau surfaces (900–1,200 m) from which rise small isolated mountain areas, the highest rising to more than 2,100 m (Mafinga Hills and Nyika Plateau) in the northeast. The plateaux are drained by the upper Zambesi, the upper and middle Kafue, the Luapula, and the Chambezi rivers which occupy extensive flood plains for much of their length. The plateaux are also drained by many tributaries of the middle Zambezi and the Luangwa rivers which flow in fault troughs (300 to 600 m) with well-defined valley walls rising steeply to heights of 600 m. Also on the plateau surfaces are several major downsags which contain the shallow lakes of Bangweulu, Mweru Wantipa, and Lukanga. In the far north, the more recent rift valleys associated with Lake Tanganyika extend into Zambia.

The climate is tropical and markedly seasonal, and all areas have a rainy season of from five to seven months and considerable rainfall variability. The northern plateaux are wettest, with a mean annual rainfall of 1,000 to 1,500 mm. This decreases steadily towards the south where it is less than 650 mm in the middle Zambezi trough (Hutchinson 1974).

The vegetation, which has been much affected by man through his use of fire and through cultivation, consists of derivatives of semi-evergreen forest in the north and dry deciduous forest in the south. Physiognomically, the vegetation consists of lightly closed to open canopy woodland with either thicket or an open shrub understorey. This understorey is replaced by grasses and sedges in those areas most affected by fire. Numerous fire-affected valley grasslands occur. Extensive areas of swamp and riverine floodplain communities occupy the lake basins and perennial river flood-plains (Fanshawe 1971).

The wild life consists of typical savanna species, mainly those of wood land habitats, though there are some open plain animals (Ansell 1978). Endemic sub-species unique to the country are Thornicrofts giraffes, Cook-sons wildebeests, and Kafue lechwe antelopes. There are major concentra-tions of other restricted species and sub-species such as red lechwes, sita-tungas, and puku antelopes. Until recently, Zambia contained some of the largest concentrations of black rhinos and elephants in Africa. It is justly famous for its concentrations of wild life typical of wet lands and of savanna woodlands.

The human population, which may have been less than one million in 1900, has now exceeded 5 million. Rural population densities are, however, frequently very low over large areas (Fig. 8.1), owing to a variety of environ-mental and historical reasons including the presence of tsetse fly. More than 40 per cent of the population lives within 40 km of the railway line from Livingstone, through Lusaka, to the Copperbelt. Migration to urban areas has been pronounced in recent years.

The Role of Wild Life Prior to Independence in 1964

Early history

Before the third to fifth centuries A.D., Zambia was occupied solely by hunter-gatherers in small dispersed family groups. They were gradually

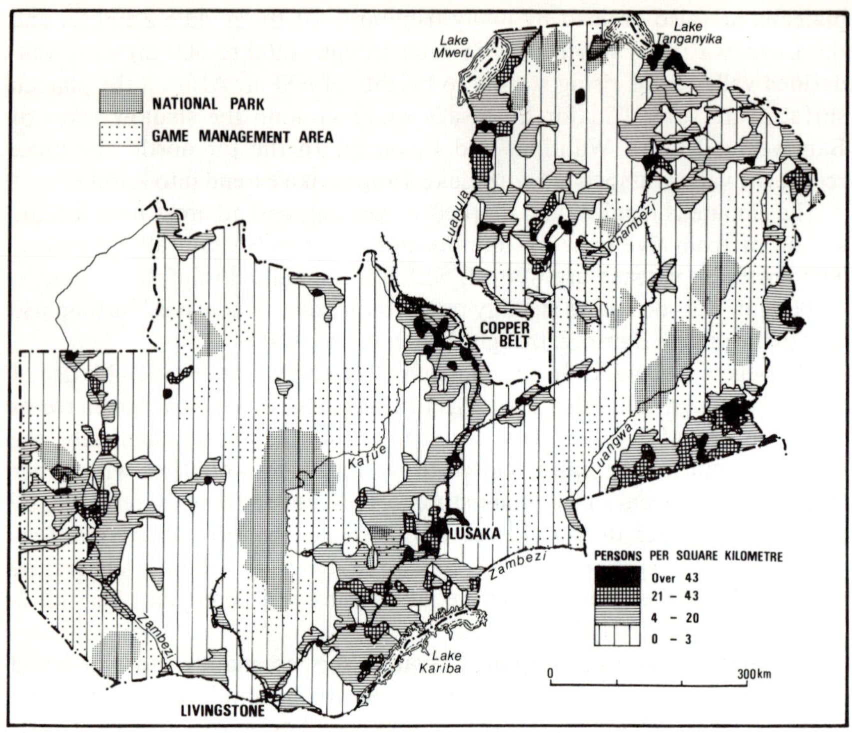

Fig. 8.1 Population density, Zambia, 1969

replaced by Iron Age farmers who also relied heavily on hunting, though some areas had domestic stock and a more pastoral economy by the tenth century (Oliver and Fagan 1975). Immigrants with new political and social structures entered Zambia from the northwest during the sixteenth and seventeenth centuries, bringing with them new crops and techniques. But they still relied extensively on wild life for meat. Long-distance trade in metals, salt, and ivory might have been carried on in Zambia in the eighteenth century, but the plateaux remained peripheral to the development of the kingdoms active in this trade.

Up to about 1840, Zambian wild life had been little affected by man's activities and was found in abundance throughout the country. Firearms were found only in small numbers and were restricted largely to the middle Zambezi valley where the Portuguese had established trading posts by 1720. Even so much of the ivory exported would have been "found" ivory at this stage. This trade was halted in 1835 when the Ngoni pastoralists emigrated across the Zambezi from Matebeleland and settled in southeastern Zambia.

The situation changed rapidly from 1850. The ivory and slave "frontiers" of exploitation moved closer to Zambia (Alpers 1975; Miller 1970), and the

Portuguese returned in greater numbers to the middle Zambezi and also penetrated into the upper Zambezi. Though David Livingstone had found few firearms, abundant wild life, and little slaving in western Zambia on his first visit in 1853 (Livingstone 1857), the situation changed rapidly in the next thirty years. As a result of his visit to the Victoria Falls in 1855 and the publication of his diaries, many traders, hunters, and tourists visited southern Zambia from 1860 (Phillipson 1975). Though tsetse fly-borne disease prevented the southern hunters from operating on horseback, they introduced more powerful firearms which were very effective against elephants (Chapman 1868).

George Westbeech (Tabler 1963) established a hunting and trading post just south of the Zambezi fly zone in 1871 and instituted a very large and remunerative trade in firearms for ivory throughout Barotseland (9,000 to 13,500 kg per year were exported). It was also used as a base for hunting parties of white adventurers from the south (Selous 1881). The first organized safari hunts were advertised for this area in the *Field*, a London sporting journal, during 1876. Further east, in the middle Zambezi, parties of up to 300 men from Portuguese trading posts penetrated into southern and eastern Zambia, and hunted the elephant with firearms. Similarly, the Mambari in the northwest and the Nyamwezi in the north, both hunting and trading in ivory, were to establish networks of local hunters trading ivory for imported goods and slaves. The Nyamwezi were followed by the Arab-Swahili slavers and traders who imported large numbers of firearms during 1870–95 (Roberts 1968). White elephant hunters followed the missionary routes into Malawi and thence into northeastern Zambia.

By the middle 1880s, the elephant population had been decimated in many areas and large numbers were now to be found only in open marshes and river floodplains where hunting with low-powered muzzle loaders was difficult. Other wild life was little affected for, though the numbers of firearms had increased very greatly, traditional hunting and trapping methods were still widely used and, in many areas, hunting pressure would have been reduced after the disturbances of slaving, disease, and famine. However, in the late 1880s white hunters using high-powered rifles were employed by the African Lakes Company based in Malawi to hunt the relict elephant populations and they penetrated far into eastern and northern Zambia (Sharpe 1888).

The BSAC and Protectorate Status, 1894–1911

The British South Africa Company (BSAC) continued exploiting the remaining ivory resources after it assumed administration of the Protectorate of northeastern Rhodesia. Prior to 1894, the amount of ivory exported from the area was not known, though much of the ivory exported from British Central Africa in 1893, amounting to 19,280 kg, came from three localities in northeastern Rhodesia (Johnston 1894). After 1894, the export of ivory was maintained initially at earlier levels, but by 1899 it had fallen to 11,747 kg. In 1901–2 the value of ivory exports was only two-thirds that of 1899 (BSAC Reports).

Data for northeastern Rhodesia were omitted from subsequent reports, which suggests that the importance of ivory was diminishing. Data for northwestern Rhodesia were often omitted, but a specific reference to ivory (BSAC 1900–2) referred to there being little ivory in the area though new districts were being exploited. However, much powder and many guns were being smuggled into the region from the north, and the ivory was being exported illegally. The enforcement of stringent laws was being considered to conserve elephants, but the difficulties of enforcement were recognized. Ivory exported from northwestern Rhodesia in 1907 was valued at £794 and karosses, made from wild life skins, at £721 (total exports were valued at £105,610; BSAC Report 1907).

Sport hunters had also followed the ivory hunters, but from 1892–96 the rinderpest virus panzootic had decimated the populations of many species of wild life. However, most districts of northeastern Rhodesia were reported to be rich in game of all kinds (Chesnaye 1901), an observation confirmed in northwestern Rhodesia from 1903 (Moubray 1912), suggesting that the effect of rinderpest had been very patchy. The sport hunters accordingly increased in numbers. The railway line from the Cape reached Victoria Falls in 1905 and was extended to the Congo frontier by 1909. This facilitated access to new hunting areas, while the Shire Valley railway (1908) facilitated access to the Luangwa Valley. The delights of trophy hunting were advertised by the BSAC from 1904 (BSAC 1908) and many safaris were undertaken (Aitkens 1906; Brunton 1909; Lyell 1910).

Game regulations were introduced in 1902 for northeastern Rhodesia, and in 1905 for northwestern Rhodesia, to replace earlier inappropriate legislation to control hunting and provide different classes of licences for revenue collection from both subsistence and sport hunters. In particular, it was hoped that these would prevent the continued decline of the elephant population.

To this end, the first game reserve was declared in 1899 and three other reserves were established during 1904–8. One of these reserves was at Victoria Falls. A curator had been appointed at the Falls in 1903, owing to the large number of people visiting the area (BSAC Reports 1903; 1904). The nearby riverine area was made into a game reserve in 1907. Indeed, as early as 1894 considerable pressure was being exerted on the BSAC to establish game reserves and introduce protective legislation (Bryden 1894).

The trade in otter skins appeared to have developed in the Bangweulu swamps in 1900 (BSAC Report 1900–1902), and by 1910 more than 10,000 were being exported (Pitman 1934). Some leopard and other cat skins were also being traded at this time. The BSAC felt it worthwhile to introduce an export duty on otter skins in 1909.

Thus wild life had moved into a phase of supporting sport hunting, skin and fur exports as well as ivory exports and the administration was receiving revenue from export duties and hunting licences. During this period, the value of wild life exports formed a large percentage of the total exports. The use of wild life for meat enabled many white settlers to exist during their first difficult years, and in spite of restrictive legislation subsistence hunting by natives remained widespread and of great importance.

The British Protectorate of Northern Rhodesia, 1911–1924

During the first three years of this period, the export of skins and ivory increased and sport hunting was popular. The amount of ivory exported during 1911 from the whole protectorate was 1,105 kg, valued at £887. The value of wild life hides and skins was £2,054. There were substantial increases in exports up to 1913 (ivory 3,018 kg, valued at £2,454; skins valued at £3,865). But these represented only 7.5 per cent of domestic exports in 1912. The value of skins exported increased still further during the First World War, but that of ivory declined (A.R. Customs 1911–23).

Sport hunting, which remained popular until 1914, probably ceased abruptly with the advent of the war and the mobilization of all white residents as fighting broke out on the northern frontier. The military supply route was pioneered from the railhead across the Bangweulu swamps and northwards along the Chambezi Valley. On the average, 72,455 carriers and 1,886 canoemen were employed each month to support about 2,500 troops (Marshall 1956; Gann 1964). Professional hunters, black and white, were employed to supply all these men with meat, and licences were granted for some hunters to trade meat for grain. Many thousands of wild animals, such as the tsessebes and black lechwe antelopes, were killed during each year of the war, especially on the lake flats. Elsewhere in northern Rhodesia, there was little hunting apart from subsistence hunting.

The Colony of Northern Rhodesia, 1924–1940

The discovery of rich copper ores in 1925 heralded a rapid increase in white immigration and employment for natives. Wild life had largely recovered from the rinderpest plague and was extensively used as a meat source by many people. Concern was soon expressed about the rapid decline in wild life numbers especially near the line of rail, and in 1931 a wild life survey was commissioned (Pitman 1934). The rapid reduction in wild life was attributed to excessive hunting by the rural native population using licensed firearms (30,000 muzzle loaders). One thousand shotguns and another thousand rifles were also in the hands of hunters, black and white, in rural areas. However, in many areas large numbers of native men were recruited for employment in the mines of both Northern Rhodesia (22,000) and other countries (43,000) (A.R. Native Affairs), and this would have reduced hunting pressure in many areas. Many more were employed in the developing urban areas and by white farmers. However, as the road network expanded and the number of vehicles grew, widespread illegal commercial exploitation of game animals is believed to have taken place (Pitman 1934). A total of 407,000 cartridges were imported in 1927, and soon pressure was applied to the British government to introduce conservation measures (Hingston 1930).

During 1924–30, exports of ivory and rhino horn increased annually, fur skins exports reached over 10,000 a year, and reptile skins became an important export. The value of all wild life exports averaged £15,800 a year, reaching a maximum of £21,000 in 1928 (this was only 4.3 per cent of total exports by value). Revenue collected from gun and game licences and export duties on wild life produce averaged £8,900 a year, of which licences contributed £5,300. In 1924 this constituted about 2.2 per cent of the total revenue. The government had little expenditure with respect to wild life as there was no Game Department. However, from 1926, some provincial administrations gave rewards for the killing of wild pigs and baboons which were becoming pests on farms. National rewards were paid for killing lions, wild dogs, and leopards. As yet no control shooting had been instituted for crop-raiding elephants, though demands were being made for this. The tsetse fly had spread in many areas and this was attributed, perhaps falsely, to an increase in buffaloes, and by 1931 native hunters were being encouraged to shoot this animal which they had formerly been forbidden to kill.

Sport hunting was unimportant during this period. Though certain species were protected under the Game Legislation (1925), there were no restrictions as to where hunting might be carried out, earlier game reserves having been decontrolled. Only after conservationist pressures were increased was the number of animals available on game licences controlled in 1931 and the first five colonial game reserves declared.

The economic depression in 1932 led to considerable white emigration and the near economic ruin of the country. Large numbers of natives returned to their rural homes, and pressure on wild life increased throughout the whole country with respect to subsistence hunting, though wild life exports dropped. The Pitman Report published in 1934 had recommended the establishment of a Game Department, but this was not implemented until much later. However, other recommendations, such as the formation of an elephant control unit, were implemented. As the price of ivory had dropped, it is doubtful whether this unit paid for itself, for Pitman had advocated the shooting of 800 a year and the most shot in any one year was 344 in 1938. The problems created by vermin species, as crop pests, increased (Colonial Office 1938) and buffaloes were declared a vermin species. Rewards for their control constituted a cost to the administration. Similarly, the first controls were applied to prevent the spread of the tsetse fly, mainly as a result of the increase in vehicular traffic, and this constituted a further expense.

However, though expenditure related to wild life had increased, the revenue derived from wild life had increased too and averaged £10,000 a year (1932–39). Game licence fees reached their highest total (£3,742) in 1932, indicating the pressure on wild life at the height of the economic depression. The average annual value of all wild life exports was £12,678, but the value in 1937 was £21,497, suggesting a much higher degree of exploitation than before the depression. About half the total was contributed by ivory exports, derived in later years mainly from elephant control hunting.

Hunting in rural areas may well have decreased in intensity as the economy recovered and men left home to work. There were also alternative domestic meat supplies for the towns and mining communities, and trading in game meat was generally forbidden. Yet as the number of white immigrants increased, so did the number of rifles. A total of 635,000 cartridges were imported in 1939.

There was, however, little development of sport hunting, though some semi-professional operators were known to have advertised safaris. The administration tried to encourage safari hunting by advertising (Yearbook 1939), but little revenue was received from this source. Likewise, though Huxley (1931) had predicted the rapid rise of tourism in Africa as air routes were developed, the country was too poorly served by its infrastructure to accommodate tourism related to wild life. The Cairo–Cape Town air service was inaugurated in 1932 and used two airports in Northern Rhodesia, but though the first camps were built for tourists at the Victoria Falls in 1925 (Suckling 1964) and game animals could be seen in the adjacent game park, the Information Office (1938) could suggest no other game viewing areas apart from the private Lochinvar Ranch.

Conservation also received little attention from the government in spite of the many recommendations by Pitman and despite political pressure from London. Some further protection was afforded by legislation to certain species and three new game reserves, including two in the Luangwa Valley, were established. Recommendations to establish a national park were not implemented.

During the period under review, wild life continued to be extensively exploited for food. The number of wild animals declined rapidly near the urban areas, along the line of rail and within the Native Reserves. There was probably little ivory poaching, however, and elephant and buffalo numbers were increasing rapidly in some areas according to farmers and Game Department officials. The expansion of both white and native farming caused displacement of animals in some areas. Many rural administrators and the public believed that wild life would, and always should, provide an assured supply of meat for subsistence hunters, and so they should exercise their traditional rights without interference. Implementation of the Game Laws was negligible, as shown by the fact that there were only 120 prosecutions for gun and game law offences by native courts throughout the country in 1937.

The exploitation of skins and ivory fluctuated and exports fell during the economic depression. They were increasing rapidly towards the end of the period, but even at this time the major importance of wild life as a resource was as a source of meat. However, apart from the revenue obtained from gun and game licences, the government did not profit from the use of wild life in this way.

Safari hunting did not develop significantly, as it did in East Africa during this period and, apart from the game park at Victoria Falls, wild life tourism was completely undeveloped.

Conservation was not seen as a major concern by the government in spite of the Colonial Office pressure with respect to the implementation in 1936 of the agreements of the Convention for the Protection of the Fauna and Flora of Africa (1933). Most residents and government officials saw wild life in the context of crop destruction and damage, which created the need for animal control hunting and for vermin rewards. The increasing threat of tsetse fly-borne disease to domestic animals, especially cattle and to a lesser extent to man, and the implication of wild life as a major host of the tsetse fly, resulted in conflicts of interests. Buffeted by these views, the administration failed to define its policy towards wild life and failed to see the significance of the ecological interaction of wild life with other forms of land use.

The Colony of Northern Rhodesia, 1940–1953

A Department of Game and Tsetse Control was constituted on 1 January 1940 and, for the first time, a clear policy with respect to wild life was presented. An essentially ecological approach was adopted for the management, conservation, and rational use of wild life. New legislation was introduced in 1941 controlling all aspects of this policy. Initially, the idea was to protect subsistence hunting interests and to this end Controlled Hunting Areas (CHAs) were declared which enabled residents to hunt in them without control, and non-residents by special licences which were to be strictly controlled. By 1953, 56 per cent of the total area of northern Rhodesia was designated as CHA land.

The total, but theoretical, protection of certain species and of all wild life within the 16 game reserves (6.7 per cent of the area of the country) established in this period was afforded by legislation. The first national park of 22,420 km^2 was established on the Kafue River in 1950. Though theoretically controlled by law, subsistence hunting increased in intensity over this period. The rural population was increasing and 42,500 muzzle loaders, 5,187 shotguns, and 1,225 rifles were registered by rural administrators in 1953. The number of convictions for game or gun offences also rose to 480, indicating a greater measure of patrol activity. Outside the designated game estate, the number of large game animals was severely reduced by hunters over this period. The white population increased to 50,000 and more than one million cartridges were imported each year.

After the end of the Second World War in 1945, control shooting by the department increased more than had been expected. Between 1943 and 1953, 4,746 elephants and more than 9,000 buffaloes were shot on control. Initially, it was intended that buffalo control should provide revenue through the sale of meat and hides, thus forming the first national attempt at cropping. But trials showed that, in the areas proposed for hunting, shooting on control interfered with tsetse control operations. Buffalo control was therefore handed over to the native authorities with a mandate to shoot 3,000 every year, but this target was never achieved.

Neglect of small animal control work during the war, together with increased hunting of leopards and other predators which had valuable skins, increased the problem of wild pig and baboon attacks on crops during this period. Provincial and national reward schemes for vermin kills were continued and two Game Department control units were established for work in the two worst hit areas. Hyenas and wild dogs were added to vermin rewards as their skins had no value and so shooting was limited, but leopards were removed from the vermin list. Jackals were also added in 1950 when a rabies threat appeared. More than 3,840 herbivores and 550 carnivores were reported killed by these units or by reward hunters.

Shooting of game animals on control also became important in the context of tsetse clearance and control schemes. Data are incomplete, but at least 23,000 animals were shot between 1944 and 1953 by hunters employed specially for this work. In addition the first game-proof fences were erected to prevent recolonization of cleared areas by game animals.

Further control measures had to be initiated with the threat of rinderpest entering the country with either cattle or game animals moving from Tanganyika where the disease was spreading south during 1941. The control measures involved digging a ditch and building a flanking brushwood fence over a distance of 250 km. Hunters patrolling this line and adjacent areas shot more than 10,000 game animals and, at times, up to 65 hunters were employed. These controls were successful and in 1953 the threat had disappeared.

The export of wild life products continued after the war. Ivory averaged 14,604 kg a year, and the value of skins exported was higher than that before the war. Certainly reptile skins were in demand in Europe during this period, and by 1953 professional crocodile hunters had reached Northern Rhodesia from Rhodesia and Botswana.

Sport hunting was unimportant initially, but in 1950 the Game Department organized its first safari. International publicity was given to this activity and 13 hunters were catered for in 1951, after which time the increasing political uncertainties prior to Federation led to a fall in numbers.

Tourism, which was now becoming very important in south and east Africa, was boosted by the appointment of a Tourist Officer in 1949, and the first game-viewing camps were opened in association with a native authority in the Luangwa Valley in 1949. The first venture was successful and profitable. After the creation of the Kafue National Park in 1950, temporary game-viewing camps were opened in 1953, two years before the initial plan had intended. Four camps provided accommodation for 50 visitors. At Livingstone, more than 4,500 cars visited the game park.

The Game Department was well established by the end of this period, and the implementation of its policy was well under way. However, far from being self-supporting as envisaged by Pitman (1934), it was financially heavily dependent on the administration. Table 8.1 gives the recurrent and capital expenditure of the department and the revenue attributable to wild life. The recurrent costs rose sharply to more than £20,000 from 1942 as a result of various control work, and from 1948 to 1953 there was a gradual increase as other activities were started and operations extended. By 1953 the recurrent costs were £138,000 and capital expenditure was nearly £60,000 a year. Total revenue was only £28,000. However, the benefits to agricultural production derived from control hunting cannot be costed, and much meat was provided free to local residents as a result of this activity. The revenue was certainly much greater than in pre-war years, and the sale of government ivory had more than doubled to £10,700. Safari hunting made a profit only in 1951 and losses occurred in other years. Rinderpest control cost £36,700, buffalo control £4,100, and tsetse control £43,100, excluding salaries and travel. Vermin rewards cost the government £24,200 over the period.

The value of wild life exports averaged about £25,000 a year, of which ivory contributed £15,500.

Tourist game viewing, while producing a profit for the low-cost Luangwa Valley scheme, was being developed mainly in the national park, and this cost £25,000 in capital expenditure in the first three years and accounted for

TABLE 8.1 REVENUE AND EXPENDITURE OF THE GAME
DEPARTMENT IN NORTHERN RHODESIA,
1940–53

	Total Recurrent Expenditure (£)	Total Capital Expenditure (£)	Revenue (£)
1940	7,200	n.a.	4,600
1941	9,100	n.a.	9,600
1942	20,200	n.a.	5,100
1943	23,300	n.a.	13,400
1944	29,600	n.a.	18,200
1945	29,200	n.a.	9,700
1946	28,200	n.a.	25,100
1947	27,700	29,800	15,400
1948	33,100	36,400	15,100
1949	53,100	40,200	18,800
1950	70,500	47,000	12,900
1951	82,600	60,000	24,100
1952	92,700	53,300	28,400
1953	138,300	60,000	n.a.

Note: Revenue includes licence fees for guns, ammunition, and
game, sale of government ivory, sale of buffalo meat from
rinderpest control, (1947–51) and revenue from conducted
hunting parties.

Source: Northern Rhodesia Colonial Office Annual Reports,
various years.

a substantial rise in total recurrent costs. To offset these and other costs,
help was sought through Colonial Development Aid from the British government with respect to a ten-year development plan (Vaughan-Jones 1945)
involving the expenditure of £732,000 (recurrent) and £95,000 (capital) in
conjunction with Nyasaland. Revised plans in 1947 allowed £502,000 to be
spent and about £400,000 had been contributed by 1954.

The period under review was one when expenditure on wild life and
associated problems and developments increased greatly and was not
covered by revenue gained. A game estate had been established which was
theoretically justified but too large in terms of available resources, and
management of this estate was negligible. Protection was afforded to the

animals in varying degrees in the game reserves, though some had to be decontrolled because of excessive illegal hunting. Clearly the legislation was not effective, and one observer reported that "nearly ninety per cent of the game has been killed in the last twenty years" (Watt 1954). Yet wild life was still abundant in many areas and the diversity of species had not declined on the whole.

The control work had prevented a rinderpest disaster and some areas had been freed from trypanosomiasis. The framework for management had been laid but not implemented and little undertaken. The period was still one of exploitation and control. The first tentative ventures into government-organized sport hunting and tourist game viewing had been started and a national park established.

Federation of Northern and Southern Rhodesia and Nyasaland

The ten uneasy years of Federation saw a rise of 24,000 in the white population to reach 77,000 in 1962, after which it declined. A census shows the black population to be 3.4 million in 1963, 75 per cent of these being rural dwellers. The area of land under cultivation increased considerably, reducing the habitat available to wild life. The number of guns also continued to grow, increasing the hunting pressure. The notable ecologist F.F. Darling was asked to visit Northern Rhodesia in 1956/57 and make recommendations with respect to wild life conservation and use (Darling 1960). A substantial increase in staff and expenditure for the Game Department in the first years was approved. However, a fall in the price of copper, the country's main revenue earner, resulted in a drastic retrenchment in 1958, to be followed by the removal of tsetse control work to the Veterinary Department and a demotion of the Game Department to the administration of the Ministry of Native Affairs. Therefore, Darling's recommendations for the establishment of a scientific department involved in ecological research and management of the game estate and wild life generally, supported by a strong unequivocal policy of the administration, were not implemented immediately. The situation was redressed in 1961 after pressure was applied by the Colonial Office (Worthington 1961). At this time, the IUCN Conference at Arusha (UNESCO 1963) identified wild life as "Africa's most neglected but potentially one of its most valuable renewable natural resources". A new policy was defined by the government and the value of wild life as a resource was recognized. The Game Department expanded but it was not made a scientific department. Preservation, conservation, cropping, and control were seen as the prime functions.

Hunting. The earlier policy of uncontrolled residential hunting in the CHAs was believed to have failed to conserve animal stocks. Therefore, new legislation was introduced in 1954 whereby all hunting in the better stocked areas (1st Class CHAs) were controlled by permits issued by the Game Department for both residents and non-residents, thus allowing white hunters better access to areas formerly reserved for local residents. This was bitterly opposed by residents and not passed until 1958.

The Game Preservation and Hunting Society had been formed in 1952 and soon had many branches with a thriving membership. Areas with poor game stocks (2nd class CHAs) were left in the control of native authorities who issued game licences and all activities of the Game Department were withdrawn. Other developments were legislation allowing Private Game Areas (PGAs) to be designated to protect game on farms. By 1963, 114 PGAs had been established but only 9 were vested in native chiefs and 3 in hunting associations. The rest were on white farmers' land. It was believed generally that during this period game stocks did increase in the national parks and those game reserves regularly patrolled by game guards. But illegal hunting increased elsewhere as the number of vehicles and roads increased and the demand for game meat in the mining townships grew.

During the last three years of Federation, when political unrest and lawlessness increased, poaching became very heavy. A measure of this can be gathered from the 2,669 elephant tusks which were surrendered during an amnesty in 1964. To combat illegal hunting, Honorary Game Wardens were appointed and there were more than a hundred by 1963. Clearly, subsistence hunting continued at high levels, and though much of this was legal there was little constraint on numbers of animals killed. Illegal hunting for profit also increased in many areas, and the overall result was the continued rapid reduction in game stocks in many areas, especially of the larger antelopes.

Sport hunting. This continued as a departmental operation, and it catered for a limit of ten hunters a season. Although profitable during the early years of Federation, increasing political unrest led to a fall in the number of hunters in the last three years when it operated at half capacity.

Control hunting. The two vermin control units killed more than 15,500 small animals and crop raiders in the first four years, but in 1958 the control units were disbanded. Rewards continued to be paid, however, and more than 36,000 animals were killed from 1954 to 1959. National rewards were dropped completely in 1961, and the responsibility for controlling pests was given to provincial authorities. From 1958, efforts were concentrated on the resettlement areas to be used when Lake Kariba was formed, and 14,500 animals were killed in these operations. The overall impact of these operations is difficult to assess and would have provided short-term relief only.

Tsetse control operations continued with hunters patrolling game fences, but elimination over larger areas was not carried out. At least 1,500 animals were killed each year, and 618 km of game fence constructed and maintained. More than 3,500 elephants were killed on control over ten years, but this was a reduction over the previous ten years. However, it should be noted that economic necessity rather than decreasing need caused the reduction.

Wild life cropping. Darling had advocated the introduction of game cropping in his report to the government, and from 1956 some small schemes were carried out, in PGAs, with departmental advice. Three larger schemes were started on a trial basis by the department in conjunction with native authorities. The first, started in 1960, was abandoned after three years because of operational and marketing difficulties. A total of £13,000 was lost on this scheme (Auditor General's Report 1966). Sustained cropping demanded scientific data, and for the second scheme, using lechwe antelopes, a research programme was carried out in 1961/62 which recommended the cropping of 4,235 animals a year on Lochinvar Ranch, to yield more than 1,000 t of meat. However, only 532 animals were cropped in the first year. The third scheme was undertaken in the Luangwa Valley where a small number of elephants were shot to provide dried meat. More than 18,000 kg were produced, and the sale of ivory allowed a profit to be made. The planned target had not been met, but it was felt that larger schemes could be successful.

Tourism. The success of the native authority camp in the Luangwa Valley was followed by the involvement of the Game Department in creating a unit to construct game-viewing roads and other facilities in the Luangwa South Game Reserve. By 1960 five camps (46 beds) were used, together with two camps (24 beds) outside the reserve. Development of the national park continued and a catering lodge was built with 36 beds and nine other bush camps with 50 beds being operational. Two airstrips were built in the park in 1962 for light planes, and air tours from Livingstone started. The number of tourists from southern Africa was gradually increasing but, at a time when air transport allowed large numbers to come to East Africa for wild life tourism, Northern Rhodesia remained isolated and distant from the major airline routes, and by 1962 political disturbances brought about a decline in the number of visitors. At this time, much of the accommodation was geared to local residents only, who could use it only during the dry season. The tourist industry had thus expanded but generated little revenue and it was costly to maintain the services.

Wild life exports. Separate figures have not been published by the Federal Government for Northern Rhodesia. Extrapolations from pre-Federation data for all three countries suggest a high level of elephant exploitation initially (1955) until the new laws became effective. Ivory exports increased by about 20 per cent. Exports of skins are not known, but Cott (1961) reported excessive crocodile hunting between 1954 and 1957 and possibly as many as 17,000 a year were shot for their skins. Crocodiles were taken off the vermin list in 1962 after fears for their survival were voiced and some controls applied.

Finance. Incomplete records do not allow a detailed presentation of revenue derived from wild life to be given, but known revenue rose from £28,000 (1954/55) to £69,000 (1962/63). Ivory sold by the government constituted 32 per cent and tourist revenue only 10 per cent of this total. Wild life contributed an insignificant amount to the total government revenue of £20.3 million.

If Northern Rhodesia contributed 70 per cent of the total ivory export of the Federation, then this would be worth about £25,000 a year, but the value of crocodile skins alone may have been worth £40,000 a year. The total export of wild life products may have been about £80,000 a year. An estimate of the annual value of meat and trophies lost to the government by illegal hunting (A.R. Game 1964) has been given as £200,000, so that the average annual production of wild life may have been between £250,000 and £300,000. Recurrent expenditure for the Game Department is available only for the first four years. It rose from £151,000 (1954/55) to £310,000 (1957/ 58), after which it was cut. More than 70 per cent was spent on wild life and the rest on fisheries and tsetse control, though expenditure on tsetse control was to increase substantially in the following years.

The average capital expenditure was £48,000 until 1958, but declined to only £38,000 during the three years 1958–60. Expenditure on the Kafue National Park was £106,000 over the period, and £47,000 was spent on the Luangwa Game Reserve. Tourist development received more than 50 per cent of the total capital expenditure.

During the period under review, Northern Rhodesia was in the forefront in trying to develop game cropping. But the government did not see wild life as having an important role in the future of the country. With increasing control hunting being necessary, together with the increasing threat of tsetse fly to commercial farm expansion, the government did not see the necessity for a scientifically based department undertaking research and applying this to wild life management and conservation. The activities of the Game Department were again hindered by an economic recession and later by national lawlessness and political uncertainties. Though a clearer government policy had been established after considerable pressure, the means to implement this were not granted and the development of tourism and safari hunting were not vigorously pursued. There was some evidence that wild life exploitation of ivory, skins, and meat increased considerably during this period to the benefit of both government and individuals. But the total wild life population continued to decrease, following the trends of earlier years. Fortunately, the size of the country and its still small population meant that the wild life population as a whole was still large and gains were made in the national parks and in some of the game reserves.

Role of Wild Life after Independence

Republic of Zambia: The First Years, 1964–1965

After the establishment of the Republic of Zambia in October 1964, the new government's policy on wild life was to accord it increasing importance as a natural resource which could be developed within the context of the long-neglected rural sector, now seen as a priority area. Conservation and utilization of the resources were to be compatible, and to this end a Department of Game and Fisheries was created in the new Ministry of Lands and Natural Resources. Certainly the time seemed ripe to take new initiatives, for total governmental revenue was to increase from £31.8 million during the last year of the Federation to £78.5 million in 1964/65.

A Transitional Development Plan was introduced early in 1965 which greatly increased both recurrent and capital funds for the Game Department. Initially, much of the development was in improving the living conditions of junior staff and extending the activities of conservation and management into the remoter and formerly neglected areas of the country. There were no distinctive changes in policy with respect to wild life. The rapid development of tourism was seen to be important, but the Director noted that "perhaps the greatest contribution that the Game Estate can make to the national economy is the production of meat" (A.R. Game 1965). Protein malnutrition was believed to be widespread, especially in those areas where domestic livestock could not be kept because of trypanosomiasis carried by the tsetse fly. It was estimated that beef provided only one-fortieth of the meat consumed in the country. Another reason for considerable concern was the increasing environmental degradation caused, it was believed, through too many animals, especially elephants and buffaloes, in the Luangwa game reserves. Cropping of these animals was seen as a means of solving or ameliorating two pressing problems at once. A research programme to investigate the domestication of certain game animals was also initiated with a view to their ultimate use in tsetse fly areas as an alternative to tsetse fly clearance.

Hunting.   Animals on the game estate were seen now to be a national asset and a resource to be used for the benefit of all people. Illegal hunting had become very widespread in the unsettled period prior to Independence, and a major initial task of the department was seen to be the immediate reduction of poaching. Game cropping trials had shown that their successful operation partly depended on the absence of illegal hunting. The President accordingly issued a national appeal for citizens to obey the game laws. The Game Department had 203 junior field staff in 1963; they were increased to 356 in 1965 and a new command structure introduced to direct operations. Estimates were made that the loss through illegal hunting of meat and ivory exceeded £200,000 per year, though the basis of the calculation was not given and there were little data available on which to base this estimate at this stage. Using the data from Pitman (1934) who estimated that the lowest number of animals poached was 90,000 a year, and assuming a mean carcass yield of 45 kg of meat, the loss through illegal hunting would have been £450,000, excluding ivory and trophies. Given that there had been a great increase in population and in the number of guns by 1965, a higher figure would not be impossible. The legal offtake of meat from CHAs by 1,800 licensed resident hunters was given as 1.24 million kg (A.R. Game 1966) and valued at more than £300,000. It is possible, therefore, that the value of meat and trophies from wild life shot each year throughout the country may have exceeded £1 million.

Sport hunting.   There was a substantial increase in the number of safari hunters from 18 (1964) to 30 (1965), catered for by two private safari companies. This was taken as an indication that Zambia was now on the "hunters map of Africa", and it encouraged the department to pay attention to the conservation of some areas specifically for trophy hunters.

Control hunting.   Very little control of animals other than elephants was recorded, and the number of elephants shot declined from previous years. Delegation of responsibility for vermin control seemed to have passed completely to local authorities and individual farmers.

Wild life cropping.   Departmental policy in these years was committed to the more extensive development of cropping, even though initial trials had not always been encouraging. In 1964, 86 large animals had been cropped in the Luangwa Valley without too much difficulty and the meat disposed of locally, and in 1965, 32 animals were cropped and the meat refrigerated and sold in urban areas, more than 600 km away. A total of 13,000 kg of frozen meat were produced at a profit of more than £1,200. As a result of this trial, the comment was made that "the time is now opportune to produce meat as a commercial venture and on a sustained yield basis for consumption in urban areas" (A.R. Game 1964). Detailed planning followed, with the intention of implementing a bigger scheme in

1966. Lechwe antelopes were cropped at Lochinvar Ranch in 1964 and the meat sold in the mining areas, but further development awaited government acquisition of the land.

Wild life domestication. Considerable interest was being shown in the possible domestication of some species of animals in Africa for use in connection with game ranching in tsetse fly areas (Bigalke 1964). A small trial programme was undertaken using lechwe and eland antelopes.

Tourism. After Independence, the number of international visitors to Zambia increased and 117,000 were recorded in 1965. About half of these were estimated to take some sort of holiday in Zambia; though many stayed with friends or relations, 22,000 were classed as tourists. The foreign exchange spending of these holiday-makers was estimated to be £443,000 (A.R. Game 1965), so that it came apparent that tourism could become an important source of revenue. However, the development of tourism related to wild life had been slow and beset with many difficulties during Federation. Upon Independence, there were only 15 camps with 150 beds and only one was a catering camp (36 beds). Political disturbances and the declaration of Rhodesian Independence in November 1965 created problems for certain areas, while unexpected environmental problems such as the flooding of the camp site on Lake Tanganyika also occurred. Therefore, the number of visitors did not change substantially on the whole, but petrol shortages were a factor accounting for a drop of more than 6,000 visitors to the Livingstone Zoo Park, a fenced park for animal viewing but without elephants and free roaming carnivores, built in 1965. Tourism related to wild life was flourishing elsewhere in Africa but, as Table 8.2 shows, Zambia could not take advantage of the changing situation.

Wild life exports. Exports of large numbers of skins continued during 1964 and 1965, though the first signs of a decline were apparent. Crocodiles (5,224 exported and 1,889 imported for re-export) and monitor lizards (3,089) formed 74 per cent of all skins exported in 1964. However, exports of skins of both animals fell by half in 1965, though python skin exports increased to 315. Otter skins also increased to more than 1,000 from 330 and other skins of zebra, antelopes, and spotted cats increased considerably in 1965. There was also a considerable importation of skins of some species, presumably in transit. Certainly, the exploitation of certain species for skins was at such a level as to cause some concern among conservationists (A.R. Game 1964).

Ivory exports fell from 28,000 kg in 1964 to 20,000 kg in 1965, though the number of tusks exported rose from 2,286 to 3,389. This apparent anomaly was due to the large number of small tusks acquired by the government as a

TABLE 8.2 NUMBER OF VISITORS TO SELECTED
NATIONAL PARKS IN AFRICA,
JULY 1964–JUNE 1965

Kruger National Park (South Africa)	225,398
Nairobi National Park (Kenya)	104,695
Tsavo National Park (Kenya)	39,943
Murchison Falls National Park (Uganda)	26,006
Queen Elizabeth National Park (Uganda)	6,611
Luangwa Game Reserve (Zambia)	2,334
Kafue National Park (Zambia)	2,171
Sumbu Game Reserve (Zambia)	434

Source: Annual Reports of the National Parks and Game Departments of the countries specified.

result of the ivory amnesty in 1964. As the ivory export for 1963 was 27,000
kg (2,492 tusks), the annual legal cull of elephants over these years exceeded
1,200 animals.

Finance.   Table 8.3 shows that, after Independence, there was a rise in
revenue attributable largely to a substantial increase from hunting after the
introduction of a new fee scale. However, the value of wild life exports fell
considerably after the Federation period as crocodile and ivory exports
could not be sustained after the severe exploitation during that period
(Table 8.4). Wild life exports made a very small contribution to total national
exports.

Recurrent expenditure for the department is not available from published
sources which are given for ministries only. However, Dodds and Patton
(1968) have given a figure of £299,849 for 1965 which is less than the figure
of £310,000 for 1957/58, a time of expansion but which also then included
tsetse control costs of £25,000. Tsetse control costs for 1964/65 are not
known, but £142,000 was spent on Disease and Tsetse Control in 1963/64.
Table 8.5 shows that increasingly greater sums were made available for
capital development. However, staff shortages and other difficulties prevented implementation of all the plans. The emphasis on junior staff housing
as the number recruited increased is apparent, and the implementation of
research programmes in the Transitional Plan of 1963–65 is an important
development. Expenditure on tourist promotion rather than on tourist
development is an interesting trend. More than £100,000 was spent on
capital expenditure relating to tsetse control during this period.

TABLE 8.3 REVENUE DERIVED FROM WILD LIFE, 1963–78

Revenue Sources	1963	1966	1969	1972	1975	1978
			(in '000 £)			
Hunting licences[1]	n.a.	31.6	49.0	99.1	452.7	647.5
Game tourism[2]	10.8	19.0	50.1	131.2	213.6	169.0
Luangwa cropping	0	31.0	113.3[3]	0	0	0
Ivory and trophies	14.3	30.3	1.3	48.7	3.8	44.6
Live animals	0	0	0	6.0	0	0
Total revenue	25.1	111.9	213.7	285.0	670.1	861.1
Court fines (wild life)	n.a.	n.a.	7.5	18.0	19.0	30.2
Wild life revenue (including fines) as % of total national revenue		0.10	0.055	0.10	0.14	0.15

Notes: [1]See Table 8.9 for specific components.

[2]See Table 8.12. After 1970 revenue from wild life tourism was derived from ZNTB and the Department of Wildlife. ZNTB data are not available for 1978 and the figures given for this year have been estimated by the author.

[3]No value was given for revenue for this year. This estimate is derived from the number of elephants culled and previous comparable revenue data.

Sources: Annual Report of the Department of Game and Fisheries, 1963–68; Wildlife Fisheries and National Parks, 1969–73; National Parks and Wildlife Service, 1974–78; Zambia National Tourist Board, 1969–72; Financial Report for the years 1963–78; Statement of External Trade, 1963–74.

TABLE 8.4 VALUE OF WILD LIFE EXPORTS, 1963–74

	1963	1966	1969	1972	1974
			(in '000 £)		
Ivory	n.a.	35.4	62.5	89.2	95.8
Crocodile skins	n.a.	28.1	16.1	0.1	4.0
Fur skins	n.a.	1.1	n.a.	1.0	n.a.
Other wild life hides and skins	n.a.	28.4	34.7	53.6	0.4
Game trophies	n.a.	1.3	2.0	n.a.	n.a.
Total	n.a.	94.3	115.3	143.9	100.2

Note: In some years, re-exports of wild life products are recorded. When this
          occurs the value of the re-exports has been deducted from the total value
          of products exported.
Sources: Annual Statement of External Trade, 1963–74.

During the period under review, the substantial increase in government
revenue and a desire to develop rural activities led to the establishment of
new priorities and the input of greater finance into wild life resource development. Tangible results, however, were slow to materialize with respect to
tourism which suffered because of many varied local problems. Direct
exploitation of wild life slowed down. Plans for increased use of the resource
on a sustained basis were initiated.

First National Development Plan (FNDP), 1966–70

The FNDP was published in July 1966. It proposed, as a major priority,
to change the rural sector through development, thus redressing the disparities of rural and urban life. This was to be undertaken, in part, through
the development of regional identities through provincial administrations.
Game animals and fish were seen to have "reached the stage where they
were poised for further exploitation which will have beneficial effects on the
nutrition standards of living and the economy as a whole" (FNDP 1966).
Through 1964/65, the number of visitors to Kenyan national parks had
increased greatly (Denney 1968), and the FNDP recognized the need for
Zambia to take part in this rapidly developing industry. Therefore, more
than 70 per cent of the capital investment allocated to wild life was ear-
marked for tourism-related developments (Table 8.6).

To further this end, the Department of Game and Fisheries was placed
under the new Ministry of Natural Resources and Tourism, thus linking the

TABLE 8.5 CAPITAL EXPENDITURE ON WILD LIFE, 1965–78

	1965/66	1969	1972	1975	1978
		(in '000 £)			
Research[1]	83.0	5.8	3.3	14.8	18.1
Management[1]	1.0	23.1	107.2	219.4	60.3
Field housing	65.2				
Kafue National Park[2]	—	22.5	—	61.3	3.7
Luangwa Game Reserves[2]	—	54.2	82.4	48.6	8.5
Cropping	—	54.0	—	—	—
Anti-poaching[3]		—	—	58.7	3.6
Training		—	—	22.8	5.8
Others[4]		—	—	—	2.6
Total	149.2	159.6	192.9	425.6	102.6
Tourism[5]	44.4	?			

Notes: [1]Research and management include expenditure on fisheries for 1964–70.
       [2]Expenditure relating to other game reserves and national parks, as well as command management, is included under management.
       [3]Expenditure relating to anti-poaching and training is included under management for 1964/65 and 1965/66.
       [4]Includes the cost of the research on wild life domestication in 1968.
       [5]Expenditure under tourist development and promotion by the Ministry of Natural Resources and Tourism 1964–68 and by ZNTB 1969–70.

Source: Reports of the Ministry of Finance, 1964–78.

overall management of tourism and wild life. In particular, further increases in staff were budgeted for, including two tourist officers. Research and management, both of which had been seriously neglected, were accorded improved status and funds. Nine new senior field staff were approved, together with three ranger pilots as the air wing was expanded.

Hunting.   Non-residential hunting in CHAs proved very popular, and all the available licences were taken up initially. However, owing to heavy poaching in many of these hunting areas, the number of hunters allowed were reduced by half by 1970. Unfortunately, the response of citizens to the appeal from the President was very poor indeed and, in those areas away from the major game reserves and the national parks which were well patrolled, the wild life estate lost "significant quantities of their stocks"

TABLE 8.6 PROPOSED EXPENDITURE RELATING TO
WILD LIFE IN THE FIRST NATIONAL
DEVELOPMENT PLAN, 1966–70

	Capital	Recurrent
	(in '000 £)	
Tourism	1,081	123
Game cropping	162	157
Research	172	67
Protection	56	67
Training	17	35
Total	1,489	448

Source: First National Development Plan, 1966.

(A.R. Game 1968). The formation of a highly mobile anti-poaching unit which was operational by late 1967, and which could call upon the help of helicopters and aircraft, resulted in an increasing number of prosecutions for wild life offences. But shortages of fuel and transport and the failure of the courts to impose deterrent fines and sentences hindered the effectiveness of anti-poaching activities, and illegal hunting continued to be widespread. Though much of the meat from animals shot by non-resident trophy hunters was consumed locally (an estimated 90 per cent was used), no revenue was derived from this source. Poaching continued to deprive the government of much licence revenue as hunting was reduced in certain CHAs to conserve the stock of animals and also to ensure that the increasingly remunerative sport hunting by visitors was successful.

Sport hunting.   Safari hunting continued to be advertised abroad, and attracted increasing numbers of hunters from all over the world. The number catered for by the private licensed safari operators rose from 40 in 1966 to 153 in 1970. In addition to the revenue received, more than 25,000 kg of dried meat and 160,000 kg of fresh meat were provided for the residents of those CHAs used for sport hunting. Rapid development of safari hunting was possible as little expensive infrastructure was required, and the success of the hunters in obtaining good trophies helped to spread the reputation of the country. In so far as this type of hunting is confined to the older male animals only, it does not compete with subsistence hunting, though subsistence hunting may disturb animals and make safari hunting unsuccessful.

Control hunting.   This work continued with respect to large and dangerous animals only. The number of elephants shot to protect crops and property varied between 334 and 381 a year, and the meat from these animals was disposed of locally and the ivory sold. Rare occurrences of lion predation on cattle near the Kafue National Park were recorded, and hippopotamuses became increasingly troublesome in some areas. Between 1,600 and 1,800 animals were shot each year by tsetse control hunters outside game fences, only a quarter of these being small animals.

Wild life cropping.   A Luangwa Unit had been established in 1965 to develop the valley and obtain scientific information. In view of the environmental degradation in the game reserves, and after the successful cropping trials in CHAs, it was decided to build an abattoir and freezing plant for game meat near the river. Cropping began in earnest in 1966 and continued through to 1969, when it was decided to continue only with hippo cropping for experimental purposes. The number of animals cropped and the revenue obtained are given in Table 8.7.

Dodds and Patton (1968) supported cropping and commended the government for taking a bold stand in the face of much hostile public opinion. However, other ecologists believed that cropping was not necessary, and that the environmental problems were caused primarily by fire and only indirectly by elephants (Lawton 1971). Local opinion became anti-cropping largely because people saw a resource being utilized outside the districts, whereas the cropping trials had provided meat for local residents, and so the scheme did not conform with the directives of the FNDP. Finally, international controversy over the merits and need for cropping, which could not be resolved because of the lack of suitable scientific information, led to the premature closure of the scheme at the end of the dry season in 1969. It is also clear that operational difficulties were becoming more apparent as time went on. The methods of hunting had to be changed and long access roads built to obtain elephants (Hanks 1979). Buffalo culling declined quickly as it became very difficult to crop enough animals without disturbance, while the presence of muscle cysts in meat was believed to be hazardous to human health. Hippos presented fewer difficulties, though there was a much greater chance of a conflict with the interests of tourist-viewing in the attractive riverine environment. It can be concluded that premature closure only hastened an event which would have been brought about by unforeseen operational and logistic difficulties.

Therefore, plans to produce 1,000 t of frozen meat each year came to nothing, and the expensive plant became idle. Supplementary plans to install a by-products plant to produce 500 t of carcass meat were not implemented, and the production of dried meat never got beyond the trial stage.

TABLE 8.7 CROPPING DATA FOR THE LUANGWA VALLEY, ZAMBIA,
1966–69

Number of animals cropped	1965	1966	1967	1968	1969
elephant	27	204	374	404	448
hippo	9	218	224	67	87
buffalo	33	100	59	12	27
impala	9	—	—	135	139
Number of carcasses condemned	0	27	35	n.a.	n.a.
Weight of frozen meat sold (tons)	12	204	244	372	n.a.
Revenue — estimated (in £)					
meat	1,265	25,520	36,000	42,640	n.a.
ivory	n.a.	2,820	3,150	21,630	n.a.
skins	n.a.	2,660	20,100	28,420	n.a.
Total	—	31,000	59,680	92,690	n.a.

Source: Astle 1971; Annual Reports of the Department of Game and Fisheries 1966–68,
Wildlife, Fisheries and National Parks 1969.

A second cropping scheme was also planned for Lochinvar Ranch using
lechwe antelopes, and the ranch was declared a Game Management Area in
1967. A small abattoir was built in 1968 with a capacity of 50 animals a day,
but a total of only 78 were cropped in 1968 and another 143 in 1969. To
attain a more balanced age and sex structure in the population, mainly old
males were cropped. However, the incidence of disease was high, and 25 per
cent of the carcasses was condemned. In the absence of a steam sterilization
plant, much produce was wasted, and it became evident too that it was not
possible to kill enough animals to keep the plant running economically.

Though further research was undertaken, cropping was not continued on
commercial lines. The cropping of black lechwe antelopes in the Lake
Bangweulu Basin was also considered feasible if poaching could be controlled. A sustained yield of 1,125 t of meat a year, valued at £230,000, was
envisaged at optimum stocking densities (A.R. Game 1966), and a research
project was initiated in 1969. It was reported that the maximum annual
sustained yield would be only 400 t (Bell and Grimsdell 1973). Cropping was

not tried. Therefore, all the initial estimates of cropping production, which were based on insufficient knowledge and understanding, were considerably over-optimistic. At the same time, it was reported that the site of experimental cropping by a local chief on Chisenga Island in the Luapula Valley, initiated in 1960, which could not be closely supervised in the last years of Federation, had resulted in the complete elimination of lechwes and a severe reduction of puku antelopes.

Domestication. The use of elands as domesticated animals had first been suggested in 1904 (BSAC 1904) and of lechwes in 1944 (A.R. Game 1944). Through the period of the SNDP, research into the use of elands and lechwes for semi-intensive management was carried out by a small unit with considerable enthusiasm. However, as various difficulties became apparent, it was concluded that "under Zambian conditions and taking into account sociological as well as economic factors, there appears to be little or no practical extension prospects for the use of these animals under semi-domestic conditions" (A.R. Game 1970).

Tourism. Zambia had only three catering lodges in wild life areas in 1966. The new international airport at Lusaka was completed in 1966 and the development of a network of internal air services was completed in 1968. These linked the catering lodges and were seen as the springboard from which Zambia could make its bid for a share of the rapidly developing international tourist trade, which was soon to use large jet aircraft on international routes.

Much time and energy were spent in the initial years in planning the new developments. But their implementation in the later years was to prove very difficult because of shortages of trained personnel, the poor supply of materials, and many operational difficulties which stretched the resources of a department committed to other forms of development as well. This led to the successive transfer of the management of tourist facilities to contractors and then to the Zambia National Tourist Board (ZNTB). The ZNTB opened New York and London offices and publicity was increased, but the slow development of city and national park accommodation together with the late opening of game-viewing areas because of heavy rains and floods produced little increase in international tourism. However, internal policies encouraged an increased use by residents of Zambia. The year 1969 was declared the "International Year for African Tourism", but though the will and the finance were available, Zambia could not take full advantage of a situation that was developing throughout the continent. In particular, political difficulties meant there was little tourism from the natural catchment of southern Africa.

By the end of 1970, Zambia had improved much of the infrastructure relating to tourism in Lusaka and at Victoria Falls, but in wild life viewing areas the number of beds had been increased by only 38 to 208 and of these only 82 were associated with hotel-type facilities. Two new camps had been built and a third converted from other uses. Official statistics showed that the number of international visitors to Zambia fell dramatically in 1969 compared with 1968, though they started to recover in 1970 (Table 8.8). However, few of these visitors went wild life viewing. There was a general decline in the use of non-catering camps rather than an overall decline in catering camp use. The percentage occupancy of beds varied greatly between viewing areas and camps, but only Luangwa Game Reserve maintained a level of more than 50 per cent.

TABLE 8.8 TOURISM IN ZAMBIA, 1963–78

	1963	1966	1969	1972	1975	1978
No. of international visitors ('000)	n.a.	108.5	37.7	61.6	51.7	53.3
No. of visitors to National Parks and Game Reserves ('000)	4.9	5.3	5.2	7.2	20.2	14.4
No. of camps and lodges open	15	15	17	17	23	23
No. of beds available	166	170	206	n.a.	455	n.a.
% occupancy	59	54	48	35	41	n.a.
No. of visitors to Livingstone Zoo Park ('000)	25.3	12.1	11.4	30.2	23.2	6.7

Sources: Annual Report of the Department of Game and Fisheries 1964–68; Wildlife, Fisheries and National Parks 1969–73; National Parks and Wildlife Service 1974–78; Zambia National Tourist Board 1969–72. Other data from the ZNTB.

Visitors to the Livingstone Zoo Park increased greatly from 11,431 in 1969 to 28,587 in 1970 following the introduction of day tours by ZNTB, thus halting a progressive decline in previous years which had reflected the decreasing number of southern visitors. The zoo park caters only for day visitors staying near Victoria Falls.

Wild life exports.   The importance of wild life exports declined over the period, reflecting a changing attitude to exploitation, a change in international demand as a result of the increasing impact of conservationist ideas, and a general decline in the number of pythons, monitor lizards, and crocodiles available for hunting outside protected areas. The number of crocodile skins exported fell from 4,345 in 1965 to 703 in 1970, and the number of other hides and skins from 12,252 in 1966 to 3,271 in 1970. However, ivory exports increased owing to the ivory made available from cropping. The average export per year was just under 30,000 kg, and represented an elephant kill of more than a thousand animals a year. In 1968, 783 were killed by the Game Department alone.

Finances. Data for all revenue sources are not available, but the major revenue earners are given in Table 8.3. Total revenue attributable to wild life increased by 92 per cent over the period, and though revenue derived from game cropping was the largest single source, all other sources increased. In particular the increase in revenue from wild life tourism came from some increase in the number of tourists, but even more from higher charges. There was little change in the later years.

Sport hunting catered for greatly increased numbers and revenue showed a progressive increase throughout the period (Table 8.9). The importance of game cropping was very great and the extent of cropping is reflected in the

TABLE 8.9 REVENUE FROM HUNTING LICENCES, 1963–78

	1963	1966	1969	1972	1975	1978
			(in '000 £)			
National game and bird licences	n.a.	4.4	1.8	10.1	59.2	48.8
Safari licences	2.5	11.7	32.2	56.3	50.2	8.8
Elephant licences	n.a.	4.3	6.5	6.7	100.3	137.0
Special game licences	— not introduced —			6.9	22.7	27.8
Supplementary game licences	n.a.	11.2	4.9	6.6	175.8	287.5
CHA/GMA permits			3.2	4.0	15.7	82.1
Hunters and dealers permits	n.a.	n.a.	0.3	1.9	4.7	5.5
Export and import licences	n.a.	n.a.	n.a.	6.6	24.1	48.1
Misc. (including photography licences)	n.a.	n.a.	0.1	—	—	1.9
Total	2.5	31.6	49.0	99.1	452.7	647.5

Sources: Annual Reports of the Department of Game and Fisheries 1963–68; Wild life Fisheries and National Parks 1969–73; and National Parks and Wild life Service 1974–78.

revenue from the sale of ivory, though difficulties in selling this commodity using local tenders resulted in much ivory remaining in stock. Revenue from court fines is given for the period following the creation of the anti-poaching unit when increased penalties became operational and fines constituted a considerable revenue source (Table 8.3).

Wild life exports declined in value initially (Table 8.4), though ivory exports increased later through the sale of ivory from cropping.

Data for current expenditure with respect to wild life are not published. Data for 1970 are given (Table 8.10), but include fisheries. An estimate of the total current expenditure related to wild life would be in excess of £500,000, and much of the expenditure on tourism by the ZNTB was related to wild life and can be estimated at £145,000.

Capital expenditure (Table 8.5) includes some related to fisheries, but an estimated £1.2 million was spent over the period in relation to wild life — its conservation, management, and use. Therefore, during this period, recurrent and capital expenditures were not far short of that proposed in the FNDP.

Clearly, there were many other indirect sources of government revenue and earnings for the country resulting from tourism, crop protection, and other aspects which cannot be costed at this level of analysis.

During the period under review, though expenditure on the wild life estate and its management and exploitation had very nearly reached the proposed targets, the objectives had not been reached. Cropping had been difficult to carry out and had not provided an economic return over the few years it was tried. The problems of too many game animals in some areas had not been solved and still required attention, while meat production targets had proved to be over-optimistic.

The development of sport hunting was very satisfactory and demanded the least capital outlay for infrastructure development. Its returns, both in revenue and the provision of meat and labour in rural areas, had increased and were encouraging. However, it was not known whether this level of hunting could be sustained or whether a marked depletion of trophy animals was taking place in the hunting areas. There was as yet little diminution in the estimated loss of wild life through illegal hunting.

After seven years of Independence, the tourist industry was still being developed and, though considerable progress had been made in the sector not related to wild life viewing, the difficulties experienced in the development of the remote game reserves were not encouraging. Though there had been a considerable capital input into this sector, use of the facilities, which were available for the dry season only, was far below capacity. There were still major deficiencies in basic information which was needed for sound planning, and little emphasis had been placed on research and the formulation of management plans prior to the initiation of development schemes.

TABLE 8.10 CURRENT EXPENDITURE ON WILD LIFE, 1970–78

| | SECOND N.D.P. | | | | | | | | |
	1970	1971	1972	1973	1974	1975	1976	1977	1978
					(in '000 £)				
Salaries[1]	425.1	513.0	549.3	590.8	612.4	460.2	595.7	565.0	467.1
Travel and maintenance[1]	194.8	243.6	156.0	163.6	205.0	189.9	278.6	121.1	118.8
Training[1]	31.3	23.2	17.4	34.5	44.2	15.5	15.5	23.1	8.1
Game conservation and control	307.7	330.0	231.4	234.7	280.1	397.0	451.4	285.1	227.6
General	68.6	108.3	80.1	86.4	94.5	99.2	56.8	41.3	43.1
Total	1,027.5	1,218.1	1,034.2	1,110.0	1,236.2	1,161.8	1,398.0	1,035.6	864.7

Note: [1]These categories contain a component relating to fisheries, 1970–73.
Source: Reports of Ministry of Finance, 1970–78.

The question remained whether, at a time when East African countries were reaping a valuable income from their wild life resources, Zambia would be able to do so in the next phase of its development.

In its own assessment of the success of the FNDP, written as a prologue to the formulation of the next planning stage (SNDP 1971), the Zambian government stated that the cropping schemes ran into unforeseen difficulties, that the anti-poaching unit operated only with reasonable success, that the research programme for the Luangwa Valley in cooperation with UNDP was initially beset with difficulties, and that wild life management duties had often been neglected as staff were directed to work on physical development projects, especially related to tourism. The training of Zambian wild life officers at all levels and an improvement of their working and living conditions had been successfully started. By the end of the period, the tourist industry was starting to recover from the slump which it experienced in the first years of Independence, and tourist visitors to the country were now estimated to be 20,000. Although some 60 per cent of the hotel visitors were expatriate residents, foreign exchange receipts from genuine "tourists" were estimated to be worth £1.5 million, though much of this may be seen as currency imported by visitors of relatives resident in Zambia rather than independent tourists without other reasons for visiting the country.

The number of international visitors who stayed in game camps or lodges is not known, but was probably about one per cent of the tourist visitors — and so foreign exchange earnings from this group were small.

Transitional Period (1971) and the Second National Development Plan (SNDP), 1972–76

A period of consolidation followed the completion of the FNDP in 1970 which lasted through 1971, during which time the price of copper fell sharply. The revenue of Zambia fell from £251 million in 1970, to £209 million in 1971, and to £183 million in 1972. As well as economic difficulties, the SNDP must be seen as a period of increasing difficulty on Zambia's frontiers on all sides, which led to unpredictable conditions with respect to land communications affecting imports and exports.

The SNDP gave as its priorities the development and transformation of agriculture, but listed as its fourth priority the intensive development of the tourist potential. This was seen as important in diversifying the national economy and making a significant contribution to rural development. The number of tourists was expected to increase from 20,000 in 1970 to 50,000 in 1976, and the development of the tourist spots, "especially the National Parks and Game Reserves", was to get priority. A total investment of £14.5 million was planned for tourism, of which £5.8 million was to come from private sources. This investment represented about 1.28 per cent of the total investment for 1972–76.

A sum of £4.5 million from the government can be identified as being directly related to the development of wild life-based tourism, largely within the Luangwa National Park for new lodges, roads, and ancillary infra-structure. A further £1.2 million was approved for the building of a new airport in the Luangwa Valley for tourist air services.

The Department of Wildlife, Fisheries and National Parks, created in 1969, was given the task of establishing a new and comprehensive system of national parks which could be properly managed on a scientific basis and adequately protected from changes of all kinds. Game Management Areas (GMAs) were to be established for the development of international safari and local hunting linked with a more thorough appraisal of the value of the animals as a resource. Game cropping would be reinstated in some areas, and the development of a large zoo park was envisaged near Lusaka. Zambianization of the department was to be completed, and the need for scientifically trained field and laboratory staff fulfilled. Two divisions were to be established: (1) Research and Management, and (2) Conservation and Development. A total financial outlay of £5 million was projected and allocated (Table 8.11).

TABLE 8.11 PROPOSED CAPITAL EXPENDITURE RELATING TO
WILD LIFE IN THE SECOND NATIONAL DEVELOPMENT
PLAN, 1972–76

		(in '000 £)
WILD LIFE: research and management		484
conservation and development		4,233
Total		4,717

Components:

Kafue National Park	684	Game cropping	292
Luangwa Valley	947	Training and public education	160
Other areas	1,740	Anti-poaching unit	72
Zoo parks	290	Aircraft	48
Other	484		

Source: Second National Development Plan 1972.

This plan can be seen as a logical continuation of the FNDP with respect to the use of wild life as a resource. Efforts were to be made to establish a substantial part of the wild life estate in a way which qualified for the designation of national park on international standards (Harroy 1971). The tourist developments were to be concentrated in the isolated Luangwa National Park which, although containing the most varied and numerous wild life, also posed major problems of management and conservation. However, the eventual establishment of the UNDP research team in the valley with Zambian counterparts in 1971, although dogged by operational and staffing difficulties, enabled prospects for the integrated development of the resources of this large area to be envisaged for the future. To this end, massive inputs were necessary to bridge the river permanently, to maintain all-season and all-weather game viewing and access roads, and to build new accommodation and a new airport. It is worth noting too that game cropping had not been abandoned as a positive management tool in spite of earlier experience, but its role as a source of protein and revenue was subsiding.

In effect, however, the political and economic forces, both internal and external, and the uncertainties and difficulties these produced, were to make development exceedingly difficult during this period. By the end of 1975, some observers thought that Zambia was on the brink of economic disaster (Legum 1976).

Hunting. The National Parks and Wildlife Act, introduced in 1972, represented a major development in legislation in accordance with the new policies of the department. Under this Act, 32 GMAs initially covering 163,853 km^2 (22.3 per cent of the national land area) and 17 national parks (59,000 km^2, 7.9 per cent of the land area) were established (Fig. 8.2). Outside these areas, there was little wild life of any size worth hunting, and the department's policy was to maintain wild life in these areas for hunting on a sustained yield basis following the dictates of management plans designed for each GMA. However, data on animal populations were not available, and management plans had not yet been drawn up. Nine GMAs were allocated initially for use by safari hunters, and fourteen were allocated for use by 93 non-residential permit holders, of which 74 were taken up.

Local residents of the GMAs could hunt on payment of a game licence, but were theoretically restricted to a specific number of species and individuals of those species. Research was undertaken in seven GMAs of the Luangwa Valley, comprising an area of 35,050 km^2 (A.R. Game 1972). One hundred were interviewed out of a total resident population of 74,000 people. Seventy-five per cent of the hunters had muzzle loaders and the average annual offtake claimed for each hunter was 2.4 animals yielding

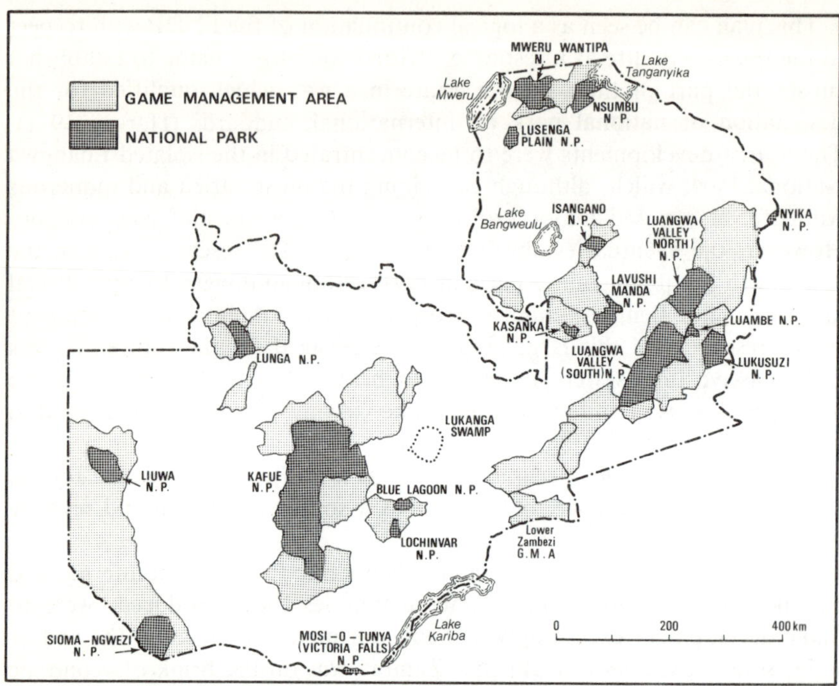

Fig. 8.2 Game management areas and national parks, Zambia

440 kg of meat. Contrary to earlier opinion, buffaloes were a favoured target, as well as impalas and warthogs. The total value of the meat so obtained throughout these GMAs was estimated at £87,000, though other data suggested an annual offtake of 947 t from this area, which accounted for only 45 per cent of the claimed offtake of the sample survey. Marks (1973) in a smaller but more complete study, in a GMA well-stocked with animals, gave a figure of 5.1 animals per hunter per year, yielding 973 kg of meat per hunter.

Mean offtake for the Luangwa Valley GMAs by hunting was estimated at 27 kg per km^2. Assuming similar yields for the national total of the Game Management Area, this would give an annual production of 4,400 t of game meat. However, the Luangwa Valley has a higher biomass than many GMAs, so that this figure may be too high. Based on a total national number of resident hunters of 1,800, the number licensed in the previous decade, the yield of the GMAs would be 800 t. Using the data of Marks (1973), the yield would be 1,750 t, a figure near that of the 1,500 t given by the Game Department (A.R. Game 1966) for the production of the Game Estate. The value of this meat, assuming a market price of £0.19 per kg, would be more than £333,500. The value of meat at the Luangwa Valley production level, giving a total production of all GMAs of 4,400 t, would be £0.9 million. The optimum sustained yield of the Game Estate in 1968 was given as 3,000 t (A.R. Game 1968). Therefore, if the skins and trophies could also be utilized, the value of the wild life production might possibly exceed £1.2 million.

Sport hunting. The number of safari hunters who came to Zambia during this period fluctuated between 93 and 119. Professional safari operators were allocated from six to eleven GMAs, usually those surrounding national parks. Here they were not troubled by other resident sport hunters, though local subsistence hunting continued. Good trophies, sometimes award winners, were obtained each year, and there seemed little to suggest that this form of exploitation could not be maintained at this level for many years to come. A later analysis (A.R. Wildlife 1978) suggested that the success rate for safari hunters for several species was lower in 1977 than it had been in earlier years, and there was also a drop in the mean weight of elephant tusk and rhino horn from trophy animals starting in 1976 and 1977 respectively.

Initially, the licence for safari hunting (£580) was a composite one which could be supplemented by the purchase of a Special Licence (£1,160) under which protected animals such as rhinoceroses, lechwes, and cheetahs could be shot. In 1975 this structure was changed to enable safari hunters to buy individual or second animals as they desired. A measure of the value of the rarer protected animals may be gathered from the fact that in 1975, out of a total of 107 safari hunters who came to Zambia, 20 bought a special licence for a rhino at £1,140 each, 87 bought an elephant licence at £182 for a male elephant, and 3 bought black lechwe special licences at £760 each. Licences for other animals had been increased by two or three times since 1972.

As in previous years, meat from animals killed by sport hunters was made available to local residents of the GMA after smoking. In 1973, 160 t of fresh meat equivalent was available, and this rose to 341 t in 1974, when it was valued at £90,000.

A study of other aspects of safari hunting (A.R. Wildlife 1972) shows that safari hunting in the Luangwa Valley supported 299 local employees, but only 47 of these were skilled to any degree and these only worked for five months of the year. The rest were casual labourers. Total wages paid were about £9,000 annually, while further indirect revenue to the government from safari hunting, e.g., income and vehicle taxes, was about £3,000.

Therefore, sport hunting, which had become well established, was able to increase its earnings greatly over this period, though there were signs of possible difficulties in maintaining the standard of animals available to hunters. Fortunately, the safari hunting areas were not particularly affected by border incidents and, because of the small number catered for at any one time, it was not difficult to ensure logistic support for these operations.

Control hunting.   The highest number of elephants shot on government control since 1957 was recorded in 1971 when 459 were killed but, following the severe transport and financial difficulties experienced after 1972, the number fell progressively until 1975 when only 174 were shot. The number killed on control in 1976 was 289. No information is available on the extent to which district or provisional control measures were taken against smaller crop pests. The hunting of game animals along game fences erected by tsetse control units continued and included the whole range of species. More than 1,500 animals were shot on an average each year between 1971 and 1975, but only 109 were shot in 1976 when this activity diminished along most fences for various reasons.

Wild life cropping.   Though envisaged in the SNDP, no cropping was undertaken during this period.

Wild life domestication and ranching.   Further trials in domestication were not undertaken, but a provincial pilot project relating to the small-scale provision of herds of impala to be kept in ranch situations was started in 1973.

Tourism.   The year 1971 was described as a most difficult year for the ZNTB which experienced severe staffing and operational difficulties. Optimism was still high that tourism could be developed into Zambia's second most important external revenue earner, after copper exports, though major deterrents to this prospect continued to be operative during this period. Through 1971, an effort was made to integrate wild life tourism in the national parks more fully into the tourist pattern through the upgrading of facilities, especially in the Luangwa National Park. Following the preliminary work of the UNDP team, plans were made for substantial developments during the SNDP to increase both visitor capacity and the infrastructure of roads, bridges, and air services in this area in an effort to not only increase the capacity of the region for tourism but also extend the period when tourist facilities were available. This programme was completed in 1976.

Before the presidential lodge in Luangwa was completed, it was decided to convert it into a luxury tourist lodge connected by all-weather roads. It provided accommodation with 30 beds and was expanded to 43 beds by 1975. Many other smaller catering lodges were expanded so that by the end of 1975, 169 beds were available in catering lodges, 116 beds in non-catering camps, 86 beds in safari camps, and 84 places in walking trial camps. Walking game-viewing safaris were pioneered in 1970, and were extended considerably during the SNDP in Luangwa Valley and also introduced in the Kafue National Park. Special safari villagers were built as base camps.

The private sector not only was involved with walking tours but also started work on two lodge developments outside the Kafue National Park near the edge of Itezhiteshi Lake, created by a new dam completed in 1977.

Thus considerable progress was made in the development of tourist capacity during a period of great difficulty. These varied difficulties resulted in a 30 per cent decline in the number of international visitors to Zambia between 1972 and 1973, though there was an improvement in subsequent years. There was also an encouraging increase in the total number of visitors to the national parks (Table 8.8), and this resulted in a doubling of visitors for the Kafue and Luangwa National Parks. However, visitors to Lake Tanganyika dropped sharply after 1973 as transportation to this isolated area became more and more difficult. Data for the origin of visitors are not available, but the southern border was closed in 1973 and this had serious repercussions on the number of visitors to the Livingstone Zoo Park (Table 8.8). This border closure, together with restrictions on travel and holidays out of Zambia, increased the proportion of Zambian residents using the facilities in the national parks. Many of the international visitors were relatives of residents.

Therefore, tourist facilities were better able to cater for international visitors during this period. But the political and economic climate of the country did not encourage international tourism. A new small national park (Blue Lagoon) near Lusaka was opened to day-visitors during 1973, but no start was made on the Lusaka Zoo Park. Instead the development of a small zoo and pleasure gardens was continued at the site of the Botanical Gardens outside the capital. A survey undertaken in the Kafue National Park of tourists in 1975 indicated that 25 per cent were international tourists and 56 per cent were temporary residents. As only 11 per cent travelled to the national parks by air, it may be assumed that many of the international tourists were visiting the country primarily to see relatives (A.R. Wildlife 1975).

Wild life exports.   Data are available for 1971–74, and these show that the number of game skins exported declined considerably with the introduction of the new Fauna Conservation legislation. Increased export duties, together with an increasing concern for conservation measures, reduced the export of crocodile skins to 181 in 1971 and 48 in 1972. The number of hides and other skins was also reduced to 364 in 1971 (3,271 in 1970). However, the figures show increases in reptile skins and fur skins in 1973 and 1974, suggesting a slight upturn in hunting activity and trade, but none of these items now formed an export of substantial value. The situation also reflected the continued reduction in all forms of wild life outside the protected game estates.

Ivory continued to be a valuable product, with the price of ivory increasing on the world market. During 1971–74, the average annual export of ivory was 21,400 kg. But much ivory remained in store within the country, owing to the unsuccessful introduction of new disposal methods. No government ivory was sold in 1975 when 2,952 tusks were in store, representing 18,000 kg of ivory. Nevertheless, with world ivory prices high, this wild life asset continued to be a valuable potential export. The Game Department became most anxious to use other wild life trophies more extensively and instituted schemes for training its staff in the recovery and processing of skins, horns, and heads. Thus there was a movement away from export of wild life products towards the establishment of small industries based on the use of wild life products for the production of tourist souvenirs.

Finances.   Data for this period are more complete than in previous years, though smaller items such as gun licence revenue are not available. Between 1972 and 1975, current expenditure on wild life (Table 8.10), including fisheries, was about £1 million a year and was kept closely within this limit. In particular, the mean expenditure on travel was held at 60 per cent of the figure for 1970/71.

Capital expenditure (Table 8.5) fell a long way behind that originally planned for the SNDP. Out of an original allocation of £4.7 million, only about £1 million was authorized and of this only £0.79 million was actually spent (TNDP 1979), owing to shortages of materials, delays in awarding tenders, and organizational problems. Calculations from figures given in Financial Reports suggest that £0.9 million was actually spent. Whichever figure is taken, it is clear that wild life development suffered greatly from capital expenditure cuts, especially in the first three years. However, considerable increases were allocated in 1975, but no progress was made with the development of Lusaka Zoo Park and there were no further developments with respect to game cropping, which may have meant a change in policy. Research and management continued to be starved of resources.

Revenue continued to increase substantially (Table 8.3). The actual revenue paid directly through licences for hunting over the period is given in Table 8.9. A very large increase in revenue in 1974 relates to the introduction of new licence fees by the government in this year which, though they greatly increased the cost of individual licences, did not result in an appreciable reduction in the level of demand for them. Thus the district game licence was increased from £0.5 to £3.35, and the national game licence was increased from £10 to £33 for residents. Supplementary game licences enabled individual animals to be bought for a specified fee which varied from £6.6 to £66 according to species. Special licence fees for protected animals were high; for example, 20 rhinos were bought for £20,000.

The amount of revenue resulting from the sale of government ivory increased threefold from the FNDP if the annual average is compared with that for the SNDP, though the number of elephant killed on control declined sharply after 1972. As the licence revenue for shooting an elephant also increased sharply during the SNDP, it is clear that elephants were still an exceptionally valuable resource. It also seems clear that more elephant licences were being sold to resident hunters during 1975 and 1976, and that there was a thriving demand for these. During 1976, 1,620 tusks were registered with the Wildlife Department, suggesting an elephant kill of well over 800 animals. Of this total, 359 were killed on control, on tsetse control, or as a result of damage feasant.

Revenue derived from game tourism is not easy to separate from other forms of tourism, in so far as game viewing complements social visits and visits to Victoria Falls. Table 8.12 gives the revenue of the ZNTB and the revenue from national parks. Revenue from other tourist sources such as the pleasure gardens and urban bus trips is not included. If the possibly unreliable figure for 1973 is disregarded, then the revenue shows a gradual increase from £182,000 to £302,000 through the period, in spite of the substantial problems posed by the Lake Tanganyika lodges. This increase, however, reflects both increased charges and possibly increased length of stay, for the number of visitors remained static over 1974–76 (Table 8.8).

It would appear that, while the running of the game lodges in the national parks continued to suffer a loss, the deficit was being narrowed (Table 8.13). Capital expenditure by the ZNTB during the SNDP amount to £1.14 million on lodges, other buildings, and vehicles, and another £3.3 million was spent on roads and bridges in the Luangwa Valley related to tourist development (Table 8.14). The new airport cost another £3 million by the end of 1976. Thus capital expenditure on tourism very nearly achieved the target proposed in 1972 of £4.5 million. The airport, however, had cost a great deal more than had been budgeted for. In keeping with the target for tourist development, there had been under-expenditure in other directions including the research and management aspects of wild life.

The values for wild life exports (Table 8.4) are incomplete. Crocodiles increased in value but other skins declined greatly. Initially, there was a new source of revenue with the export of animals to zoos, but difficulties prevented this trade continuing. Ivory showed a higher value than for the average figure of the FNDP, and the figure is consistent with the increased number of licences available for elephant hunters.

TABLE 8.12 REVENUE DERIVED FROM WILD LIFE TOURISM THROUGH THE NATIONAL PARKS AND THE LIVINGSTONE ZOO PARK

	SECOND N.D.P.								
	1970	1971	1972	1973	1974	1975	1976	1977	1978
				(in '000 £)					
ZNTB bookings	—	4.8	106.7	336.7[1]	149.4	199.9	225.0	n.a.	n.a.
Wild life service bookings and entry[2]	51.9	76.3	24.5	5.8	13.8	13.7	9.9	19.1	13.6
Total	51.9	81.1	131.2	342.5[1]	163.2	213.6	234.9		
ZNTB total revenue from tourism[3]	6.3	50.4	175.0	215.5	259.4	369.9	461.6	265.0	n.a.

Notes:　[1]This figure was given by the Wildlife Department and probably includes all ZNTB income.

[2]This figure also included entry fees to the Livingstone Zoo Park.

[3]A small proportion of this total may be regarded as additional wild life revenue, being derived from bookings by ZNTB offices overseas and the sale of publicity material and tourist goods in shops, some of which will relate to wild life.

Sources:　Annual report of the Department of Wildlife; Fisheries and National Parks 1971–73; National Parks and Wildlife Service 1974–78; Zambia National Tourist Board 1970–72; Reports of the Ministry of Finance 1970–78.

TABLE 8.13 COMPARISON OF REVENUE AND EXPENDITURE ON WILD
LIFE VIEWING AREAS BY THE ZAMBIA NATIONAL TOURIST
BOARD, 1974–77

		1974	1975	1976	1977
		(in '000 £)			
Kafue and Luangwa National Parks	revenue	102	122	181	90
	expenditure	171	195	191	254
Lake Tanganyika National Parks	revenue	37	61	44	2
	expenditure	33	56	43	44
Total	revenue	159	183	225	92
	expenditure	204	251	234	298

Source: Financial Reports of the Ministry of Finance 1974–77.

TABLE 8.14 EXPENDITURE ON THE LUANGWA VALLEY RELATED TO
TOURIST DEVELOPMENTS CONCERNING WILD LIFE, 1971–76

	1971	1972	1973	1974	1975	1976
	(in '000 £)					
Accommodation	227[1]	170	111	73	109	12
Roads bridges	—	—	611	963	1,241	491
Airport	—	21	580	524	929	882
Total	227	191	1,302	1,560	1,279	986

Note: [1]This figure includes £214,000 spent on the President's Luangwa Lodge in 1970
and 1971. It was converted to a luxury tourist lodge and opened in 1975.

Source: Reports of the Ministry of Finance, 1971–78.

Conclusions. The period was one where distinct progress was made with respect to the development of the wild life-based tourist industry, and a considerable number of major capital ventures were completed in the Luangwa Valley by way of lodges, roads, bridges, and an airport. Further development awaited the completion of the new major lodge with extensive accommodation which would make the high capital expenditure on roads and the airport acceptable. However, it is not easy to compute precisely the value and cost of tourism in Zambia. Estimated foreign currency earnings related to tourism have been given as follows: 1970, £1.74 million (SNDP 1971); 1972, £3.07 million (ZNTB, A.R.); and 1976, £6.07 million (TNDP 1979). But it would seem that during this period very few of the goods and only some of the services required by the tourist industry could be supplied by Zambia. Much had to be imported so that the benefits of high foreign currency earnings from tourists were restricted. During this period, the tourist industry was allowed to work out many of its initial problems in the context of the large expatriate community who were denied holidays elsewhere in eastern and southern Africa. Circumstances beyond the immediate control of the nation prevented Zambia from taking a more important role within the context of wild life tourism in Africa.

The development of sport hunting continued, and this largely catered for foreign hunters. The national income derived from this activity increased fourfold during the period and did not necessitate the large capital investment accorded to tourism. Safari hunting also generated foreign currency, amounting to £1 million in 1976 (A.R. Wildlife 1978). In addition, 346,000 kg of fresh meat were supplied to local residents of the GMA in 1976 and employment for 275 people was created.

The export of wild life products can be seen to be declining in importance as far as skins and hides were concerned, though ivory remained very important. Little wild life now remained outside the game estate which could be profitably used, and efforts were made for the greater local utilization of wild life products as the basis for tourist souvenir craft industries.

It is difficult to assess the continuing need for control hunting over this period. The number of animals shot on control declined but travel was difficult and finance restricted. More particularly, it was believed that illegal hunting continued without restraint in many areas owing to the frontier difficulties, while illegal hunting in other areas was always seen as a major continuing problem, which the department was less able to control.

As the number of people buying hunting licences increased during the last years of the period, so must the number of animals killed as shooting over and above licence limits became easier with less chance of detection.

Systematic cropping of wild life did not take place, though one of the alternative strategies proposed for the management of the Luangwa Valley was an immediate reduction of the elephant population by cropping, to be followed by sustained cropping in subsequent years. Costing showed that a profit was possible only if hides and ivory were used. The use of the meat as well would possibly only just cover the cost of recovery (Naylor *et al.* 1973).

One small game ranching project was initiated. The use of wild life as an alternative, or complement, to cattle on low potential grazing on established ranches was pioneered in Rhodesia in 1960 (Dasmann 1964). Further experiments and schemes in East and southern Africa have been reviewed by Field (1979). Game ranching continued to expand in Rhodesia, and the number of permits issued for cropping on such ranches increased from 49 in 1964 to 108 in 1967 to 179 in 1973. Estimated revenue from game ranching was £147,000 in 1964, £227,000 in 1968, and £163,000 in 1973 (A.R., National Parks and Wildlife Management, Rhodesia). In Zambia, however, large game animals rarely exist on ranch lands, but the Rhodesian experience suggests that the potential development through restocking would be worth investigating in various areas where protection of the animals could be assured.

The Period after 1976

Though 1976 had seen a resurgence in the Zambian economy as exports increased again, the costs of imports increased in response to inflationary trends throughout the world, while the international situation on many of Zambia's frontiers remained extremely difficult. The currency was devalued and government expenditure severely curtailed.

In this situation, the number of tourists declined, internal travel became more and more difficult, and many activities of the Department of Wildlife came to a virtual standstill.

Anti-poaching activities were reduced, and the number of arrests, which had increased from 944 in 1974 to 1,830 in 1976, declined progressively to 1,454 in 1978. The internal situation meant that many parts of the wild life estate could not be patrolled at all, and the introduction of a ban on beef imports in 1977 increased the incentives to hunt wild life illegally. The number of modern firearms in the country had increased greatly over recent years, making it easier to kill the larger animals. Indeed the anti-poaching unit alone confiscated 198 elephant tusks and 130 pieces of ivory in 1978, and this must have represented a small part of the total illegal offtake. As part of the campaign to make the illegal export of ivory more difficult, the export of unworked ivory was banned in 1977, and it was hoped that Zambian carvers and curio markers would use all the ivory produced.

Sport hunting was little affected by the overall situation, though some GMAs could not be used. The number of safari hunters increased to 125 a year, and their economic significance grew. It was estimated that the foreign currency earnings from this source alone were £1.2 million (1977), and £300,000 was paid to 350 employees (A.R. Wildlife 1978). As was the custom, 350,000 kg of fresh meat were provided for local consumption by residents of the GMAs. There is little doubt that safari hunting in Zambia increased during 1973–77 because of the hunting ban in Tanzania (which ended in 1978) and its introduction in Kenya in 1977. Both these countries were facing severe pressure from conservationists as a result of declining wild life which was the result of not only commercial hunting activities but also much illegal hunting generated by a great demand for tourist wild life products and the high price of ivory and horn in Asian markets

Hunting by residents of Zambia also appears to have increased during these years. The total revenue from hunting was about £650,000 a year, a substantial increase from previous years. A total kill, under licence, of 3,320 animals was made in the GMAs in 1978. This included 365 elephants, 372 zebras, 244 buffaloes, 168 crocodiles, and 116 leopards. Not all these were trophy standard animals, however, as hunting for meat was a profitable activity for hunters even with the increased licence fees.

Illegal hunting in both the national parks and GMAs increased, and it was believed that a tightening of security in the East African wild life areas, together with declining wild life there, had forced international poaching gangs to move into Zambia, with the result that rhinos and elephants were severely threatened.

Wild life tourism was also badly hit by the national situation, and the number of visitors to the national parks dropped from 20,058 in 1976 to 14,454 in 1978. The remote lodges on Lake Tanganyika were badly hit by supply problems and restrictions on airflights, while Livingstone Zoo Park visitors declined from an estimated 20,000 in 1976 to 6,717 in 1978 as a result of border incidents with Rhodesia. Some progress was made with extending the length of the game-viewing season in the Luangwa Valley as the new bridges and roads allowed ''green season'' visitors. Yet in 1978, the occurrence of unprecedented floods damaged the roads and delayed the opening of the viewing facilities, emphasizing once again the environmental hazards associated with this large river basin in which there is uncontrolled land use in many watershed areas.

The response of the government to the worsening financial situation was a sharp cutback in capital expenditure from £261,000 in 1976 to £102,000 in 1978 (Table 8.5), while current expenditure was trimmed from £1.4 million to £0.9 million (Table 8.10). Replacement of field staff (rangers, scouts, and guards) was especially difficult, and the department was 22 per cent under-staffed. Some national parks and GMAs could not be patrolled because of the national security situation, and illegal hunting undoubtedly increased as a result. Many other critical activities relating to research and management were curtailed or not undertaken. But in the long term, this could result in increasing difficulties and must mean a complete reassessment of the wild life situation.

Because of restraint in spending, revenue attributable to wild life nearly equalled expenditure. The revenue from tourism is not known, and must have declined considerably but was offset by increasing licence fees for hunting. The sale of ivory steadied near the value at the beginning of the decade and fines resulting from game offences were maintained at more than £30,000 because of increased penalties, although there was a drop in the number of convictions.

The annual grant to the ZNTB to run its operations was increased (Table 8.15) and expenditure on tsetse fly control remained at a high level, and field schemes including fencing, air and ground spraying, and vegetation clearing cost more than £800,000 in 1977 (A.R. Vet. and Tsetse Control 1977).

TABLE 8.15 EXPENDITURE ON TOURISM BY THE
ZAMBIA NATIONAL TOURIST
BOARD, 1970-78

Year	Government Grant	Total Recurrent Expenditure
	(in '000 £)	
1970	158	149
1971	454	485
1972	464	675
1973	550	718
1974	610	894
1975	597	1,014
1976	666	1,128
1977	666	1,033
1978	522	n.a.

Note: The difference between the government grant and
         the total recurrent expenditure was derived from
         income.

Source: Reports of the Ministry of Finance 1970-78.

No data are available for wild life exports, but it is doubtful whether these were now of any importance outside the export of trophies by safari hunters. The value of export permits for this activity is given in Table 8.9.

No further attempts were made to investigate the possibility of game cropping or to expand the work in game ranching. With little new information on game stocks, and field research greatly hampered, the progress made with management plans for the game estate was poor.

Therefore, the financial and political situation prevented the natural conclusion of the SNDP, and in such abnormal circumstances it is not possible to evaluate the effect of the expenditure on revenue earners such as wild life tourism. The impression given is one of drawing on the national stock of wild life to maintain revenue. But with increased illegal exploitation of wild life for meat, teeth, and horn, and in the absence of on-going wild life enumeration, such a policy was clearly dangerous and could only be maintained in the short term.

An Assessment

Zambians have used wild life as an important resource providing meat, skins, fat, bone, horn, and sinews throughout their history. Wild life has been an essential component of the subsistence food supply, and was of particular significance where tsetse fly-borne trypanosomiasis prevented the rearing of domestic animals. In these areas, fish were the only alternative major source of animal protein to wild life meat.

Subsequently, wild life has been used in various ways, and these have been described in this paper for different periods in the country's history. The importance of different uses has varied greatly through time, and they have not all been economically successful. Nevertheless, the resource has always been valuable to the country.

Wild life is a renewable resource if the offtake is kept at or below the reproductive potential of the wild life population. In many areas, the un-controlled exploitation of wild life has resulted in the elimination of many species, especially the larger mammals. In other areas, changes in habitat, competition from man, and disturbance by man have reduced or eliminated wild life. Therefore, the total national production from the wild life resource must have decreased at an accelerating rate. The question remains whether wild life should continue to be replaced by alternative forms of land use, or whether it is both economically and socially desirable to maintain those areas which remain and which today are designated as national parks and Game Management Areas. If wild life remains a valuable resource, the question might also be asked as to whether new areas of wild life estate should be designated.

An assessment is made here of each of the uses of wild life.

Ivory.   Apart from traditional subsistence uses, ivory has the longest history of use of a wild life resource in Zambia. The early export of ivory enabled men to acquire trade goods, including guns and powder, and this trade, together with the additional hunting of elephants by foreigners, led to a great reduction in the number of elephants by 1890. Subsequently, the export of ivory was continued by the African Lakes Company and the British South Africa Company, and successively by the Colonial, Federation, and Republican administrations from ivory acquired by them. Revenue was obtained from the sale of elephant hunting licences and from ivory export duties. There has been a general increase in the level of exploitation for ivory, though there have been many fluctuations as a result of economic depressions and falls in the price of ivory, increased control hunting, and both internal and external political disturbances. There has also been a continuous and fluctuating level of illegal hunting which cannot be measured. As the price of ivory has increased more than tenfold since 1969, so the level of illegal elephant hunting is believed to have increased greatly over this period and there can be little doubt that the number of elephants has declined greatly throughout the country.

It has been estimated that the Luangwa Valley has lost half its population of 100,000 elephants during the past seven years as a result of illegal, highly organized poaching by international operators (Brannen 1979). Clearly, if ivory is to remain a valuable resource, the illegal export of ivory will have to be stopped. A £1.5 million international protection and conservation programme for elephants and rhinos has been launched in Zambia to contain illegal hunting for ivory and horn. If protection can be achieved, Zambia will continue to have an ivory resource from tusks found in national parks and from control shooting on farmlands and on the game estate.

Zambia banned the export of unworked ivory in 1977, believing that it should use its ivory for the production of tourist consumer goods and art objects. Whether it can use all its normal supply of ivory in this way is debatable, and it may certainly be more profitable to sell ivory abroad officially unless tourist demand increases greatly. However, ivory exports of any kind do enable illegal exports to be made with less chance of detection.

Data available from 1970 to 1977 given an average annual official offtake of ivory for this period at 20,000 kg, valued at more than £80,000.

Other wild life products.   Hides, horns, and skins have varied in importance and though otter, monitor lizard, and python skins have been important items of export in the past, and antelope, buffalo, zebra, and spotted cat skins of minor importance, only crocodile skins have ever achieved an annual value greater than the value of ivory. This was for a short period only, and the export of all skins is now small and may be expected to decline further as public opinion and prohibition on imports limit the countries where they can be sold. There may, however, be considerable room for an increase in the use of skins for tourist and local consumer sales, both as items of household use (rugs) and in the context of handicraft. Efforts are being made to train craftsmen to deal with the processing and use of skins. Yet without the elimination of illegal hunting, the use of skins in this way encourages poaching.

Therefore, ivory and other wild life products, apart from meat, will continue to be important in Zambia only if the tourist industry increases and provides a market for goods which can be sold internally or by the government abroad. The supply of wild life products will depend on the ability of the Wildlife Service to implement plans for protection conservation and different forms of management. Wild life products will probably become progressively minor earners of revenue for the government.

Subsistence hunting.   The ability of traditional farmers to supplement their food supply with much-needed protein from game has diminished over large areas of Zambia as a result of increased subsistence hunting in the past, as rural populations have increased, and as illegal commercial hunting has also taken place. Certainly, the larger antelopes have disappeared from much of their former range though many smaller species remain. The Zambian will use most small animals including mice and birds. Subsistence hunting on a traditional level may now be confined to the GMAs where wild life continues to make a substantial contribution to the nutrition of the local people, supplemented by meat provided by safari hunters.

Therefore, the government is faced with providing alternative sources of protein for the increasing rural population as well as the urban population. In previous decades, this has been achieved through the marketing of dried fish, mainly from Lake Tanganyika and Lake Bangweulu, the Lukanga Swamps, and major river fisheries, especially the middle Kafue River. However, there is a real danger of overfishing and resource depletion. In recent years, Zambia has not been self-sufficient in beef and has imported some from Botswana. Attempts to increase the production from the national beef herd did not meet with great success until 1975 when there was a considerable increase in the number slaughtered (A.R. Veterinary 1966–77). However, in those areas affected by tsetse fly the provision of alternative protein sources is very expensive in terms of tsetse clearance schemes, trypanosamiasis prophylaxis for cattle, or meat importation.

Clearly, an increase of wild life protein at sustainable levels is highly desirable. But with more land being required for cultivation by both traditional and high technology farmers, and with access to the rural areas improving with the creation of new arterial and feeder roads, the containment or elimination of illegal hunting is unlikely. GMAs could probably make an increased contribution to protein supplies if illegal poaching were eliminated and careful management of these areas undertaken to increase their game stocks. This would involve either allowing subsistence hunters to sell meat to drying or freezing plants, or the introduction of controlled cropping schemes.

Therefore, there may still be a role for subsistence hunters in the future if the wild life stocks in GMAs can be protected, increased, and managed to provide a sustainable yield.

In view of the increasing interest in allowing indigenous residents to remain in national parks to follow their existing lifestyles and economies (IUCN meeting, Kinshasha 1975), the possibility might be considered of allowing low-level controlled hunting to be gradually introduced into some national parks on a subsistence basis. With adequate supervision, hunters would use animals which stay in the parks and cause habitat deterioration. It would create little disturbance and present no more of a problem than ration hunting which formerly took place in the parks to supply game staff. In so far as illegal hunting has taken place in most parks in the past and so provided a measure of control to animal population growth, permitted hunting would regulate an existing situation and enable more control to be exercised.

The provision of a marketing infrastructure for game meat provided by resident subsistence hunters in GMAs would enable non-resident hunting to be phased out if this was desirable, while the loss of revenue that this would entail could be collected from the resident hunters as they are drawn into the money economy.

Safari hunting.   Safari hunting has a long history in Zambia but did not reach substantial levels, with appropriate revenue returns, until the last decade. The nature of the environment has frequently made sport hunting more difficult here than in East Africa. It is significant, however, that as the opportunity for this type of sport declines throughout the world, very substantial revenue can be received from small operations with strictly limited expenditure. There would seem to be no sound reasons why safari hunting for trophy animals by both residents of Zambia and aliens should not continue to provide revenue, employment, and meat in the future if game stocks, with suitable old male animals, can be maintained by protection and management. This form is compatible with other uses in GMAs and is flexible in terms of location and logistics. Such use could provide much of the finance needed to implement longer term strategies as well as control poaching.

The major problem remains to ensure that the offtake of trophy animals is at the correct level for a sustained yield. A detailed study from recent hunters' returns (A.R. Wildlife 1978) has shown that success rates are declining for many species of animals and that tusk and horn weights are also generally declining. Only the provision of detailed information on numbers and population structures will enable the continued use of the wild life resource by safari hunting to continue and possibly expand.

Control hunting.   The killing by government hunters of dangerous crop or stock raiding animals has featured increasingly as the areas available to wild life have decreased and as man's activities have created conditions or ecological imbalance either through the destruction of natural predators or through increased competition for grazing, browsing, and water. However, as the rural population grows still greater and the rural infrastructure increases, so the areas where control hunting has to take place should decrease. The transformation of the land by agriculture will ensure this in many areas. However, in other areas the process will take place very slowly, especially in those areas where the soils are infertile, the rainfall irregular, and tsetse fly present. Here the cost of such control should diminish except near those national parks with growing wild life populations, where emigration is taking place. However, implementation of management in national parks should restrict such movements and the damage caused by such animals.

Effective control of smaller vermin will follow increased occupation of farm land as has taken place in Nigeria and other African countries.

Game cropping.   Systematic controlled cropping of wild life, as opposed to commercial slaughter of large numbers in a short time, has proved to be difficult in Zambia but might be tried again should illegal hunting be controlled and the stock of wild life increase in GMAs and national parks. The high capital costs which earlier schemes demanded could be avoided if small-scale units can be operated with a view to supplying local rather than national markets, including possibly the tourist market. This seems to have been successful in the Kruger National Park (South Africa).

Game ranching.   Though a programme of semi-domestication of two antelopes was tried, without success, Zambia has done little research into the use of wild life in ranching situations in spite of its success in Rhodesia since 1964. Recently, impalas have been kept under such conditions in one experiment. The use of wild life in this way has much to recommend it, especially in tsetse fly areas. Even if wild life is absent today, the translocation of selected species from national parks can be undertaken. Indeed, Zambia successfully translocated 102 black lechwe antelopes from the Bangweulu Swamps to the Chambeshi flood plain and in 1978, 500 animals were surviving. Such introductions demand careful planning in the selection of sites suitable not only from an environmental but also from social and economic viewpoints. The schemes need protection and rural education in the immediate areas where they are tried. Research and application could prove rewarding, and the short-term elimination of tsetse fly might then await the application of new research methods which would possibly be less costly and be free of environmental damage from toxic substances.

Tourism.   Tourists were visiting the Victoria Falls more than a hundred years ago. But wild life tourism was initially neglected by successive governments, and development then came too late to take advantage of the increased use of air travel before better placed countries had increased their tourist facilities. The development of tourism since Independence was affected seriously by political decisions initially and then by the very limited infrastructure that was available for use or could be provided quickly. Subsequent economic and political difficulties have prevented an economic evaluation of what has been done to redress the situation. However, even before the developments of the past seven years, Dunlap (1973) had forecast that it might be a long and difficult task to reach the point where tourism might make a significant contribution to the economy of the nation. The reasons for this are varied and many. It is salutary to compare the situation in Zambia with that of Uganda in the early 1970s where Jahnke (1974) concluded, after a detailed analysis of the economics of wild life development, that "tourism is simply not profitable enough to be expanded". He highlighted the fact that Uganda's neighbours had tourist attractions other than wild life, in particular the tropical coasts, and that wild life tourism in Uganda was always "short stay" and located at some distance from the major airports. This situation would appear to be similar to that in Zambia, and with the projected development of wild life tourism in Mozambique in association with nearby tropical coasts (Hanlon 1980), Zambia may well soon have neighbouring countries as serious rivals for international tourists.

It is clear that national tourism should be developed together with short-stay tourism from visitors who come to Zambia for other reasons. In this context, it would seem that the major need is to establish zoo parks near Lusaka and on the Copperbelt, for day visitors only. Large areas are required but no new accommodation need be provided. The value of such zoo parks has been demonstrated at Livingstone and more particularly by the Nairobi National Park. The establishment of these parks would mean the translocation of all the game animals, fencing, and tsetse control, but all these are feasible today.

Conservation.   The conservation of wild life in Zambia has a history dating back to BSAC administration, and the first game reserve was designated in 1899. Undoubtedly, game legislation and the demarcation of protected areas ensured the continuance of wild life to the present day, though the successive administrations were only passively engaged in conservation until the formation of a Game Department in 1940. From this date, active conservation through protection and later through management has led to an increasing expenditure on wild life to maintain the integrity of the game reserves and national parks. This has been accomplished with varying degrees of success, but the question must be asked whether the country can now afford the increasing cost of protection and management of the very large game estate. It is unlikely that there is going to be immediate financial support from tourist revenue related to wild life for some years to come. National parks can supply animals to the surrounding GMAs, but they cannot generate hunting licence revenue.

Tourism has never been envisaged for some national parks and the future of these, especially those near the international frontier and therefore liable to outside interference, may be questioned. Williams (1976) estimated that the value of the wild life in the parks if it were hunted was more than £3 million. Sooner or later, the cost of keeping the national parks will be questioned unless they can generate more revenue. One alternative is to seek external financial help from conservationist organizations or some other external international source. It would now seem unrealistic to ask a developing country to conserve vast areas which replicate others within that country or in other countries. Fortunately, at the present time pressures from alternative land uses are occurring only in association with some small national parks such as Lochinvar or sections of some of the larger ones. The main threat to many of the larger parks would come if the threat of tsetse fly were removed, or could be removed easily from these areas.

If the tsetse fly remains associated with wild life in the national parks, then the threat of tsetse encroachment into nearby cleared areas will persist, though control by game fences and through correct land use permitted in these border zones could be effective. This will involve further expense initially until the ecological balance is achieved.

At the present time, the GMAs are not under specific threat from alternative land uses. But that time may come in the near future. Then a decision will have to be made with respect to the importance of their role in facilitating the integrity and effective management of the national parks, and whether this continued designation for wild life use is rational in an economic sense. Decisions will have to be taken and choices made as to which areas can be so used. Before this can be objectively undertaken, much research and data collection will be required.

The Future

The wild life estate in Zambia is large and of international as well as national importance. The cost of maintaining and managing this estate will increase with time. It is doubtful whether Zambia will ever be able to keep this estate without it contributing substantially to the nation's expenditure. Unless tourism, hunting, and wild life products can generate that revenue, the wild life estate will gradually degenerate through neglect or the estate will have to be reduced. The generation of revenue through tourism would seem to be still associated with many problems and has probably yet to show a profit except for the Livingstone Zoo Park. The continuance of safari hunting at its present level is uncertain, and its very existence is subject to protests from conservationists today. In the present difficult financial context, it is clearly very important and can be justified. Local hunting can be justified only if there is effective control on the number of animals shot and a move made from pure subsistence use to an economy whereby more revenue can be generated from this type of hunting.

Though it may be desirable that new areas should be added to the wild life estate, on conservation or management grounds, these are most likely to reflect specific protectionist interests rather than broad ecosystem needs. There would seem to be every reason to encourage research and small varied trial schemes related to game ranching of some species chosen to suit a range of environments. Ultimately, this may enable the costs of tsetse fly control and eradication to be limited as well as contributing substantially to protein supply.

All alternative plans will come to naught unless immediate protection can be afforded to the remaining wild life in the game estate, to be followed immediately by the implementation of effective management in the light of ecological needs and uses both within the estate and in surrounding areas.

The Third National Development Plan (TNDP 1979) has been operating for one year. Its stated objectives are to increase the capital expenditure associated with tourism to "reap the maximum benefit from this unlimited potential". Marketing campaigns will be mounted at home and abroad and development given a high priority. The expressed hope is that "the earnings from tourism will defray a growing proportion of public expenditure on wild life conservation". The projected foreign currency earnings that will be generated are between £12.7 and £20 million (compared to £4 million in 1977). How much of the revenue generated will be profit remains to be seen and, until Zambia can supply the many needs of the tourist industry from its own industries, much of the foreign currency earnings could be used on imports related to tourism.

Wild life is still a valuable resource. It can either be exploited, as in the past, and fade away, or it can be used rationally but at considerable initial cost. Much research needs to be done, but time is running out either to find ways of making wild life self-supporting or to generate international concern and, therefore, international financial support to protect wild life and manage the estate until the answers are found.

Acknowledgements

I am indebted to the Director of the National Parks and Wildlife Service, Zambia for his help. However, I am solely responsible for the opinions expressed in this paper.

I am also grateful to the University of Liverpool for financial support and to the Department of Geography of that university for secretarial, cartographic, and photographic help. My gratitude goes also to Dr Brenda McLean for commenting on an early draft of the script and for subsequent helpful discussions.

Currency

Zambia changed its currency from pounds sterling to kwacha in 1968 and from this date all statistics in government publications are given in the new currency. To maintain continuity, all financial data have been converted into pounds sterling at the following conversion rates (Whittakers Almanacs). The value of one kwacha in 1968–72 was £0.58. The value for subsequent years was:

1973	£0.65	1977	£0.73
1974	£0.67	1978	£0.62
1975	£0.76	1979	£0.58
1976	£0.74	1980	£0.52

REFERENCES

Aitkens, S.F. 1906. *The Sportsmans Guide to North-Western Rhodesia*. London: Commercial Intelligence Publishing Co.

Alpers, E.A. 1975. *Ivory and Slaves in East Central Africa*. London: Heinemann.

Ansell, W.F.H. 1978. *The Mammals of Zambia*. Government Printer, Lusaka.

Astle, W.L. 1971. "Management in the Luangwa Valley". *Oryx* 11:135–39.

Bell, R.H.V., and Grimsdell, J.J.R. 1973. "The Persecuted Black Lechwe of Zambia". *Oryx* 12:77–92.

Bigalke, R.C. 1964. "Can Africa Produce New Domestic Animals?" *New Scientist*, no. 374: 141–45.

Brannen, J. 1979. "Poachers Wipe Out Wildlife". *The Observer*, 16 Dec. 1979.

Brunton, J.D. 1909. *The Sportsmans Guide to North East Rhodesia*. London: Scientific Press.

______. 1912. *Big Game Hunting in Central Africa*. London: Melrose.

Bryden, H.A. 1894. "The Extermination of Great Game in South Africa". *Fortnightly Review* 62:538–51.

BSAC. 1908. *Rhodesia: Information for Tourists and Sportsmen*, 4th ed. British South Africa Company, London. (Issued annually from 1904 to 1910.)

Chapman, J. 1868. *Travels in the Interior of South Africa*. London: Stanford. (Also an edited version by E.C. Tabler, Balkema [Cape Town], 1971.)

Chesnaye, C.P. 1901. "Notes on the Fauna and Flora of North-Eastern Rhodesia". *Reports on the Administration of Rhodesia, 1898-1900*. British South Africa Company, London.

Colonial Office. 1938. *Report of the Commission Appointed to Enquire into the Financial and Economic Position of Northern Rhodesia*. Colonial Report no. 145. London.

Cott, H.B. 1961. "Scientific Results of an Enquiry into the Ecology of the Nile Crocodile (*Crocodilus niloticus*) in Uganda and Northern Rhodesia". *Transactions of the Zoological Society*, London, vol. 29:211-356.

Darling, F.F. 1960. *Wildlife in an African Territory*. London: Oxford University Press.

Dasmann, R.F. 1964. *African Game Ranching*. Oxford: Pergamon.

Denney, R.N. 1968. "The Case for Intensive Wildlife Management". *East African Agricultural and Forestry Journal*, Special Issue, pp. 118-32.

Dodds, D.G., and Patton, D.R. 1968. *Wildlife and Landuse Survey of the Luangwa Valley*. UNDP no. TA 2591 FAO. Rome.

Dunlap, R.C. 1973. *A Tourism Plan for the Luangwa Valley*. Document no. 4. Luangwa Valley Conservation and Development Project, DP/Zam/68/510, FAO. Rome.

Fanshawe, D.B. 1971. "The Vegetation of Zambia". *Forest Research Bulletin*, no. 7. Ministry of Rural Developmnet. Lusaka.

Field, C.R. 1979. "Game Ranching in Africa". *Applied Biology* 4:63-101

FNDP. 1966. *First National Development Plan, 1966-70*. Office of National Development and Planning. Lusaka.

Gann, L.H. 1964. *A History of Northern Rhodesia*. London: Chatto and Windus.

Hanks, J. 1979. *A Struggle for Survival*. London: Country Life Books.

Hanlon, J. 1980. "Can People Live with Elephants?" *New Scientist* 88:630-32.

Harroy, J.P. 1971. "The Criteria for Selection Established by the International Committee on National Parks". *United Nations List of National Parks and Equivalent Reserves*. 2nd ed. Brussels.

Hingston, R.W.C. 1930. "Report of a Mission to East Africa for the Purpose of Investigation Suitable Methods of Ensuring Preservation of Its Indigenous Fauna". *Journal of the Society for the Preservation of the Fauna of the Empire* 15:21-57.

Hutchinson, P. 1974. *The Climate of Zambia*. Occasional Study no. 7. Zambia Geographical Association. Lusaka.

Huxley, J. 1931. *Africa View*. London: Chatto and Windus.

Information Office. 1938. *Government Handbook, Northern Rhodesia*. Livingstone.

Jahnke, H.E. 1974. "The Economics of Controlling Tsetse Flies and Cattle Trypanosomiasis in Africa". *Forschungsberichte der Afrika*, no. 48. Ifo-Institut fur Wirtschaftschung. Munich.

Johnston, H.H. 1894. *Report on the First Three Years' Administration of the Eastern Portion of British Central Africa*. Parliamentary Paper c7504. London.

Lawton, R.M. 1971. "Destruction or Utilization of a Wildlife Habitat?" In Duffey, E., and Watt, A.S. (eds.), *The Scientific Management of Animal and Plant Communities for Conservation*. Oxford: Blackwell.

Legum, C., ed. 1976. "Zambia". *African Contemporary Record, 1975-76*. New York: Africana Pub. Co.

Livingstone, D. 1857. *Missionary Travels and Researches in South Africa*. London: Murray.

Lyell, D.D. 1910. *Hunting Trips in Northern Rhodesia*. London: Cox.

Marks, S.A. 1973. "Prey Selection and Annual Harvest of Game in a Rural Zambian Community". *East African Wildlife Journal* 11:113-28.

________. 1976. *Large Mammals and a Brave People*. Seattle: University of Washington Press.

Marshall, H.C. 1956. "Water Transport in the Bangweulu Swamp". *Northern Rhodesia Journal* 3:189–91.

Miller, J.C. 1970. "Cokwe Trade and Conquest in the Nineteenth Century". In Gray, R., and Birmingham, D. (eds.), *Pre-Colonial African Trade*. London: Oxford University Press.

Moubray, J.M. 1912. *In South Central Africa*. London: Constable.

Naylor, J.N.; Caughley, G.J.; Abel, N.D.; and Liberg, O. 1973. *Game Management and Habitat Manipulation*. Working Document no. 1. Luangwa Valley Conservation and Development Project. DP/ZAM/68/510, FAO. Rome.

Oliver, R., and Fagan, B.M. 1975. *Africa in the Iron Age*. Cambridge: Cambridge University Press.

Phillipson, D.W. 1975. "The Victoria Falls and European Penetration of Central Africa". In Phillipson, D.W. (ed.), *Mosi-oa-Tunya*. London: Longman.

Pitman, C.R.S. 1934. *A Report on a Faunal Survey of Northern Rhodesia*. Government Printer. Livingstone.

Roberts, A. 1968. "The Nineteenth Century in Zambia". In Ranger, T.O. (ed.), *Aspects of Central African History*. London: Heinemann.

Selous, F.C. 1881. *A Hunter's Wanderings in Africa*. London: Bentley.

Sharpe, A. 1888. "Elephant Shooting in East Central Africa". *The Field* 71.

SNDP, 1971. *Second National Development Plan*, January 1972– December 1976. Ministry of Development Planning and National Guidance. Lusaka.

Suckling, N.J. 1964. "The Victoria Falls Trust". In Fagan, B.M. (ed.), *The Victoria Falls*. 2nd ed. Commission for the Preservation of Natural and Historical Monuments and Relics. Northern Rhodesia, Livingstone.

Tabler, E.C. 1963. *Trade and Travel in Early Barotseland*. London: Chatto and Windus.

TNDP. 1979. *Third National Development Plan 1979–83*. National Committee for Development Planning. Lusaka.

UNESCO. 1963. "Conservation of Nature and Natural Resources". *Modern African States*. IUCN, New Series no. 1. Morges.

Vaughan-Jones, T.G.C. 1945. *Game and Tsetse Development Plan 1945–55*. Government Printer. Lusaka.

Watt, N. 1954. "A Game Ranger in Northern Rhodesia". *African Wildlife* 8:39–49.

Williams, V.C. 1976. "Infrastructural Development for Tourism Based on Wildlife". *Proceedings of the Fourth Regional Wildlife Conference for Eastern and Central Africa*, Zambia, July 1976. Ministry of Lands, Natural Resources and Tourism. Lusaka.

Worthington, E.B. 1961. *The Wild Resources of East and Central Africa*. Colonial Office Report no. 352. London.

Yearbook and Guide of the Rhodesias and Nyasaland, 1938–39. Salisbury.

List of Government and Official Publications from which data have been derived.

North-Eastern and North-Western Rhodesia

Directors Reports and Accounts. Year ending 31 March . . . British South Africa Co. (1889–1914).

Reports on the Administration of Rhodesia. British South Africa Co. (1889–92, 1900–02).

Protectorate of Northern Rhodesia

Customs Department, Annual Report (1911–23).

Colony of Northern Rhodesia

Annual Statement of Trade (1948–52).
Blue Book for the year ending 31 December . . . (1924–39).
Customs Department, Annual Report (1924–53).
Department of Animal Health, Annual Report (1931–40).
Department of Game and Tsetse Control, Annual Report (1940–52).
Department of Native Affairs, Annual Report (1929–52).
Financial Report for the year ending 31 December . . . (1947–52).
Report on Northern Rhodesia for the year . . . (1928–60).
Statistics of External Trade (1933–48).

Federation of Rhodesia and Nyasaland (Northern Rhodesia)

Annual Statement of Trade (1953–63).
Department of Game and Tsetse Control, Annual Report (1953–58).
Department of Game and Fisheries, Annual Report (1959–63).
Department of Native Affairs, Annual Report (1953–63).
Financial Report for the financial year ending 30 June . . . (1953–63).
Department of Veterinary and Tsetse Control Services, Annual Report (1959–63).

Republic of Zambia

Auditor General, Annual Report (1966–).
Annual Statement of External Trade (1964–).
Department of Game and Fisheries, Annual Report (1964–68).
Deprtment of Veterinary and Tsetse Control Services, Annual Report (1964–).
Department of Wildlife, Fisheries and National Parks, Annual Report (1969–73).
Financial Report for the year ending . . . (1964–).
National Parks and Wildlife Services, Annual Report (1974–).

9
The Grazing Resources of the Sudan

MUSTAFA M. KHOGALI

Livestock occupies an important position in the economy of the Sudan. An estimated 15 per cent of the total population of the country are nomadic, engaged in and depending primarily upon the raising of animals for their livelihood.[1] Furthermore, all the settled cultivators, except in certain unfavourable localities, raise animals to supplement their income and as a source of milk. The livestock sector contributes about 18 to 19 per cent to the gross national product (Cur. Agric. Stat. 1980:31), and it provides all the meat consumed in the Sudan. In addition, many of the planners and administrators think that the livestock sector can play yet a bigger role in the economy of the country, especially in the export sector which is heavily dominated by cotton, groundnut, and gum Arabic. Such planners think that the grazing resources of the Sudan have not yet been exhausted and that the Arab countries provide good potential markets for Sudan meat.

This paper describes the grazing resources of the Sudan and focuses upon some of the important problems that hinders the future development of the livestock sector.

The Physical Environment

The livestock of the Sudan subsists mainly on natural and unimproved grazing resources. This makes it necessary to discuss the natural vegetation which in turn is very much correlated with the distribution of rainfall in the country. The Sudan extends from about 4 °N to about 22 °N. It has a tropical continental climate characterized by high temperatures throughout the year, and summer rainfall that decreases in amount and duration northwards. The northern part is desert with very scanty rainfall (Wadi Halfa: 0.1 mm), while in the southwestern part rainfall totals exceed 1,500 mm and the rainy

---

[1] The number of nomads in the Sudan is not known exactly; the censuses of 1955/56 and 1973 do not give precise data on the number and percentage of nomads to the total population.

326

season lasts for nine to ten months of the year. Between these two zones, there is a narrow semi-desert zone and different savanna types. The narrow Red Sea coast receives most of its rains in the winter months, but the annual total remains small, not more than 150 mm.

The distribution of rainfall tends to be zonal (Fig. 9.1); to some extent this is also the case with the vegetation belts. However, a number of factors modify the distribution of vegetation. These include altitude, slope, soil types, and river water. The Sudan is mostly a flat country with only a few areas of high lands. Where such areas exist, e.g., Jebel Marra (3,100 m),

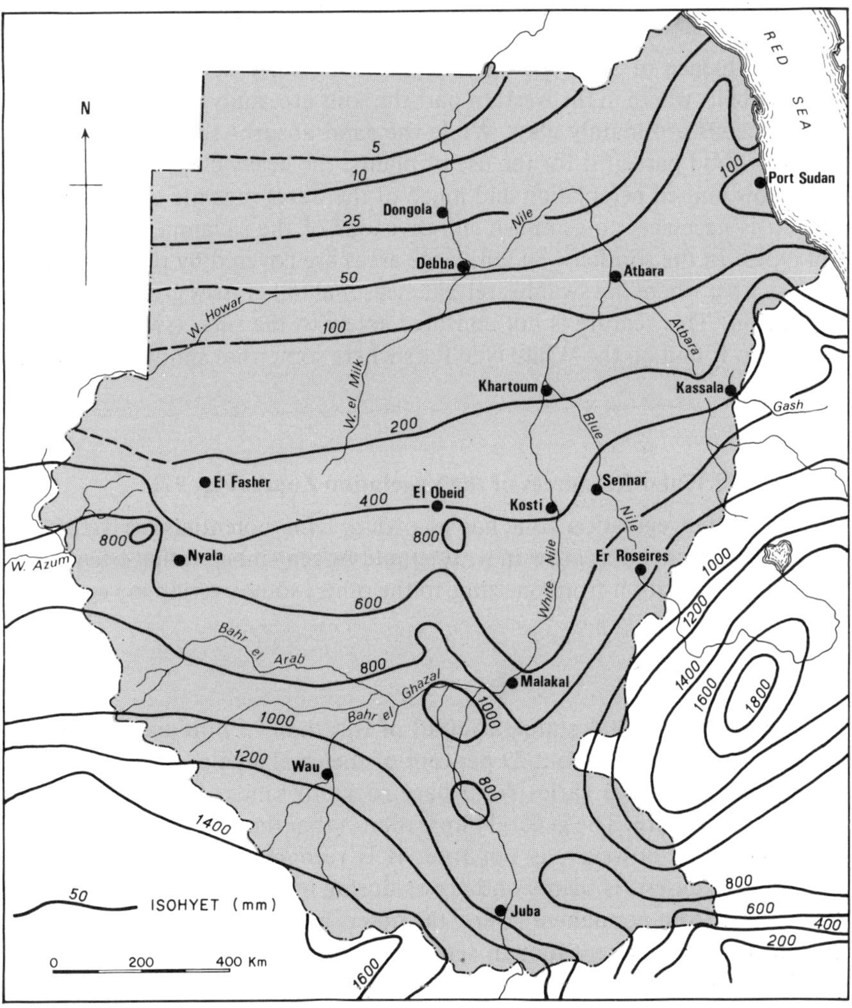

Fig. 9.1 Average annual rainfall of the Sudan, 1921–62

montaine vegetation dominates. The Red Sea Hills are not high enough to change the vegetation, but because of the winter rainfall, a difference exists between the type of vegetation on the hills and that of the plateau further westwards where the scanty rainfall occurs in summer. However, the most important influence of the Red Sea Hills is in the fact that the numerous water courses concentrate the rain water along their channels and therefore have a richer vegetation than would be expected if only rainfall totals are considered. It is therefore customary to consider most of the area of the Red Sea Hills as having a semi-desert rather than a desert vegetation. The other areas that are located on the same latitudes in the country have a desert vegetation.

The influence of soil type on vegetation is clearly demonstrated in the savanna belt, where in the western part the soils are sandy and in the eastern part the soils are mainly clay. While the sand absorbs the rain water and retains a good part of it for the use of plants, the heavy clays of the eastern part allow limited percolation and much of the water runs off or evaporates. Thus it is necessary to establish sub-divisions of the savanna according to soil types. In the southern Sudan, large areas are covered by river and rain-fed swamps; where the swamps retreat, seasonal tall grasses grow during the dry season. This feature is not confined solely to the *sudd* (swamp) region but is also found on the White Nile flats where somewhat similar conditions prevail.

Pasture and Water Resources of the Vegetation Zones (Fig. 9.2)

Each of the vegetation zone has its own grazing potentials derived from the types of plants that grow in it. It should be remembered that because of the gradual transition from one zone to the other, some overlap in vegetation occurs.

The Desert

The zone has a total annual rainfall of less than 75 mm and an area of about 725,000 km² or about 29 per cent of the total area of the Sudan. The nature of the surface varies from bare rocks to sands. Soils, where they exist, are immature. The general impression is that the desert has no grazing potential. This, however, is not true. It is recognized that vegetation, if found in the desert, is scanty and grows during a very short period of time. Also, apart from permanent rivers, the desert has no surface water and has not attracted permanent human settlement. Yet the nomads of the Sudan use the desert to raise a large number of animals. The nomads recognize two types of desert grazing. In years of good rainfall, the desert may receive some showers and, consequently, water collects in the natural depressions

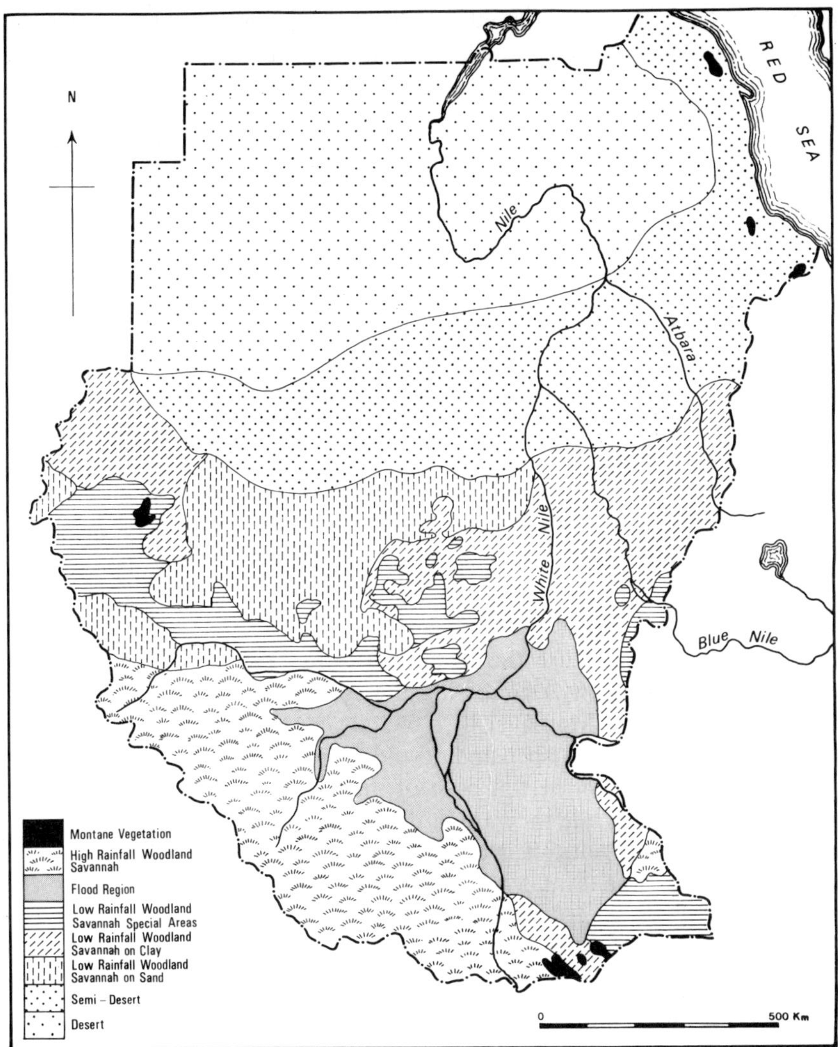

Fig. 9.2 Vegetation of the Sudan

and plants, mainly grasses and herbs, burst into life. The nomads who graze their animals along the desert margin take the opportunity to use these short-lived grazing and water resources. In bad years the desert grazing fails. The Bayouda desert as far north as the latitude of Debba is utilized in this way. In the second type of grazing, the *jizu* of Wadi Howar and the adjacent areas along the Sudan/Chad borders is extensively utilized by the nomads.[2] The *jizu* is of special interest to the nomads who send large numbers of their animals to it. The *jizu* plants are mainly herbs and grasses that grow in sandy depressions after the end of the rainy season and when the cool northerly winds blow from November to April. Usually the *jizu* succeeds after a good rainy season. The *jizu* plants include *Neurada procumbens* (*Ar. Saadan*), *Aristida papposa* (*Ar. Nissa*), *Indigofera brachteolata* (*Ar. Derma*), and a number of other similar plants. All of them are grazed by the animals and they have two important merits. First, they are very nutritious, with the *Saadan* especially so, and second they are succulent plants and the animals that subsist on *jizu* do not need water for as long as they eat it. It is not uncommon for animals to spend five months grazing solely on *jizu* plants.

However, the *jizu* does not grow every year, but only in years of good rainfall. It did not grow during the Sahelian drought of 1969–74, but appeared in 1975 when the author had the chance to fly over Wadi Howar and the adjacent areas. It was estimated in November 1975 that at least 50,000 heads of camel and a similar number of sheep and goats were actually in the *jizu*, and another 10,000 camels were trekking towards it. Although the *jizu* supports a large number of animals for three to five months a year, it cannot be depended upon to grow in any one year.

Although cattle can use the *jizu* plants, they cannot trek to the *jizu* because of the difficulty of crossing the rocky desert region that comes before the *jizu* area. Moreover, life in the *jizu* is extremely hard for the herdsmen as the region is open and devoid of trees or any other shelter. The nomads are exposed to the full blast of the northerly winds and the bitter cold of the desert. While they are in the *jizu*, the nomads subsist on milk, porridge, and date fruits. It is estimated that the carrying capacity of the desert ranges between 0 and 5 animal units per km[2].

The Semi-Desert[3]

The semi-desert in the Sudan is a narrow zone with an annual rainfal of between 75 and 300 mm. It covers an area of about 490,000 km[2]. Rainfall is usually not sufficient for cultivation. Nevertheless, some quick-maturing crops are grown. According to Koppen's classification, this zone should be part of the desert (Bw). But because the scanty rainfall is sufficient to support vegetation on which a large number of animals subsist, the zone is considered semi-desert (Lebon 1965:54–57). The main occupation away

---

[2]*Jizu* is a collective name for certain desert plants, and is also the name of the desert area where such plants grow. In the past, the nomads of the semi-desert used the Wadi Howar and the Dongola *jizu* areas. For various reasons, the Dongola *jizu* failed and is now not used.

[3]This is also known as the semi-desert acacia scrub and short grass lands of north-central Sudan.

from the irrigated areas along the river is animal raising, mainly under nomadic conditions. The vegetation of the semi-desert consists of herbs, annual and perennial grasses, and bushes and trees of the *Acacia* genus. The slope and the type of soil play an important role in the dominance of grass/ herbs and trees. Thus, in the *wadis* of the Red Sea Hills *Acacia* trees dominate, while herbs and grasses are more common on the plateau and on hill tops. *Acacia* trees also dominate in low depressions where the soil is silty. In the Butana, grasses dominate and trees are found only along the beds of the water courses. Apart from such areas, grasses and scattered *Acacia* grow together in western Sudan. The grasses are usually short and consist of a number of trees, mainly of *Aristida* genus, especially *Aristida plumosa* (*Ar. Bayad*), and *Blepharis spp.* (*Ar. Siha*). *Siha* was originally dominant over the Butana and western Sudan, but because of its high nutritional value it has been severely over-grazed and in many places it has been replaced by less palatable or unpalatable species such as *Cyperus conglomeratus* (*Ar. Ushar*). The grasses become dry shortly after the rainy season and thus lose part of their nutritional value. The *Acacias*, especially *A. tortilis*, are important sources of grazing since the leaves are readily browsed by camels and goats, and the trees bear seed-pods that appear during the dry season, March-April, at a time when there is a severe shortage of other grazing. These seed-pots are eaten by all animals, especially sheep and goats.

Away from the riverine lands, the main problem is the shortage of water since rainfall is seasonal and occurs only during the three months July to September. But wells dug in the river beds and the vicinity of the water courses supply water most of the year. In recent years, some deep bores have been drilled and some *hafirs* (surface water reservoirs) have been dug. A few dams have also been constructed across some seasonal water courses, such as Wadi el Milk. However, the geological formation is not everywhere suitable for wells or *hafirs*. For this reason, settlement is restricted to the few areas where water is found throughout the year. The semi-desert is therefore used mainly by nomads who raise hardy animals, camels, sheep, and goats that can afford to remain without water for days, and can walk long distances between grazing areas and watering centres. It is estimated that the semi-desert of the Sudan accommodates about 7 million sheep, 5 million goats, 2 million cattle, and 2.5 million camels.

The *dars* (homes) of the camel nomads are in the semi-desert, around the permanent water points. The nomads use extensive areas in the low woodland savanna, in the desert, and in the semi-desert, and in this way they maximize the number of their animals. At the onset of the rainy season, the nomads move southwards to the northern part of the low woodland savanna to make use of the green pasture of the region. As the rains are established in the semi-desert margin, the nomads move northwards to make use of the new grazing and of the water in the natural pools. The grazing resources of the *dars* are thus spared, to be used during the dry season. During the cool part of the dry season, many of the nomads of northern Kordofan and north Darfur journey to the *jizu* areas. It has also been observed during the past few years, especially during drought periods, that an increasing number of the camel nomads began to appear during the dry season in the areas of the low woodland savanna and as far south as Bahr el Arab, an area of Baggara nomads.

The carrying capacity of the semi-desert is estimated to be between 5 and 10 animal units per km².

Low Rainfall Woodland Savanna[4]

The annual rainfall in this zone is between 300 and 800 mm, and the zone covers an area of about 690,000 km². It is customary to divide the zone into at least two main sub-divisions: one on clay, covering the clay plain along and east of the White Nile; and the other on sand, west of the White Nile, in Kordofan and Darfur. In both sub-divisions, rain-based cultivation is possible and is widely practised, although towards the northern edge cultivation becomes more risky. Animal raising is also an important activity, especially in the southern area where most of the cattle of the Sudan are raised by the Baggara tribes.

The vegetation of the low woodland savanna on clay is very much dominated by *Acacia mellifera*. In areas of medium rainfall (400–600 mm), the *Acacia* forms dense thickets. Grasses have the chance to dominate only when old trees die or when fire kills the trees. *A. mellifera* has seed-pods that can be grazed, but when the vegetation is dense it is difficult for the livestock to penetrate it. The grasses that grow are *Sorghum purpureo*, *Cymbopogon nervatus*, and *Brachiaria obtusiflora* — all these are rather poor grazing. On the other hand, there are short grasses associated with *Acacia mellifera* where the bush is not thick enough to prevent grass growth. These grasses, which include *Sehima ischaemoides*, *Tetrapogon spathaceus*, and *Setaria verticillata*, are eaten by animals and are nutritious.

As the rainfall in the southern part of this sub-zone increases, *A. mellifera* gives way to taller trees, mainly *A. seyal* and *Balanites aegyptiaca*. Grasses are tall and fibrous during the dry season, but these can be grazed by animals when the grasses are young. At the southern margins of this area, annuals disappear and give way to perennial grasses which on burning give regrowth that can be used by the grazing animals. A major problem of this southern zone, however, is that stomoxys and other biting flies spread during the rainy season and force the nomads to move northwards to areas of less rainfall and, consequently, of fewer flies. This is the main reason for

---

[4]This is known also as the low woodland savanna of Central Sudan.

the nomadic movements northwards. As the dry season sets in, the nomads return southwards in the direction of Mashar Marches in the southern Gezira and to southern Butana.

The low woodland savanna on clay was originally a land of farmers and nomads. Irrigated cultivation was restricted to a narrow zone along the riverine lands of the White and Blue Niles and River Atbara, while rain-based cultivation was practised away from the rivers. The nomads used the rivers for watering their animals, but for most of the year were away from the banks of the rivers. Since 1925, when the Gezira scheme was started on the central clay plain, the area under irrigated and rain-fed cultivation has continuously expanded at the expense of the range land. It is estimated that since 1925 about 9 million feddans[5] have been converted to crop land. The Gezira scheme alone covers 2 million *feddans*, and the mechanized crop production schemes in the Gedaref and the southern Gezira account for another 4 million *feddans*.

The low woodland savanna on sand is a belt in Kordofan and Darfur provinces. While sand dominates there are clay depressions, some of them large. However, towards the southern margin the sands are replaced by the wide clay plains of Bahr el Arab and Bahr el Ghazal regions.

The type of vegetation found here depends on the water available and on the type of soil. Generally, the *qoz* has a mixture of *Acacia* trees and grasses while in the clay depressions trees are more common. The dominant tree of the *qoz* is *A. senegal* from which gum Arabic is obtained. This zone is the world's main source of gum Arabic. The tree is linked with the agricultural rotation, which is why it is often found in pure stands on the sandy soils. The *tebeldi* (the Baobab, *Adansonia digitata*) is also found on the silty part of the sand, but it is thought that this tree owes its existence here to a former more humid period. The leaves of the *A. senegal* trees are browsed by camels and goats, but not those of the *tebeldi* tree.

The grasses of the zone include *Aristida pallida*, *Andropogon gayanus*, *Blepharis linarifolia*, *Zornia diphylla*, *Cenchrus biflorus*, and *Eragrostis tremula*. These grasses are palatable to various degrees, but the vegetation of the sandy areas is generally deficient in minerals, especially salt. This is one of the reasons the nomads do not stay throughout the year on the *qoz*. However, on the lower slopes of the *qoz* some salty (*taim*) grasses grow, and these include *Sporoblus marginatus*, *Brachiaria* spp., *Chloris* spp., and *Digitaria gayana*. These grasses are much sought after by the cattle owners.

The best grazing of the zone is thus found at the junction of the *qoz* with the clay plain. The zone of contact between the sand and the clay is not a

---

[5]A *feddan* is a measurement unit very slightly greater than an acre, but for practical purposes it is taken to be equal to an acre.

continuous belt but an area where sand ridges alternate with clay depressions. The ridges gradually become smaller and more isolated, and finally disappear. This zone of contact is where the Baggara nomads are found.

The vegetation of the clay depressions and plains in this sub-zone is not very different from that of the area east of the White Nile. There, *Acacia seyal*, *A. mellifera*, and *Lanna humilis* dominate, while the grasses include *Aristida* spp. and others. The slightly higher areas of the clay from where water runs off are usually bare.

The dominant economic activities of this sub-zone are rain-based cultivation and livestock raising, usually under nomadic conditions. The northern part of the zone, which is dominated by sand, is a land of traditional cultivators. The main crops are *dura* (sorghum), *dukhn* (millet, pennisetum typhoideum), and groundnuts as well as gum Arabic. In addition, the cultivators raise livestock under settled conditions, although some animals may be sent away with the nomads. The southern part of the sub-zone is of more varied soils where sand ridges alternate with clay depressions. Clay soils become dominant as Bahr el Arab is approached. This southern part is the domain of the Baggara who raise cattle, sheep, and goats. The nomads use a wide zone that stretches from just south of Bahr el Arab, from about latitude 8 °N to about latitude 13 °N. The area of Bahr el Arab is clayey, and with rain the mud becomes breeding grounds for stomoxys and other biting flies, and so the nomads are forced at the beginning of the rainy season to move northwards to the *qoz* land where rainfall is lower and biting flies fewer. But at the end of the rainy season, water becomes scarce and this, together with the saltlessness of the grazing on the *qoz*, force the nomads to move back to Bahr el Arab where water can be obtained. The burning of the grasses initiates some green regrowth which is used as fodder. Through their movements, the Baggara nomads are thus able to exploit extensive areas in a number of different ecological sub-zones and, consequently a large number of animals can be raised. It is estimated that the Baggara of the low woodland savanna west of the White Nile possess 5 million heads of cattle, 2 million heads of sheep, and 3 million heads of goats. These Baggara supply about 80 per cent of all the beef consumed in the Sudan.

Partial settlement of the nomads has taken place in this sub-zone. It became common for a great number of families to split, and for part of each family to settle and engage itself in both subsistence and commercial cultivation based on *dura*, *dukhn*, and groundnuts. The other part of the family moves and cares for the livestock. But cultivation and raising of animals are part of one and the same economy with the animals subsisting on the agricultural residues. However, the settlement and engagement in cultivation has caused a reduction of the range land.

The carrying capacity of the low woodland savanna varies widely, between 10 and 25 animal units per km².

High Rainfall Woodland and Savanna[6]

The annual rainfall is between 800 and 1,500 mm, and the total area of the zone is about 347,000 km². This is an area of tall trees and tall perennial grasses which include *Ctenium eligans*, *Hyparrhenia rufa*, and *Sprobolus pyramidalis*. These grasses become woody when they approach maturity and thus they cannot be used by animals. The animal owners therefore burn the grass so as to initiate new green growth. The northern end of this zone is an area where the Baggara nomads and Nilotic animal-owners meet.

A major problem in the zone is that biting flies appear in it during the wet season, and a great part of the zone is infested by the tsetse fly. Therefore, in such areas the population depends on crop production and animals are not reared.

The Flood Region

The annual rainfall in this zone varies from 300 to 800 mm. But the rainfall is not the decisive factor in the vegetation since the rivers overflow their channels and cause widespread swamps which are enlarged by rain water that cannot drain into the rivers. The total area of the swamps is about 260,000 km². The swamps can be divided into two types: permanent and seasonal. It is the latter that is important for grazing because it is in the areas from which water seasonally recedes that annual and perennial grasses grow. The perennial grasses are of the *Echinochloa* genus, namely, *E. stagnina* and *E. pyramidolis*. These make good grazing and are very important to the Nilotic pastoralists because the grasses grow during the dry season at a time when the other grazing areas are barren (Lebon 1965:39). Other vegetation types depend on the soil and on whether the area is flooded or under permanent swamps. The papyrus sedge, plants of the permanent swamps such as *Cyperus papyrus*, are not grazed by the livestock except when there is no other grazing available.

The swamp region (*sudd*) is inhabited by Nilotic tribes who are mainly pastoralists but who also practise primitive cultivation and fishing. The Nilotes have their permanent huts on the slightly higher grounds that are not inundated, and it is here that they spend the rainy season, cultivating their crops and caring for their animals. Because such higher land is small in area, and because part of the land is under cultivation, there is insufficient grazing for the livestock, especially at the end of the rainy season. As the flood begins to recede and vegetation grows on the uncovered land, the Nilotes move with their animals from the higher to the lower areas for grazing. The journey may involve a distance of a few to as much as 120 km. Biting flies are endemic to the flood region, but the livestock appear to have

---

[6]This is known also as the deciduous woodland savanna of southern Sudan.

acquired some immunity to them. The Nilotes keep the cattle inside their huts at night and use smoke to keep flies away.

The southern Sudan, that is, the high rainfall woodland and savanna, and the flood region has a large animal population composed, in 1978, of 6,257,967 heads of cattle, 3,370,656 heads of sheep, and 2,173,162 heads of goats. But not much economic use is made of this wealth, as the Nilotic and other associated tribes are reluctant to part with their livestock, especially their cattle, which have a kind of spiritual significance to them. Furthermore, the animals of the southern Sudan suffer from a number of diseases and are therefore not very marketable.

Problems of the Grazing Resources

The Sudan has extensive areas of range land, the utilized parts of which come to about 60 million *feddans*. Large sections of the Sudan population, the Hamites, the Nilotes, the Arabs, and Arabicised stocks are known for their love of their livestock and their long history of contact with raising camels, cattle, sheep, and goats. The availability of resources has enabled the population to raise large and growing numbers of animals (Table 9.1).

There are three main reasons for the increase in the livestock numbers: (1) the increase of the rural population and the desire of each household, nomadic as well as settled, to raise as many animals as possible; (2) the improvement in the provision of water supply; and (3) the introduction of veterinary services. Since the end of the Second World War, an active programme for the provision of water was initiated in the low rainfall wood-

TABLE 9.1 NUMBER OF ANIMALS IN THE SUDAN, 1924–79
(million head)

	1924	1944	1964	1974	1979
Cattle	1.5	3.1	10.0	12.9	17.3
Sheep	1.9	4.8	10.0	3.1	17.2
Goats	1.8	3.4	7.0	11.5	12.2
Camels	0.4	1.1	20.0	8.8	2.6

Sources: 1924 and 1944 figures are from *Report by His Majesty's Agent and Consul General on the Finance and Administration of the Sudan*, London, 1925, 1946 respectively. Other figures are from Ministry of Planning, *Economic Surveys*, 1965, 1975, and 1979 respectively. The figures for 1979 are based on the animal survey carried out in 1976, by using air photos and ground work.

land savanna, with the intention of enabling the livestock to delay their passage to their summer grazing grounds to the south by about two months so that the grazing there will be better able to last out the summer (Jefferson 1956:16). At a later stage, the purpose of the programme as stated by the then Director of the Land Use and Rural Water Corporation was to shorten nomadic movements with a view to further settlement (Shawki 1968:13). The result of the water programme is that animals today rarely die of thirst, though many may die of hunger.

Important progress was also achieved in the area of combating animal disease such as *rinderpest* (cattle plague) and contagious bovine pleuro-pneumonia. This was done through vaccination. As far back as 1924, anti-*rinderpest* vaccine was developed and produced in the Malakal Veterinary Laboratories. However, it was only immediately after the Second World War that an active policy of veterinary services was adopted. The main animal diseases, especially the *rinderpest*, which used to kill large numbers of livestock, are now under reasonable control.[7] In fact by 1951 it was recognized that although not all animal diseases were under complete control, the Sudan livestock population had emerged from the survival stage to the quantitative development stage (Bisschop 1951).

Despite the good development of the livestock sector, there are problems involving the grazing resources of the Sudan. Some of these are related to the nature of the resource themselves, while others arise from man's management of the resources. Three main problems are discussed here.

First, there is the problem of the seasonality of the pastural resources. Rainfall is seasonal all over the Sudan, but the duration of the wet season varies from nine months in the extreme southwestern Sudan to three months in the semi-desert, while in the desert rain may not fall for five continuous years. Consequently, water is not available for periods that vary between three and nine months, depending on the latitudinal location of the region. Furthermore, the nutritive value of the pasture drops markedly during the dry season. In the savanna areas, the population tries to overcome the problem by burning the grasses to get rid of the old growth and to initiate green regrowth. But this measure, important as it is, has limited effect as not all areas are covered by perennials; and as the dry season approaches its end, this practice becomes less fruitful. The problem is more acute in the semi-desert as perennials cover small areas and burning does not have the same effect as in the savanna region. In fact during the dry season, the vegetation of the semi-desert becomes dry and the grasses and herbs wither away. Therefore, the non-nutritious fodder that is available is insufficient to meet

---

[7]In certain cases, *rinderpest* could decimate a herd by 90 per cent. In the 1970s the Organization of African Unity took an active role in the fight against the disease by organizing a continental campaign involving countries such as the Sudan, Chad, and Ethiopia.

the requirements of the livestock. The lack of nutritive grazing during the dry season has many adverse effects. The growth of the animals is checked and many animals become weak and some die. This, in turn, leads to late age at first calving, high mortality, slow and uneven growth, decrease in live weight during the dry season, late slaughtering age, low dressing percentage, and poor carcass composition (Hassan 1979:63, 65).

Another adverse effect is that the supply of animals and meat to the local markets and for export also becomes seasonal. The producers sell the surplus of their animals during the rainy season when these animals are at their best. But during the dry season the livestock are in poor condition and prices are low. Consequently, the producers hesitate to sell even if there is a risk that some of the animals may die. Traditional producers have the hope that their animals can survive until the rainy season. Furthermore, the livestock merchants who work between the main producing areas (western Sudan) and the consuming centres (mainly the towns of the Nile valley and eastern Sudan as well as some of the neighbouring countries) hesitate to buy animals during the dry season and trek them to the consuming centres as at that time there is very little water and pasture on the route.[8]

The second problem is the inability to use all the pasture resources. The agricultural lands of the Sudan are extensive, but only a small proportion is used for crop production and for grazing. The lands that can be utilized but so far have not yet been used total about 181,315,000 *feddans*, and these are considered cultivable waste. If infrastructural facilities become available and some problems are solved, a great part of this land can be used for pasture. In fact it has been estimated that out of a total area of 251 million ha, 140 million ha are suitable for crop and livestock production (Table 9.2).

The inability to use all the pasture land of the Sudan is due to a number of factors — shortage of water during the dry season, biting flies in the clay areas of the savanna where the annual rainfall is more than 500 mm, seasonal fires, and the fact that large areas are covered by non-palatable plants. Although there has been important improvements in the water supply situation since 1946 when the rural water programme started, large areas still remain without water during the dry season. Furthermore, at the end of the dry season when the livestock are hungry, they become too exhausted to walk long distances from the watering centres to the grazing areas just to feed on fill-up stuff. In addition to this, the pasture of the *qoz* areas of the savanna is deficient in salt and is not used throughout the year. On the other hand, the clayey areas of the southern parts of the savanna cannot be used by the nomads during the wet season when the vegetation is

---

[8] In the Sudan, animals are usually transported on the hoof. The railway system, being inefficient and lacking the necessary facilities, transports only a negligible percentage of the marketed livestock.

TABLE 9.2 PRESENT AND POTENTIAL LAND USE IN THE SUDAN

	Estimated Area (million ha)	%
Present land use:		
Crop production, irrigated and rain-fed	6.3	2.5
Livestock production	42.5	16.9
Potential Land use:		
Crop production	33.7	13.4
Livestock production	106.3	42.4
Other land uses	62.2	24.8
Total area of Sudan	251.0	100.0

green and tender and at its best because of the biting flies. Flies also play an important role in reducing the available grazing. This is specially so in the savanna zone where the vegetation cover is fairly dense and the areas that are covered by grasses become dry some weeks after the end of the rainy season, and thus catch fire easily.

The natural vegetation as everywhere is composed of grazable and non-grazable species. But under conditions of over-grazing and over-cultivation, the non-grazable species tend to increase and, as a result, wide areas become of little use as range land. So far no study has been made to estimate how much of the 181,315,000 *feddans* that are classified as cultivable waste are covered by unpalatable species, but the impression gained from observation in the field is that the percentage is quite high.

It should also be noted that the expansion in the crop area referred to earlier was achieved at the expense of grazing lands. Moreover, further expansion of the area under cultivation is the aim of the planner, and this would encroach further on the area of grazing land and cause a good deal of frustration to the nomads who are the traditional users and whose migration routes would be disrupted. However, it must be mentioned here that the conversion of land to produce crops does not necessarily mean a total loss of grazing since crop residues constitute an important feed to livestock, especially those of the sedentary population. It has been estimated that the crop residues in the Sudan can support 7,253,000 animals units. Most of these residues come from sorghum (*dura*), millet (*dukhn*), groundnuts, sesame, cotton, tubers, and sugar cane (Hassan 1979:50; Arab Organization for Agricultural Development 1974).

The third problem is over-grazing. Despite the fact that only between 30 and 35 per cent of the land potentially suitable for grazing is actually used for raising livestock, over-grazing is an observed and a serious problem in the Sudan. Many scholars think that nomadism, under which most of the Sudan livestock are raised, has a number of advantages as a strategy for managing the grazing resources. One of these merits is that the nomads by their seasonal movements use the pastoral resources in a rotational way, thus avoiding over-grazing and deterioration of the resources. In a sense, the nomadic use of resources amounts to rotational use. But over-grazing can occur as a result of any of the two processes: concentration of animals in one area throughout the year, thus giving no chance to the vegetation to regenerate, or seasonal concentration of large number of animals in one area. It is the latter process that characterizes the nomadic use of the resources.

In the past, the number of animals raised by the nomads was small and a sort of an equilibrium was established between the number of animals and the grazing resources. As Lebon observes: "there seems to have been little evidence of over-use of land until about 1930, except in the immediate neighbourhood of the small towns and villages where trees were denuded for firewood and all plants except *Calatropis procera* consumed by goats in their daily coming and going" (1965:158). But with the introduction of veterinary services and the provision of rural water supply the number of animals began to increase rapidly (Table 9.1). The ecological balance was upset and severe over-grazing began to occur. By 1954 when Harrison wrote his report, all ecological zones where animals were raised except the flood region in the Upper Nile Province were overstocked (Table 9.3). The situation concerning overstocking and over-grazing seems to be far worse than many authorities think. One scholar's view was that there is overstocking in the Sudan during all the three climatic seasons, including the *Kharif*, the rainy season (Bakhit 1977). His calculation is as follows:

Season	Overstocking by Animal Units
Rainy season	1,558,710
Winter season (early dry period)	4,472,402
Summer season (late dry season)	10,321,344

Note: The description of early and late dry season has been added by the present author.

TABLE 9.3 SUMMARY OF STOCKING CONDITIONS IN THE NOMADIC AND PASTORAL REGIONS OF THE SUDAN, 1955

Region	Estimated Animal Units in 1954	Stocking Rates (Animal Units per sq. ml.)	Stocking Position in 1954
Semi-desert Beja country	440,000	7.3	At least 50% overstocked
Semi-desert, including all animals using the Butana in the rainy season	340,000	11.3	50% overstocked
Semi-desert northern Kordofan and north Darfur	1,000,000	14.3	30% overstocked
Low rainfall woodland savanna, western parts and Baggara	2,200,000	22.0	30% overstocked
Low rainfall woodland savanna on clay	1,100,000	15.0	20% overstocked
Toposa, especially area of low rainfall woodland	176,000	13.0	20% overstocked
Flood region, Upper Nile Province	1,100,000	22.0	50% understocked
Flood region, Upper Bahr el Ghazal	1,100,000	44.0	25% overstocked
High rainfall wood-land savanna	41,000	Most areas have no animals because of tsetse flies	

Source: After Harrison 1955.

The result of overstocking is widespread over-grazing, which manifests itself in the following ways:

(1) In many places, some of the most palatable and nutritive plants, such as *Siha* and *Bogheil*, both of *Blepharis* spp., have disappeared and have been replaced by less palatable or by unpalatable species. *Siha* used to dominate in the semi-desert, especially in the Butana and Kordofan and Darfur, but it has been replaced by *Nal* (*Cymbopogan nervatus*) and *gao*

(*Aristida* spp.). *Bogheil* was an important plant in the savanna of Kordofan and Darfur, but it has disappeared except in some isolated locations. The savanna areas of the western Sudan are being invaded by quite undesired plants such as *Hirab Hawsa* (*Acanthospermum hispidum*) and *gao* (*Cassisa* spp.).

(2) In many places in the semi-desert and the savanna region, some perennial plants have disappeared and have been replaced by annuals.

(3) In some places, especially around water points, the vegetation cover has been removed altogether. As a result, the unprotected top soil is being eroded, causing mobile sand dunes or sand sheets that cover cultivated areas or natural vegetation.

The Future

The grazing resources of the Sudan play an important role in the Sudan economy. These resources have been exploited in raising livestock to satisfy almost all the domestic demand for meat and, in addition, a surplus is generally available for export. It is the intention of the Sudanese planners that such a surplus should be increased to earn more foreign exchange for the country. The important question is how far the grazing resources can allow the livestock sector to continue playing its present role and the role the planners intend to assign to it. Two important factors have to be borne in mind when considering this problem.

First, the margin between meat production and local consumption is narrow and thus there is little surplus available for export (Table 9.4).

Second, the human population in the Sudan is increasing at a rate of at least 2.8 per cent a year. Therefore, if the number of the livestock remains

TABLE 9.4 MEAT PRODUCTION, CONSUMPTION, AND EXPORT, 1977/78

Meat	Total Production (000 tons)	Total Consumption (000 tons)	Export (000 tons)	Percentage of Export to Total Production
Beef and veal	193	188	5	2.6
Mutton and lamb	123	118	5	4.0
Goat	27	25	2	7.1
Total	343	331	12	3.5

Source: Hassan 1979:77.

at its present level during the next few years, no meat will be available for export unless local consumption is reduced. But the livestock population has not been static; instead it is continuously increasing, as shown by Table 9.1. However, with the current overstocking and over-grazing, it is hard to envisage further increases in livestock numbers without causing extremely serious environmental problems. It becomes of vital importance, therefore, to think in terms of improving both the grazing resources and the management of resources. So far no improvement has been made to the pasture resources, although some improvements were achieved in the provision of water supplies. The pastures have also suffered degradation as a result of over-grazing and, consequently, their carrying capacity has declined. Moreover, crop residues are not being used to the full, especially in the eastern Sudan in the area of the mechanized cultivation where more of the owners of the farms have few or no animals. The nomads would be in the southern areas when the crops are harvested and the residues become available for use. On the other hand, as seen earlier, vast areas classified as cultivable waste are used neither for cultivation nor for grazing.

It seems, therefore, that there is still scope for increasing the carrying capacity of the land, making more use of crop residues and for bringing more land under range management. But this needs planning and costs money. Planning is especially needed for the proper distribution of water points, as grazing around the new water points would otherwise be destroyed. If the distribution is not sound, large areas will continue to be unused. Re-establishing the nutritive plant species that have disappeared because of over-grazing, and introducing other grazable species that have proved themselves in similar environmental situations, are among the measures that should also be taken.

The foregoing suggested measures by themselves are not enough; they should be accompanied by sound management of resources. The nomads now use the grazing resources in a communal way. What is important to them is to satisfy the immediate grazing needs of their livestock, and no consideration is given to the preservation of the resources. Furthermore, the nomads are usually not in a hurry to take their animals to the markets. This is because (1) the value their livestock as a form of wealth that they know best and can manage; (2) the pasture and to certain extent water are free and thus the upkeep of an animal for a number of years does not cost much; and (3) the material needs of the nomads are few. It is suggested, therefore, that the system of the ownership of grazing resources be re-examined so that a sense of personal ownership be developed. One can foresee the political and social difficulties that will be encountered if such a suggestion is to be carried to the letter, that is, each household to have a piece of grazing land. Fewer problems will be created if each nomadic tribe is to be divided into small grazing groups of households, *Khashm beit*, and each group is to use and be responsible for one or more pieces of grazing land. In this way, it may be hoped that a sense of responsibility for the resources be developed.

Although most the Sudanese cultivators raise animals, livestock is not really integrated into the farming system, except in as far as animals are allowed to feed on some of the crop residues. In fact, the authorities in the large-scale irrigated schemes such as Gezira and Khashm el Girba have shown marked hostility to the presence of animals within the boundaries of their schemes. This hostility has been legalized by the issuing of local ordinances prohibiting the entry of animals to the schemes during the cultivation season, July–April, except for one milking animal unit per tenancy. It is time that a policy of integrating livestock in the agricultural systems be developed for the benefit of both the farmers and the livestock. Legume crops can be grown as part of the rotation system and used for fodder. Some feed lots based on some crops and their by-products such as oil-cakes from pressed cotton seeds and groundnuts can also be used. In fact, a beginning has already taken place in the White Nile region by former nomads who are selling milk for cheese making. They have realized that the natural grazing is not sufficient for their livestock and so they supplement it by feeding them oil-cakes. What is important is to encourage such practices not only among nomads but also among cultivators so that some traditional crop production areas can be turned into fattening zones. It is only then that the animal wealth of the Sudan can pass from the quantitative development stage to the qualitative development stage where more animals of better quality can be raised in a unit area in conjunction with field crops.

REFERENCES

Arab Organization for Agricultural Development. 1974. *The Technical and Economic Feasibility Study for the Meat Production in the Sudan*. Khartoum, pp. 63–78 (in Arabic).

Bakhit, A.K.M. 1977. "Characteristics of Meat Production in the Sudan". Sudan Veterinary Association, Proceeding of the Eighth Veterinary Conference, *Livestock and Animal Production Development in the Sudan*. Khartoum.

Bisschop, J.H.R. 1951. Detailed Report on the Animal Industry in the Medium Rainfall Area North of Bahr el Arab-Sobat. Khartoum, p. 11 (typed script).

Democratic Republic of the Sudan, Department of Economic Research, Ministry of Finance and National Economy. 1976. *The Economic Survey*. Khartoum.

———, Statistical Section, Department of Agricultural Economics, Ministry of Agriculture, Food and Natural Resources. 1977. *Sudan Yearbook of Agricultural Statistics*. Khartoum.

———, Statistical Section, Department of Agricultural Economics, Ministry of Agriculture, Food and Natural Resources. 1979. *Current Agricultural Statistics*. Khartoum.

Harrison, M.N. 1955. *Report on a Grazing Survey for the Sudan* (Khartoum). Mimeographed and better known as Harrison Report.

Hassan, Mohammed Hassan. 1979. *Prospects for Livestock Development in the Sudan.* Khartoum.

Jefferson, J.H.K. 1956. *Hafirs or Development by Surface Water Supplies in the Sudan.* Khartoum, p. 16.

Lebon, J.H.G. 1965. *Land Use in the Sudan.* London.

Shawki, M.K. 1968. *The Role of Rural Water in the Economic and Social Development in the Sudan.* Khartoum.

Shepherd, W.O. 1968. *Report to the Government of the Sudan on Range and Pasture Management, FAO.* Rome.

10
Agricultural Land and Man-Land Ratio in India: An Analysis of Change

E. DAYAL

Agricultural land is perhaps the most vital natural resource modified by man to provide him with food, fibre, and other agricultural raw materials. Although the economic prosperity of nations can never be gauged in terms of their land resources, it is nevertheless not too unfair to say that the high standards of living in some wealthy countries, such as the United States, Canada, and Australia, can be partly attributed to the enormous land resources they possess (Barlowe 1972:2). In a country as large and as densely populated as India, the importance of agricultural land is obvious. It is regarded as the most influential source of economic, social, and political power. In Hindu mythology, land — which in an agrarian society implied agricultural land — was looked upon with great respect as "Mother Earth".

The agricultural land in this study has been defined as land that is normally used for agricultural production, and it includes net sown area, current fallow, other fallow, and land under tree crops. A similar definition of agricultural land has been used by Chopra in the *Gazetter of India* (Chopra 1975:46). Pasture and grazing land are often not included in the agricultural land in the tropical countries, as livestock raising is not an organized activity.

The areal frame of this study comprises 309 districts for which data were available for the selected years. All data for this study were obtained from government publications (Government of India 1954, 1967, 1977, 1978, 1979). The data on the items included in the definition of agricultural land are generally reliable as they are collected on a field-to-field basis by the village revenue officers. Spot checks of these data are made by higher officers of the revenue department and further cross-checking is carried out by the canal officials in the irrigated areas (Singh 1975:3). The base year 1950/51 marks the beginning of planned agricultural development in India and 1969/70, the final year, was the latest year for which published data were available when the study commenced.

This study attempts to examine regional trends in the expansion of agricultural land and man-land ratio in India over a period of twenty years. An attempt has also been made to identify and explain the under-utilization of agricultural land in terms of selected variables. The percentage changes for districts could not be worked out because 1950/51 district data are not comparable to 1969/70 data owing to changes in the boundaries of several districts during that period.

Extent and Quality of Agricultural Land

Agricultural land in India covers a total area of 167 million ha. The proportion of total area at present used for agricultural production in India is 54 per cent, which compares quite favourably with other densely populated countries of Asia and Europe. However, there are considerable regional variations in the quality of agricultural land. In some regions, although the extent of agricultural land is large, its productive capacity is limited because of poor quality, for example, in eastern Rajasthan, eastern Madhya Pradesh, the Malwa Plateau, and the Deccan Plateau.

The variations in the quality of agricultural land in India, which determines its production capability, stem from differences in rainfall and access to other sources of water supply, topography, soil characteristics, and location relative to market. The last characteristic is less important in India because the large population creates a ready market almost everywhere. But the constraints imposed by the three other characteristics limit the supply of good agricultural land.

Perhaps the most serious limitation on the supply of agricultural land is inadequate and erratic rainfall. The rainfall over an area of roughly 97 million ha (one-third of the total area of India) is less than 800 mm, which in a hot tropical climate is not sufficient for crop production. The areas of inadequate rainfall in India are Punjab, Haryana, Rajasthan, about two-thirds of Gujarat and Mysore, about one-third of Maharashtra and Andhra Pradesh, and about one-fifth of Uttar Pradesh. In these areas, rainfall is not only insufficient but also uncertain owing to high variability. Some areas of insufficient rainfall have high agricultural productivity owing to good irrigation facilities, but the production in unirrigated tracts fluctuates considerably. About 47 million ha of agricultural land is in the low rainfall zone (Chopra 1975:210).

The mountainous topography in the north and parts of the Peninsula is another obstacle that limits the supply of agricultural land. The Himalayas in the north occupy about 500,000 km^2. Land here is too steep, rough, and inaccessible for agricultural production. The topography of the peninsular region, the Western Ghats, the Aravalli Range, the Bundelkhand Plateau, the Vindhyan Scarplands, and the Bghclkhand region is also unsuitable for agriculture. Large tracts in the Chambal Basin have been rendered useless for agriculture owing to severe gully erosion.

The most productive agricultural land in India is found in the alluvial tracts of the northern plains and the coastal plains. The alluvial soil tracts occupy about 44 per cent of the total area of the country, and almost everywhere more than 70 per cent of the alluvial soils are used for agricultural production. The alluvial soils themselves are also not uniform, as their productivity depends on their texture, plant nutrients contained in them, and their moisture-holding capacity. Often the productivity of alluvial soils is hindered by the presence of hard pans of lime (*kankar*) at certain depths, which prevent root growth and percolation of water.

In general, however, the alluvial soils are very productive because they contain a large variety of salts derived from different rocks. The agricultural land in the alluvial tracts has the added advantage of the ease of tillage and construction or irrigation channels. The Sutlej-Ganga Plain has the largest concentration of canal and well irrigation. The development of irrigation, which made the Ganga Plain a prosperous agricultural region, has however created some problems of salinity and alkalinity (Worthington 1978:281). In the heavily irrigated areas of the Punjab and western Uttar Pradesh, about 5 per cent of the previously cultivated land (about 7 million ha) has been rendered useless owing to the accumulation of salts (Dayal 1977:108).

The productivity of alluvial soils has gradually declined because they have been cultivated for several centuries. In some parts of the northern plains, agriculture was established as early as 3000 B.C. (Basham 1954:194).

The agricultural land in the black soil region of the Deccan Plateau is also quite productive, where the soil is deep. The black soils occupy about 15 per cent of the total area of the country, and more than 50 per cent of the black soils area is used for agricultural production. The black soils are heavy and difficult to work. When too dry they crack and are hard to break, and when too wet they stick to the implements and make ploughing difficult. However, the black soils are capable of high productivity, where they are deep and properly irrigated. They are rich in potash, phosphorus, and have a good capacity for holding moisture.

Elsewhere, the agricultural land is spread over the variety of soil types of which the red soils are most extensive. Their red colour is due to diffusion of iron in them. They vary greatly in colour, depth, and fertility. In the highlands the red soils are generally thin, gravelly, and poor; but in the lowland regions they are deep, rich, and fertile. However, relative to other soil zones, smaller proportions of area under them are used for agricultural production. The red soils occupy another 15 per cent of the total area of India and are in Tamil Nadu, eastern Madhya Pradesh, Orissa, Mysore, northeast Andhra Pradesh, southern districts of Uttar Pradesh, and parts of Rajasthan and Assam. The quality of agricultural land in the red soil regions is not consistent.

The quality of agricultural land in India has gradually deteriorated because it has been cultivated without much use of fertilizers for many centuries. In several areas, the quality has suffered from soil erosion owing to massive destruction of natural vegetation, over cultivation and grazing. These are the result of rapidly increasing pressure of population on land resources.

Agricultural Land: A Historical Perspective

The archaeological excavations at Mohenjo-daro and Harappa have provided sufficient evidence to support the view that agriculture was established in northwest India as early as 3000 to 2500 B.C. It has also been established that agriculture in the Indus Valley at that time was fairly advanced in technology, using ploughs and irrigation. Therefore, it was probably in the northwest that land was first developed for agriculture. From the Indus Valley, agriculture spread into Peninsular India and then into the northern plains. The establishment of agriculture and permanent settlement is reported to be earlier in the Peninsular India than in the northern plains. The dense natural forests in the northern plains delayed the spread of agriculture for quite some time. Agriculture appears to have been established in the Malwa Plateau and the Deccan by the sixteenth century B.C. (Schwartzberg 1978:159).

The size of agricultural land in ancient India is difficult to estimate, but some idea can be gained by the spread of excavation sites, which occupy over 1.3 million km² in the Indus region alone. Fa-hsien, a Chinese traveller who visited India in the fifth century A.D., reported large tracts of forests and empty land, indicating the small extent of agricultural land in the northern plains (Farmer 1974:7). Even by the end of the sixteenth century A.D., the Tarai forests in the Ganga Plains extended quite far to the south. The upper Ganga Plains were heavily cultivated, but the extent of agricultural land decreased towards the east into Bihar. One characteristic feature of agriculture and settlement at that time was the existence of nuclear population concentrations separated by large tracts of forest and waste land. The expansion of agricultural land in the Ganga Plains, as perhaps in other parts of India, has been the result of a continuous fight against nature. Forests and waste land have recovered and retreated several times throughout Indian history (Farmer 1974:6). The reversion of agricultural land to waste land was due to excessive taxation by local rulers, and also due to wars, pestilence, and famines. The reoccupation of abandoned land was slow and piecemeal, occasionally achieved by drafting people from overcrowded areas and sometimes by providing such incentives as free seeds and work animals to farmers and giving tax concessions for the first few years.

During the British period, India recorded an unprecedented growth of population and a corresponding increase in the area of cultivated land. During the early years of British Raj, the ancient pattern of retreat and recovery of waste and forest land continued. The fluctuations in the extent of agricultural land during the British period were also due to changes in government policy concerning revenue assessments. The pressure for higher revenue often led to a shrinking of agricultural land (Klein 1974:196). The population increase, however, changed the old pattern into one where waste and forest land were developed for agriculture, never to be allowed to revert back to waste. This critical point was reached at different time periods in different regions of India, depending probably on the population densities.

The introduction of plantation agriculture opened up new land for agriculture in Assam, north Bengal, Tamil Nadu, and Kerala; and thus increased the extent of agricultural land in India. Assam was virtually a forest up to 1826 (Farmer 1974:13). The introduction of tea plantations not only established plantation agriculture in the state but also attracted large numbers of peasants from overcrowded Bengal, who occupied unsettled waste for subsistence farming.

During the Second World War, the "Grow More Food" campaign brought additional land under cultivation through government incentives. For example, in 1942/43 the government sanctioned about 2 million rupees to the provinces for food crop cultivation on non-food crop land and for the development of irrigation. Under the "Grow More Food" campaign, significant areas were improved and brought into cultivation in Uttar Pradesh, Mysore, Madras, and Assam states.

The most outstanding achievement of British India in land development was the establishment of canal colonies. In the canal colonies (now mostly in Pakistan) 2.2 million ha of agricultural land were developed through the provision of canal irrigation. Most of this land was previously waste land or occasionally used for nomadic pastoralism.

Although British efforts resulted in significant additions to agricultural land, there is strong evidence that several regions of India such as the northern plains were heavily populated as early as the mid-nineteenth century. This population pressure led to the extension of cultivation in areas which were previously considered barren, prone to flooding, and infested with malaria (Klein 1974:194). The pressure of population on land was perhaps greatest in the northern plains where a high percentage of the total area had been brought under cultivation. In eastern Uttar Pradesh, population densities had reached 1,800 to 2,330 persons per km^2 in several districts. In most parts of north India cultivation expanded between 1840 and 1870, but after that it levelled off and there was little relative change in the extent of agricultural land. In some districts of eastern Uttar Pradesh, the farthest limits of cultivation had already been reached by 1880 (Klein 1974:198). By 1900 the possibilities of expansion of agricultural land had been nearly exhausted even in the western districts of Uttar Pradesh. Klein also mentions that several low-lying areas, which were avoided for cultivation between 1850, were brought under cultivation by 1900, despite fears of crops being submerged under water during the monsoons.

For about fifteen years prior to becoming independent, India faced continuous food shortages which became more serious after partition, because a share of the food surplus areas went to Pakistan. Hence the most pressing problem facing independent India was to make the country self-sufficient in food. A high priority was, therefore, placed on the improvement and expansion of agricultural land through extension of irrigation and a programme of land reclamation. Considerable success was achieved in the expansion of agricultural land in new areas in the 1950s and 1960s. Most of the increase in food grain production during this period came through the expansion of area sown to crops (Able 1970:7; Malenbaum 1971:147).

In the early 1950s, some exaggerated claims were made about the amount of land that could be cultivated by reclaiming culturable waste. The term "culturable waste" applies to land that might be tilled but that has never been tilled, despite high population pressure. Although agricultural statistics of India indicate some decline in the culturable waste, much of the expansion of agricultural land occurred in areas which were previously under forests, were too dry, or were under tree crops. For example, large tracts in Haryana and Rajasthan were unavailable for cultivation because they were too dry. The annual rainfall in these two states is less than 300 mm, and there are practically no rivers which could be used for irrigation. The underground water table is so low that well irrigation is only a remote possibility, and then only possible if the water is not brackish. The extension of irrigation through the construction of Rajasthan canal has opened new land for agriculture in southern Haryana and northern Rajasthan. The Rajasthan Canal that draws water from Sutlej and Beas rivers now irrigates several thousand hectares in these two states. On completion the 445 km long canal will create an irrigation potential of 1.9 million ha. The whole project, including the construction of 3,000 km of distributaries, is expected to be completed by 1985/86.

In the Tarai region of Uttar Pradesh and Bihar too, some land, previously under forest and waste, has been developed for agriculture. The Tarai is a narrow belt of flat land along the foothills of the Himalayas in Uttar Pradesh and Bihar. On reaching the plains, the rivers drop much of their heavy load of eroded material and form a boulder and pebble country, but a few kilometres farther south the structure of soil is fine enough for agriculture. The Tarai region suffers from excessive moisture and swampy conditions in some parts during the monsoons. Malaria, tall grass forests, and wild animals discouraged settlement. Until recently, Tarai was inhabited only by some tribal people, but increased population pressure and the introduction of D.D.T. has pushed cultivation deep into the Tarai region. Considerable land has been developed for agriculture in Nainital, Kheri, and Pilibbit districts of Uttar Pradesh and some in the Tarai region of Assam.

Dandakaranya, Chambal Valley, and Malwa Plateau are some other major areas where previously uncultivated land has been developed for agriculture. Furthermore, state and central government mechanized farms have made additional contributions to the size of agricultural land in India. Through these land development schemes about 11 million ha were added to the agricultural land of India between 1950 and 1975 (Table 10.1).

Expansion of Agricultural Land in Modern India

Despite considerable efforts to increase the extent of agricultural land in India, the achievements were not very encouraging. Agricultural land recorded a small increase over a period of twenty-five years. Between 1950/51 and 1974/75, it increased from 166.7 million ha to 167.5 million ha, an increase of only 10.4 per cent (Table 10.1). The population during the same period increased by 66 per cent from 361 million to 598 million (Government of India 1977:6). This rather less than impressive agricultural land increase does not reflect lack of enthusiasm or effort but the limitations of terrain, rainfall, and other sources of soil moisture on the expansion of agricultural land in India. Spate wrote in the early 1950s that rural population has settled down everywhere to a density appropriate to local land resources and that the best of the possibilities of expansion of agricultural land have already been taken up (Spate and Learmonth 1967:280). Owing to a rapid increase in the population, India's potential for increasing agricultural land diminished towards the end of 1950s (Veit 1976:73). During 1951 to 1961, the net area sown to crops increased by 11.9 per cent or at the rate of 1.2 per cent per year; but during the next fifteen years it increased by 1.5 per cent or at the annual rate of 0.15 per cent. If we only take the change in net cropped area it reflects a certain measure of success of land development schemes, which reduced some area under cultivable waste and fallow land (Dasgupta 1977:17). The agricultural land, however, remained unchanged. Nath also arrives at a somewhat similar conclusion in his study of agricultural growth rates (1970:409).

The net sown area and fallow land, two major components of agricultural land, are subjected to annual fluctuations mainly due to variations of rainfall. In bad monsoon years, farmers leave larger areas under fallow to reduce the risk of losing their investment on inputs. Four-year moving averages were, therefore, computed to examine the actual trends in the three

TABLE 10.1 AGRICULTURAL LAND AND ITS COMPONENTS, 1951–74

Years	Net Sown Area	Fallow Land	Area Under Tree Crops	Agricultural Land	Man-Land Ratio (ha per head)
	Million hectares				
1951	118.75	28.12	9.83	156.70	0.46
1952	119.40	28.96	7.89	156.24	0.42
1953	123.39	36.35	7.65	167.39	0.45
1954	126.77	24.73	5.69	157.20	0.41
1955	127.78	24.93	5.63	158.34	0.41
1956	128.77	24.45	5.63	158.84	0.40
1957	129.16	23.67	5.95	158.77	0.39
1958	129.08	25.32	6.09	160.49	0.38
1959	131.83	23.71	6.01	161.54	0.38
1960	132.94	23.00	5.82	161.76	0.37
1961	133.20	22.82	4.46	160.48	0.36
1962	135.40	21.63	4.50	161.53	0.36
1963	136.34	21.25	4.56	162.15	0.35
1964	136.48	21.29	4.38	162.15	0.34
1965	138.12	20.36	4.11	162.59	0.33
1966	136.15	22.44	4.13	162.72	0.32
1967	137.21	22.60	4.09	163.91	0.32
1968	139.75	20.83	4.07	164.65	0.31
1969	137.51	23.16	3.92	164.59	0.30
1970	139.07	21.80	4.04	164.90	0.30
1971	140.40	19.74	4.37	164.51	0.29
1972	140.86	20.56	4.35	165.77	0.29
1973	140.22	25.27	4.62	170.11	0.29
1974	136.91	20.45	4.16	167.52	0.28

components of agricultural land, that is, net sown area, fallow land, and area under tree crops. The graphs based on moving averages reveal some interesting features. It is quite clear from the graph that the net sown area increased significantly up to 1962, but thereafter it demonstrates little or no change (Fig. 10.1). Fallow land and the area under tree crops both declined

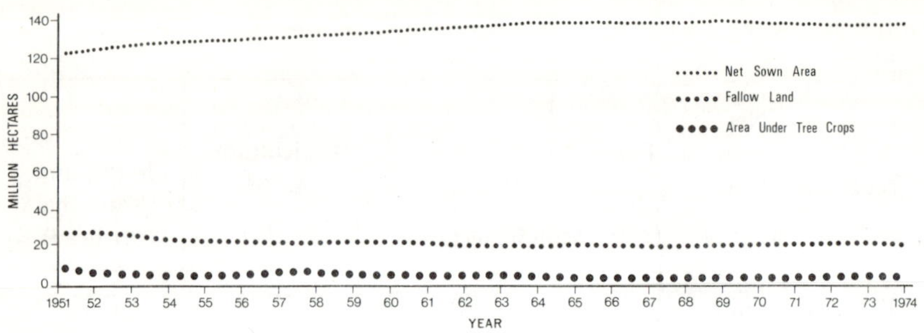

Fig. 10.1 Major land use trends (four-year moving averages)

up to 1962 and thereafter remained practically unchanged. It implies that much of the increase in the area sown to crops was due to a decline in fallow land and tree crops areas. This inference is further confirmed by examining the trend of agricultural land, which depicts much less change over time than its components. The area under tree crops declined from 19.8 million ha to 4.1 million ha. Moreover, fallow land declined by about 8 million ha, accounting for about a 15 per cent increase in the net sown area and only a 0.4 per cent increase in the agricultural land. Although efforts were made to increase agricultural land between 1950 and 1974, and some success was achieved in a few areas, the overall increase in agricultural land was quite nominal. All that has really happened was a transfer of land from the area under tree crops and fallow to the area sown to crops. This clearly illustrates that there is little scope for the further expansion of agricultural land in India and that most of the land suitable for agriculture is already under cultivation.

Distribution of Agricultural Land

There is a high concentration of agricultural land in two clearly noticeable regions — the Sutlej-Ganga Plains and the northwestern Deccan Plateau, where more than 80 per cent of the total area is used for agricultural production (Fig. 10.2). From these two core areas, the density of agricultural land decreases in different directions. Gujarat, Tamil Nadu, and Kerals are some other areas of relatively high density of agricultural land. The high percentage of total area used for agriculture is generally in the regions of fertile soils and high population density. In the mountainous regions such as Kashmir, Himachal Pradesh, Southern Bihar, Assam Hills, northeast India, and in the hilly areas of Peninsular India, the density of agricultural land for obvious reasons is low. Also, in extremely arid regions such as western Rajasthan, and in heavily forested areas such as Chatisgarh Plains, the density of agricultural land is quite low. The number of districts where less

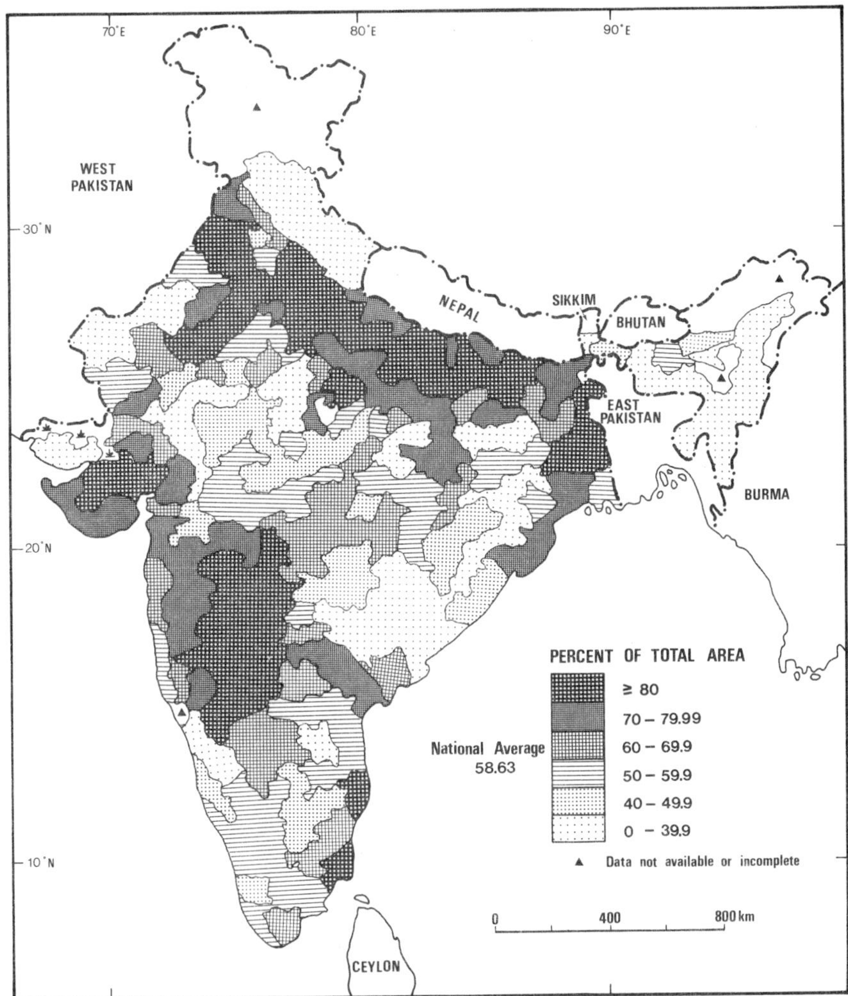

Fig. 10.2 Agricultural land, 1951

than 40 per cent of the total area is used for agriculture is small in India. In such districts, the physical constraints are formidable; the scope for the expansion of agricultural land is therefore slim.

The distribution pattern of the density of agricultural land has not changed much between 1951 and 1970. The most noticeable change during that period is a reduction in the area in the lowest class interval (Figs. 10.2 and 10.3). There is also some increase in the density of agricultural land in scattered districts, largely in response to increases in population density. One

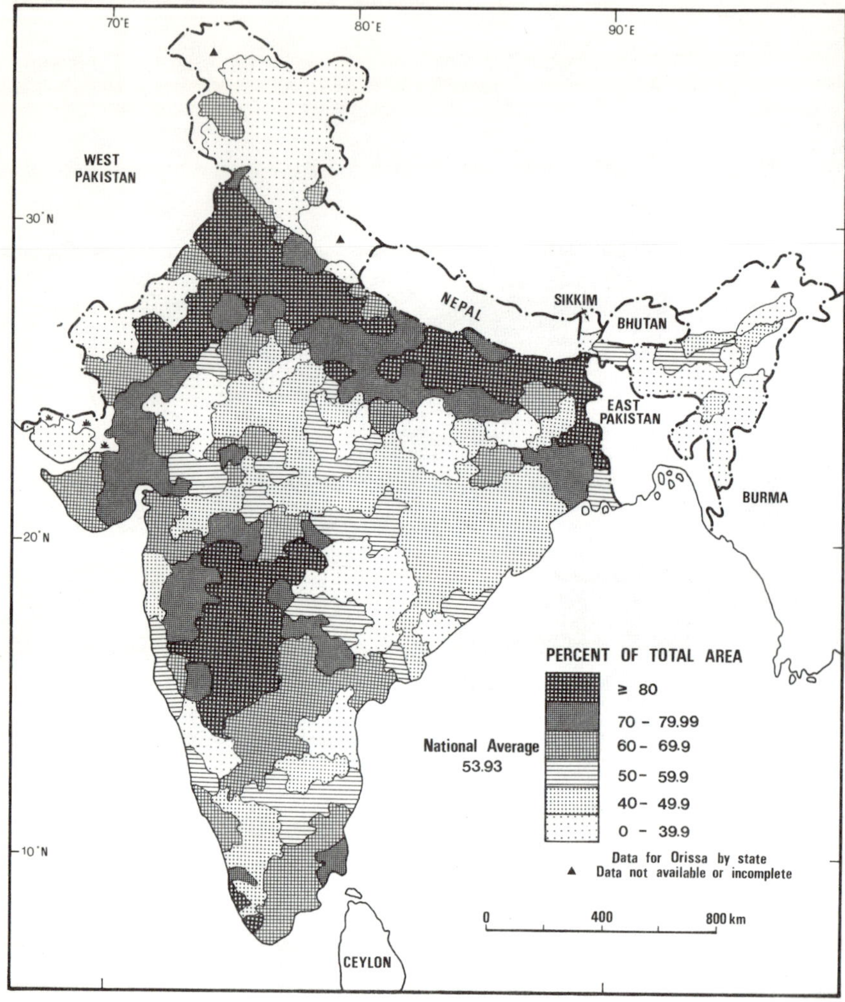

Fig. 10.3 Agricultural land, 1970

such area is the Assam Valley, where there is some increase in agricultural
land. However, on the whole the national trend has been maintained even at
the regional level showing little change in the percentage of total area used
for agriculture, with some exceptions. The overall pattern has become more
consolidated due to the filling up of some gaps in the pattern. In a few isolated districts in the northern plains, Gujarat, and Tamil Nadu the density of
agricultural land also declined, perhaps because of the expansion of non-
agricultural activities, such as urban encroachment on agricultural land and
expansion of roads and industries.

There appears to be a strong positive correlation in the distribution of agricultural land and population density. The spatial accord between the two patterns is because both are governed by the same set of ecological variables, that is, land productivity, terrain, and climate. Population pressure compels people to bring more land under cultivation, leading to higher densities of agricultural land in more populated regions.

Man-Land Ratio

As stated earlier, the population has increased at a much faster rate than agricultural land in India during the period 1950 to 1975. Consequently, the man-land ratio has declined markedly. According to Tiwari, such a trend had begun much earlier, and cultivated land could not keep pace with population, at least in one state of India, since 1930/31 (1970:73). The agricultural land per head of total population presents a dismal picture in India. The agricultural land per head of total population has never been large. Tiwari states that in Uttar Pradesh the man-land ratio was only 0.3 ha as early as 1905 (1970:70). Agricultural land was 0.45 ha per person in 1951 but decreased to a mere 0.28 ha per person in 1974 (Table 10.1). Agricultural land per head of working agricultural population (cultivators and agricultural labourers) also declined from 2.07 ha to 1.31 ha during this period. This was also apparently due to agricultural land not having increased in the same proportion as the agricultural population. The position would have been much worse if significant out-migration from the rural sector had not taken place. As can be seen from Table 10.1, the supply of agricultural land remained virtually unchanged during 1951 and 1974. The often stated increases in the acreage of cultivated land during the 1950s and early 1960s were simple increases in the area sown to crops, which increased at the expense of fallow land and land under tree crops (Veit 1976:73; Malenbaum 1971:147; Able 1970:7).

There are significant regional differences in the man-land ratio as one would expect for a country as large and varied as India. Low man-land ratios (agricultural land per head of agricultural population) are clearly concentrated in well-watered and productive areas. The agricultural population is heavily concentrated in the large alluvial tracts of the northern plains, the Assam Valley, and the river basins and deltas on the east coast. The agricultural land per head of agricultural population is less than one hectare throughout these regions, leading to small holdings and consequently to highly labour-intensive farming. There is, therefore, good spatial accord in the low man-land ratio and agricultural intensity. Obviously, because farmers have small holdings they use them very intensively to produce enough for their requirements. In the hilly tracts such as Kashmir, Himachal Pradesh, Meghalaya, Manipur, and Mizo Hills, the density of agricultural population relative to other parts of India is quite low, but the pressure of agricultural population is quite high owing to the scarcity of agricultural land. In such regions, the percentage of total area available for agriculture is much less than 30 as compared to 70 and 80 in the northern plains and other alluvial tracts (Fig. 10.4).

The high man-land ratio, more than 1.5 ha of agricultural land per head of agricultural population, is confined to the Thar Desert, Chambal badlands, the Malwa Plateau, the Deccan Lava Plateau, and the Karnataka Plateau. In these regions, the quality of agricultural land is low because of aridity and poor soils; hence there is a low pressure of agricultural and total

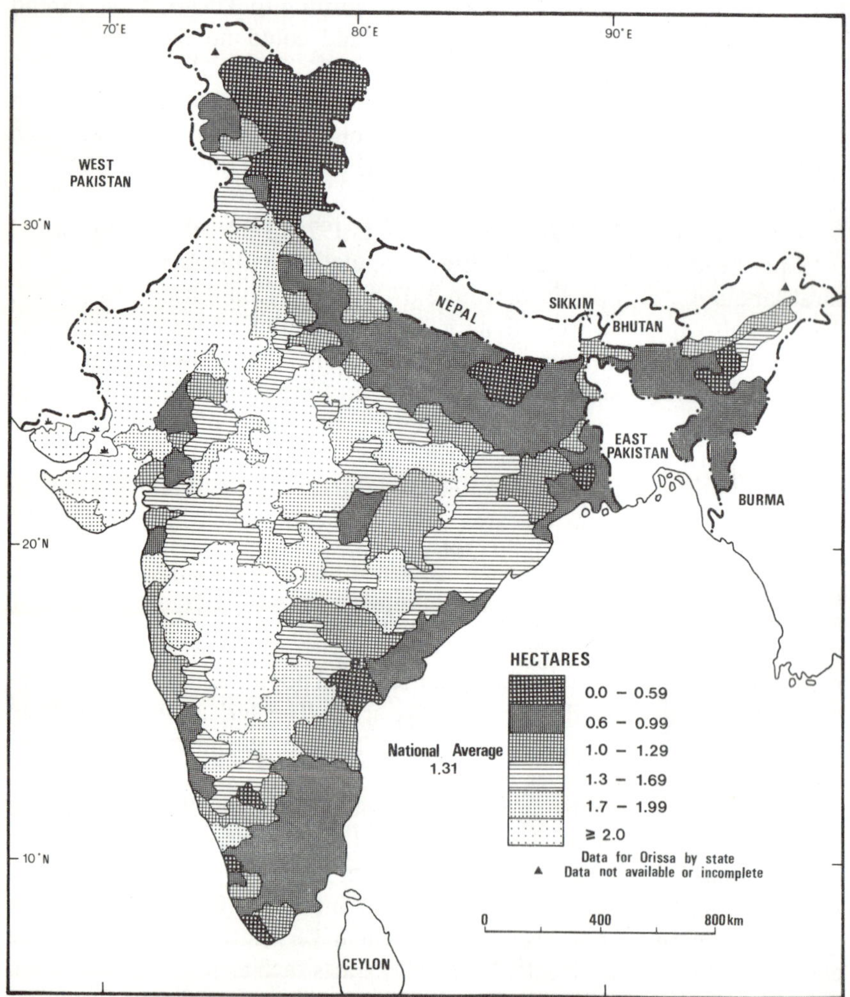

Fig. 10.4 Agricultural land per head of agricultural population, 1970

population on land resources. The low carrying capacity of land necessitates larger landholdings for subsistence.

An inspection of all-India figures confirms that agricultural land per head of agricultural population has declined steadily over the past twenty-five or thirty years. Between 1951 and 1974, it declined by almost 37 per cent (Table 10.1).

The regional trends of decline of agricultural land per head of population are even more obvious. Between 1950 and 1970, a dramatic encroachment of population on agricultural land in marginal areas appeared to have occurred. In 1950 there was a large area in India, spread over most of Peninsular India, the Punjab Plains, Rajasthan, and Gujarat, and the hilly districts of northeast India, where agricultural land per head of agricultural population was more than 2 ha. But in 1970 this large area had shrunk to a fraction of its former size (see Figs. 10.4 and 10.5). Over Peninsular India, the man-land ratio declined considerably on the west coast, in the Narmada Valley, the Gujarat Plains, parts of Chambal Basin, the Chattisgarb Plains, the Dandakaranya Plateau, and parts of the Telangana Plateau. The ratio also declined in the Punjab Plains, the Ganga Plains, the Assam Valley, Manipur, Tirpura, and Mizo Hills during the same period. The decline was due to population increase and little change in the quantity of agricultural land. The increased pressure of population on land resources has compelled people to invade marginal and less productive land. The regional disparities in man-land ratio are thus gradually decreasing. This, however, does not mean that the man-land ratio in the more productive areas is constant or increasing. The ratio is declining even in the more productive northern plains, but because of the already tight situation there the deterioration of the ratio is not as significant as in the other parts of India. In other words, equally low man-land ratios are also found in several other parts of India, although not as yet over the large areas such as the northern plains.

The analysis of change of agricultural land per head of agricultural population indicates the development of a grim situation in Indian agriculture, which will adversely affect agricultural productivity. Although agricultural productivity is a function of several variables, the size of the operational holding is one important variable affecting productivity. A declining size of operational holding reduces the risk taking capacity of the producers and discourages innovation. It also limits the ability of farmers to invest in inputs. A consistently declining man-land ratio is also likely to result in diminishing returns in densely populated areas.

Man-Land Ratio and Agricultural Intensity

This situation in India directs attention to the concept that increasing population pressure on land resources leads to more frequent use of land

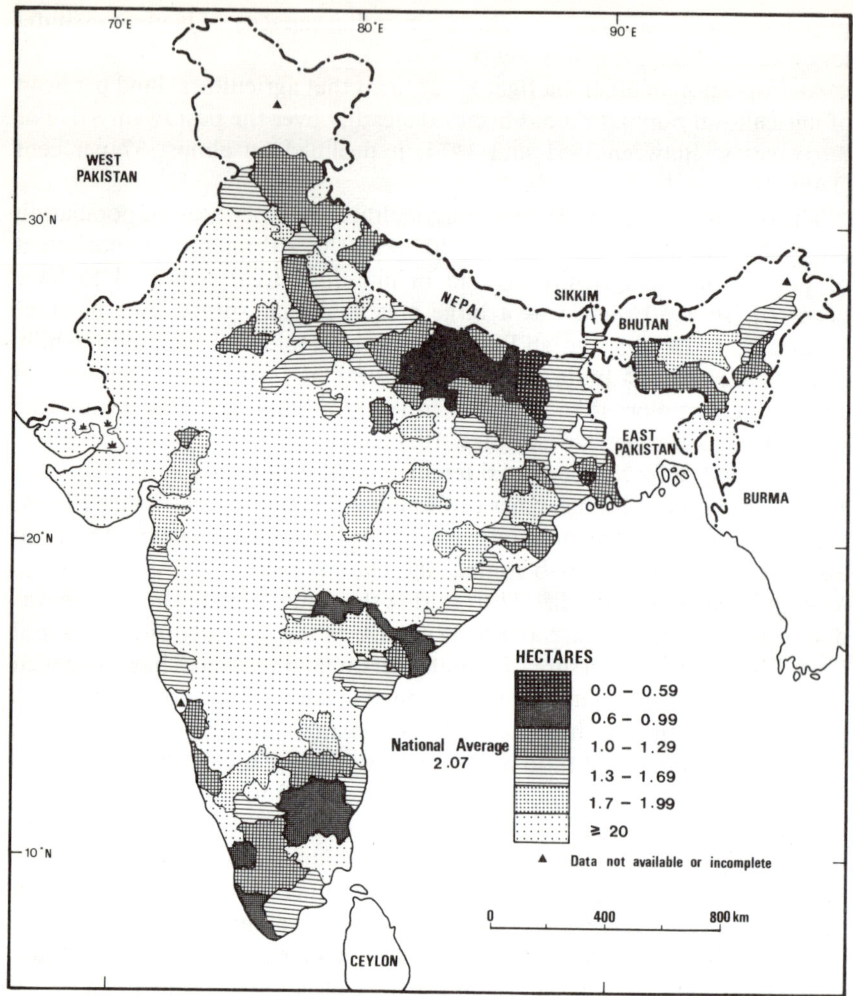

Fig. 10.5 Agricultural land per head of agricultural population, 1951

and consequent shortening of fallow. The amount of land left fallow is one
indication of agricultural intensity. Several scholars have examined the
influence of population pressure on agricultural land (Brookfield 1962:
242–52; 1972:30–48; Boserup 1975; Verneer 1970:299–314; Grigg 1976).
Boserup puts the relationship between population pressure and agricultural
intensity in a theoretical framework. Her thesis is based on the premise that
subsistence farmers are labour efficient, and that they choose the level of
agricultural intensity that will satisfy their demand with minimum input of

labour. It also implies that as the population pressure on agricultural land increases the farmers shorten the length of fallow and begin to use land at shorter intervals (Boserup 1965:16). In sparsely populated areas where land suitable for agriculture is freely available, population growth leads to an expansion of agricultural land. But in India such possibilities had been exhausted by 1900, and further increases in population has led to an intensification of agriculture. Therefore, the prevailing situation in India provides an excellent opportunity for testing Boserup's theory.

An inspection of national figures for agricultural land per head of population and land under fallow reveals a close positive relationship, indicating that as the man-land ratio deteriorates owing to increasing population pressure, the amount of land under fallow declines. It also implies that farmers have resorted to more frequent cultivation of available agricultural land because of the increasing pressure of rural population and the limited opportunities for the expansion of agricultural land. The coefficient of correlation between man-land ratio and area under fallow is very high ($r =$ 0.87) when computed for all-India over a period of twenty-five years from 1950–74.

In spatial terms, it may be quite appropriate to assume that in the regions where population pressure on agricultural land is high the proportion of land under fallow will be low. A comparison of the patterns of fallow land and population density for 1951 and 1970 seems to provide some support for the hypothesis. For example, over much of Uttar Pradesh, west Bengal, and Kerala, the percentage of agricultural land under fallow was below the national average in 1951. These are also the most populated regions of India (Fig. 10.6). The higher percentages of fallow occurred in south Bihar, northern Madhya Pradesh, Rajasthan, Telangana Plateau, and the southern part of Karmataka Plateau, where population densities are low. The change in the proportion of fallow land between 1951 and 1970 was mainly from high to low, particularly in the areas that recorded significant increases in population, namely, Assam, West Bengal, Punjab, Haryana, western Madhya Pradesh, Maharastra, Andhra Pradesh, and parts of Tamil Nadu (Figs. 10.6 and 10.7 and Johnson 1979:166). For the country as a whole, the percentage of agricultural land under fallow declined from 16.87 to 13.2. In terms of four-year averages, the percentage declined from 17.30 to 12.80 between 1951 and 1974. There is thus a clear indication of increase in farming intensity, if measured in terms of fallow land as a percentage of agricultural land in the regions that recorded significant increases in population during 1950 and 1970.

However, over considerable areas the relationship between fallow and population density is not so clear, for example, in northern Bihar, eastern Maharashtra, and parts of Andhra Pradesh and Tamil Nadu. This may be

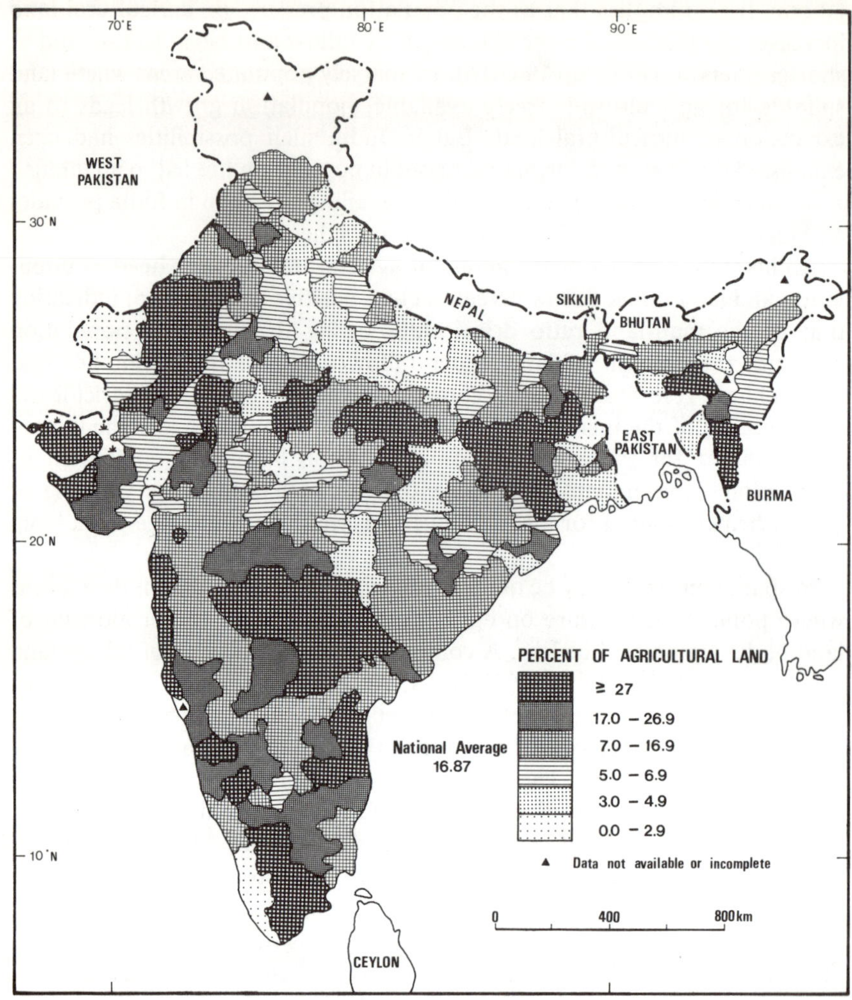

Fig. 10.6 Fallow land, 1951

because farmers leave land fallow for a variety of reasons and population
pressure may be just one of the several factors affecting the regional variations of the proportion of fallow land. For example, the distribution of
fallow land may be affected by the amount of rainfall received in a given
year. It may also be affected by the availability and non-availability of
irrigation. Normally, in the areas of high rainfall variability, farmers may
be encouraged to leave more land under fallow because of the uncertainty
involved in the farm business in such areas. They would try to reduce the
risk of losing investment in inputs by cultivating only a small proportion of

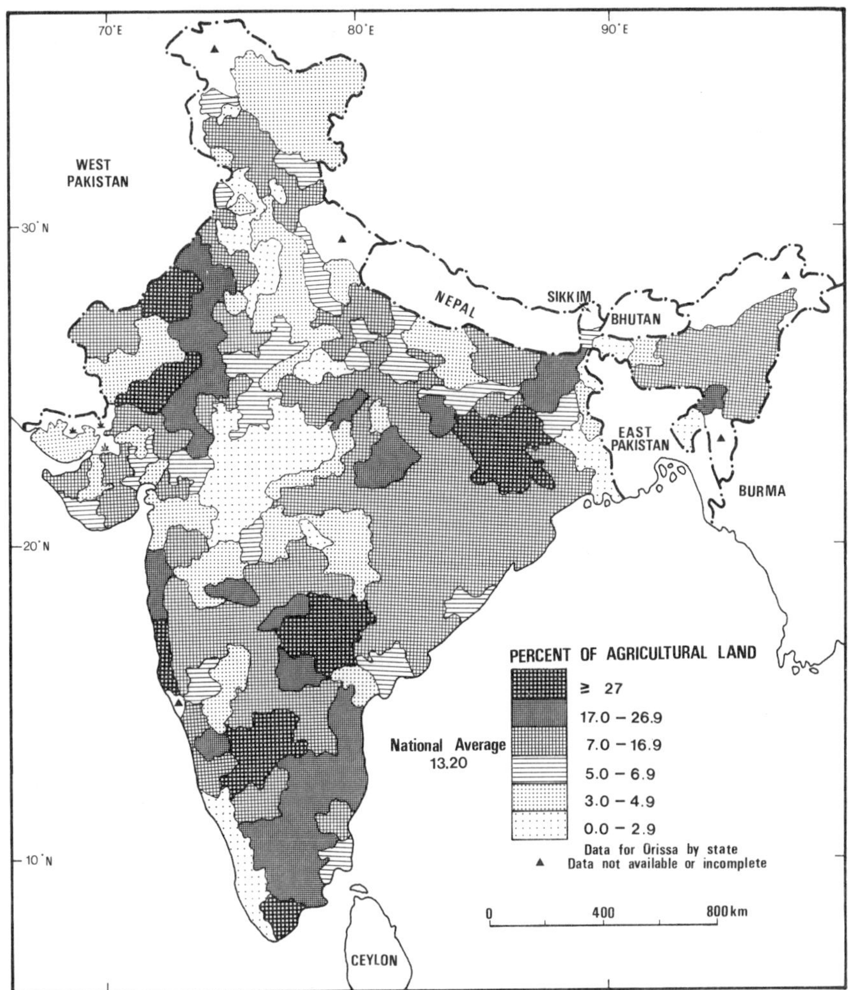

Fig. 10.7 Fallow land, 1970

agricultural land. As the risk of losing the crop is much less in irrigated
regions, the farmers would be expected to leave less land under fallow. In all
the major irrigated regions, the farmers would be expected to leave less land
under fallow. This conclusion is fairly well supported by the distribution of
fallow land in such major irrigated areas as the Punjab, Haryana, western
Uttar Pradesh, the deltas of Mahanadi, Krishna, Godwari, and Cauvery
where the percentage of land under fallow is quite small (less than 5). The
upper Cauvery region of Tamil Nadu is an exception.

The statistical test also indicates a significant but low correlation between population pressure and fallow land ($r = 0.312$), which explains only 10 per cent of the variation in the spatial distribution of fallow land. The low value of correlation coefficient is a reflection of an unsystematic fluctuation of fallow land in different regions as it is affected by several variables. Nevertheless, with few exceptions, the distribution of fallow land appears to have a significant negative relationship with the intensity of agriculture.

The area under fallow provides an indication not only of the intensity of cultivation but also of the under-utilization of available agricultural land. The annual and regional fluctuations in agricultural production are thus also partially due to variations of fallow land. There is, therefore, a case for examining the variables that appear to affect the variations in the area under fallow. As stated earlier, farmers may leave land under fallow for various reasons. A multiple regression model based on four explanatory variables has been employed to seek a more satisfactory explanation for the spatial variation of area under fallow in different parts of India. To save time in the collection of data on the four explanatory variables for all the 309 districts of India, it was decided to take a random sample of 66 districts or 22 per cent of the population.

The first explanatory variable is the man-land ratio. It is assumed that in the areas where man-land ratio is low because of high population pressure, farmers will leave a minimum area under fallow. Hence a positive correlation between man-land ratio and area under fallow is expected. The second variable is population density. It is assumed that high population density will lead to more intensive cultivation and, consequently, to a smaller area under fallow. A negative correlation is, therefore, expected between fallow land and population density. The third and fourth variables are rainfall and irrigation. In areas with adequate and reliable rainfall, and where good irrigation facilities exist, there is little risk of crop loss and farmers may be encouraged to use all available land. In such areas, the land under fallow will be quite small. Hence a negative relationship is expected between these two variables and the area under fallow.

The simple correlation analysis indicates that three out of four explanatory variables, namely, man-land ratio, population density, and rainfall are significantly related with the dependent variable (area under fallow). The man-land ratio has the strongest positive relationship with the area under fallow (Table 10.2). The direction of relationship for each variable is the same as anticipated in the hypotheses, indicating the validity of the assumption in the formulation of those hypotheses. There is, however, no significant relationship between the area under fallow and irrigation, because in some irrigated regions such as Tamil Nadu and coastal Andhra Pradesh a considerable area of land is left fallow, whereas in some non-irrigated areas

TABLE 10.2 SIMPLE CORRELATION COEFFICIENTS OF FOUR EXPLANATORY VARIABLES AND MULTIPLE REGRESSION COEFFICIENTS OF SIGNIFICANT VARIABLES REGRESSED AGAINST PERCENTAGE OF TOTAL AREA UNDER FALLOW

Explanatory Variables	Simple Correlation Coefficient	Multiple Regression Coefficient (β)	Percentage Contribution to R^2	R
Man-land ratio	0.312*	0.172*	9.75	0.312
Population density	-0.308*	-0.232*	3.24	0.360
Rainfall	-0.245*	-0.130	1.95	0.387
Percentage of NSA irrigated	-0.060	0.073	0.39	0.392

N = 66
*Significant at 0.05 level.

there is little land under fallow. But in the major irrigated areas in Punjab and western Uttar Pradesh, the percentage of fallow is quite small.

In the stepwise multiple regression model too, the three variables, namely, man-land ratio, population density, and rainfall, emerge as significant explanatory variables, each making a significant contribution to the variation of fallow land (Table 10.2). The three explanatory variables together explain about 15 per cent of the spatial variation in the proportion of land under fallow. Much of the spatial variation of fallow land remains unexplained by the model. Several other institutional variables, such as size of holding, nature of cropping pattern, credit facilities, and borrowing capacity of farmers may also affect the area under fallow, but district-wise data for such variables could not be obtained. However, the cartographic and statistical analyses provide enough support for the main idea imbedded in Boserup's theory. The temporal change in area under fallow has a high negative correlation with change in population pressure (rural population density on agricultural land), and spatial variation of fallow also has a significant negative correlation with rural population density.

Conclusion

This study directs attention to temporal and regional changes in agricultural land, man-land ratio, and fallow land. In reviewing the historical development of agricultural land in India, it was noted that the possibilities of expansion of agricultural land, according to some writers, had been exhausted by 1900, and that little change in the size and regional pattern of agricultural land has occurred since then. The population increased rapidly after India became independent and the need for increasing the size of agricultural land became crucial. Several land development and reclamation programmes were launched during the 1950s and 1960s. However, despite these efforts, the extent of agricultural land remained unchanged since 1950. The often stated increases in the area under cultivation were simply increases in the area sown to crops, which recorded some increase at the expense of a decline in the area under fallow and under tree crops. In regional terms too, there was only a nominal change in the percentage of total area under agricultural land. In fact the percentage of total area under agricultural land declined between 1950 and 1970, but that could be due to more accurate land utilization statistics.

Because the extent of agricultural land did not increase as much as the rural population, the man-land ratio deteriorated between 1950 and 1974. This is a serious development and calls for urgent measures to reduce the pressure of rural population on agricultural land. This apparently can be done by transferring population from the rural sectors to the non-rural sectors of the economy through the development of small-scale industries in the rural areas. This, however, is not an easy task. It will require considerable improvements in the accessibility of rural areas. Some progress in this direction has been made through the Industrial Estate Programme. The man-land ratio has reached a serious level, and any further deterioration in the ratio is bound to result in diminishing returns with grave consequences, particularly in the northern plains of India.

One of the ironies of Indian agriculture today is that while the agricultural land per head of agricultural population is too low, many farmers are not using their land intensively enough to maximize agricultural production. The area under fallow is about 13 per cent of the total area of agricultural land, and in some regions it is more than 25 per cent. The statistical analysis reveals that the area under fallow is negatively related to the population density and positively related to the man-land ratio. It means that in the sparsely populated regions, where the agricultural holdings are large, farmers do not use their land intensively enough to maximize agricultural production. The results of the analysis of fallow land provide good support for Boserup's theory that increasing pressure of population on land resources leads to more frequent use of agricultural land and a consequent shortening of fallow.

REFERENCES

Able, Martin E. 1970. "Agriculture in India in the 1970's". *Economic and Political Weekly* 5:A5–A14.

Barlowe, R. 1978. *Land Resource Economics*. Englewood Cliffs, N.J.

Basham, A.L. 1954. *The Wonder That Was India*. London.

Boserup, E. 1965. *The Conditions of Agricultural Growth*, London.

Brookfield, H.C. 1962. "Local Study and Comparative Method: An Example from Central New Guinea". *Annals of the Association of American Geographers* 52:242–52.

__________. 1972. "Intensification and Disintensification in Pacific Agriculture". *Pacific Viewpoint*. 13:30–48.

Chopra, P.N., ed. 1975. *The Gazetteer of India*. New Delhi.

Dasgupta, B. 1977. *Agrarian Change and the New Technology in India*. U.N. Research Institute for Social Development, Geneva.

Dayal, E. 1978. "A Measure of Cropping Intensity". *The Professional Geographer* 30:389–96.

__________. 1977. "Impact of Irrigation Expansion on Multiple Cropping in India". *Tijdschrift voor Economische en Sociale Geografie* 68:101–9.

Farmer, B.H. 1974. *Agricultural Colonization in India*. London.

Government of India. 1979. *India 1979: A Reference Manual*. New Delhi.

__________. 1977. *Statistical Pocket Book* New Delhi.

__________. 1967. *Statistical Abstracts of India*. New Delhi.

__________. 1978. *Indian Agricultural Statistics, 1969–70*. New Delhi.

__________. 1954. *Indian Agricultural Statistics, 1950–51*. New Delhi.

Grigg, D.B. 1976. "Agricultural Responses to Population Pressure". *Progress in Geography* 8:147–75.

Johnson, B.C.L. 1979. *India: A Geography of Development*. London.

Klein, I. 1974. "Population and Agriculture in North India". *Modern Asian Studies* 8:191–216.

Malenbaum, W. 1971. *Modern India's Economy*. Columbus.

Nath, V. 1969. "The growth of Indian Agriculture". *Geographical Review* 59:348–77.

Singh, J. 1974. *An Agricultural Atlas of India*. Kurkshetra-Haryana, India.

Spate, O.H.K., and Learmonth, A.T. 1967. *India and Pakistan*. London.

Schwartzberg, J.E. 1978. *A Historical Atlas of South Asia*. Chicago.

Tiwari, R.N. 1970. *Agricultural Development and Population Growth*. S. Chand and Co., Delhi.

Verner, D.E. 1970. "Population Pressure and Crop Rotational Changes among the Tiv of Nigeria". *Annals Assoc. American Geographers*, 60:299–314.

Veit, L.A. 1976. *India's Second Revolution*. New York.

Worthington, E.B., ed. 1978. *Arid Land Irrigation in Developing Countries*. Oxford.

III

RESOURCE MANAGEMENT AND DEVELOPMENT

11
Resource Inventory and Conservation Aspects in Tropical Vegetation

A.N. GILLISON

Over the past decade, the momentum of resource exploitation in the tropics has increased, in many cases to an unacceptable level. The problem concerns the physical removal of non-renewable resources. A conspicuous feature is the removal of tropical woody vegetation, sometimes involving the complete and irreversible destruction of a regional biome.

To effectively manage the resource there must be a continuing effort to inform political decision-makers. The extent of the resource must be determined, the primary environmental determinants of the resource identified, and the best way of managing the resource established.

This paper sets out to define the important terms and concepts used in relation to the tropical vegetation resource, to review briefly the existing methods of field estimates, and to present a case for an improved approach to field survey and vegetation classification methods. Some case studies are drawn from the Australian region.

Definitions

The term *resource* refers to a material quantity which may be exploited to satisfy either material or non-material (aesthetic) requirements (Usher 1973; Allaby 1977). Qualifying terms, such as "finite", "renewable", "potential", and "diminished", tend to be used in a uniform way in literature and require no further definition here.

Inventory is a term used to define limits to a resource. Resource inventories are a function of the parameters and attributes used to describe them, and because these vary from place to place, the final data may not always provide inventories which can be uniformly compared. Description and classification are integral parts of the inventory process. The descriptive phase is the recording of data, and classification is the ordering of such data into meaningful patterns. The last usually employs analytical methods which vary according to circumstance.

Allaby (1977) has defined *conservation* as "the planning and management of resources so as to secure their wise use and continuity while maintaining and enhancing their quality, value and diversity. . . . Nature conservation is the application of this concept to fauna, flora and physiographical features'' (see also Macfadyen 1963; Usher 1973:204).

In this paper, conservation is used in the above sense, as distinct from "preservation" or "protection", which may imply certain exclusions of manipulation or management — that is, where the resource unit is left "intact". Usher (1973:213), on the other hand, regards preservation as central to conservation.

For the biologist and the geographer, such terms as *tropical* are unduly restrictive, there being many environmental elements in both temperate and tropical zones which will be similar or dissimilar depending on local geoclimatic variability. The geographical imprecision of the term "tropical" has been discussed by Oliver (1979). For Australia the implications for vegetation have been dealt with by Gillison and Walker (1981), Gillison (1982), and Walker and Gillison (1982). In this paper, "tropical" will refer to the area between the two Tropics, but within that context, differences arising from variation in temperature and annual moisture regimes will be defined by the terms "megatherm", "mesotherm", and "microtherm". The rationale for these terms is derived from a measure of plant growth indices based on specific global thermal regimes (Nix 1982). Applications of this are dealt with further in Gillison (1982) and Walker and Gillison (1982).

Table 11.1 outlines some provisional bioclimatic regions for Australia and the island of New Guinea, based on these optima and a seasonality term. The definition of bioclimate as a primary environmental determinant has important implications for resource inventory and management. Figure 11.1 outlines, for example, the major bioclimatic regions and provinces that largely influence the distribution of the vegetation resource in Australia and New Guinea. With further validation this approach, based on a broad understanding of plant/environment interaction, may well be applicable to the classifications of vegetation on a global level as it offers an objective and quantitative primary framework for a functional resource definition.

Figure 11.2 indicates other areas of the southwest Pacific referred to in this paper.

Vegetation is simply the total assemblage of vascular plant life in any given area. It is usually described via measures of floristics (species), as well as structure and physiognomy.

Concept of Natural Vegetation

The question of whether a vegetation type is natural is an important aspect of resource inventory, and this has concerned ecologists for some time. Yet there are no generally accepted criteria for "naturalness". Natural vegetation is, however, usually that which is regarded as "intact", "un-

TABLE 11.1 PROVINCIAL BIOCLIMATIC REGIONS AND PROVINCES FOR AUSTRALIA AND PAPUA NEW GUINEA

Province No.	Code*	Bioclimatic Region	Bioclimatic Sub-Region	Thermal† Optima	Gradient‡ Range
1	AIS	Megatherm	I	28/35	0.1–0.2
2	AIIS		II	28/35	0.2–0.4
3	AIIIS		III	28/35	0.4–0.5
4	AIVS		IV	28/35	0.5–0.7
5	ABIS	Megatherm	I	28/19	0–0.1–0.15
6	ABIIS	Mesotherm	II	28/19	0.15–0.3
7	ABIIIS		III	28/19	0.4–0.4
8	ABI	Megatherm	I	28/19	0.1–0.15
9	ABII	Mesotherm	II	28/19	0.15–0.3
10	ABIII		III	28/19	0.1–0.3
11	ABIV		IV	28/19	0.2–0.3
12	ABV		V	28/19	0.25–0.45
13	ABVI		VI	28/19	0.15–0.75
14	ABVII		VII	28/19	0.5–0.75
15	BIS	Mesotherm	I	19	0.15–0.3
16	BIIS		II	19	0.3–0.45
17	BIIIS		III	19	0.45–0.5
18	BI	Mesotherm	I	19	0.15–0.3
19	BII		II	19	0.3–0.45
20	BIII		III	19	0.45–0.5
21	CI	Microtherm (Aust.)	I	12	0.3–0.5 (approximate only for PNG)

Note: *A province is seasonal (S) if cv% ≥ 60. Seasonality as indicated by coefficient of variation per cent of water availability (weekly mean values) throughout the year.

†Thermal optima for plant growth set at megatherm (28° and 35 °C), mesotherm (19 °C) and mlcrotherm (12 °C).

‡Scale of 0–1 growth index for plant growth within a specified thermal optimum, taking into account known radiation and moisture levels.

Source: Gillison 1981.

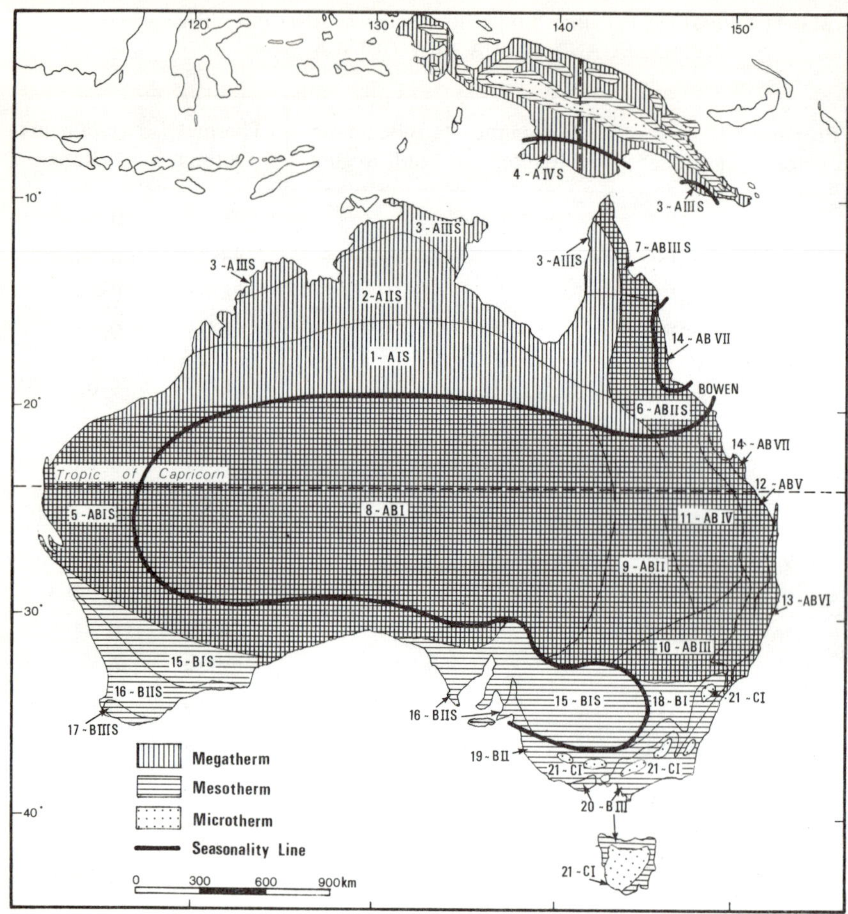

Fig. 11.1 Provisional map of bioclimatic regions of Australia and the island of New Guinea
(Source: Gillison, 1981)

disturbed", "virgin", "climax", and so on. To date there are no available objective means of determining these "states"; even mature rainforests with trees hundreds of years old can be shown to be in a state of flux with local changes in "gap-phase" reproduction where trees fall to create openings and thus influence subsequent regeneration patterns (Hopkins 1981).

For most practical purposes, it can be argued that such perturbations are minor and can be accepted within a low amplitude of change. The more open woody communities, which may owe their origin to fire, are usually described as "disturbed" or "artificial", as man is often regarded as the

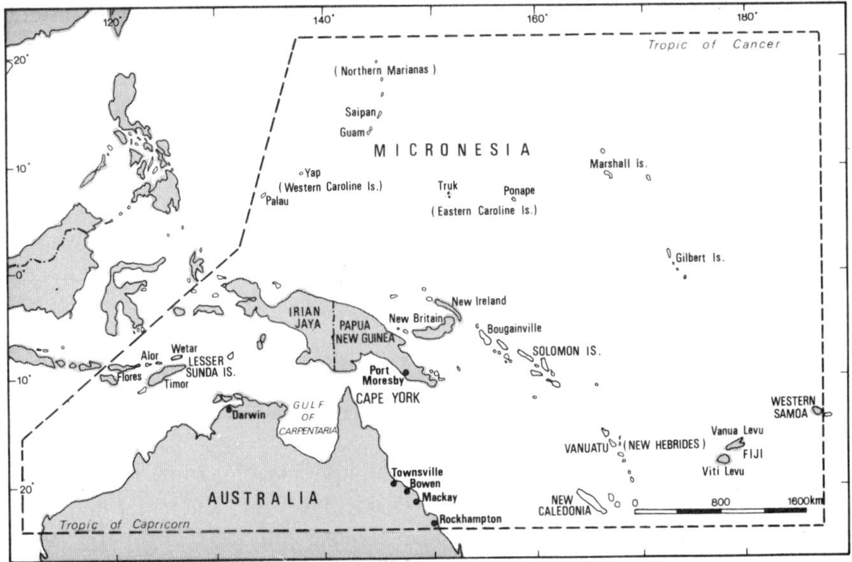

Fig. 11.2 Map of southwest Pacific area under study

destructive agent. In Australia, man has undoubtedly been associated with fire from at least 32,000 years B.P. (Barbetti and Allen 1972), and it is likely that in tropical Australia much of the original vegetation was "monsoon" forest that retreated in the face of fire (Stocker 1981; Gillison 1982). On the other hand, within-horizon (30 km radius) lightning (ground) strikes of up to 19,000 a month have been recorded in seasonal northwest Australia (Hooper 1979). Shaw (1968) recorded the effects of lightning strikes in Papua New Guinea on plantations of *cacao* and copra and I have observed fires caused by lightning in Papuan savannas. This phenomenon, together with the likely presence of man in both Australia and Papua New Guinea for 30,000 to 40,000 years, as either hunter and/or subsistence farmer, has undoubtedly influenced the evolution of vegetation patterns reflected in millenia of fallow cycles and hunting fires. It is certain that similar influences have existed throughout the old and new-world tropics.

For the plant ecologist investigating these systems, it is therefore difficult not to regard such vegetation as natural, as with the "climax rainforest". There are evident periodic changes in environment with which are associated dependent biota. White (1975) and Gillison (1975) have also pointed out that for Papua New Guinea, continual landslips, and in some cases wind and water damage, are a common feature where successional stages are well geared to such events. For the Solomon Islands, both Whitmore

(1974; 1975) and Rollet (1978) recommend that silvicultural methods should take into account the periodicity of hurricanes so that fast-growing, short-term crops can be established.

For this paper, vegetation that is subject to these periodic phenomena is therefore regarded as "natural". "Unnatural" vegetation is regarded as that which has been disturbed by man involving an actual *removal* of the resource, e.g., either very intensive subsistence gardening or levels of gràzing outside the evolutionary history of the vegetation, or else by wholly destructive events such as wood-chipping or intensive logging.

Tropical Natural Vegetation Resource: Definition and Extent

To determine the extent of a vegetation type assumes, *a priori*, that the type in question is readily classifiable.

Vegetation classification methods are highly variable, some being based on or combined with climatic criteria (Holdridge 1947; Küchler 1949), and others on structure and physiognomy (Dansereau 1951; Fosberg 1970; IUCN 1973; and UNESCO 1973). Of these authorities, Fosberg (1970) emphasizes physiognomy and structure as the main criteria, followed by functional or physiologically-based attributes. Floristic and other ecological terminology are then used as additional information. More recently, according to Letouzey (1978), Brünig (forthcoming) has produced a classificatory system based on a hierarchy of characters which employ physiognomy, structure, and physiology related to changes along humidity gradients. The last is mentioned here as it will be shown later in this paper that there is much to recommend the use of physiological as well as structural criteria (*per se*) in classifying vegetation according to its apparent environmental relationships.

For the tropics, global inventories have concentrated mainly on the tropical moist forest (TMF) biome as the major source of biomass, with the drier woodland savannas and open grasslands decreasing in importance. The focus in this paper is the TMF, as it epitomizes both the promises and problems attached to the exploitation of a natural tropical resource.

Several workers, in particular Sommer (1976) and Myers (1980), have outlined the difficulties in determining the extent of the forest resource. Basic problems arise through differences in inventory method (including differences in terminology), in the political effects of having data published, and in the often severe logistic constraints and methodological limitations involved in monitoring levels of change in tropical moist forests. Sommer (1976) has provided several useful figures, including an estimate of the actual and potential TMF climax area of the world (Table 1.2), and an approximate list of countries with reported decreasing rates in the TMF area

TABLE 11.2 SUMMARY TABLE OF TOTAL LAND AREA, MOIST FOREST CLIMAX AREA, ACTUAL FOREST LAND AREA AND ACTUAL MOIST FOREST AREA
(in million ha)

Sub-Continent	Total Land Area	Total Moist Forest Climax Area	Percentage of Total land Area	Percentage of World Moist Forest Climax Area	Total Actual Forest Land Area	Actual Moist Forest Area	Percentage of Total Land Area	Percentage of Total Forest Area	Percentage of Total Moist Actual Forest Area	Regression of Climax Area 2–6	Regression of Percentage of Climax Area
	1	2	3	4	5	6	7	8	9	10	11
East Africa	236	25	10.6	1.6	55	7	3.0	12.7	0.7	18	72.1
Central Africa	408	269	65.9	16.8	176	149	36.5	84.6	15.9	120	41.1
West Africa	356	68	19.1	4.2	103	19	5.3	18.4	2.0	49	72.1
Total Africa	1,000	362	36.2	22.6	334	175	17.5	52.4	18.7	187	51.1
Latin America	1,401	750	53.5	46.9	864	472	33.7	54.6	50.5	278	37.1
Central American Caribbean Region	166	53	31.9	3.3	100	34	20.5	34.0	3.6	19	35.1
Total Latin America	1,567	803	51.2	50.2	964	506	32.3	52.5	54.1	297	37.0
Pacific Region	374	48	12.8	3.0	78	36	9.6	46.2	3.8	12	25.0
Southeast Asia	448	302	67.4	18.9	268	187	41.7	69.8	20.0	115	38.1
South Asia	348	85	24.4	5.3	71	31	8.9	43.7	3.3	54	63.5
Total Asia	1,170	435	37.2	27.2	417	254	21.7	60.9	27.2	181	41.6
Total Humid Tropics	3,737	1,600	42.8	100.0	1,715	935	25.0	54.5	100.0	665	41.6

(Table 11.3). Other syntheses of data of TMF have been provided by Lanly and Clement (1979) and Myers (1980). Brünig (1977) summarized a broader situation as follows: The area covered by tropical closed forests and open woodland is roughly 20 million km². These forests represent about 50 per cent of the forested land, 60 per cent of the forest growing stock, and 69 per cent of the forest productivity of the world. Half of this area is closed forest; of this 5.4 million km² is in America, 1.9 million km² in Africa, and 2.7 million km² in the Asia Pacific region.

In these inventories, the most effective breakdown of figures is related to the most simplified classification of TMF into essentially evergreen and deciduous (or semi-deciduous) components. Sommer (1976) has provided figures for these in the context of the UNESCO international system (1973) which employs the following terms and definitions:

Ombrophilous: mainly broad-leaved evergreen trees; forest canopy remains green all year though a few individual trees may be leafless for a few weeks.

TABLE 11.3 COUNTRIES WITH REPORTED
DECREASING RATES IN THE TROPICAL
MOIST FOREST AREA

Country	Area of Forest Reported To Be Lost Per Year (million ha)
Bangladesh	0.01
Colombia	0.25
Costa Rica	0.06
Ghana	0.05
Ivory Coast	0.4
Lao	0.3
Madagascar	0.3
Papua New Guinea	0.02
Philippines	0.26
Thailand	0.3
Venezuela	0.05 (?)
North Vietnam	0.01
Malaysia	0.15
Total	2.16

Evergreen seasonal: mainly broad-leaved evergreen trees; foliage reduction during dry season is noticeable.

Semi-deciduous: most of the upper canopy drought deciduous; many of the understorey trees and shrubs evergreen.

Deciduous: majority of trees shed their foliage simultaneously in connection with the unfavourable season and foliage is shed regularly every year.

Exploitation

Traditional methods of selective logging, originally determined by specified lumber and cabinet wood requirements, have changed in recent years to be intensive-logging or clear-felling that has become associated with the wood-chip industry. Technological development has kept pace with demand, to the extent that standing timber previously unacceptable because of short fibre length or other physical or chemical problems has now become acceptable for commercial exploitation. One of the major problems that besets the conservationist of the broad natural resource, as well as the lumber man, is the level of diminishing nutrients in the system. Whole trees removed with bark from site may represent a loss of up to 80 per cent of available nutrients in some rainforest systems. There is little reliable information on natural regenerative systems in most tropical moist environments, and there is still no certainty that a resource can be maintained under many increasingly severe levels of exploitation. Lanly and Clement (1979) provide a measure of the changing levels of timber resource (Table 11.4), but this does not take into account the long-term effect on ecosystem management which may well be problematical (see also Pringle 1976).

The continued decrease in readily available natural vegetation has meant that more attention is being paid to the establishment of a "sustained-yield" rather than the "cut-out and get-out" philosophy that has been the hallmark of much exploitation in the tropics.

Neither are developing countries exclusive victims to this kind of exploitation. According to Routley and Routley (1977) Australian forests, which were a small proportion of the land area before settlement, have since been reduced by two-thirds and now constitute only 5 per cent of the land area. Of this, L.J. Webb (pers. comm.) estimates that less than one per cent is rainforest (i.e., non-sclerophyll or non-eucalypt-dominated). Myers (1980: 63) has said that of the 7,000 km² of forest lands controlled by the Queensland Forestry Department, most are still forested, although somewhat disturbed. All but 660 km² have been logged at least once since European settlement. In some of the remaining areas that are being threatened with logging, there is probably no more than about twelve to fifteen years of

TABLE 11.4 ESTIMATED AREAS OF NATURAL FORESTS FROM 1975 TO THE YEAR 2000 (thousands hectares)

Regions	1975 Hardwood Closed and Softwood			1980 Hardwood Closed and Softwood			2000 Hardwood Closed and Softwood		
	Productive	Unproductive	General total	Productive	Unproductive	General total	Productive	Unproductive	General total
Tropical America	513,131	146,176	659,307	495,845	145,325	641,170	441,390	146,060	583,450
Tropical Africa	134,183	69,470	203,653	130,487	69,470	199,957	119,310	69,480	188,790
Tropical Asia and Far East	196,637	106,480	303,117	185,351	105,546	290,897	150,910	102,920	253,830
Total Tropical Countries	843,951	322,126	1,166,077	811,683	320,341	1,132,024	711,610	314,460	1,026,070

Source: Data extracted from Lanly and Clement 1979.

exploitation of the natural resource left. Most estimates of future logging potential in the southwest Pacific suggest the currently available resource will be fully exploited by 1990 (see also Myers 1980:169).

In his summation of differentiated conversion rates for TMF, Myers (1980:168) has argued that elimination of TMF is not a phenomenon common to all parts of the biome. While most of the world's TMF may be drastically affected by the year 2000, areas with abundant mineral resources, for example, Central Africa, may not have the same need to use the TMF. Thus the final result may be remnants of relatively intact forest in those areas rich in other resources.

Inventory

Requirements

To be effective, any inventory process must take into account the aims and objectives in solving the problem at hand. In this respect, inventories conducted for merchantable volume may have little relevance to specific questions being asked by a biologist. Lanly (1976:42) has said that for industry "the aims of a forest inventory should reflect the level and scope of planning under consideration, the nature and size of the unit of management and the stage reached in decision making". He has further pointed out that a changing technology usually requires a change in the inventory process. Here, communication may become a problem, as with new developments there is often resistance to change from traditional inventory or classificatory techniques. Cameron (forthcoming) has emphasized the difficulty of communication between classifiers, let alone classifier and user.

The scale and purpose of the objectives should therefore control the development of an inventory system.

Common Inventory Methods

Inventories covering large areas usually operate with small mapping scales (e.g., 1:60,000). Initially, very general survey parameters are used to seek out the course pattern. Within this framework samples are often taken at meso-scale ($>$1:25,000), as at this level the pattern becomes more readily interpretable for a wider range of purposes, particularly for the extrapolation of productivity measures from sample points.

Visual interpretation or "eye-balling" of imagery is usually the first step in deciding how to sample an area. Recent development in techniques for analyzing data from satellites is receiving much attention and the literature in this respect is considerable.

Sayn-Wittgenstein (1978:1187) has emphasized that application of satellite remote sensing to tropical forest inventories requires efficient methods to collect more detailed information. He has pointed out that there are many cases where large-scale aerial photography provides an immediate and effective means of interpretation. Aldred (1976) has concluded that the best measurements of trees and crowns came from photography at scales of from 1:2,000 to 1:4,000. In Australia, in the CSIRO Division of Land Use Research, aerial photography of tropical rainforest and wood lands has shown that effective crown identification can be achieved at 1:5,000 (Paijmans [1975] has produced a vegetation map of Papua New Guinea largely on the basis of determination of crown type from aerial photographs).

Cannon *et al.* (1978) concluded from a comparison of aerial photographic versus LANDSAT data that the latter can provide immediate national data and a monitoring base, whereas the former must be relied upon for management data. Morain and Klankamsorn (1978) found that for forest mapping and inventory techniques, "the discovery of rapid forest depletion in eastern Thailand and the mapping of watershed classes and shifting agricultural practices in the north are prime examples of the multitude of resource management problems approachable by LANDSAT".

In the Australian region, application of LANDSAT data to inventory is problematical where there are seasonal changes in the soil moisture patterns and subsequent vegetation regeneration. Also in some localities, changing reflectance signature requires considerable "ground-truthing" to provide an adequate predictive base. It is nevertheless likely that with increased resolution of pixel size in the forthcoming generation of satellites, vegetation inventory via satellite will become increasingly important.

Sampling

Following initial determination of photo-patterns, samples are usually made of each pattern for representative "ground-truth". In the tropics for heavily forested natural areas more than 1,000 km², it is none the less unusual for the on-ground sample to exceed one per cent of the area. Subsequent extrapolation is therefore highly dependent on sampling methods, the attributes used, and the level of variance in the biological system. Forestry surveys in Papua New Guinea usually employ airborne assistance from helicopters. Sample transects are made on the ground through discrete air-photo "forest types", with counts of merchantable species more than 10 cm d.b.h. within about a 20 m radius every 200 m. Data on terrain, soil type, and access are similarly recorded. There is rarely any collection of other information of broader ecological importance. Unknown species are usually collected and later identified where possible. From these surveys, estimates of merchantable volume and cost-benefit studies of harvesting are made. This type of ground survey or "cruising" is generally applicable in the TMF, although satellite remote sensing data have been employed to identify commercial forest types as well, for example, in the northern Amazon region.

Follow-up monitoring surveys of natural forest are not common, mainly because of high logistic costs and often lack of staff.

Typology

At a global level, methods of forest typology are by no means uniform. Forest typing for commercial inventory, apart from using structural data for volume estimation, is usually assisted by the addition of species names. At a generic or family level, taxonomic data can be useful, for example, *Anisoptera*, *Drybalanops*, *Hopea*, and *Shorea*, all members of the Dipterocarpaceae, are common in parts of Malaysia and Papua New Guinea, where such forest often has a typical structural character.

For broader purposes of conservation, other methods may be used, depending on the purpose at hand. The "Yangambi" classification in Africa, the UNESCO (1973) and IUCN (1973) classifications, and the classification of Fosberg (1970) are methods which are essentially structural/physiognomic in nature and, for simplicity, employ a limited range of attributes. For forests of Australia and the southwest Pacific, Webb *et al.* (1976) have, in contrast, used an extensive *proforma* that includes life-form, leaf-size classes, and structural and physiognomic features such as crown type, bark type, and so on. The Webb method has been used successfully in broad surveys to classify rainforests at the "climax" level, but is not designed for heavily disturbed areas or regrowth situations. For disturbed areas and for open vegetation, Webb *et al.* (1976) suggest floristics will provide the most useful diagnostic.

In Vanuatu (the New Hebrides), Gillison (1975a) found that by incorporating additional attributes for graminoids, the Webb *proforma* could be extended to include savanna formations. The additional attribute of bark colour (presumably as a function of stem exposure to light), also provided assistance in identifying seral or successional stages. In the Australian region, the classifications of Specht (1981), Beard and Webb (1974), and Carnahan (1976) are largely based on height and foliage projective cover of the dominant stratum. Large areas of Australia are currently being mapped via this method at a 1:250,000 scale.

Limitations of Present Methods

Tropical vegetation is not static. As mentioned earlier, even in mature forest stands it can be argued that there is constant change. In stands which have been logged-over, it is unlikely there will be a return to the former conditions. In Malaysia, some of the silvicultural systems have been criticized by Meijer (1970) who maintains that many removal procedures are pointless. He also says that the degree of disturbance is critical in determining the final composition of the rainforest. He was able to detect stands dominated by the secondary species *Anthocephalus chinensis* in areas logged more than forty to forty-five years before. Other disturbance effects have been discussed by various authors (Richards 1955; 1966; Jones 1956; Conklin 1957; Budowski 1965; 1970; Fox 1973; Gillison 1975a, b; Hopkins and Graham 1981), from which it can be concluded that the advanced growth of secondary species makes it difficult to distinguish between old secondary forest and primary or climax forest. Gillison (1975b) has emphasized that if one is to "restore" an altered tropical forest ecosystem, it is desirable that some classification of the former unaltered status is available as in many cases restoration may simply mean a return of tropical forest to an advanced seral stage — not that this is necessarily undesirable. An indication of a typical example of forest succession following disturbance is described by Hopkins (1981) for a TMF in Queensland (Fig. 11.3).

From the above, it should be evident that because of the calibration of many inventory processes to static rather than dynamic systems, there is much meaningful information lost, particularly where disturbance or perturbation has taken place. The kinds of successional stages described by Symington (1933), Budowski (1965; 1970), Gillison (1970), Gomez-Pompa and Vasquez-Yanes (1974), Fontaine and Gomez-Pompa (1978), and Ashton *et al.* (1978) suggest there are distinct behavioural and strategical characteristics of plant response to changing environment that can be used to identify characteristic states of forest type. To some extent this has already been considered by several authors, for example, the "nomad" and "dryad" species of van Steenis (1958), and certain "vital attributes" which have been suggested by Noble (1981).

Other limitation of classificatory systems are the rigidity or arbitrariness of classes and more particularly the subjectively based criteria for attribute selection. Very rarely have attributes been experimentally validated for use in forest classification. Rollet (1978) and others have discussed at length the difficulty of employing the recognition of strata or layers within rainforest as classificatory attributes, and it is certain there are major difficulties both in layer recognition and in subsequent numerical analyses of the data. Even terms such as *sclerophyll*, *xerophyte* and so on are elusive of adequate quantitative definitions.

In this respect, the extension of life-form classes proposed by Ellenberg (1956) and Ellenberg and Mueller-Dombois (1974) are limited when used in classifying, for example, complex forest types. Further, the system of drawing forest profiles is a useful descriptive technique but data from this source can rarely be used sensibly in numerical analyses. Finally, the appli-

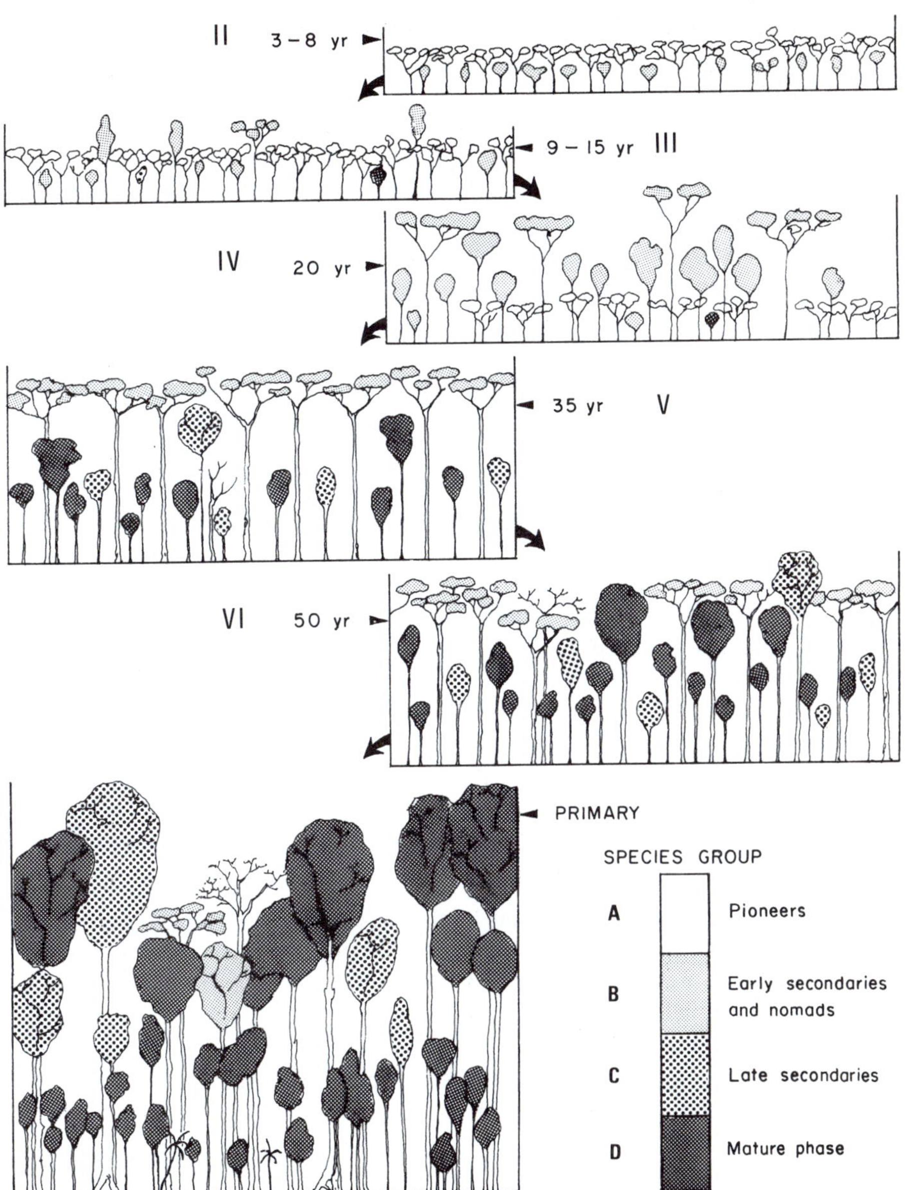

Fig. 11.3 Schematic representation of reconstructive secondary succession in complex notophyll vine forest
(Source: Hopkins 1981)

cation of height and cover classes, often used in forest typology, will differ in meaning depending on age structure and other features of the forest environment. Cameron (forthcoming) has already suggested a change of classes from the Specht system, for forestry purposes in Australia. There are other limitations to current classificatory processes, and a fundamental and continuing problem is that it seems there will always be as many systems as there are ecologists to provide them. However, with changing demands and technology this may be inevitable.

It remains to mention the sampling philosophy behind inventories. It is usual for some stratification of photo pattern to take place for subsequent sampling — usually by transect, though it is not usual to seek out locally intensive environmental discontinuities but rather to avoid them. Such discontinuities usually reflect the steepest environmental gradients and thereby the focus of greatest biological diversity in any one area. The avoidance of such gradients to obtain a wide dispersion of sample points may extend the logistic demand well beyond that required for adequate survey and may not necessarily represent the best way of sampling for management or conservation purposes.

Problems of Vegetation Classification

It is generally accepted that the distribution of vegetation pattern tends to reflect environmental change. For this reason, and because vegetation patterns tend to be conspicuous both in the field and in imagery, the analysis and interpretation of this medium has become widely employed in land resource surveys. Despite this, problems associated with the application of vegetation classification are rarely considered in depth and tend to be subjugated. Some crucial problems confronting the land resource surveyor are:

1. At continental and regional levels there is a very limited functional basis for using vegetation parameters to predict physical environment and land-use capability. Below the regional level, a useful functional relationship is not yet apparent.
2. Because land-use problems differ, some flexibility in the application of different vegetation classification methods is required; classifications by definition are inflexible in themselves.
3. There is rarely a clear relationship between vegetation classification and purpose.
4. Different classificatory methods produce different results.
5. Most classifications and sampling methods are without critical validation and are difficult to test.
6. Different terminologies and the highly subjective nature of most classifications makes uniform application difficult.

7. There is no currently acceptable method of classifying the disturbed or seral conditions that probably occur in about 70 per cent of world tropical moist forest biomes.
8. There is often irreconcilable conflict in scale between vegetation parameters used in the field and those resolvable in final mapping from available imagery. Large-scale imagery is commonly expensive and usually difficult to obtain *ad hoc*.

Possible Developments

Application of Functional Attributes

The decrease of the tropical vegetation resource and the inapplicability of traditional methods of inventory to perturbed systems suggests there is a requirement for more efficient (and more sensitive) methods so that different developmental vegetation stages can be identified and hopefully forecasted in advance of manipulation. Such development requires an adequate study of the function and behaviour of ecosystems — a cry that rings through most UNESCO, UNEP, FAO, and IUCN publications. Because TMF ecosystems are complex, this is an extremely daunting task. It is not proposed to attempt here a review of such important features as nutrient cycling or fluxes in carbon balance, but rather to examine briefly some broader aspects of functional attributes, a plant functional attribute being defined as that which behaves in a demonstrable and predictable way with a change in physical environment (Dansereau 1951; Fosberg 1970). Although physiology is important, for present purposes ecological rather than physiological response is reflected in a functional attribute which, depending on scale, may be described in terms of structure, morphology, or anatomy. There is no explicit relationship between a functional and floristic attribute.

General assumptions for land resource purposes

(a) "That characters of many land attributes are interdependent and tend to occur in correlated sets;
(b) That every land-use is constrained by the combined and interacting effects of several land attributes" (Austin and Basinski 1978).

Austin (1978) defined several purposes for providing a vegetation description for a region, of which one is "to use vegetation as an indicator of environment". These and other assumptions implicit in vegetation description suggest there is an underlying theoretical basis for using vegetation as an indicator of environment. The basis is far from clear when one examines available literature, although in the past decade there have been considerable advances in the field of ecophysiology. Lewis (1972) has summarized it thus:

> More recently, important quantitative relationships have been established between the structure and arrangement of leaves and their physiological behaviour. Variation in leaf size, shape, thickness, stomatal structure and frequency has a direct and profound influence on the transfer rates of heat energy, water vapour and carbon dioxide between the leaf and the environment. Improved understanding of the physical principles involved has provided the basis for an analytical as distinct from a descriptive approach to these problems.

In the following discussion, some common functional terminology is considered, together with functional aspects, broad (morphological) and fine (anatomical) structure. This is followed by a brief discussion of environmental relationships and related evolutionary and adaptive traits.

Terms such as *Sclerophylly* and *xerophily, xerophyte, xeromorph, etc.*, have strong environmental connotations but are rarely described on a quantitative basis. *Sclerophyll* usually indicates a (usually small) thickened, evergreen leaf found in environments with a marked seasonal water deficit (Schimper 1903; Cooper 1922; Adamson and Osborn 1924; Grieve 1955; Ferri 1961; Mooney 1973; Webb *et al.* 1976; Medina *et al.* 1978). Loveless (1961; 1962) attempted to place sclerophylly on a quantitative basis, where a sclerophyllous leaf was characterized by "increased cutinization and sclerification" and the degree of sclerophylly expressed by the ratio:

$$\frac{\text{crude fibre dry wt.} \times 100}{\text{crude protein dry wt.}}$$

In defining this relationship, Loveless assumed and later provided evidence that there was a nutritional (functional) basis for sclerophylly where such leaves can tolerate low levels of phosphate.

A *Xerophyte* is typically regarded as a plant growing in a dry place (Schimper 1903; Schouw 1922 — cited by Kummerow 1923; Adamson and Osborn 1924; Maximov 1931; Patton 1933; Thoday 1933; Ferri 1961; Easu 1965:277; Gindel 1969; Grieve and Hellmuth 1970; Lewis 1972; Medina 1975; Lloyd and Woolhouse 1978). Other related but more specific attributes are succulence and crassulacean acid metabolism (CAM) (Medina 1975) and transpiration features (Maximov 1931; Gindel 1969). Because of the controversy surrounding the term, Thoday (1933) went so far as to propose that the term "xerophyte" "should not carry any particular functional or structural implications". Nevertheless it remains in use in that sense today. This and related terms are further reviewed in Gates (1914), Cooper (1922), Shields (1950), Grieve (1955) and Esau (1965:277). Walter (1979) believes that *xeromorphosis* is a direct result of a growth deficiency and therefore better called *peinomorphosis* (Greek *peino* = hunger).

To date, *xero-* and *meso-* plants remain variously defined but never on a quantitative basis. For this reason, they cannot be satisfactorily regarded as functional terms; neither can they be used efficiently in diagnostic field proformas for forest typing.

Foliage characteristics.   Horn (1970) showed that for some North American forests it was possible to provide a theoretical basis for the development of foliar geometry of trees related to the foliar interception of light. He developed the monolayer versus multilayer concept, and showed that trees with a monolayer foliage shell developed later in a successional sequence and were most common in climax forets, while the converse was true of multilayered trees. In tropical forests, my observations suggest there may be a reverse trend from mono- to multilayer types to that suggested by Horn.

The quantity and structure of leaves of a desert shrub *Encelia farinosa* was shown by Cunningham and Strain (1969) to be controlled by the moisture stress of the environment. Their field observations also showed the influence leaf quantity and structure exert on CO_2 exchange capacity and water status of the shrub. In Australia the development and use of foliage projective cover (f.p.c.) as developed by Specht (1970; 1981) is in wide use by Australian vegetation classifiers. Specht related the measure of C.A.G.I. (current annual growth increment) of a plant to its projective foliage characteristics based on physiological parameters but failed to take into account the variability encountered in different crown densities and seasonal variability. This aspect has been investigated more recently by Walker and Tunstall (unpub.) who have shown there are predictable relationships between foliage type and density, and light transmission.

Functional attributes of leaves.   There has been a recent emergence of considerable physiological information that deals with the functional relationships of certain attributes. Of these the following have received most attention: size, shape, indumentum, thickness, dissection of margin, stomatal characteristics and other anatomical features, spectral properties, and phytochemistry.

These have recently been reviewed by Gillison (unpub.) and by Karlsson (1981), and although the subject matter is too extensive for inclusion in the present paper, there are several relevant points. Recent workers (Webb 1959; Gentry 1969; Brünig 1970; 1978; Dolph 1978) have found a general link between leaf size and environmental gradients such that size decreases with increasing harshness of environment. According to Webb (1959), for Australian rainforests, the climatic formations of tropical, sub-tropical, and temperate rainforests are "characterized under optimum conditions by mesophyll, notophyll, or microphyll sizes respectively". Leaf Area Index (L.A.I.), that is, the total leaf area of a plant as a ratio of the area of ground covered by the plant, has been examined by several workers as an important index to light-demanding plant processes. More critical examination has come from Mooney (1972), Larcher (1975), and Wassink and van den Noort (1976). In Australia, Carbon *et al.* (1979) have developed a visual method for estimating leaf area to estimate L.A.I. problems of field estimation remain.

Givnish and Vermeij (1976) consider that leaf size is functionally related to shape. Work by Raschke (1969), Gates (1968), and Taylor (1975) examined boundary layer effects in relation to heat transfer and Taylor calculated a "characteristic dimension" related to energy transfer for basic leaf shapes, across a range of environments. This aspect of leaf form has been further reviewed in depth by Givnish (1978). Increasing leaf pubescence was found by Clausen *et al.* (1940) to be directly correlated with increasing dryness. Further work on leaf indumentum has been undertaken by Ehleringer *et al.* (1976) and Baruch (1979). Leaf margin (dissection) was found by Bailey and Sinnott (1916) to vary with environment where leaves and leaflets with entire margins are found mainly in the lowland tropics. The stomatal structure of leaves has long fascinated ecologists. Various measures of stomate frequency, density, stomatal index, and aspects of pore dynamics have shown a number of environmentally related trends (Cooper 1922; Lewis 1972; Larcher 1975; Clay and Quinn 1978; de Michele and Sharpe 1978; Sharpe and Hsin-Iwu 1978; Hsin-Iwu and Sharpe 1979).

Anatomical features have important compensatory effects on stomate frequency, for example, the total pore area per leaf, the kind of stomate apparatus — whether sunken or raised, etc. — which is one reason why simple stomate indices often do not relate simply with environmental measures.

Leaf spectral properties are known to be associated with environmental change (Gates *et al.* 1975; Mulroy 1977) where, for example, glaucousness is associated with very high reflectance of UV radiation. Mulroy (1979) indicated that high UV reflectance is ecologically significant in reducing damage to dehydrated leaves from visible and UV-b radiation. Brünig (1970) has further discussed changes in reflectance with changes along a humidity gradient.

Figure 11.4 illustrates in a very general way how some of these functional attributes are related to environmental change.

Apart from leaf function, patterns of plant behaviour in relation to environment are evident in other organs, particularly those underground, and in the position of the perennating buds.

ATTRIBUTE	ATTRIBUTE STATE		
Visible reflectance	>B	<A	>A
Absorbtance and Transmittance	<B	>A or B	<A or B
Vegetation profile A = rainforest B = open sclerophyll C = mixed sclerophyll	A	B	C

Stomate distribution	hypostomatous	isostomatous	isostomatous
Stomate density	low	high	intermediate
Leaf cuticle	thin	thick	mixed
Leaf size	'mesophyll'	'microphyll'	'notophyll'
Leaf angle	lateral (radiation +)	pendulous (radiation -)	composite (radiation ±)
Leaf water potential	low (<B)	high (>A or C)	medium (<B)
Leaf type	'mesic'	'sclerophyll'	'sclerophyll'
Life forms	complex	simple	simple
Plant biomass	high	medium	<B
Density	high	low	medium
Available water	high	low	>b
Soil depth	deep	shallow	>B
Nutrient store	high	low	>B
Soil type	Um Uf	Uc	Gn
Land use capability class (American system)	2 - 3	7 - 8	4 - 5

Fig. 11.4 A hypothetical relationship between some vegetation, environmental, and land use attributes

Case Studies in the Use of Functional Attributes in Tropical Vegetation Classification

There is little published work in this area. The philosophical approach of using functional attributes is not new, but in the past problems have existed with validating attributes used, and the absence of a computer technology has made difficult the analytical process. Because of their relevance to dynamic rather than static systems, it is useful to comment on some field trials that were undertaken using a proforma of the kind shown in Appendix 11.1. These trials were conducted at a range of scales and environments where sites were ordered along the steepest conspicuous environmental gradients. The assumption that data collected along such gradients are likely to reflect greatest diversity of pattern in any one area has been recently tested (Gillison and Brewer, unpub.) and found to be highly significant for a meso-scale (1:25,000) pattern in Australian woodland.

In Australia, unpublished work by Gillison and Goodwin has shown that for classification at a major geographic level in far north Queensland, most attributes could be reduced to a set that involved an estimate of leaf size, leaf angle, leaf type and the "life-form" to which the leaf was attached. These features, together with an overall estimate of foliage projective cover (Specht 1981) and height of the tallest layer, were sufficient to provide an adequate classification of vegetation for most inventory purposes. Further work by Gillison and Bennett in northwest Australia and by Gillison and Nix in the Eungella rainforest area of north Queensland indicate that these attributes are also useful correlates with aspects of the physical environment and with habitat features that are important to avifauna. This work is, as yet, unpublished. The Eungella study used the three leaf characters in combination with "life-form" so that unique combinations occurred when drawn from attributes in a finite set (see Appendix 11.1). Such a combination is termed a *modus*, for example, a meso-latero-ortho-plan is an individual with mesophyll-sized leaves that are laterally disposed, "orthodox" in type (i.e., dorsiventral or bilateral) and supported by a phanerophyte. Further details on the application of this method can be found in Gillison (1981). When used in a range of disturbed rainforest sites, the method of using *modi* to describe the character of the vegetation is sufficiently sensitive to identify a range of seral regrowth stages. This method is more sensitive in that regard than the structural typological technique developed by Webb *et al.* (1976). Published applications of the technique in sub-tropical and temperate Australia can be found in Gillison (1981) and Fox and Fox (1981). In the former, a gradient-based sample was taken in a grazed wood land in southwest Queensland. The results indicate that the *modi* used produce a useful vegetation classification in a way that is closely correlated with a number of soil features, in particular soil salinity, soil depth, and infiltration capacity (Fig. 11.5). The study by Fox and Fox (1981) showed that in a heath-land community, the method when combined with samples taken along a transect ordered along a edaphic gradient, produced a means of classifying vegetation that effectively described the distribution pattern of the ground-dwelling mammal species.

Each of these studies suggests that the further refinement of an approach using parameters that are closely and measureably linked with the physical environment will be one of the more profitable ways to classify vegetation in a dynamic context.

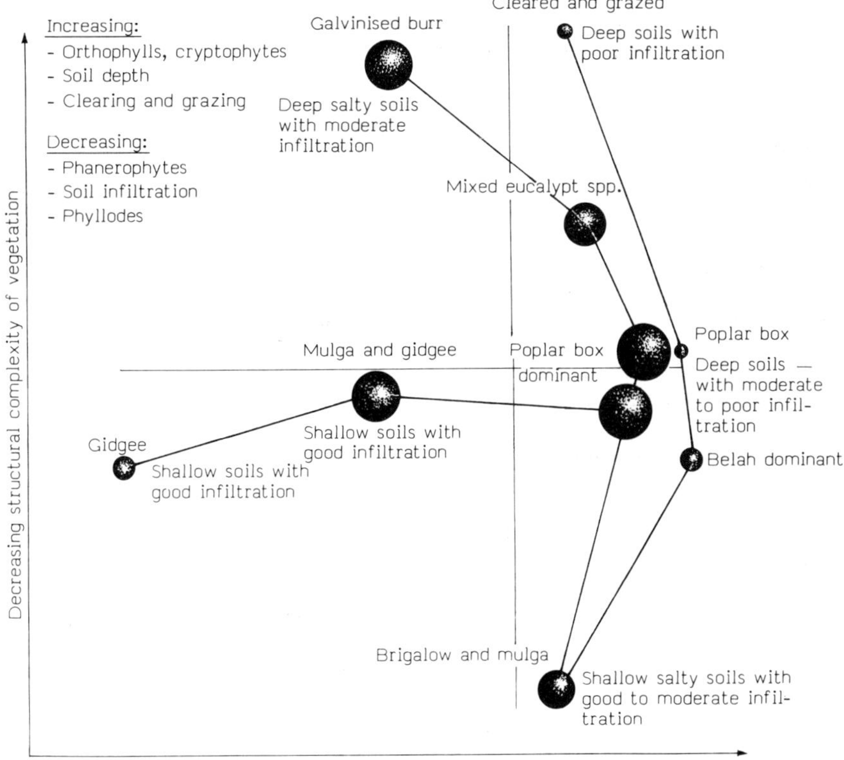

Fig. 11.5 *Minimum spanning ordination of functional attributes from sub-tropical semi-arid woodland system in southern Queensland. Edaphic features are superimposed.*
(Source: Gillison, 1981)

Conclusions

Recent work in tropical countries suggests that there are measurable responses by plants to environment which can be correlated with fine-scale morphological attributes. These attributes can be regarded as functional. In cases where disturbance has occurred, or is likely to occur, such attributes can perform a useful role in classifying different vegetation types or in monitoring vegetation changes within one location. The attributes so used can be employed as core attributes, to which additional attributes can be attached for specified purposes.

The use of quantitative data of this kind and the selection of environmental attributes for which there is a measurable response in vegetation should facilitate management and planning. The crucial development in the future will be the identification of those assemblages of attributes which together provide the necessary information.

REFERENCES

Adamson, R.S., and Osborne, T.G.B. 1924. "The Ecology of the Eucalyptus Forests of the Moutn Lofty Ranges (Adelaide District), South Australia". *Transactions Royal Society of South Australia* 48:87.

Aldred, A.H. 1976. *Measurement of Tropical Trees on Large-Scale Aerial Photographs. Information Report FMR-X-86.* Forest Management Institute. Ottawa.

Allaby, M. 1977. *A Dictionary of the Environment.* Southampton: Camelot Press.

Ashton, P.S.; Hopkins, M.S.; Webb, L.J.; and Williams, W.T. 1978. "The Natural Forest: Plant Biology, Regeneration and Tree Growth". In *Tropical Forest Ecosystems: A State of Knowledge Report.* UNESCO-UNEP. France.

Austin, M.P., and Basinski, J.J. 1978. "Bio-Physical Survey Techniques". In J.J. Basinski (ed.), *Land Use on the South Coast of New South Wales.* I: *General Report.* CSIRO Melb.

Bailey, I.W., and Sinnott, E.W. 1916. "The Climatic Distribution of Certain Types of Angiosperm Leaves". *American Journal of Botany* 3:24–39.

Barbetti, M., and Allen, W. 1972. "Prehistoric Man at Lake Mungo, Australia, by 32,000 B.P.". *Nature* (Lond.) 240:46–48.

Baruch, Z. 1979. "Elevational Differentiation in *Espeletia Schultzii* (Compositae), a Giant Rosette Plant of the Venezuelan Paramos". *Ecology* 60:85–98.

Beard, J.S., and Webb, M.J. 1974. Vegetation Survey of Western Australia. Great Sandy Desert 1:1,000,000 Vegetation Series, Explanatory Notes to Sheet 2. Part 1. *The Vegetation Survey of Western Australia, its Aims, Objects and Methods.* University of Western Australia Press.

Brunig, E.F. 1967. "On the Limits of Vegetable Productivity in the Tropical Rain Forest and the Boreal Coniferous Forest". *Journal of the Indian Botanical Society* 46:314–22.

______. 1969. "On the Seasonality of Droughts in the Lowlands of Sarawak (Borneo)". *Erdkunde,* 23:127–33.

______. 1970. "Stand Structure, Physiognomy and Environmental Factors in Some Lowland Forests in Sarawak". *Tropical Ecology* 2:26–43.

______. 1973. "Some Further Evidence on the Amount of Damage Attributed to Lightning and Wind-throw in *Shorea Albida* — Forest in Sarawak". *Commonwealth Forestry Review* 52:261–65.

______. 1977. "The Tropical Rain Forest: A Wasted Asset or an Essential Biospheric Resource". *Ambio* 6:187-91.

Budowski, G. 1965. "Distribution of Tropical American Rain Forest Species in the Light of Successional Processes". *Turrialba* 15:40-42.

______. 1970. "The Distinction between Old Secondary and Climax Species in Tropical Central American Lowland Forests". *Tropical Ecology* 11:44-48.

Cameron, D.M. 1979. "Review of Forest Land Classification in Tropical and Sub-Tropical Australia". In *Forest Land Assessment and Management for Sustainable Uses.* Unpublished report from conference sponsored by the Environmental and Policy Institute of the East West Centre, Hawaii, 19-28 June 1979.

Cannon, T.K.; Ellefsen, R.A.; Carib, K.B.; and Crespo, J. 1978. "The Application of Remote Sensing Techniques to Foreign Vegetation Surveys in Tropical Areas and Urban Fringe Land Use Problems in Costa Rica". In *Proceedings of 12th International Symposium on Remote Sensing of the Environment, Manila, Philippines,* 1978. Centre for Remote Sensing Information and Analysis, Environmental Research Institute of Michigan, Ann Arbor, Michigan. Vol. 3, pp. 2081-90.

Carnahan, J. 1976. "Natural Vegetation". In *Atlas of Australian Resources.* 2nd series. Dept. National Resources, Canberra.

Clausen, J.D.; Keck, D.; and Hiesey, W.M. 1940. *Experimental Studies on the Nature of Species.* I: Effect of Varied Environments on Western North American Plants. Carnegie Institute Washington Publication 520.

Clay, K., and Quinn, J.A. 1978. "Density of Stomata and Their Responses to a Moisture Gradient in *Danthonia sericea* Populations from Dry and Wet Habitats". *Bulletin Torrey Botanical Club* 105:45-49.

Conklin, H.A. 1957. "Hanunoo Agriculture: Report on an integral system of shifting cultivation in the Philippines". *FAO Forest Development Paper* no. 12.

Cooper, W.A. 1922. "The Broad-Sclerophyll Vegetation of California". Carnegie Institute Washington Publication 319.

Cunningham, G.L., and Strain, B.R. 1969. "An Ecological Significance of Seasonal Leaf Variability in a Desert Shrub". *Ecology* 50:400-408.

Dansereau, P. 1951. "Description and Recording of Vegetation upon a Structural Basis". *Ecology* 32:172-229.

De Michele, D.W., and Sharpe, P.J.H. 1978. "A Parametric Analysis of the Anatomy and Physiology of the Stomata". *Agricultural Meteorology* 14:229-41.

Dolph, Gary E. 1978. "Variation in Leaf Size and Margin Type with Respect to Climate". *Couse Forschung — Institut Senckenberg* 30:153-58.

Ehleringer, J.; Bjorkman, O.; and Mooney, H.A. 1976. "Leaf Pubescence: Effects on Absorptance and Photosynthesis in a Desert Shrub". *Science* 192:376-77.

Ellenberg, H., and Mueller-Dombois, D. 1967. "A Key to Raunkiaer Plant Life Forms with Revised Sub-Divisions". *Berichte Geobotanische Institut ETH Stiftung Rubel, Zurich* 37: 56-73.

Easau, K. 1965. *Anatomy of Seed Plants.* New York: John Wiley.

FAO. 1976. *Forest Resources in the Asia and Far East Region.* Paris.

Ferri, M.G. 1961. "Problems of Water Relations of Some Brazilian Vegetation Types, with Special Consideration of the Concepts of Xeromorphy and Xerophytism". In *Plant-water Relationships in Arid and Semi-arid Conditions.* UNESCO Paris. *Arid Zone Research,* 16:191-97.

Fontaine, R.G.; Gomez-Pompa; and Ludlow, B. 1978. "Secondary Succession". In *Tropical Forest Ecosystems: A State of Knowledge Report.* UNESCO-UNEP Paris.

______, and Lanly, J.P. 1978. "Inventory and Survey: International Activities". In *Tropical Forest Ecosystems: A State of Knowledge Report.* UNESCO/UNEP/FAO. France.

Fosberg, F.R. 1970. "A Classification of Vegetation for General Purposes". In G.F. Peterken (comp.), *Guide to the Check Sheet for I.B.P. Areas*. I.B.P. Handbook 4.

Fox, J.E.D. 1973. "Dipterocarp Seedling Behaviour in Sabah". *Malaysian Forester* 36:205–14.

Gates, D.M. 1968. "Energy Exchange in the Biosphere". In F.E. Eckhardt (ed.), *Functioning of Terrestrial Ecosystems at the Primary Producing Level*. Proceedings of the Copenhagen Symposium. Paris. *UNESCO Natural Resources Research* 5:33–43.

Gates, F.C. 1914. "Winter as a Factor in the Xerophilly of Certain Evergreen Aricads". *Botanical Gazette* 57:445–89.

Gentry, A.H. 1969. "A Comparison of Some Leaf Characteristics of Tropical Dry Forest and Tropical Wet Forest in Costa Rica". *Turrialba* 19:419–28.

GIllison, A.N. 1970. "Structure and Floristics of a Montane Grassland/Forest Transition, Doma Peaks Region, Papua". *Blumea* 18:71–86.

———. 1975a. "A Review of Problems and Techniques in Restoring the Tropical Forest Ecosystem Once It Has Been Altered". In UNESCO M.A.B. report of symposium on *Ecological Effects of Increasing Human Activities on Tropical and Subtropical Forest Ecosystems* pp. 51–71. University of Papua New Guinea and Australian Government Publishing Service.

———. 1975b. "Vegetation Types of the New Hebrides, with Particular Reference to the Northern Islands". Mimeo.

———. 1981. "Towards a functional vegetation classification', in A.N. Gillison & D.J. Anderson (eds.), *Vegetation Classification in Australia*. CSIRO/ANU Press (Canberra) pp. 30–41.

———. 1982. "Tropical Savannas of Australia and the South-West Pacific". In F. Bourliere and D.W. Goodall (eds.), *Tropical Savannas*. Ecosystems of the World Series. Elsevier, Amsterdam. In Press.

Gillison, A.N., and Walker, J. 1981. "Woodlands of Australia". In a R.H. Groves (ed.), *Australian Vegetation*. Cambridge: Cambridge University Press. In press.

Gindel, I. 1969. "Stomatal Number and Size as Related to Soil Moisture in Tree Xerophytes in Israel". *Ecology* 50:263–67.

Givnish, T.J. 1978. "Ecological Aspects of Plant Morphology: Leaf Form in Relation to Environment". *Acta Biotheoretica, Supplement: Folia Biotheoretica* no. 7, pp. 83–142.

Givnish, T.J., and Vermeij, G.J. 1976. "Sizes and Shapes of Liane Leaves". *American Naturalist* 110:743–78.

Gomez-Pompa, A., and Vasquez-Yanes, C. 1974. "Studies on the Secondary Succession of Tropical Lowlands: The Life Cycle of Secondary Species". In *Proceedings of the First International Congress of Ecology*. The Hague.

Gomez-Pompa, A.; Vasquez-Yanes, C.; and Guevara, S. 1972. "The Tropical Rain Forest: A Nonrenewable Resource". *Science* 177:762–65.

Grieve, B.J. 1955. "The Physiology of Sclerophyll Plants". *Journal Royal Society Western Australia* 39:31–45.

Grieve, B.J., and Hellmuth, E.O. 1970. "Eco-Physiology of Western Australian Plants". *Oecologia Plantarum* 5:33–67.

Holdridge, L.R. 1947. "Determination of World Plant Formations from Simple Climatic Data". *Science* (N.Y.) 105:367–68.

Hooper, R. 1979. "Fires in the Northern Territory Environment and Implementation for Rehabilitation Programmes". Paper presented to the Northern Australian Mine Rehabilitation Workshop, Darwin, N.T.

Hopkins, M.S. 1981. "Disturbance and Change and the Resulting Problems of Functional Classification". In A.N. Gillison and D.J. Anderson (eds.), *Vegetation Classification in Australia*. Canberra: CSIRO/Australian. National University Press.

Hopkins, M.S., and Graham, A.W. 1981. "Structural Typing of Tropical Rainforest Using Canopy Characteristics in Low-Level Aerial Photographs: A Case Study". In A.N. Gillison and D.J. Anderson (eds.), *Vegetation Classification in Australia*. Canberra: CSIRO/ Australian National University Press.

Horn, H.S. 1970. *The Adaptive Geometry of Trees*. Princeton: Princeton University Press.

I.U.C.N. 1973. *A Working System for Classification of World Vegetation*. I.U.C.N. Occasional Paper no. 5, Morges, Switzerland.

Kuchler, A.W. 1949. "A Physiognomic Classification of Vegetation". *Annals of the Association for American Geographers* 39:201–10.

Kummerow, J. 1973. "Comparative Anatomy of Sclerophylls of Mediterranean Climatic Areas". In F. Di Castri and H.A. Monney (eds.), *Mediterranean Type Ecosystems: Origin and Structure*. New York: Springer.

Lamb, D. 1977. "Conservation and Management of Tropical Rain Forests: A Dilemma of Development in Papua New Guinea". *Environmental Conservation* 4:121–29.

Lanly, J.P. 1976. "Tropical Moist Forest Inventories for Industrial Development Decisions". *Unasylva* 28:42.

Lanly, J.P., and Clement, J. 1979. "Present and Future Forest and Plantation Areas in the Tropics". FO:MISC/79/1. Food and Agriculture Organization. Rome.

Larcher, W. 1975. *Physiological Plant Ecology*. Berlin: Springer-Verlag.

Letouzey, R. 1978. "Floristic Composition and Typology". In *Tropical Forest Ecosystems: A State of Knowledge Report*. UNESCO/UNEP. France.

Lewis M.C. 1972. "The Physiological Significance of Variation in Leaf Structure". *Science Progress*, (Oxford) 60:25–51.

Lloyd, N.D., and Woolhouse, H.W. 1978. "Leaf Resistances in Different Populations of *Sesleria caerulea* (L.) Ard". *New Phytologist* 80:79–85.

Loveless, A.R. 1961. "A Nutritional Interpretation of Sclerophylly Based on Differences in the Chemical Composition of Sclerophyllous and Mesophytic Leaves". *Annals of Botany New Series* 25:168–84.

Loveless, A.R. 1962. "Further Evidence to Support a Nutritional Interpretation of Sclerophylly". *Annals of Botany New Series*. 26:551–61.

Macfadyen, A. 1963. *Annual Ecology: Aims and Methods*. London: Pitman.

Maximov, N.A. 1931. "The Physiological Significance of the Xeromorphic Structure of Plants". *Journal of Ecology* 19:273–82.

Medina, E. 1975. "Dark CO_2 Fixation, Habitat Preference and Evolution within the Bromelicaceae". *Evolution* 28:677–86.

Medina, E.; Sobrado, M.; and Herrera, R. 1978. "Significance of Leaf Orientation for Leaf Temperature in an Amazonian Sclerophyll Vegetation". *Radiation and Environmental Biophysics* 15:131–40.

Meijer, W. 1970. "Regeneration of Tropical Lowland Forest in Sabah, Malaysia, Forty Years after Logging". *Malaysian Forester* 33:204–29.

Mooney, H.A. 1973. "Vegetation in Mediterranean Climate Regions. Sect. III: Introduction". In F. di Castri and H.A. Mooney (eds.), *Mediterranean Type Ecosystems: Origin and Structure*. New York: Springer.

Morain, S.A., and Klankamsorn, B. 1978. "Forest Mapping and Inventory Techniques through Visual Analysis of Landsat Imagery: Examples from Thailand". In *Proceedings of 12th International Symposium on Remote Sensing of the Environment, Manila, Philippines, 1978*. Vol. 1. Centre for Remote Sensing Information and Analysis, Environmental Research Institute of Michigan, Ann Arbor, Michigan.

Mulroy, T.W. 1979. "Spectral Properties of Heavily Glaucous and Non-Glaucous Leaves of a Succulent Rosette-plant". *Oecologia*, 38:349–57.

Myers, N. 1980. *Conversion of Tropical Moist Forests*. National Academy of Sciences. Washington.

Noble, I.R. 1981. "The Use of Dynamic Characters in Vegetation Classification". In A.N. Gillison and D.J. Anderson (eds.), *Vegetation Classification in Australia*. Canberra: CSIRO/Australian National University Press.

Oliver, J. 1979. "A Study of Geographical Imprecision: The Tropics". *Australian Geographical Studies* 17:3–17.

Paijmans, K. 1975. *Explanatory Notes to the Vegetation Map of Papua New Guinea*. CSIRO Australian Land Research Series no. 35.

Parkhurst, D.F., and Loucks, O.L. 1972. "Optimal Leaf Size in Relation to Environment". *Journal of Ecology* 60:505–37.

Parkhurst, D.F. 1978. "The Adaptive Significance of Stomatal Occurrence on One or Both Surfaces of Leaves". *Journal of Ecology* 66:367–83.

Patton, R.T. 1933. "Ecological Studies in Victoria: The Cheltenham Flora". *Proceedings of the Royal Society of Victorian New Series*, vol. 45, Pt. 2.

Poore, Duncan. 1976. "The Values of Tropical Moist Forest Ecosystems and the Environmental Consequences of Their Removal". *Unasylva* 28:127–46.

Pringle, S.L. 1976. "Tropical Moist Forests in World Demand, Supply and Trade". *Unasylva* 28:18–25.

Raschke, K. 1969. "Heat Transfer between the Plant and the Environment". *Annual Review of Plant Physiology* 11:111–26.

Richards, P.W. 1955. "The Secondary Succession in the Tropical Rain Forest". *Science Progress* 43:45–47.

Richards, P.W. 1966. *The Tropical Rain Forest*. Cambridge: Cambridge University Press.

Rollet, B. 1978. "Organization". In *Tropical Forest Ecosystems: A State of Knowledge Report*. UNESCO-UNEP. France.

Routley, R., and Routley, V. 1977. "Destructive Forestry in Australia and Melanesia". In J.H. Winslow (ed.), *The Melanesian Environment*, Canberra: Australian National University Press.

Sayn-Wittgenstein, L. 1978. "New Developments in Tropical Forest Inventories". In *Proceedings of the 12th International Symposium on Remote Sensing of the Environment, Manila*, Philippines, *1978*. Centre for Remote Sensing Information and Analysis, Environmental Research Institute Michigan.

Schimper, A.F.W. 1903. *Plant Geography on a Physiological Basis*. Hist. Nat. Class., vol. 2. English Translation by Codicote, Wheldon and Wesley (repr. Weinheim, Cramer, 1964).

Shaw, D.E. 1968. "Lightning Strike of *Cacao* and *Leucaena* in New Britain". *Papua New Guinea Agricultural Journal* 20:75–84.

Shields, L.M. 1950. "Leaf Xeromorphy as Related to Physiological and Structural Influences". *Botanical Review* 16:399–487.

Sommer, A. 1976. "Attempt at an Assessment of the World's Tropical Moist Forests". *Unasylva* 28:5–24.

Specht, R.L. 1970. "Vegetation". In *The Australian Environment*. 4th ed. Melbourne: CSIRO and Melbourne University Press.

Specht, R.L. 1981. "Foliage Projective Cover and Standing Biomass". In A.N. Gillison and D.J. Anderson (eds.), *Vegetation Classification in Australia*. Canberra: CSIRO/Australian National University Press. In press.

Steenis, C.G.G.J. van. 1959. "Tropical Lowland Vegetation: The Characteristics of Its Types and Their Relation to Climate". In *Proceedings of the Ninth Pacific Science Congress*, Bangkok.

Stocker, G.C., and Mott, J.J. 1981. "Fire in the Tropical Forests and Woodlands of Northern Australia". In A.M. Gill, R.H. Groves, and I.R. Noble (eds.), *Fire and the Australian Biota*. Canberra: Australian Academy of Science.

Symington, C.F. 1933. "The Study of Secondary Growth on Rain Forest Sites in Malaya". *Malaysian Forester* 2:107-17.

Taylor, S.E. 1975. "Optimal Leaf Form". In D.M. Gates and R.B. Schmere (eds.), *Perspectives of Biophysical Ecology*. Ecological Studies, vol. 12. Berlin: Springer.

Thoday, D. 1933. "The Terminology of 'Xeromorphism'." *Journal of Ecology* 21:1-6.

Usher, M.B. 1973. *Biological Management and Conservation*. London: Chapman and Hall.

Walker, J., and Gillison, A.N. 1981. "Australian Savannas". In B.J. Huntley (ed.), *Dynamics of Savanna Ecosystems*. UNESCO, Scope. Pretoria.

Walter, H. 1979. *Vegetation of the Earth*. 2nd ed. New York: Springer.

Wassink, E.C., and Van Den Noort, M.E. 1976. "Note on Leaf Area Index in a Solitary Plant". Medelingen landbouwhogeschool, Wageningen, Nederland, 76-13.

Webb, L.J. 1959. "A Physiognomic Classification of Australian Rainforests". *Journal of Ecology* 47:551-70.

Webb, L.J.; Tracey, J.G.; and Williams, W.T. 1976. "The Value of Structural Features in Tropical Forest Typology". *Australian Journal of Ecology* 1:3-28.

Webb, L.J. 1977. "The Dynamics of Tropical Rainforests". *The Forestry Log* 10:11-18.

White, K.J. 1975. *The Effect of Natural Phenomena on the Forest Environment*. Department of Forests. Port Moresby.

Whitmore, T.C. 1974. "Change with Time and the Role of Cyclones in Tropical Forest of Kolombangara, Solomon Islands", *Commonwealth Forestry Institute Paper* no. 46, 73.

Whitmore, T.C. 1975. *Tropical Rain Forests of the Far East*. Oxford: Claredon Press.

APPENDIX 11.1
VEGETATION STRUCTURE
FIELD SHEET (A) STRUCTURAL MODI

Site No. .... Lat. .... Long. .... Alt. .... (m) Date (   )

STRATUM	HT m	F.P.C. %	CORE UNIT				SPECIES	CODE
			LEAF SIZE	LEAF ANGLE	LEAF TYPE	STRUCT. UNIT		
Em								
PRIM.	18	75	meso	latero	ortho	phan	Alstonia sp.	ALS. SP.
			platy	"	"	phan	MALLOTOS MOLISSIMUS	MALMOL
			"	"	"	"	MALARANGA ALEURITOIDES	MALALE
S1	2	25	NOTO	latero	ortho	REPTO PHAN	LANTANA LAMARA	LANCAM
S2								
S3								
TOTAL spp 4	TOT. WOODY 4		TOT. HERB —		TOT. GRAM —		TOT. MODI 3	spp/MODI 4/3	BASAL AREA 10

REMARKS REGENERATING RAINFOREST SITE FOLLOWING LOGGING   W.F. = (1)

Slope.......... 8° .... Aspect......... S.W. .... Soil Type... KRASNTOZEM ... Rock... BASALT ....

APPENDIX 11.1 (*cont.*)

Field Sheet (A) *STRUCTURAL MODI FOR CORE UNIT*

*Leaf size**

1. pico (up to 0.01 cm^2)
2. lepto (0.10 – 0.25 cm^2)
3. nano (0.25 – 2.5 cm^2)
4. micro (2.5 – 20 cm^2)
5. noto (20 – 45 cm^2)
6. meso (45 – 150 cm^2)
7. platy (150 – 750 cm^2)
8. macro (750 – 1640 cm^2)
9. mega ($>$1640 cm^2)

Leaf types†

1. ortho — simple bilateral
2. iso — isobilateral
3. phyllodo — phyllode or phyllode-like (e.g. *Melaleuca viridiflora*)
4. clado — caldode
5. gramo — graminoid or grass-like
6. erico — ericoid
7. solido — of solid form e.g. *Callitris*
8. tereto — cylindrical in X-section (takes precedence over solido if used exclusively)
9. suculo — succulent leaves
10. aphyllo — aphyllous

STRUCTURAL UNITS

1. -phan — phanerophyte‡
2. -cham — chamaephyte‡
3. -hemicrypt — hemicryptophyte‡
4. -crypt — cryptophyte‡
5. -there — therophyte‡
6. -pter — pteridophyte
7. -bryod — bryophyte
8. -thallod — thallophyte

prefi types

1-epi-(epiphyte)
2-para-(parasite)
3-sapro-(saprophyte)
4-repto (reptant)
5-rosu (rosulate, e.g., palm)
6-liano (vine-like)

Notes: *see diagram for field comparison (mature ls. only)

-suggic "ic" is added if leaflet (not used in analysis)

†scale leaves are ignored

‡after Raunkiaer

12
Desertification of the Sahel

D.O. ADEFOLALU

"The desert strikes again"; "Death as Sahara Creeps on". Such were the headlines (in the wake of the 1972 drought) during the middle of 1973, of articles and radio programmes describing the drought across a belt of West Africa, 4,000 km long and up to 1,000 km wide, where 25 million people depend upon regular rainfall for agricultural practice. E.G. Davy (1974)

The 1972/73 drought in West Africa, now referred to as the "Sahelian" drought, brought disaster to at least six million people in that zone of the tropical sub-region called the Sahel (Fig. 12.1). The drought was believed to have started in 1968 (Glantz 1975). Six countries were said to be affected — Senegal, Mali, Mauritania, Upper Volta, Niger, and Chad. Focusing attention only on the climax year 1973 has resulted in attention being given to some countries which perhaps were suffering from perennial economic problems that were aggravated by the drought and no attention being given to other countries within the same climatic region.

The immediate concern was to provide relief measures for the most affected countries, but no attempt was made to trace the root causes of the

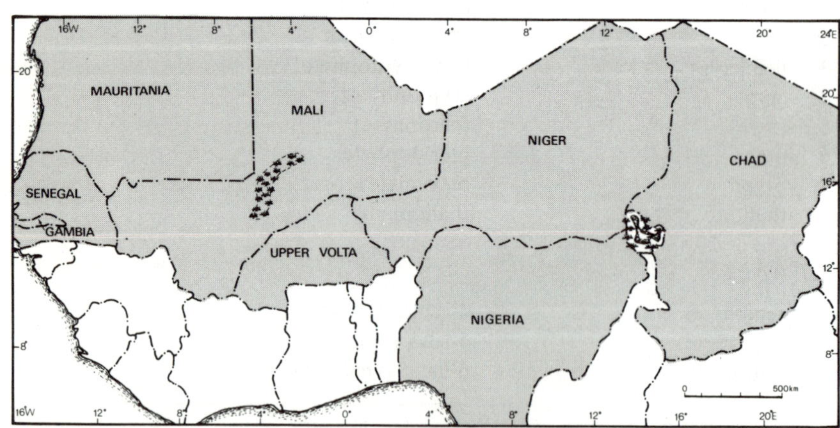

Fig. 12.1 Countries in the Sahel-Sudan zone of West Africa

402

drought. It was not until later (1974/75), when there was a reversed climatic situation (with above normal rainfall) in the same area and reports of flooding in countries which were still struggling with the aftermath of the 1972/73 drought, that there was general awareness of the linkages between these climatic events and the deterioration of the land resources of the Sahel countries (Wade 1974).

Rainfall and Drought

What first aroused interest in the Sahelian drought was a drop in agricultural production and the consequent socio-economic impact. As there was a corresponding drop in rainfall, it was easy to blame the failure of the monsoon for, rather than associate it with, the diminished productivity (Winstanley 1973a; Davy 1974).

Glantz (1976) has enumerated "nine fallacies of natural disaster" with respect to the Sahel. One of them is:

When the rains come, everything will return to normal.

The failure of the land to respond to the abundant 1974/75 rains that fell in the Sahel illustrates the pertinence of Glantz's observations. As Baker (1974) has pointed out: "There is . . . a great danger that when the rains come in the Sahel, and the millet grows again, the 'problem' will be considered over until the next time."

The emphasis has always been on rainfall deficiency as the cause of drought and of poor crop yields. Desertification is thought of as being a physical advancement of the existing Sahara desert. The 1974/75 situation which saw rainfall above normal but poor response by the overburdened vegetative cover indicates that the Sahel seems to be undergoing some irreversible desertification process not connected with the lack of rainfall. This is illustrated in Figure 12.2 for Niamey in Niger. It is obvious that considered either by monthly or annual values, the 1972/73 episode was not the worst in history and was only one of many such drought situations. The only significant difference between this episode and other extreme situations such as those of 1944 and 1963 is the duration of the drought.

Socio-Political Implications

There is doubtless a link between drought occurrence, food production declines, and general socio-economic instability in Africa south of the Sahara in the past two decades. To the extreme east of the Sahelian zone, Ethiopia overthrew its monarchy by a military coup d'etat in 1976 after what has been described as "political instability, economic ruin, social disorganization, abject poverty among 99 per cent of the people . . ." (Obisesan 1981). A similar take-over had occurred in the Sudan in 1969

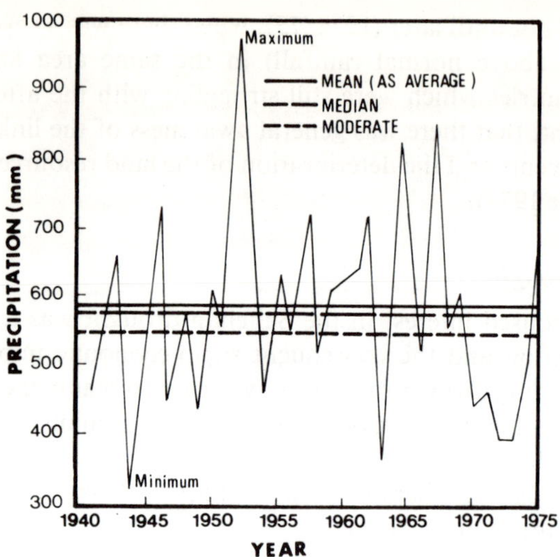

*Fig. 12.2 Annual precipitation at Niamey-Niger
(from Glantz, 1975).*

when the basic food crops such as millet and sorghum were in short supply
due to the emphasis on cotton production. Such agricultural policies affected
other countries such as Senegal, Mauritania, Mali, Niger, and Nigeria
where efforts were concentrated on groundnut and soybean production for
export. Nigeria's abandonment of basic agriculture may be attributed to its
oil boom (Adefolalu 1976).

The Place of Research

Fundamental research conducted in other areas has revealed the intricate
and complex nature of climate-land-man relationships. The approach to the
problems of the Sahel should be through such research and not through
seeking short-term solutions. In this respect, the World Meteorological
Organization (WMO) and other United Nations agencies and scientific
groups have played important roles in bringing to focus the effect of weather
and climate on plant crops and animal production in tropical countries. In
particular, the WMO has, through its Commission for Agrometeorology,
suggested that long-term fluctuations of climate have a greater impact on
agricultural production than short-term ones. The point has also been made
that there are basic differences between drought and desertification, especially in semi-arid regions where the latter is symptomatic of serious land-
soil degradation. This viewpoint is supported by other organizations such as
the International Federation of Institutes for Advanced Study (IFIAS)

which are seeking a lasting solution to the Sahelian zone desertification problem.

The Sahel

Definitions of the Sahel have been given by various authors. These are based on its political location, physiography and climatology. The Sahel is in general a zone which lies almost entirely between latitudes 23½ °N and 10 °N, extending from the west coast of Africa to central Africa and covering an area of 5 million km² (Fig. 12.1).

The word "Sahel" is now commonly used to mean that zone most affected by the 1972/73 drought. Except for Nigeria, the countries shown in Figure 12.1 are known as the "Sahel States" (Glantz 1976). They are all Francophone countries with a common lingua franca. More than 100,000 people and up to 40 per cent of livestock died during and immediately after the 1972/73 drought in these countries.

In this arid sub-region of Africa, cattle is the main export commodity. The pre-independence political situations in these countries encouraged nomadism, and there was no restriction on the movement of people (Grove 1967). But after most of these countries gained independence, they began to foster national consciousness and a sense of belonging to a particular country rather than a race.

It must be noted that two main ethnic or tribal groupings are dominant in the sub-region — Hausas and Fulanis. Both are herdsmen leading a nomadic type of life (Murdock 1959). With independence, the unrestricted movement of people and livestock came to an end, mainly for population control, revenue collection, and national boundary adjustments for exploitation of the land and its resources.

The immediate effects of such restrictions were border clashes and struggles for fertile areas. Nomads and herdsmen were now compelled to become sedentary crop farmers, keeping their herds in the vicinity of cultivated lands. They have had to abandon their former pattern of cattle-rearing (Boudet *et al.* 1976).

The main crisis confronting the so-called Sahelian States was therefore political instability (Lofchie 1971) caused by nomads and the sedentary farmers competing for meagre resources for their cattle. This in time has led to overgrazing, recognized as one of the most important factors in the desertification of the Sahel.

Recognizing the problem, the United Nations General Assembly by resolution 3337 (XXIX) commissioned studies that led to projects on the management of cattle and rangelands to combat desertification in the Sudano-Sahelian region. One such project has been code-named SOLAR

(Boudet *et al.* 1976). But when the Inter-Governmental Permanent Committee of the States (Chad, Mali, Mauritania, Niger, Senegal, Sudan, and Upper Volta) was established after the 1972/73 drought, its major task was to seek drought control measures. The committee was concerned with anti-desertization rather than anti-desertification.

Hence, while the U.N. Project was aimed at an idealized solution for the entire sub-region, the regional committee appeared not to have grasped the basic problem. It instead sought foreign aid for immediate relief. However such a measure was a total failure as pointed out by Wade (1974). It only worsened the political instabilities as some political leaders became corrupt. As Glantz (1976) has noted: ". . . when corruption in his administration had been exposed by a reporter, President Tumbalbaye of Chad (since deposed) said that if he had to take insults in order to get American aid, then he would not eat. What he meant was that if American reporters continued to expose corruption in his government (corruption in which his wife was heavily implicated), then he would cut off relief food shipments to his people, and therefore, his people would not eat."

It is obvious that the approach adopted could not contribute to a meaningful assessment and solution as desertification continued unabated. But it pleased the Western countries as they were able to get rid of their surplus grain. In many cases, the donor countries also supplied the ships and aircraft for transporting the food to the drought-stricken countries. All these had been paid for by the U.N. agencies and other organizations who acted as intermediaries (IFIAS 1977). Compared with the total estimated cost of US\$22 billion for the five-year SOLAR project and other related activities, the relief aid could have cost close to that amount.

Physiographic Classification

Classification of the Sahel from the point of view of exploitation of natural resources for agriculture should be based on physical, biological and human activities (Davy *et al.* 1976:6). This is especially relevant for the West African sub-region where emphasis has hitherto been placed on isohytal sub-divisions with annual rainfall as the only common denominator. This is further used to delineate various sub-zones in terms of eco-climate.

Rainfall Zonation

The procedure is illustrated in Figure 12.3. Here the Sahel is defined or classified as the sub-region of West Africa which receives about 300 to 650 mm mean annual rainfall (Boudet *et al.* 1976, Davy *et al.* 1976; Glantz 1975; 1977). It is bordered to the north by another zone of sub-desert (200 mm) and to the south by the Sudan which receives between 600 and 900 mm rainfall annually. It should be noted that north of the 500 mm isohyet only

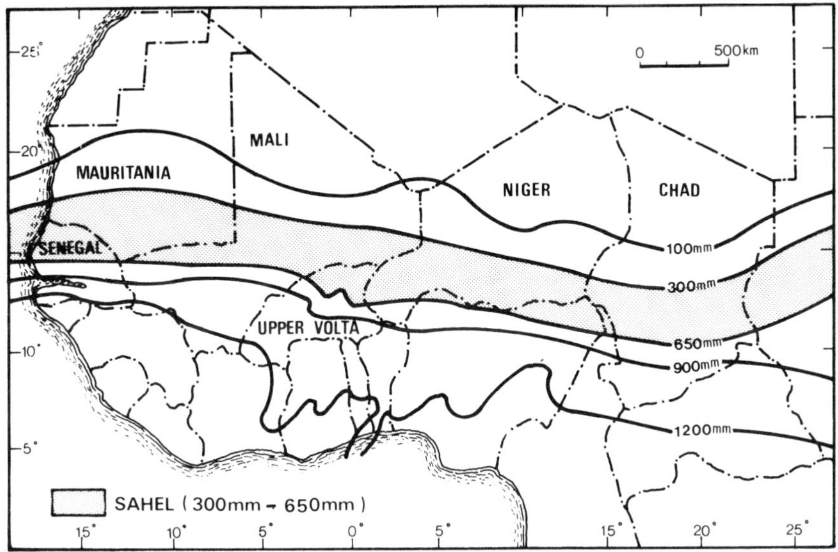

Fig. 12.3 Classification of the Sahel based on precipitation

irrigated crops can be grown. The patterns of rainfall distribution during the rainy season has been ignored here. This is recognized as a major problem in rainfall zonation in semi-arid zones (Kowal and Adeoye 1973; Gibbs 1975; Odingo 1976; Adefolalu 1977).

As Horowitz (1972) has stated, rainfall may be evenly distributed throughout the four months of the rainy season in the Sahel (June to September/October), or it may be concentrated in a few intense periods. For instance, in 1968 and 1969, rainfall amounts of 550 to 600 mm were about average in Niamey (see Fig. 12.4), but earlier than normal rains led the grass to germinate early. The immature seedlings wilted and were subsequently scorched by the sun when the rains ended prematurely.

It can be seen that rainfall amount is not ideal for classification (FAO 1967) as there is no direct correlation between the response of rapidly deteriorating and highly depleted soil and vegetation cover and mean rainfall, especially in semi-arid regions. The pattern rather than amount of rainfall affects the type of soil and vegetation in any region (Odingo 1976:5). Super-imposed on this is the constraint posed by increasing population which, as in the Sahel, depletes the natural vegetation to satisfy energy requirements and thus cause the land-soil surface to become exposed.

Vegetation

Natural vegetation is influenced by climate but human intervention has caused major modifications to the environment (Grove 1967:95; Davy *et al.*

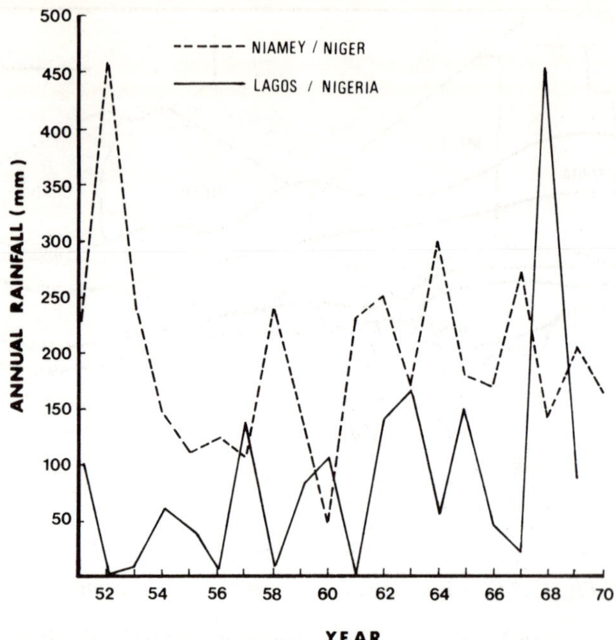

*Fig. 12.4 Monthly rainfall in August at Lagos and
Niamey (from Adefolalu, 1973)*

1976). Using rainfall amounts, Trochain (1940; 1952) classified the Sahel as belonging to a Sudan region. There are three sub-zones: the Sahel-Sahara (see Fig. 12.5), the Sahel-proper, and the Sahel-Sudan. A previous classification by Chevalier (1933) recognized only two broad zones — the Sahel and Sudan. Modification of the ecosystem may lead to the Sahel-Sudan sector becoming Sahel-proper (Le Houerou 1976).

The Sahel-Sahara has been described as sub-desert with open steppe. Spiny trees and shrubs are widely spread out but patches of herbaceous species are frequent in sandy soils. Grasses (which are seasonal) flourish for a few weeks after rain. Rainfall is generally between 100 and 200 mm annually.

The Sahel-proper is described as wooded steppe — a transition zone between the desert and sub-desert types to the north and moister woodland to the south, respectively. Taller, widely scattered plants or patchy closed woodlands abound. Grasses are also annual but constitute the bulk of rangeland production. The steppe vegetation is xerophilous with thorn-bearing woody species. Precipitation may range between 200 and 600 mm a year.

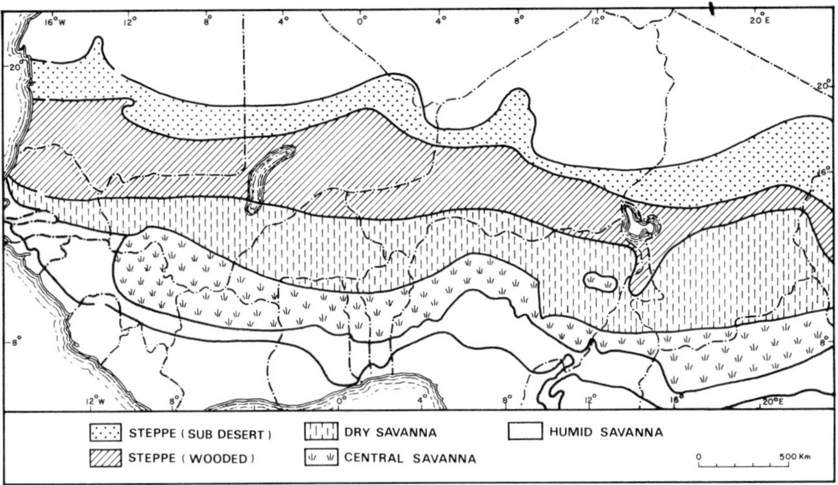

Fig. 12.5 Classification of the Sahel, based on vegetation

The Sahel-Sudan zone is, in general, a closed woodland with little grass (Schell 1970). But there are wide areas without trees in the whole zone. It has been suggested that the wooded savanna has given way to this abundant grass-growth as a result of bush fires and the use of wood for domestic fuel (Grove 1967). The vegetation is mesophilic steppe in which the annual grasses have broad flat leaves and the woody species are thornless, broad-leaved and deciduous. Rossetti (1965) found that, although they belong to the Sahel-Sahara and Sahel, some areas in Mauritania and Mali which have coarse sandy soils (dunes) have dry savanna vegetation. Here precipitation ranges from 400 to 600 mm a year, but where the drainage is poor, wooded savanna may prevail.

It is recognized that the natural vegetation has been considerably altered in the recent past (Odingo 1976:10). Human population in this zone continues to deplete the land productivity potential. This compounds the problem posed by geological substrate, topographic relief, and limited water resources. It has also been pointed out that deterioration of the land resulting from overuse has contributed to soil erosion with a consequent increase in albedo, a major factor in desertification processes (Otterman 1974).

That the present vegetation in the Sahel zones is poorer than that existing fifty years earlier (Chevalier 1933; Thompson 1965) is therefore not sur-prising. If the trend is not checked, the present wooded savanna may turn into Sahel-proper before the end of this century.

Soils of the Sahel

Classification of the Sahel into zones based on soil types is not easy. In fact, it is not possible to make distinctions in a zonal sense. According to earlier classifications (see Trewartha 1961; Chevalier 1933), the Sahel, lying between latitudes 10° and 23°N, belongs to desert and sub-tropical soil regions. Desert soils, made up mainly of sands, have a highly coarse texture and low fertility. Owing to extensive wind erosion, the surface layers are subject to lateral movements, resulting in the formation of sand dunes. This type of soil can only support scattered herbaceous plants although further south patched woodlands appear. The sub-tropical soil type to the south is in areas with higher rainfall. Such soils have a high alkaline content. They support grasses which are taller and more extensive where the rain is heavy. The above classification was valid only up to the early part of this century as over-use and over-exposure have led to significant changes.

The results of recent in-depth studies as documented by the FAO (1974) in its *Soils Map of Africa* are summarized in Figure 12.6. The complexity of the soil pattern is obvious. Prominent among the soils are the three main types of desert soils. The first is Lithosols (I), found north of latitude 20°N. These are shallow soils with hard rock. The broken-down rocks which form sandy soils are known as Yermisols (Y) and Regosols (Re). The Yermisols are without organic substance and have little or no development. The Regosols have some loose materials overlying the core but also have little or no development. These two soils are typically found in northern Niger and Chad (18° to 23°N). Within the same latitude band, Mauritania and

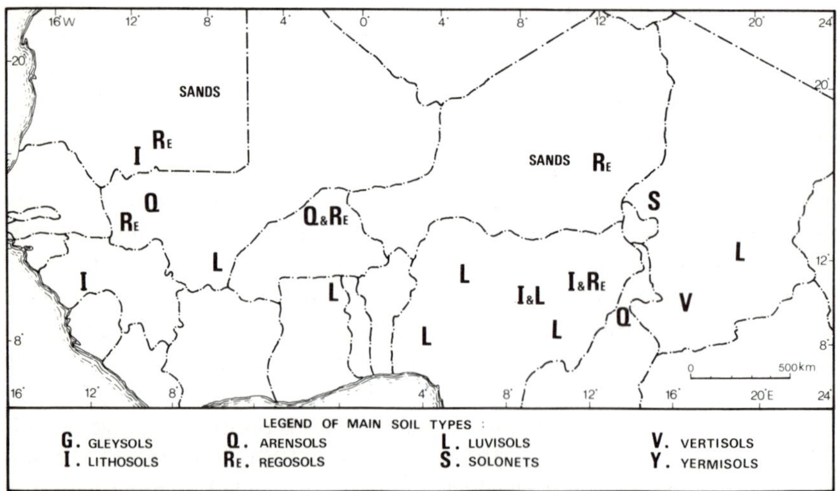

Fig. 12.6 Classification of the Sahel, based on soils

portions of Western Mali are characterized more by sand dunes of finer sand particles but are of the same Y and Re basic structure.

The second main soil type which has a more or less unbroken zonal distribution within latitudes 12° to 16°N are Arensols (Q) found along the area of transition between Sahel-desert and Sahel-Sudan environments. Such soils are coarse in texture with weak development. Between latitude 16° and 18°N vertisols (V-soils generally rich in clay at 1 to 2 m depth) are found in the central parts of Mali, Niger and Western Chad. While most of the Q-soils are found in association with patches of sterile Regosols (especially in Chad), the typical areas of fertile Gleysols (G-soils, with temporary or permanent excess of water) are found in the inner delta of River Niger. Degraded saline soils (Solonets — S) are typical of areas of transition as on the shores of Lake Chad.

The third main soil type which is found between about 10° to 14°N is the Luvisols (L) commonly formed as an association between poorly-drained Lithosols and Regosols. They are therefore subject to leaching. They have a light texture on the surface and either red-clay mottles in depth or a high iron content (Ferric Luvisols). These types are common in Senegal and Gambia.

One of the extremely sensitive areas in the Sahel is Chad where the most infertile soils are found, while the richest alluvium deposits are in the Niger along the river Niger basin. Yet the Chad basin is the subject of inter-governmental agricultural development. The question that needs to be answered is, "Is agriculture really possible in the Sahel?" Agricultural yields were low during the 1972–73 period of drought as well as the 1974/75 period of flood. Destructive agricultural techniques for large scale cash crop farming with foreign aid have led to an irreversible trend in the carrying capacity of the land. They have led to soil erosion, land deterioration and depletion of sub-surface water content on the one hand (Boudet *et al.* 1976) and higher albedo on the other hand. These have contributed to desert encroachment by systematically increasing the aridity of the Sahel (Otterman 1974). What then is drought and how is it different from desertification?

The 1972/73 Drought

The dry spell in the Sahel zone which occurred between 1968 and 1972 was a period when the mean annual rainfall was about half the value for the period between 1908 and 1956 (Lamb 1979). This is an indication of climatic change due to "increasing drought" which might have started as far back as 1955 (Winstanley 1973a; b). However, the pattern of rainfall in Figure 12.2 (for Niamey—Niger Republic) from 1940 to 1975 does not provide support for that supposition. Rather it emphasizes the point made by Odingo (1976) that variability in amounts received is large and that "even for the small amounts expected, the probability that they will not be obtained every five years out of ten is very high".

To assert that a dry spell constitutes a drought situation, it is necessary to define "drought". It will then be possible to ascertain what really happened in 1972/73 and if this was indeed a drought period.

Drought: Definition and Indicators

A general definition of drought is "a situation when the demand for water exceeds the supply". It is thus a supply and demand phenomenon. There are at least three types of drought (Palmer 1964; Saarinen 1966; Gibbs 1975). *Meteorological* drought is defined as occurring when the amount of rainfall received is less than a percentage of a long-term average. As indicated in Figure 12.2 for Niamey, all years with rainfall below the mean may be termed drought years. *Agricultural* drought relates to seasonal vegetation development directly. It may be defined as a situation when the demand for water by plants is not met. *Hydrological* drought relates to underground water. If there is a situation whereby infiltration of water to replenish the water table — perhaps already depleted due to excessive use — does not take place to an extent that the water table continues to fall, a hydrological drought may be said to occur.

Of these meteorological drought is easiest to explain because of the use of simple statistical values such as the mean, median, and mode. From the mean, anomalies can be calculated as deviations from the mean and can be used to illustrate the distribution pattern of rainfall over a period. In the case of normal distribution (and rainfall is not, being non-continuous in time and space), it is possible to estimate dispersion such as variance and standard deviation and thus describe the distribution. With these two parameters, it is also possible to calculate the probability of occurrence of values within any given range. However, rainfall is treated as obeying Gausian normal distribution if annual totals are used. Gibbs and Maher (1967) have advocated a system which uses the limits of each ten per cent (decile) to the distribution. Thus the first decile is that rainfall amount which is not exceeded by ten per cent of totals and so on. The fifth decile or median then represents the amount not exceeded on 50 per cent of occasions. Decile values give a good description of the rainfall pattern especially with respect to departure from the average. For example, decile range one suggests an abnormally dry spell, decile range 10 gives abnormally wet conditions (Gibbs 1975:15).

Agricultural drought suggests that reduced availability of water may result in a modification of demand. There are basically two requirements with respect to plants — water demand to avoid wilting (i.e., survival level) and water demand to produce maximum yield. Rainfall which may produce poor crop yield obviously results in drought, but the amount below which a crop is considered drought-stricken will depend on the degree to which reduced yield can be tolerated. In such a case, the ability of plants to resist drought is crucial. May and Milthorpe (1962) have classified plants into three categories for this purpose:

Drought escape	—	ability to complete the life cycle before being subjected to serious water stress.
Drought endurance	—	with high water content implies capacity to survive drought by virtue of a deep root system or reduced transpiration.
Drought endurance	—	with low internal water content during periods of drought but ability to recover and grow rapidly when soil water is replenished.

For cultured plants such as millet and sorghum, the most important factor in drought resistance, especially in seasonally arid areas such as the Sahel, is the ability to mature before the onset of drought during the rainy season. For perennial natural vegetation, it is a combination of the other two drought-resistance attributes that is responsible for survival. These plants are acclimatized and are usually not drought-prone.

But once the perennial vegetation has been cleared for cultured vegetation, which may (in the case of sorghum or millet) or may not (in the case of cotton) be able to adjust to extreme weather events, the area becomes much more drought-prone. Prolonged dry periods within a rainy season adversely affects cultivated crops or cultured plants more than perennial vegetation in general.

Water needs in semi-arid and arid regions are acute because of low precipitation and excessive evapotranspiration. There is thus a deficit in available soil moisture for plants and water for animals. Deep wells and bore holes are drilled to bring underground water to the surface for use. With time and in persistent meteorological drought situations, the amount of water extracted from sub-surface storage could become excessive and big surface reservoirs such as lakes may lose water to replenish the underground storage in such areas. This has been the case in Chad where Lake Chad has shrunk to its lowest size in the past 22,000 years owing to the depletion of underground water and consequent lowering of the water table in areas surrounding the lake. Compared with its areal extent covering 10 °N, 16 °E, 13 °N, 11½ °E, 18 °N, 14 °E, and 18 °N, 20 °E about 22,000 years ago, its present size of less than 200 km² is only a small fraction (Grove 1967) and can only suggest that the Lake Chad area is undergoing a desertification process.

Therefore, "desertification" as distinct from "desertization" (result of persistent drought) is more or less an irreversible reduction of the vegetation cover due to a series of human activities. Persistent drought implies the disappearance of vegetation resulting from an increase in the aridity of climate (Boudet *et al*. 1976). A common feature of all types of drought is the lack of moisture by way of atmospheric precipitation. Meteorological drought, by definition, is repetitive in nature; its influence is therefore reflected by the length of individual dry spells or dry years. Continuous dry spells will only result in agricultural and hydrological drought.

What Happened in 1972/73

During normal rainy seasons (May to October), moisture-bearing south-westerly winds from the Atlantic Ocean and the Congo basin push the surface Inter-Tropical Discontinuity (ITD) over the Sudan-Sahel zone to its most northly position (22° to 25°N) in August. The instability to the south of the ITD associated with travelling perturbations (cyclonic circulation systems) causes the rain on which the entire area depends. The mean monthly rainfall patterns in West Africa show that the bulk of the annual rainfall is received during those months and as illustrated by Table 12.1 maximum precipitation is received in August (Oguntoyinbo *et al*. 1978).

In 1972, total rainfall amounts were, on the average, 50 to 75 per cent of the annual. But these are to be expected as the zone termed Sahel-proper lies within the broken lines where the 80 per cent probability values are only half and three quarters of normal or average annual rainfall (Fig. 12.7). In particular, the 75 per cent line to the south runs almost among the 500 mm annual average isohyet in Senegal in the west and Chad in the east. In eastern Mali and western Niger, it is nearer the 400 mm annual average isohyet, indicating a lower risk of seriously dry years in the centre of the Sahel zone, compared with the west and east.

Although the lowest percentage of 44 was estimated for areas north of 13° to 14°N, it was about 68.5 per cent at 12°N and 80 per cent at 11°N. Further south of 10°N, rainfall ranged between 80 and 100 per cent of the average. These suggest that there was no serious deviation from expected departure from normal in the Sahel where in every five out of ten years, a departure of 50 per cent from expected mean rainfall is a normal situation. Thus relying on the amount of rainfall recorded could, in the case of 1972/73, lead to the conclusion that the rainfall did not fail to such a degree as to be responsible for the general drought conditions. In that case, it was simply not a mere meteorological drought situation.

It was found that in 1972 the onset of the rains north of about 12°N did not differ from average dates (Kowal and Adeoye 1973; Apeldoorn 1977). However, the retreat at the end of the rainy season started earlier than

TABLE 12.1 COMPARISON OF RAINFALL OF FIVE "SYNOPTIC-TYPES A1 IN AUGUST 1958 WITH THE TOTAL MONTHLY AND ANNUAL RAINFALL AT FIVE STATIONS WITHIN THE SAHELIAN ZONE

Synoptic Type A1	Rainfall (MM)				
Period	Stations				
Days	Fort Lamy	Kano	Niamey	Bamako	Dakar
31 July – 4 Aug.	43	04	71	24	56
2 Aug. – 7 Aug.	43	28	21	35	103
10 Aug. – 15 Aug.	33	41	48	09	26
15 Aug. – 20 Aug.	26	24	39	36	95
18 Aug. – 23 Aug.	25	37	42	11	89
Total	170	134	221	116	369
Total for August	194	258	243	405	493
Mean (Annual)	650	873	584	750	540
Percentage of Rainfall due to Synoptic-Type A1	87	51	90	28	74
Percentage of Annual Rainfall in Aug. 1958	30	30	41	53	90

Mean percentage of the monthly rainfall due to synoptic type A1 in the sub-region is 66 per cent and that of August in relation to annual and amount is 49 per cent. . . .

normal, thus reducing the length of the rainy season (usually referred to as length of the growing season). This is illustrated in Figure 12.8. Another finding was that the rate of retreat, estimated as 8.5 days per degree latitude (or 14 km per day), was faster than the mean 10 days per degree (or 11 km per day) as estimated from previous studies (Clackson 1958; Adejokun 1966; Dhonneur 1971). This implies that for a place such as Kano (12° 08 ′N) the normal rainy season in addition to being shorter by about two weeks (Table 12.2) was less effective due to the faster retreat of the ITD.

Mean patterns of longest and shortest rainy (growing) seasons are illustrated in Figure 12.9. Compared with the situation in 1972 (Fig. 12.8), they confirm that the situation in 1972 was worse than the expected worst situation, especially in areas north of the 130-day isoline in Figure 12.8. This is indicative of the severity of the drought in this part of Sudan-Sahel zone of northern Nigeria. Figure 12.10 illustrates the extreme situation when the latest and earliest dates of onset and cessation of the rains are used, respec-

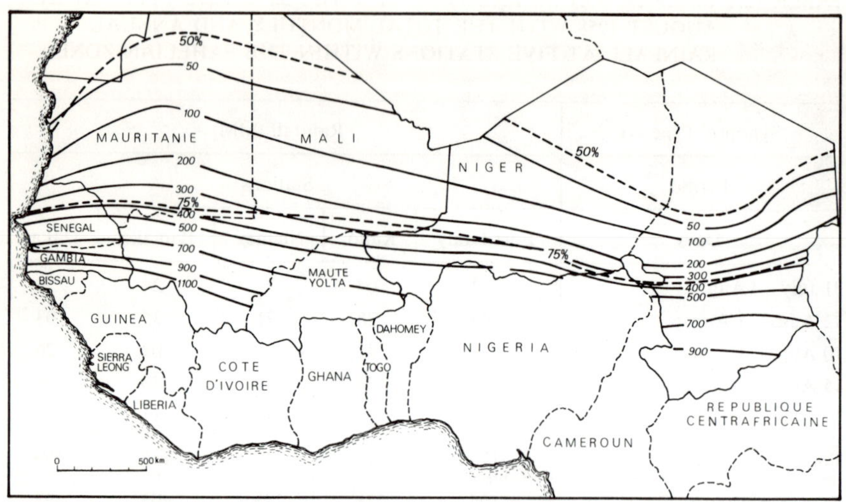

Fig. 12.7 Isolines of rainfall amounts exceeded in 80 per cent of all years. The two broken lines run through all places at which such amounts are 50 per cent and 75 per cent of 'normal' rainfall
Source: Boudet et al., 1976; p. 42.

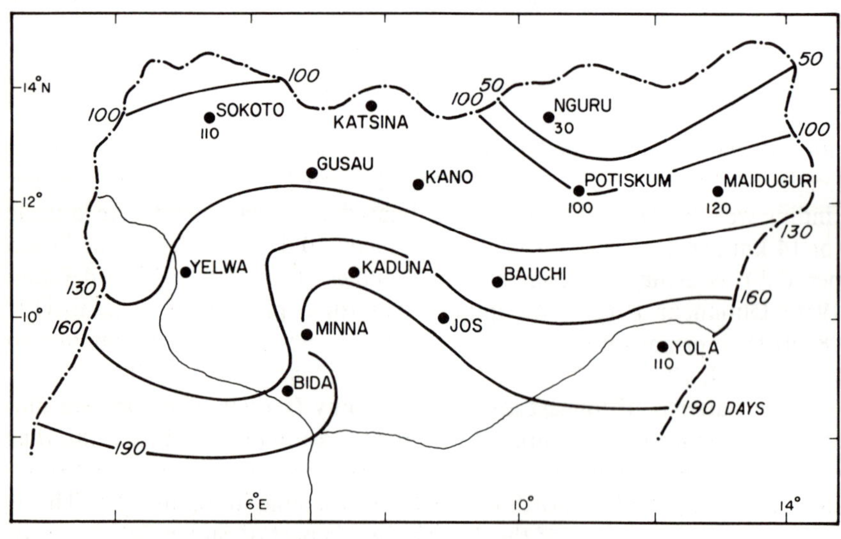

Fig. 12.8 Length of growing season in the Sahel Zone for 1972
(after Kowal & Adeoye, 1972)

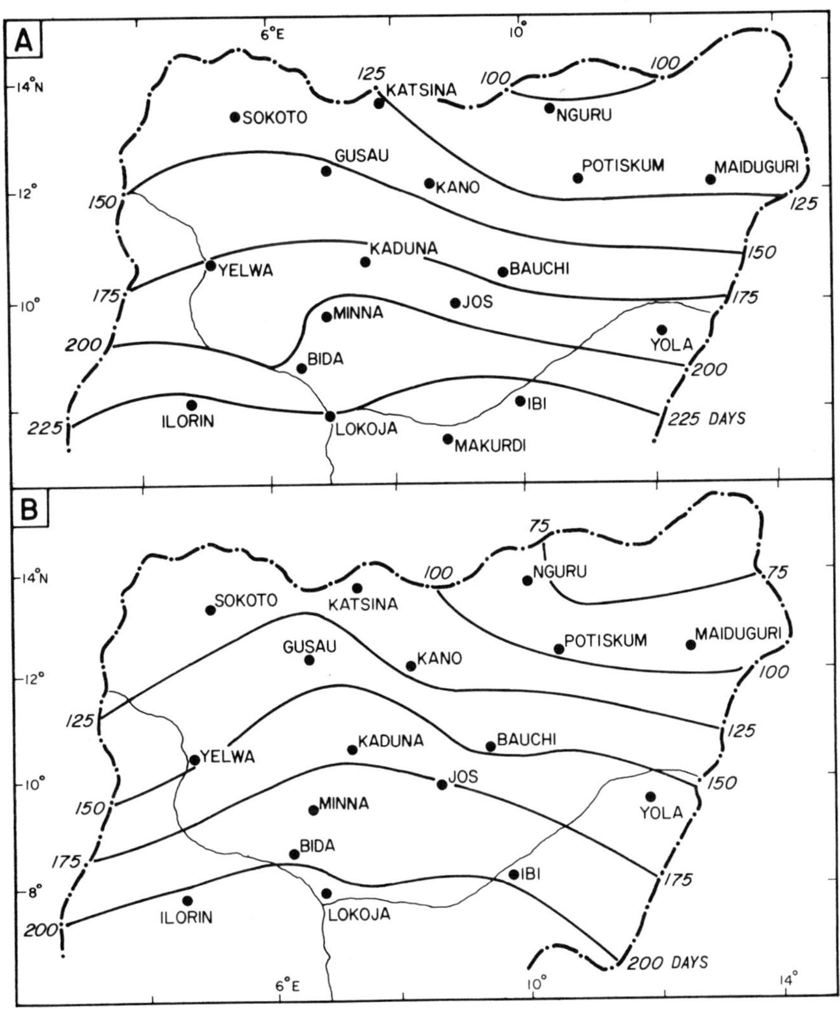

Fig. 12.9 A. *Longest growing season deduced from 50 years rainfall data by the application of standard error technique.*
B. *Shortest growing season deduced from 50 years rainfall data by the application of standard error technique*

TABLE 12.2 SHORTEST AND LONGEST LENGTH OF THE RAINY (GROWING)
SEASON IN FIVE-DAY PENTADES AT SELECTED NIGERIAN
STATIONS BETWEEN 12° AND 14°N

Year	Stations							
	Sokoto		Gusau		Katsina		Kano	
1968	28	59	21	25	—	—	23	53
1969	31	56	29	58	32	53	32	58
1970	35	55	33	54	33	49	34	54
1971	35	52	33	53	29	54	34	55
1972	30	57	25	58	29	54	34	55
1973	31	51	34	54	34	53	37	54
Mean	30.2	55	29.1	55.3	31.1	51.8	32.3	55.0
50-year mean	30.4	55.2	27.8	54.8	31.5	53.8	30.9	53.9

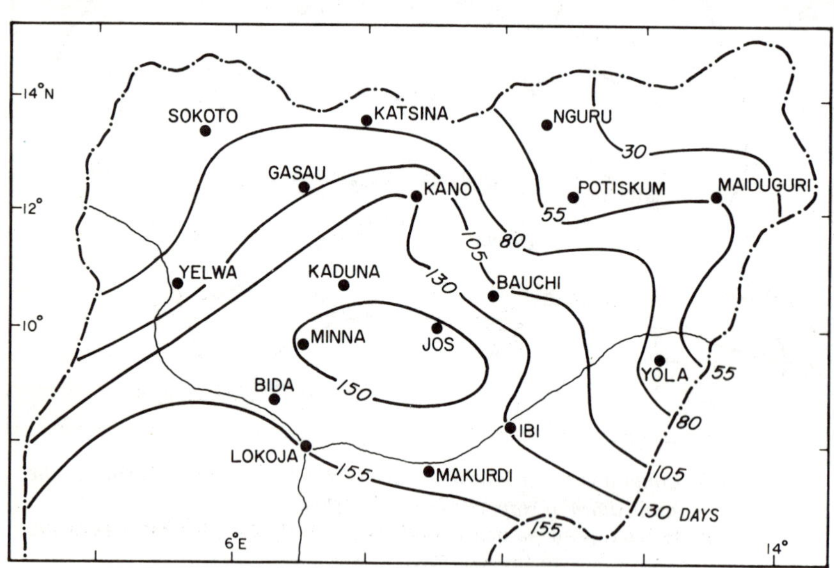

Fig. 12.10 Shortest growing season based on extreme values of onset and
cessation dates of the rainy season

tively. It reveals another salient feature in the problem of drought in the Sahel. Since these extreme patterns have been observed previously, the likelihood of such an occurrence in the future is high, and with it may come a worse drought situation.

Lastly, 1972 was characterized by uneven distribution of rainfall. Table 12.3 illustrates this feature with Kano, Potiskum, and Maiduguri exhibiting important zonal variations even though they are at about the same latitude. Of significance is the large difference of 303 mm between Sokoto and Nguru, both stations being also at about the same latitude. Thus, the Sahel experienced not only temporal but also spatially uneven rainfall distribution in 1972. It is important to note that, as illustrated in Figure 12.11 the 750 mm and 500 mm rainfall isohyets for average conditions and for 1972 are almost coincident at 12° to 13°N. This suggests that in 1972, apart from uneven spatial distribution, there seemed to have a two degree latitude equatorward shift of the axis of active weather zones in relation to the monsoon flow.

TABLE 12.3 UNEVEN ZONAL RAINFALL DISTRIBUTION BETWEEN 12° AND 14°N IN 1972 AT FIVE NIGERIAN STATIONS

Stations	Lat. (N)	Long. (E)	Total Rainfall Annual (in mm)		
			Mean (mm)	1972 (mm)	% of mean
Sokoto	13.01	05.15	731	551	75
Kano	12.03	08.32	869	596	70
Potiskum	11.42	11.02	787	690	83
Nguru	12.53	10.28	569	248	44
Maiduguri	11.51	13.05	653	440	67

From the above it is evident that while the low rainfall amounts in the Sahel during the 1972/73 drought were within the limits of tolerance for the sub-region, the shorter length of the rainy (growing) season, the uneven distribution in time and space of the amount received, and the definite shift to the south of the mean axis of active weather were more significant factors. As these could aggravate a bad situation, especially in relation to the availability of water in sufficient quantity and at the times of greatest need, the 1972/73 drought could be classified as agricultural drought. It should also be noted that the prolonged dry spells, which appeared to have started as far back as 1968, became progressively worse and reached a climax only in 1972/73. Even though it may be reasonable to suggest that hydrological

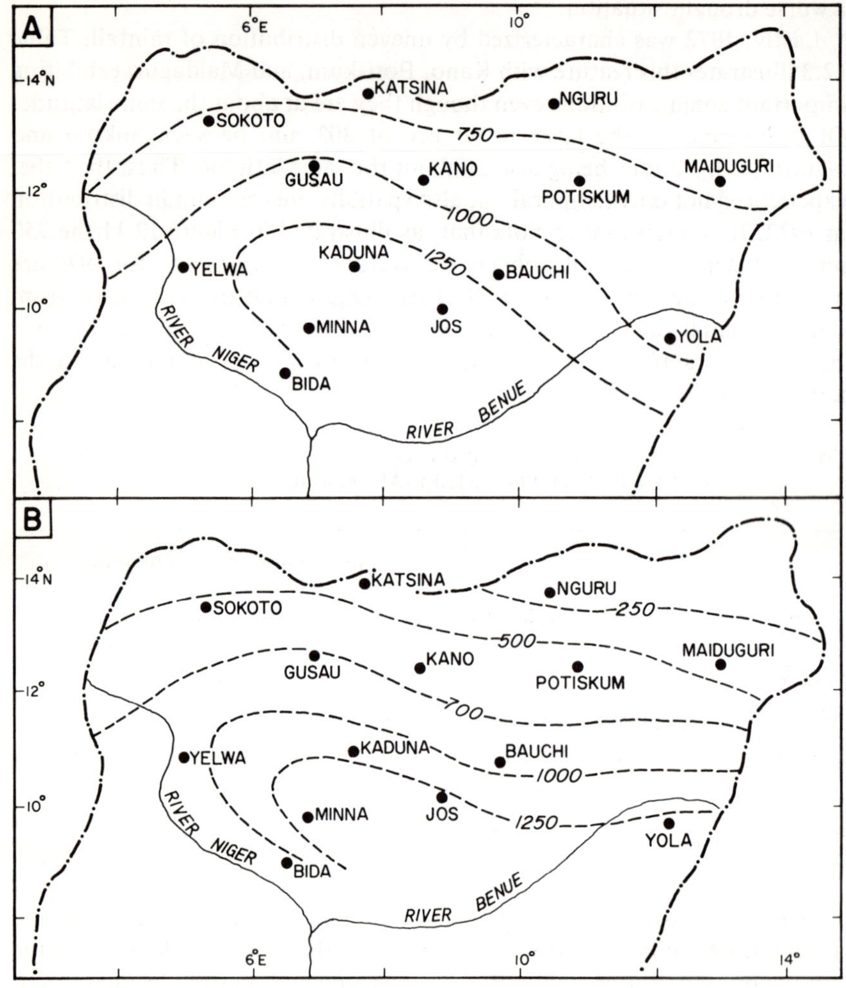

Fig. 12.11 Average rainfall (in mm) over the 'front-line' stations in Nigeria
A. Average
B. 1972
"Front-line" is used here to denote those stations bordering the real
Sahel Zone, i.e., stations north of 10°N but south of 14°N Lat.

drought could have set in, the duration of the dry-years (five years) does not justify classifying what occurred as climatic change. Thus the failure of the monsoon during those five years did not by itself lead to desertification of the Sahel.

Climatic Variability and Desertification

It has been recognized that the geographical distribution of plants (especially perennial, native plants) is closely correlated with climate. As explained by Pigott (1975), if a specie is restricted by climate, then it is probably sensitive at the edge of its range to exceptional conditions, so that it is useful to analyse its responses to the variations of climatic conditions in different years. He found that variations in temperature tend to affect cultured or cultivated plants more than the natural plant population.

Duckham (1974) has also emphasized the role of climate and weather on soil and natural vegetation. He suggested that five main climatic parameters are essential to plant growth:

(1) solar energy receipts, ET, measured as mm of potential evapo-transpiration (mean per annual thermal growing season (TGS);

(2) precipitation, P, in mm (mean per year);

(3) effective transpiration, A, that is, Thornthwaite's (1962) actual evapo-transpiration, in mm (mean per annual TGS);

(4) duration of the thermal or hydrologic (HGS) growing season;

(5) climatic uncertainty taken as coefficient of variation of mean annual precipitation.

A global investigation by Duckham and Masefield (1970) has established that a hydro-neutral zone, that is, a zone of optimum crop yield, exists in areas where ET per TGS approximately equals mean P per year and where A is relatively high. For semi-arid zones, ET increasingly exceeds P and A decreases. In such areas, yield falls. It was found that this characteristic is related to the high salinity and high pH of semi-arid soils. These inter-relationships are illustrated in Figure 12.12 for tropical latitudes. In semi-arid areas, owing to high ET and low P, the hydrologic growing season (HGS) tends to be very short.

The above are termed "local physical factors" which influence the degree of desertization in a place. Another kind of local effect which has gained prominance in recent times is the sensitivity of local moisture balance to changes in albedo (Charney *et al.* 1977). Numerical experiments indicate that over the semi-arid areas of the tropics an increase in albedo tends to lead to lower rainfall values as a result of a decrease in net radiative input at the surface and consequent lower evaporation rates (Pearce 1980).

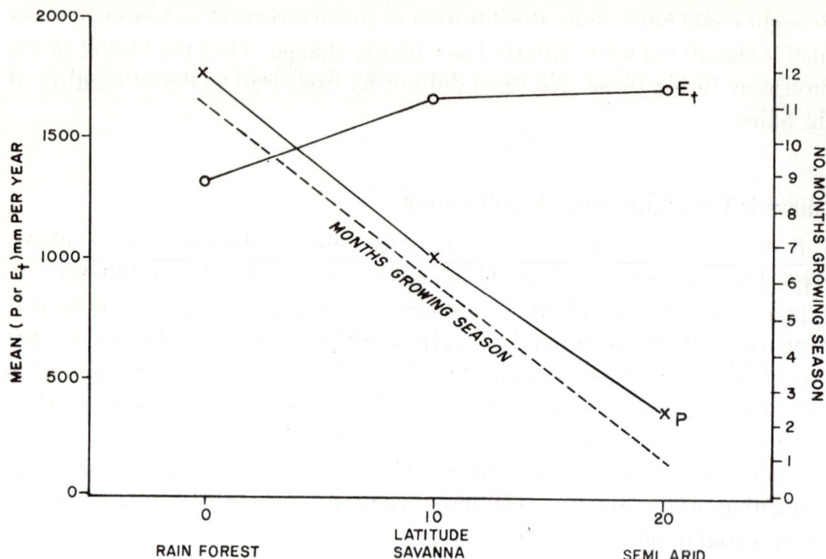

*Fig. 12.12 Relation in tropical areas between latitude, mean annual precipitation
(P) mean annual potential transpiration (E$_t$) (Penmans' method) and
length of hydrologic growing season (HGS)
Source: data from Elston 1974*

Variability of large-scale circulation has been demonstrated to have a
devastating influence on the tropical rainfall patterns. Krishnamurti (1970;
1980) computed the values of the velocity potential at the 200 mb level over
the tropics and mapped the results for 1967 (normal year) and 1972 (drought
year). He found that the main anti-cyclonic outflow at 15 °N over Southeast
Asia was displaced eastwards from over Burma in 1967 to just north of the
Philippines in 1972. From stream-function analyses, it was also found that
the upper level strong easterlies, known as the Tropical Easterly Jet (TEJ) at
about the 200 mb level was only half its normal intensity in 1972.

These large-scale features are translated into meaningful functions such
as mean specific humidity, q, over the tropical zone. A characteristic of q is
that it is often very low in dry years and high in wet years. For example,
Newell *et al.* (1972; 1974) have found that the Sahara has a very low annual
value of 2 gm kg^{-1} at 850mb level, while it is approximately 4 gm kg^{-1} and
8 gm kg^{-1} at about the surface in winter and summer respectively. Compared with the mean oceanic value of 14 gm kg^{-1}, these values cannot be
expected to result in appreciable precipitation.

The specific humidity is closely related to the static stability function Q$_t$
(static energy) of the atmosphere. The latter is given as:

$$Q \cong Cp\, \Theta_e = Lq + CpT + gZ$$

Where Cp = specific heat at constant pressure

      Θ_e = equivalent potential temperature

      L = latent heat

      q = specific humidity

      T = temperature

      g = acceleration due to gravity

      Z = geopotential height.

Since Cp is a constant, a profile of Θ_e is representative of Q_t and this has been used to show that the tropical troposphere is unstable in the mean (Garstang 1967; Adefolalu 1972). Normal years (wet) in the Sahel zone are characterized by high Θ_e values especially during periods of moisture convergence in the region (Obasi 1964) at mid-tropospheric levels. It is important to mention here that large values of q are directly related to the northward penetration of the monsoon into the Sahel zone — the farther north it penetrates, the greater the precipitation at places between latitude 10° and 20°N (Lamb 1979:54).

Subsequent discussions in this section deal with each of the parameters already outlined as they relate to what happened in 1972/73 drought period with a view to ascertaining the role (if any) climatic variability played in the progressive deterioration of the vegetation cover in the Sahel.

Local Factor of Climatic Variability

Potential evapotranspiration (ETP). Potential evapotranspiration (ETP) is a function of incoming solar radiation (R). Table 12.4 shows ETP and Table 12.5 shows R for different rainfall zones in the Sahel. The highest values of both are during the drier and hotter months while the low values are at the peak of the wet period. There is also a sharp north-south gradient

TABLE 12.4 ETP VALUES FOR DIFFERENT MONTHS AND RAINFALL
SUB-ZONES, IN MM DAY $^{-1}$

Rainfall Sub-Zones	January	March	August	October	Year
100–200	4.1	6.2	5.6	5.6	5.5
400–500	5.5	7.3	5.2	6.0	6.3
700–900	4.8	6.4	4.6	4.8	5.3
1100–1300	4.6	6.1	4.1	4.7	4.9

Source: Davy *et al.* 1976.

TABLE 12.5 GLOBAL INCOMING RADIATION IN SELECTED MONTHS
AND RAINFALL SUB-ZONES, IN CAL. CM.^{2} DAY^{-1}

Rainfall Sub-Zones	January	March	August	October	Year
100–200	454	520	496	525	507
400–500	436	490	490	502	493
700–900	448	519	460	501	485
1100–1300	454	490	424	466	465

Source: Davy *et al.* 1976.

as a result of increasing cloudiness from the south (areas of higher precipitation) which reduces incoming radiation. As explained by Davy *et al.* (1976), two local factors affect the ETP. There is the annual march of the sun to the extreme north of the zone in June which would be reflected by maximum energy input due to higher incoming solar radiation. High ETP in midsummer should result but for a second controlling factor — the annual march of the Inter-Tropical Discontinuity (ITD) to about 25°N in late August. The northward retreat of the dry harmattan wind and its replacement by the cloudy, more humid monsoon would result in a steady decrease of ETP values. These are illustrated in Figure 12.13 for Fort-Lamy and Ba-illi. Compared with the monthly rainfall patterns as illustrated in Figure 12.3, it is obvious that northwards from the 900 mm to the 300 mm isohyet, decreasing rainfall occurs in areas with increasing potential losses of water from the soil to the atmosphere because as ETP exceeds precipitation, transpiration from plants decreases (Duckham 1974). This is yet another reason why it is not proper to use the total rainfall amount alone as a basis of classification of the Sahel. Riou (1975) has hinted that the variability of the date of onset of rains is reflected in high variability of ETP values which is a common feature in the Sahel. Variation of ± 10 per cent could be expected at least once every five years.

It is obvious from the above that the Sahel is characterized by high values of the ETP, and these values will be higher during periods of lower rainfall. This is critical at the early part of the growing season when young plants are exposed to most frequent stress from both lack of soil moisture and heavy evaporative demands of the atmosphere (Kowal 1972).

Albedo effects.   Albedo is total reflectivity of the incoming solar radiation in the earth-atmosphere system. For semi-arid regions, the albedo from

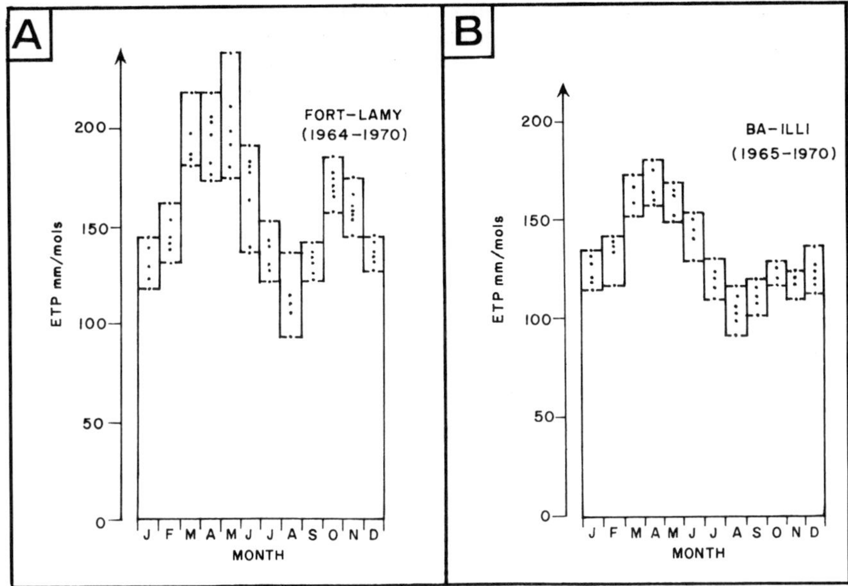

Fig. 12.13 Variation of monthly evapo-transpiration years
A. Fort Lamy
B. Ba-illi

bright desert sandy soil has been estimated at 0.37 (Ashburn and Weldon 1956). Using a contrast ratio 1.5 of upward reflectivity, the albedo of the same soil with appreciable vegetation cover is about 0.25.

Such large differences in surface reflectivities have important environmental implications. Otterman (1974) discovered a difference of 5 °C in the Sinai-Negev desert between the denuded high albedo region of the Sinai (lower temperature) and the relatively "green" Negev (warmer region). One major implication is that where the ground temperature is higher, there is more heating of the atmospheric boundary layer and the air has a greater tendency to rise leading to condensation and formation of clouds (Malkus and Stern 1953). The amounts of such cloud cover and rainfall are therefore higher in areas of green vegetation where the effect could be of the order of hundreds of millimetres of precipitation a year.

This difference in surface temperatures is well marked in the Sahel region, especially between the 100–300 mm and 500–700 mm rainfall zones (Table 12.6). Here the temperature difference is 4 °C to 6 °C from west to east, being warmer in the 500–700 mm rainfall zone. The cooler conditions in the 1100–1300 mm rainfall zone is due to the interplay of annual cycles of solar radiation and the rainfall seasons on the one hand and relative influence of maritime effects on the other hand.

TABLE 12.6 MONTHLY AVERAGE
TEMPERATURES IN THE SAHELIAN
SUB-REGION OF WEST AFRICA

Rainfall Sub-Zone	Day Maxima (°C)		
	West	Central	East
100–300	Highest 42	40	39
	Lowest 28	28	28
500–700	Highest 42	40	41
	Lowest 34	32	33
1100–1300	Highest 40	37	39
	Lowest 28	30	29

Source: Davy *et al.* 1976.

Denuded soil surfaces are also exposed to another risk. In the dry season, strong winds steadily erode surfaces, causing a shifting of soil particles and preventing the establishment of young plants and burying existing ones. They may even lead to a complete loss of soil. Davy *et al.* (1976:271) reported that in 1969 an estimated 60 million tons of soil were carried by winds from West Africa to the Atlantic Ocean during the summer. Compared with the losses from the persistent harmattans during winters (Burns 1960; Adefolalu 1968; Aina 1972; Kalu 1978), that amount must be a small fraction of the total annual soil losses. With such dust veils, the lower layers of the atmosphere are also prone to further reduction of surface temperatures. A decrease in lifting is associated with these conditions.

Climate, as discussed here, is affected by soil surface conditions. By implication, a situation which leads to the exposure of high albedo soils is a major factor in the downgrading of the ability of the environment to support life (Davy *et al.* 1976). In the Sahel, aridity results indirectly from soil exposure and from overgrazing, leading to increased albedo. The scanty rainfall, which depends largely on circulation patterns, is thus a function of local conditions. The extent to which a continuing decrease of surface temperatures will affect such circulation patterns is discussed below. There is no doubt that when changes in albedo are continuous as must have been the case in the Sahel during the 1972/73 period, desertification is at an advanced stage and any reversal of the process is going to be difficult if not impossible. Once land-use patterns are such that the vegetation (either perennial or cultured) cannot recover even when normal years of precipitation follow drought spells, the process of desert formation may be said to have been completed. The situation can only worsen as soil degradation is at an irreversible stage.

Large-scale effects. Return periods of two to three and ten years have been found in rainfall patterns in Africa (Landsberg 1975; Rodhe and Virji 1976; Ogallo 1977; Obasi *et al.* 1980). While the shorter periodicity is linked with the quasi-biennial oscillation, the ten-year cycle is associated with the sun-spot cycle. Although doubts have been expressed about their usefulness in prognosis (Gibbs 1975), what these cycles demonstrate is the oscillatory pattern of rainfall in the region. Such oscillations are linked to or associated with changes in the mean circulation over West Africa.

By virtue of the land-sea distribution and the general circulation of the atmosphere, maximum penetration of moisture-laden southerly winds are expected in summer, as postulated by many authors (Bergeron 1960; Dhonneur 1971; Dhonneur *et al.* 1978). During wet years, the low level flow also reflects the surface condition as the transient disturbances are cloud fed with adequate moisture for formation and release of precipitation (Adefolalu 1974). The depth of moisture could be up to the 700 mb level (Fig. 12.14A). The situation is different in dry years when the moisture-laden southerlies (monsoon) is shallow. The depth of moisture then falls below the 700 mb level and the vortices, even at the peak of the monsoon (August), take on the structure as depicted in Figure 12.14B for the October composite model. Here the flow to the south of the vortex is more easterly than southerly, and while convergence may still be pronounced, precipitation is limited to small areas.

It has been found (Laseur and Adefolalu 1975; Krishnamurti 1980) that the east-west axis of the monsoon trough is important in the availability of moisture in the vicinity of the travelling disturbances. Dry years correspond to an equatorward shift in the axis, while the axis is further north in wet years. In 1961 and 1967, both normal (or above normal) years in terms of total rainfall amounts, monthly mean equivalent potential temperature (Θ_e) patterns in Niamey showed that the low tropospheric minima had a distinct structure in summer different from the pattern in 1963, a dry year (compare Fig. 12.15A and B with Fig. 12.15C). In those two years, the Θ_e isolines were convex from the surface up to about 700 mb during the summer, whereas in 1963 it was concave downwards in general. These suggest higher values in the lower troposphere as well as in the middle layers during wet years and much lower values in the mid-troposphere in dry years (Adefolalu 1972; 1973). The patterns in the wet years demonstrate the

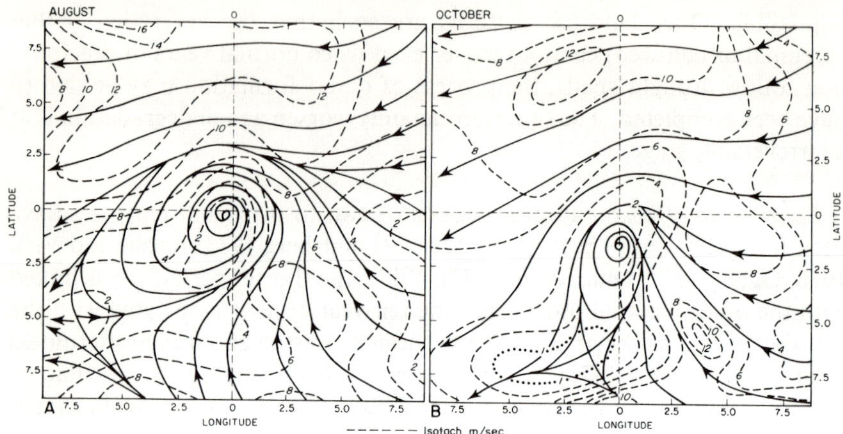

Fig. 12.14 Streamline and isotach analysis for the composite wave-vortex at 700 mb
A. August 1954 (01–042)
B. October 1954 (01–042)

"hot-tower" energy transport mechanism as postulated by Riehl and Malkus (1961) and others, that is, the maintenance of the general circulation in the tropics (the Hadley circulation) is through the tall cumulonimbus clouds (hot towers). This is not possible when moisture is absent, thus showing that drought situations are positively correlated to weak Hadley circulations (Adefolalu 1980).

The Sahel, with its low moisture content coupled with lower surface temperatures (due to high albedo) can thus be expected to continue to have decreasing precipitation if nothing alters the present trend. It can therefore be hypothesized that human pressures on the land are the causes of drought, including meteorological drought.

Human Interference and Desertification

In the previous sections, most or all of the problems resulting from rainfall deficiency have been analysed and discussed. The sum total of this is that climatic variability is a feature of the climate itself and periodic drought situations are normal, just as are periodic wet spells.

The vegetation cover of the Sahel is steadily disappearing and the situation was aggravated by the prolonged 1968 to 1973 drought. It must be recognized that semi-arid zones at the periphery of deserts are more prone to vegetation and soil degradation. In these regions, the ecological balances are so delicate that man's activities are likely to determine whether or not present steppe land will become desert (Davy *et al.* 1976). This, as explained by Le Houerou

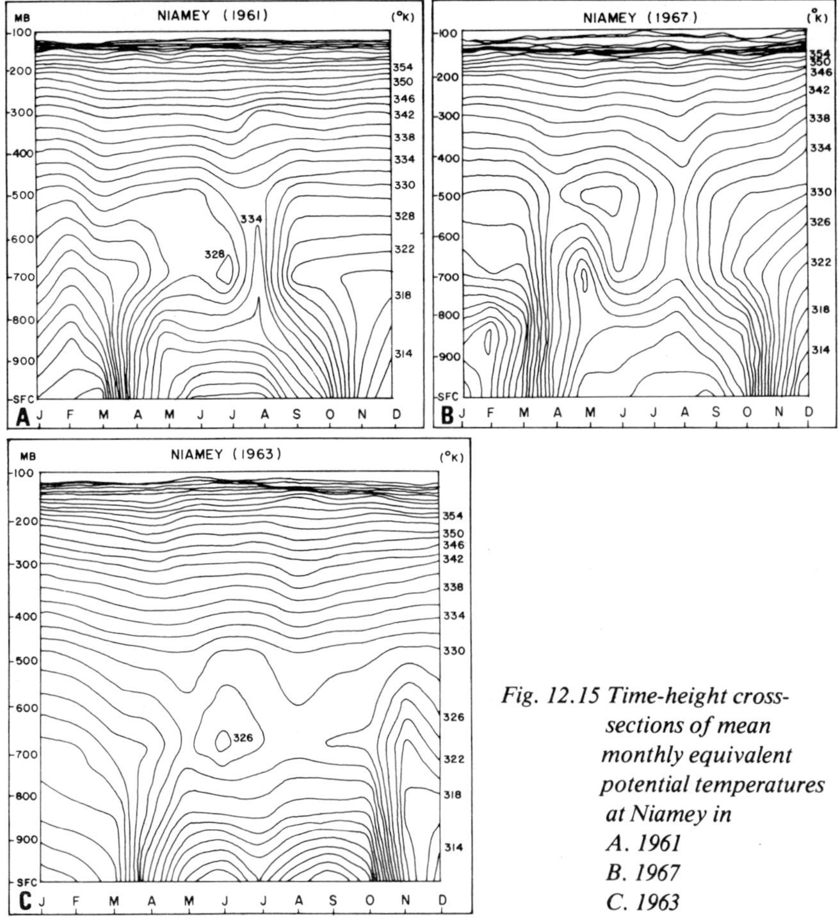

Fig. 12.15 Time-height cross-sections of mean monthly equivalent potential temperatures at Niamey in
A. 1961
B. 1967
C. 1963

(1976), is because rangeland misuse leads to such problems as a reduction in plant cover and biomass, an increase in erosion, a reduction in the productivity or rangelands, an increase in the member of unpalatable as well as of annual species of plants, and the rare-faction and occasional disappearance of valuable species and of life.

These imply that human interference of or interaction on the ecosystem results in unintended and generally undesirable side-effects which are usually local in extent but are rapidly (at least for the Sahel) increasing in scale and significance to such a degree as to cause desertification of the region.

Two basic activities in the Sahel region of Africa which will ultimately lead to the total annihilation of plant and animal life are cash crop agriculture and livestock rearing. They have climatic implications which go beyond the destruction of the surface vegetation. The climate can be modified as surface heating could be reduced by as much as 35 per cent in the Sahel. This will result in anticyclonic subsidence in the atmosphere, and what Charney (1977) has described as a positive feedback mechanism which tends to intensify and perpetuate aridity.

Cash Crop Agriculture

In 1929, the French forced farmers in the Chad Republic to grow cotton together with their food crops of millet and sorghum (Grove 1967). This was to ensure sufficient raw materials for the factories in France. Small-scale irrigation projects were constructed to use the waters of Lake Chad and some of the rivers (Hellen 1969). Similar schemes were introduced along the fairly rich alluvium basin of the Niger river in Mali and the Republic of Niger.

In Chad cotton, which provided 80 per cent of the country's export earnings, was grown on over 200,000 ha of land, while the cotton area in Mali was projected to cover between 800,000 and 1,200,000 ha. As Grove (1967:99) has pointed out, the soils in Mali are not as fertile as those of the Nile Valley; nor are climatic conditions so favourable for cotton.

These schemes necessitated the construction of dams and water reservoirs and irrigation channels along natural river basins, deforestation of the natural vegetation, and acquisition of marginal areas which could barely support crops whose water and nutrient requirements are less demanding than cotton.

Therefore, traditional short-cycle crops such as millet and sorghum were replaced with the more demanding cotton plant. The length of the growing season in the Sahel range between 60 and 75 days (Davy *et al.* 1976), but cotton requires up to 180–190 days from germination to picking of the cotton lint. Irrigation was therefore necessary for its cultivation.

Cotton is not food and its production was meant to boost the earnings of the poor farmers. But it is both labour and capital intensive. Reports showed that the farmers' profit margins were small and they had to rely on governmental subsidies so that total expenditure was more than total revenue.

Over the years, the soil had become exhausted as a result of cotton cultivation. Irrigation, dependent on the Niger river and Lake Chad, had lowered the water level in both natural reservoirs. This resulted in the general lowering of the water table in the sub-region. Such a conclusion is supported by findings in similar areas such as the 5.5 million ha of irrigated semi-arid land in central Asia. There, the rivers Amu Darya and Syr Daria are used for irrigation. It has been estimated that, as a result of high evaporation rates (about half of their waters are lost before reaching the Aral Sea), the Aral Sea will disappear completely in about fifty years from now if alternative sources of water are not used for the irrigation schemes (Lamb 1979). That Lake Chad is presently less than one per cent of its size 22,000 years ago is therefore not surprising; it may even dry up within the next fifty years. The situation in the Niger river is not better. Indiscriminate damming along its course and diversion of its water for agriculture and other uses have reduced it to a combination of ox-bow lakes and meanders, especially downstream in Nigeria. The volume of water required for the multi-million Kainji Electricity Project in Nigeria has dropped as a result of the 1972/73 drought.

Livestock Rearing

In the Sahel, the pastoralist raises livestock for two reasons: for food and to earn some money from the sale of cattle. This was still possible for as long as the nomad remained a nomad as he was able to move from one place to another in search of fodder and water for his herd. Thus the rhythm of human activities was in harmony with the seasonal variations of weather and water resources in the Sahel.

In the Sahel-Sahara zone, the herbaceous biomass attains a maximum of 400 to 500 kg/ha on scattered strips. In the Sahelian sector, the maximum herbaceous biomass varies from 500 kg/ha on skeletal land to 2,000 kg/ha on sandy peneplains. In the Sahel-Sudan sector, the maximum herbaceous biomass ranges from 500 to 3,000 kg/ha. All of these could become available to herds at appropriate times, but the raising of large numbers of cattle in smallholdings where sedentary agriculture is also practised has led to overgrazing and desertification.

According to Boudet *et al.* (1976:22), the Sahel region can be divided into three parts for combating desertification due to overgrazing:

1. The climatic zone lying between the 400 and 800 mm rainfall isohyets, where the desertification process is most advanced owing to the combined effect of itinerant agriculture and transhumance.

2. The stretches of sandy loess bearing non-harmful grasses, between the 200 and 400 mm isohyets. These are rangelands where the increase of livestock pressure has led to the dominance of short, thorny, almost leafless plant species, and a reduction in the possibility of grazing in-between seasons.

3. The desert Sahel interface delimited by the 100 and 200 mm isohyets, where a period of drought such as that from 1968 to 1973 in association with increased livestock pressure could lead to real desertification.

It has been recognized that until tsetse flies (the vectors of trypanosomiasis) are eradicated in the more humid savanna, the natural limit of free movement of herds is the 1,000 mm isohyet. The natural limit has contributed to overgrazing. This fact is clearly demonstrated by the patterns of livestock distribution before and after the 1972/73 drought climax.

In 1970, before the drought climax, the overall livestock population of the Sahelian countries was 50,330,000 cattle, 77,791,000 small ruminants, and 4,270,000 camels making a total of 51,459,000 TLU (Tropical Livestock Unit), distributed as follows:

> Sahel Zone: 38 per cent (6.9 herd/km^2)
> Sudan Zone: 58 per cent (4.5 herd/km^2)
> Guinea Zone: 4 per cent (0.3 herd/km^2)

Their distribution after the drought was as follows:

> Sahel Zone: 27 per cent (4.0 herd/km^2)
> Sudan Zone: 68 per cent (4.0 herd/km^2)
> Guinea Zone: 5 per cent (4.3 herd/km^2)

It has been explained that the 6.9 herd/km^2 in the Sahel in 1970 corresponded to an overall load of 14 geographic hectares per animal, which is generally accepted as the upper limit beyond which the pressure on the environment will contribute to desertification. Under normal conditions, that is, a no-drought situation, a herd of 100 cattle, 200 sheep, 50 goats, 30 camels, and 10 horses require 414 hectares of land (Le Houerou and Hoste 1976). But according to these authors and others (Condon 1968; Breman 1973; 1975) rainfall seems to be the most important factor of pasture productivity. It will seem proper, therefore, to relate pasture productivity to the lowest possible expected rainfall because it is not the normal years but the lean years that need to be forecast. In such a case, rainfall itself may not be the most important factor but the cause or causes of such dry or wet years and their return times. In other words, overuse during normal or wet years could lead to dry years. Otterman (1974) has explained that the natural unit time for such a cycle would be the age of litter-bearing of goats; a population explosion in grazing herds during the "seven fat years" is the cause of the "seven lean years" (Biblical quote: Genesis 41:17–18).

Man and the Trophic Levels

From the foregoing, it is apparent that man is responsible for most of the problems of desertification of the Sahel. Considered from a general viewpoint, of the total 3,000 kCal/day of total incident solar energy, only 0.15 kCal/day is available to man as a consumer organism (Simmons 1966). Yet man plays a key role in the actual transformation of the incident energy. He is the one who clears the vegetation for cultured plants, a process which consumes most of the 15 kCal/day available for the rural vegetation. He attends to the lower carnivore for whom 1.5 kCal/day is available, and eventually obtains in return goods and services for such efforts.

A conceptual representation of this is given in Figure 12.16 which shows man as a consumer as well as director. Present-day advances in technology and chemical use will result in further modifications to the environment. The effect on the semi-arid zones will be further desertification, unless man returns to the land what belongs to it or at least refrains from further interfering with it.

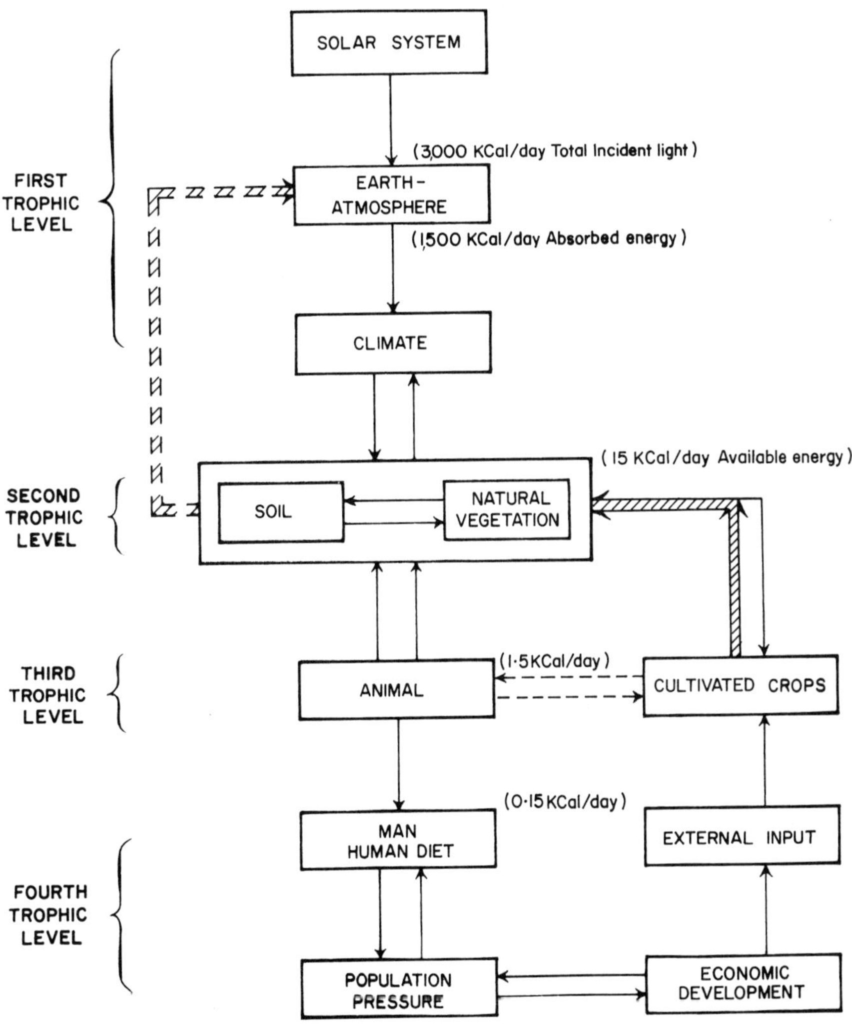

Fig. 12.16 Inter-relationships between climate, vegetation and man
Source: adapted from Simons, 1966 & Duckham, 1974

Conclusion and Recommendations

Climatic variability as manifested by dry and wet years in the Sahelian zone of Africa has been discussed as a normal aspect of climate. The contention that dry years are drought years has also been discussed in the context of the three definitions of drought. These led to the suggestion that deficient rainfall (and hence the failure of the West African monsoon) in the region can lead to directly is desertization, and prolonged drought such as the 1968–73 drought could lead to aridity owing to excessive exploitation of the diminishing natural resources of this semi-arid environment.

A variety of factors can therefore lead to the process of desertification, the most important of which is the influence of man and his activities on marginal lands bordering existing deserts. The unintended decimation of the natural vegetation of the Sahel and the careless utilization of water resources in the region can lead to the irreversible total destruction of vegetation, which may in turn lead to the annihilation of all living organisms in the sub-region. If such a situation is to be averted, there is no alternative but to manage the land and other resources of the area on a sound ecological basis. This is essential as future drought will aggravate an already serious situation. To this end, the following measures (among others) seem appropriate:

1. Policies to improve the agricultural productivity of the countries in the Sahelian region should be tied to the conservation of the biomass. For example, replacing the native millet, maize, and sorghum (all low cash-yielding commodities) with revenue-earning cotton has led to poor soil conditions in Mali, Niger, and Chad. Future "expert" advice on related matters need to be viewed with caution.

2. A drastic review of the factors which encourage and lead to excessive densities of both people and herds and restrict the movement of pastoral nomads in the semi-arid and arid zones is necessary. This restriction on movement coupled with the higher animal population have led to over-grazing, the major cause of desertification.

3. Research emphasis should shift to applied and functional studies of drought. Determination of such phenological parameters as length of the growing season, periods of break within a monsoon, and the heat sum requirements of plants as against available values are more meaningful than mere description of cycles of drought. Such quantities are related directly to soil carrying capacities and the programmes of land and water resource management. Why should "expert" suggestions for planting cotton between February and November be ever considered, let alone implemented if it is known that potential evapotranspiration is highest during the drier months and the irrigation of cotton fields using artesian wells as well as surface reservoirs will only lead to the lowering of water tables and the depletion of the level of water in running rivers and lakes?

4. It is obvious that the states of the Sahelian zone, because of their impoverished environment, cannot by themselves combat the problems of desertification. Previous efforts had been mere palliatives. For a lasting solution, all the countries (English and French-speaking) in the area should form a task force with aid and expertise from such international organisations as the World Meteorological Organization (WMO), the Food and Agricultural Organization (FAO), the International Federation of Institutes for Advanced Studies (IFIAS), and many others. Inter-governmental bodies such as the Organization for African Unity (OAU), can only pass resolutions but not solve the problem.

This paper has attempted to establish that the problem to resolve is not the Sahara spreading southwards but whether human interference or intervention in the Sahel is leading to desertification of the area. Many previous works have established the fact that desertification of the Sahelian zone has reached an advanced stage. If climatic variability is a normal feature of climate, then it needs to be explained that the increase in such variability is a result rather than the cause of depleting the vegetation cover. This has been done and man has been cited as the root cause of it all. It is he who must find the solution.

Acknowledgements

My thanks go to Prof. Barry Floyd for his interest in this work and to my colleagues for their helpful comments. I am most indebted to Mr A.J. Umoh, Secretary of the Department of Geography, University of Calabar, who typed the paper. The cartographic staff under Mr E.J. Umoh of the same department drafted all the maps and diagrams and I greatly appreciate their efforts. Finally, I wish to express my gratitude to Chief (Prof.) E.A. Ayandele, Vice-Chancellor, University of Calabar, for approving my participation in the Workshop on Natural Resources in Tropical Countries organized by the Commonwealth Geographical Bureau.

REFERENCES

Adefolalu, D.O. 1968. "Two Case-Studies of the Vertical Distribution of Dust during Occurrence of Harmattan Haze over Nigeria". *Technical Rep.* Meteorological Department, Lagos.

______. 1972. "On the Mean Equivalent Potential Temperature as a Means of Measuring the Structure of the Troposphere over West Africa". *Quarterly Meteorological Magazine* (QMM). Meteorological Department, Lagos.

______. 1973. "Further Studies on the Mid-Summer Drought in West Africa". *Quarterly Meteorological Magazine* 3:110–18.

______. 1974. "On Scale Interaction and the Lower Tropospheric Summer Easterly Perturbation in Tropical West Africa". Ph.D. dissertation, Department of Meteorology, FSU, Tallahassee.

______. 1976. *New Discoveries in Tropical Meteorology and Nigeria's 3rd National Development Plan, Zaria-Nigeria.* Annual Conference Nigeria Science Association.

______. 1977. *The Sahel Drought of 1972: Causes and Effects*. International Workshop, International Federation of Institutes for Advanced Studies, Geneva.

______. 1980. *Study of Drought in the Semi-arid Zone of the Monsoon region of West Africa: Towards Achieving a Complete Data Base*. WMO Report of Expert meeting on semi-arid zones, Geneva.

______. 1981. "on the Length of Growing Season in Nigeria and Crop Adaptation". *Nigerian Journal of Tropical Geography* (submitted).

Adejokun, J.A. 1966. "A Three-Dimensional Study of the ITD". *Technical Note no. 39.* Meteorological Department, Lagos.

Aina, J.A. 1972. "A Contribution to the Forecasting of Harmattan Dust Haze". *Quarterly Meteorological Magazine* 112. 2:77–90.

Apeldoorn, G.J.V. 1977. *Drought in Nigeria* 1/2. Centre for Social and Economic Research, Ahmadu Bello University, Zaria.

Ashburn, E.V., and Weldon, R.G. 1956. "Radiation in the Atmosphere". *Journal of Optical Society of America* 46.

Baker, R. 1974. "Information, Technology Transfer and Nomadic Pastoral Societies". Seminar and Overseas Development Institute, Reading University.

Boudet, G., *et al.* 1976. *Management of Livestock and Rangelands to Combat Desertification in the Sudan-Sahelian Regions*. SOLAR, U.N.

Bremen, H. 1973, 1975. "L'amenagement e cologique des paturages: la capacite de change maximale pour le Mali". *Collection de Biolugie*. Mali.

Burns, F. 1960. "Dust Haze in Relation to Pressure Gradients". *Technical Note no. 11.* Meteorological Department, Lagos.

Charney, J.; Quirk, W.J.; Chow, S.H.; and Kornfield, J. 1977. "A Comparative Study of the Effects of Albedo Change on Drought in Semi-Arid Regions". *Journal of Atmospheric Sciences* 34:1366–85.

Chevalier, A. 1933. "Le territoire geobotanique de l'Afrique tropicale nord-occidentale et ces Division". *Bulletin Societie Botany de France*.

Davy, E.G. 1974. "Drought in West Africa". *WMO Bulletin* 123:18–23.

______; Mattei, F.; and Solomon, S.I. 1976. "An Evaluation of Climate and Water Resources for Development of Agriculture in the Sudano-Sahelian Zone of West Africa". *WMO*, no. 459.

Dhoneur, G. 1971. *General Circulation and Types of Weather over Western and Central Africa*, Annex IV. GARPGATE, WMO Report, vol. 2.

______ *et al.* 1978. *The West African Monsoon Experiment.* (WAMEX), WMO/GARP REP. no. 21.

Duckham, A.N. 1974. "Climate, Weather and Human Food Systems: A World View" *Weather*, pp. 242–51.

______, and Manefield, G.B. 1970. *Farming Systems of the World*. London and New York.

FAO/UNESCO/WHO. 1967. *A Study of the Agroclimatology of the Semi-Arid Area, South of the Sahara in West Africa*. FAO, Rome.

______. 1974. *Soil Map of Africa in Soil Map of the World*. FAO, Paris.

Garstang, M.N.E. La Seur, and Aspliden, C.I. 1967. *Equivalent Potential Temperature as a Measure of the Structure of the Tropical Atmosphere*. U.S. Army Report no. 67–10.

Gibbs, W.J. 1975. *Drought*. WMO no. 403.

______, and Maher, J.V. 1967. "Rainfall Deciles as Drought Indicators". *Bulletin no. 48.* Meteorological Bureau, Melbourne.

Glantz, M.H. 1975. *Grazing in the Sahel: A Background Paper*. National Center for Atmospheric Research, Boulder, Colorado.

______. 1976. "Le Sahel: Nine Fallacies of Natural Disaster". IFIAS special publication in *Politics of Natural Disaster*. New York.

__________. 1977. "The Value of a Long Range Weather Forecast for the West African Sahel". *Bulletin, American Meteorological Society* 58.

Grove, A.T. 1967. *Africa-South of the Sahara*. Oxford.

Hellen, J.A. 1969. "Colonial Administrative Policies and Agricultural Patterns in Tropical Africa". In *Environment and Land Use in Africa*. Methuen.

Horowitz, M. 1972. "Ethnic Boundary Maintenance among Pastoralists and Farmers in the Western Sudan". *Journal of Asian and African Studies* 7:105-14.

IFIAS. 1977. International Workshop on Drought and Man: The 1972 Case History. Geneva.

Kalu, A.E. 1978. *The African Dust Plume: Its Characteristics and Propagation across West Africa in Winter*. Wiley.

Kamm, H. 1974. "Sub-Saharan Lands Are Hopeful of a Reasonable Harvest Soon". *New York Times*.

Kowal, J. 1972. "Radiation and Potential Crop Production at Samaru, Nigeria". *Savanna* 1, no. 1:89-101.

__________, and Adeoye, K.B. 1973. "An Assessment of Aridity and the Severity of the 1972 Drought in Northern Nigeria and Neighbouring Countries". *Savanna* 2, no. 2:145-58.

Krishnamurti, T.N. 1970. "Observational Study of Tropical Upper Tropospheric Motion Field during Northern Hemisphere Summer". *Technical Report 70-4*. Florida State University, Tallahassee.

__________, and Kanamitsu, M. 1980. "Northern Summer Planetary-Scale Monsoons during Drought and Normal Months". *Monsoon Dynamics*. Cambridge University Press.

Lamb, H.H. 1979. *Climate, Present, Past and Future*, vol. 2.

Landsberg, H.E. 1978. "Sahel Drought: Change of Climate or Part of Climate? *Archieves Meteorology, Geophysics and Biology*. Series B, no. 23, 193-200.

La Seur, N.E., and Adefolalu, D.O. 1975. *A Multiple-Scale Research Program in Tropical Meteorology*. U.S. Army Rep. no. ECOM-69-0062F.

Le Houerou, H.M., and Hoste C.H. 1976. "Nature and Desertification". Consultation CILSS/UNSO/FAO on the role of vegetation in programmes of rehabilitation of the Sahel, Dakar.

Lo Fichie, M. 1971. *The State of Nations: Constraints and Development in Independent Africa*. University of California, Los Angeles.

Maher, J.V. 1973. *The Environmental, Economic and Social Significance of Drought*.

Malkus, J.S., and Stern, M.E. 1953. "Climatic Effects of Surface Temperature Changes". *Journal of Meteorology* 10:30.

May, L.H., and Milthorpe, F.L. 1962. *Drought Resistances in Plants*. Commonwealth Bur Pastures and Field Crops, pp. 171-79.

Murdock, G.P. 1959. "Staple Subsistence Crops of Africa". *Geographical Review* 50:523-40.

Newell, R.E.; Kidson, J.W.; Vincent, D.G.; and Boer, G.J. 1972. *The General Circulation of the Tropical Atmosphere*. Massachussetts Institute of Technology.

Obasi, G.O.P. 1964. "Thermodynamic and Dynamic Transformation over Ikeja". *Technical Note no. 33*. Meteorological Department, Lagos.

__________; Mulero, I.; and Omolayo, G. 1980. *Rainfall of Nigeria in Pre WAMEX Sumposium*. Leo Press, Lagos.

Obisesan, P. 1981. "Africa's Sit-Tight Rulers". *Sunday Sketch*.

Odingo, R.S. 1977. *Systems of Agricultural Production in the African Areas of Drought Hazard with Special Reference to the Sahelian Zone of West Africa*. SIES, Sweden.

Ogallo, L.A. 1977. "Periodicities and Trends in the Annual Rainfall Series over Africa". M.A. thesis, Meteorological Department, University of Nairobi.

Oguntoyinbo, J.S.; Areola, O.O.; and Filani, M. 1978. *A Geography of Nigerian Development*. Heinemann.

Otterman, J. 1974. "Baring High-Albedo Soils by Overgrazing: A Hypothetical Desertification Mechanism''. *Science* 186:531–33.

Palmer, W.C. 1964. "Meteorological Drought". *Science*. Weather Bureau Research Paper no. 45.

Pearce, R.P. 1980. Review on "The Large-Scale Factors Contributing to Tropical Droughts". CAS/IME-TPM/Doc. 5, WMO, Geneva.

Pigott, C.D. 1975. "Experimental Studies on the Influence of Climate on the Geographical Distribution of Plants". *Weather*, pp. 82–90.

Riehl, H., and Malkus, J.S. 1961. "Some Aspects of Hurricane Daisy of 1958". *Tellus* 13:181–213.

Riou, C. 1975. "Le Determination Pratique de l'Evaporation: Application a l'Afrique centrale". Memoire ORSTDM no. 80, Paris.

Rokhe and Virji 1976. "Trends and Periodicities in the African Rainfall Data". *Monthly Weather Review* 104.

Resetti, C. 1965. *Ecological Survey: Mission to West Africa.* FAO-UNDP Desert locust project UNSF/DL/ES/5.

Saamnen, T.F. 1966. *Preception of the Drought Hazard on the Great Plains.* Chicago University Research Paper no. 106, pp. 144–48.

Schell, R. 1970. *Introduction a La Phytogeographie des Pays Tropicaux.* Gauthier-Villars Ed., Paris.

Simmons, I.G. 1966. "Ecology and Land Use". *Transactions and Papers* 38.

Stebbings, E.P. 1935. "The Encroaching Sahara". *Geographical Journal* 86:5.

Tapela, T.N. 1981. *Desertification: A Man-Induced Disaster* (forthcoming).

Thompson, B.W. 1965. *The Climate of Africa.* Oxford.

Trewartha, G. 1961. *The Earth's Problem Climates.* University of Wisconsin Press, Madison.

Thornthwaite, C.W. 1962. *Average Climatic Water Balance Data of the Continents.* Publication in Climatology, Centerton Lab. of Climat., U.S.A.

Trochain, J. 1940. *Contribution a l'Etude de la Vegetation du Senegal.* Memoirs-IFAN, vol. 2.

Wade, N. 1974. "The Sahelian Drought: No Victory for Western Aid". *Science* 185.

Winstanley, D. 1973a. "Recent Rainfall Trends in Africa, the Middle East and India". *Nature* 243:464–65.

———. 1973b. "Rainfall Patterns and General Atmospheric Circulation". *Nature* 245: 190–94.

LaSeur, N.E., and Adefolalu, D.O. 1975. *A Multiple-Scale Research Program in Tropical Meteorology.* U.S. Army Rep. M. Elom 69-0062F.

Le Houerou, H. 1975. "Ecological Management of Arid Grazing Lands Ecosystems". *Politics of Natural Disaster*, pp. 267–81.

13
The Environmental Consequences of Water Resource Development in the Tropics

C.J. BARROW

Seldom does nature provide man with adequate supplies of water of suitable quality, where and when it is needed. Man has long tried to improve natural supplies by either regulating or storing runoff, for example, flood control, irrigation, hydro-electric generation, and domestic and industrial consumption by means of dams, weirs, and barrages or by tapping ground water using wells or boreholes. The water so obtained may be distributed by canal, quanat, aquaduct, or pipeline; channels may be constructed or natural water courses altered to improve flow, reduce flooding, improve drainage, or even aid navigation. Increasingly, large quantities of water are required to dilute and disperse sewage and industrial effluents.

Irrigation techniques and dams were initially developed in the tropics (Toran 1973), with irrigation being sustained for millenia in, for example, Sri Lanka, South America, and in the Nile, Tigris, and Euphrates valleys. Although some systems have remained functional for very long periods (Obeng 1980a), most have failed because of bad maintenance, political disruption or social change, climatic fluctuations, and other factors (Biswas 1978). History cautions that water resource development involves a complex interrelationship between social and environmental processes that can, unless carefully managed, result in ruined soils and lost water supplies. This interrelationship is particularly complex and delicate in the tropics. It is only recently that sustained research has been conducted on the structure and dynamics of tropical ecosystems. The International Hydrological Decade and International Biological Program have helped increase our understanding of these ecosystems. Although environmental understanding has increased, it may still be argued that institutional structures and other factors may frustrate the use of such knowledge to benefit modern water resource development (Fig. 13.1). As Dasmann *et al.* (1973) have said: "There is good reason to think that development projects are spreading faster than efforts to anticipate their full consequences."

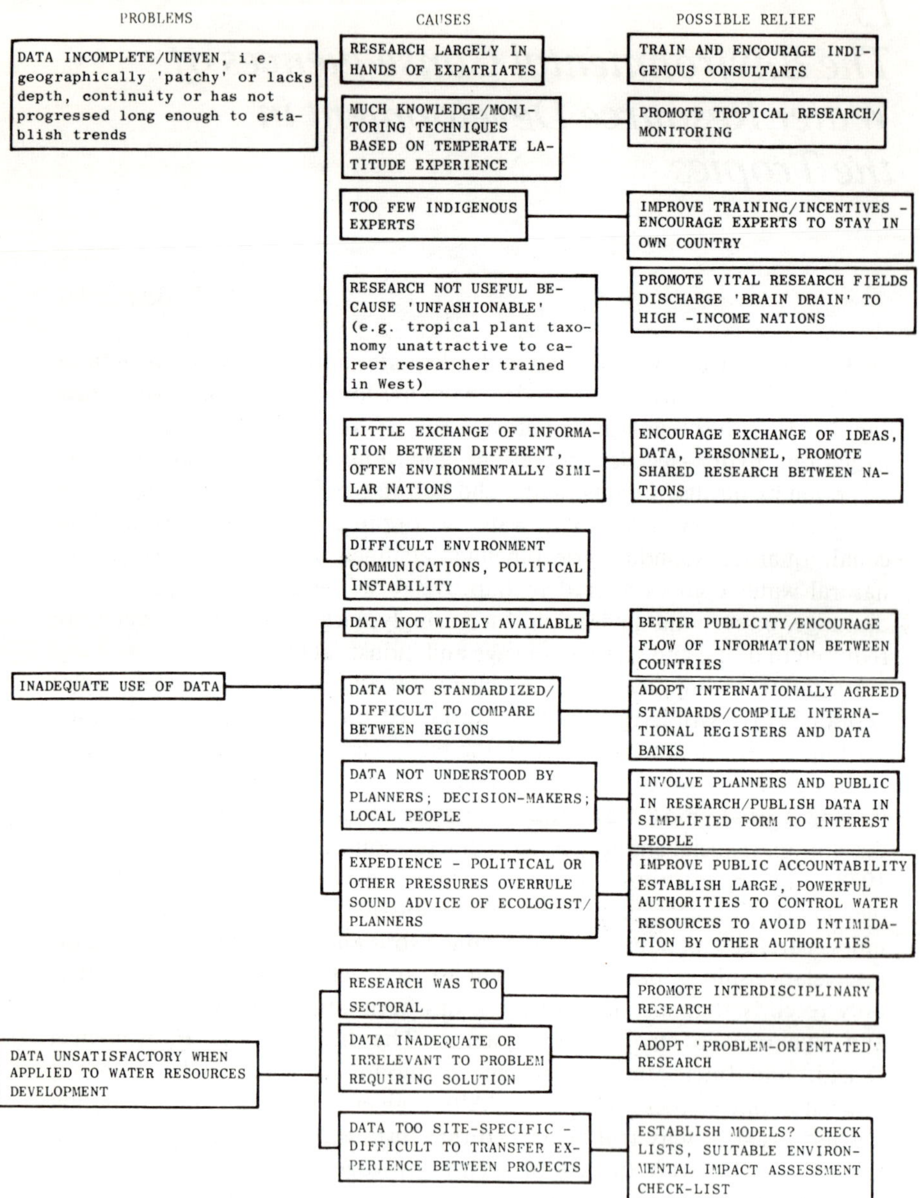

Fig. 13.1 Tropical water resource development: problems, causes, and possible relief

The Tropical Environment

The "tropics" is defined here as the region of the earth lying between the Tropic of Cancer and the Tropic of Capricorn. Between these limits, average annual air temperatures do not, under normal circumstances, fall below 20 °C (Harrison 1979; Jackson 1977). Within the tropics, marked differences in soil, altitude, and rainfall are reflected in vegetation cover, which in turn affects runoff and infiltration.

Because there is such diversity within tropical (especially humid tropical) environments, the results of development in, say, the American humid tropics may, therefore, be quite unlike those generated by developing an apparently similar habitat in Africa. If water resources developers in the tropics are to avoid environmental problems, they must ensure that careful studies of local flora, fauna, soils, etc., are carried out so that planning is not based on mere assumptions. Ideally, case studies of health and human ecology in stable and disturbed tropical environments should be executed and the results disseminated (Milton 1975:209).

Three generalized types of continuous plant cover may be recognized in the tropical environment:

Tropical evergreen rainforest. Such rainforest formerly covered most of lowland (below 1,000 m) Africa, Madagascar, southeast Africa, Central and South America, Indonesia, and New Guinea, but human activity has reduced this biome to the extent that it now covers about one-quarter of the total tropical land area — estimated in late 1975 to be 1,166 million ha (Lanley and Clement 1979).

The environment is characterized by small daily and seasonal temperature variation. Rainfall varies between 200 and 800 cm a year, although there are typically seasonal rainfall maxima. At no time does the soil dry out enough to stop plant growth. This pattern is complicated in Southeast Asia and southern Asia by the monsoons and in South America by the moist, seasonal trade winds. Precipitation may be very heavy (43 cm/day or more) and intense (15 cm/hour or more).

Runoff is moderated under natural vegetation cover but flooding may still be a problem in such regions. Streams flowing from undisturbed forest areas tend to carry little silt — the Amazon River has deposited only about 64,750 km^2 of alluvial (*várzea*) soils, though it drains about 1,295,001 km^2 and has about one-seventh of the world's total river flow (Sternberg in Dasmann 1973:56). Little nutrient leaches into streams because they are "tied" to vegetation rather than lying in the soil, tropical luxuriance being "a facade, an outward magnificence cloaking the poverty beneath" (Brennan *et al.* 1973:1–8). Within the humid tropics bedrock, edaphic and other factors may conspire to produce oligotrophic (nutrient-poor) "black water" streams.

In general, the tropics await suitable agricultural techniques for stable, long-term development, although paddy production is a stable system (heavily dependent on regulated water supply) which has evolved where the humid tropics have sufficient rainfall variation to suit the growing requirements of rice.

Tropical savannas. About half the land area of the tropics is grassland (Eckholm 1978:136). Savanna grassland, open woodlands or scattered scrub typically cover regions receiving 76–380 cm of rain a year. The climate is quite changeable, with a marked dry season during which fire often threatens vegetation. Typically, a cool-dry season is followed by hot-dry, then hot-wet, seasons with mean monthly temperatures ranging from 15° to 30°C. Rainfall intensity in savanna regions may be high; for example, in northern Nigeria 90 per cent of all rain falls in storms of 25 mm an hour or more intensity. Consequently, infiltration is reduced, resulting in strong runoff and heavy silt loads.

Some planners envisage savannas as regions of vast agricultural potential, given a suitable supply of irrigation water and adequate drainage to prevent water-logging (Grigg 1970:227). Water resources developments resulting from improved technology will doubtless result in the agricultural development of large areas of savanna, and already a high proportion of tropical water resource development schemes have been located in savanna or semi-arid environments.

Tropical shrub and deciduous forest. Occupying a place between the rainforest and savanna are biomes adapted to intermediate conditions. The lower limit of rainfall for the support of deciduous forest is roughly 100 cm a year. A marked dry season is a characteristic feature of the rainfall regime. Typical deciduous forests are the dry miombo and mopane forests of east and southern Africa, dominated by *Brachystegia*, *Berlinia*, and *Copaifera*, and in South America the zebil forest dominated by *Piptadenia*. Drier scrub vegetation is typified by the caatinga scrub of northeast Brazil and by *Acacia* and *Euphorbia* species-dominated associations in Africa and southeast Asia. Dipterocarpaceae characterize the monsoon deciduous forests of southern and Southeast Asia (Troll 1963).

Demands on Water Resources

Poverty and underdevelopment are generally attributed to the impact of colonialism, neo-colonialism, and an "unfair" world economy. Seldom is there any serious consideration of environmental problems. While gross environmental determinism is today rightly dismissed as too extreme an outlook, it must be acknowledged that the environment either supports or inhibits man's development. Little attention has been given to the relationship between the levels of economic development and climate (Biswas 1979a:239), although Harrison (1979) suggests that low national incomes are perhaps related to high average temperatures. Developing nations are almost by definition tropical nations, probably because tropical conditions hinder the attainment of an adequate agricultural surplus to support development.

An adequate supply of water is undoubtedly a major requirement for development. In the tropics, the constant solar radiation ensures that given a regular water supply plant production is potentially prodigious. However, despite considerable runoff in equatorial South America, in Africa (where almost 50 per cent of the continent's total water resources are in the Congo Basin [Biswas 1979b]) and other tropical regions, the water supply must be regulated if it is to be used to aid development. A variety of factors, including increasing population and the world economic situation have caused many Third World nations to face increasingly severe food shortages and unemployment; for example, a decade or so ago Nigeria exported groundnuts and Sri Lanka rice, but today both import these foodstuffs. The management of water resources for large-scale irrigation can help to combat such problems (Diamant 1980), particularly in regions where rainfall is seasonally concentrated or unpredictable from year to year.

Water resource development is often initiated with a view to achieving specific goals — such as irrigation to increase agricultural production or to support people resettled from overcrowded regions, hydro-electric generation, flood control, improvement of navigation, and to supply domestic water to expanding urban populations. The interaction between social, economic, and environmental factors can be complex, with water resource planners being obliged increasingly to consider regional income redistribution, or to give development priority to special interest groups (Barkin and King 1970). A planner or decision-maker may intend to interpret development as the alteration of the environment for the benefit of man with the minimum upset to natural equilibria, but the need to maximize social welfare may frustrate such an intent (UN 1972). There may be an element of national prestige involved in a major water resource project such as a large dam or an irrigation scheme. The Kariba, Volta, Aswan, Kainji, and Kossou schemes were all directly associated with national leaders (Scudder in Williams and Howard 1977; SCOPE 1972:12).

Within the tropics, socio-economic, administrative, and logistics problems which accompany large-scale water development compound problems of an already difficult natural environment. The most expertly designed water resource projects deteriorate if they are not well maintained, and maintenance depends on a government's ability to control actions within a catchment (UN 1970:2). Communications may be poor and governments' goals and methods change especially rapidly in many tropical countries.

The UN Conference on the Human Environment held in Stockholm in 1972 declared that in developing countries most environmental problems were caused by underdevelopment. The poor stewardship of natural resources which results in environmental problems is understandable when priority has to be given to the speedy reduction of human deprivation. The large, grandiose schemes carried out in a hurry by nations desperate for water supplies to improve agriculture, aid industrialization, and encourage regional development all too often result in undesirable environmental impacts. The sheer magnitude and complexity of interaction makes it almost impossible for water resource planners to predict all difficulties. Even a multi-disciplinary planning team may fail to coordinate and synthesize their knowledge sufficiently to foresee potential problems, especially social problems which are particularly unpredictable (Barrow 1981).

The broad prediction of the environmental impacts of dams, irrigation projects, drainage schemes, etc., should now be possible. Avoidance of environmental impacts is, however, more uncertain (Obeng 1980a:126). Environmental disruption may be generated by: (1) inadvertance (the lack of knowledge and expertise or inadequate application of expertise); (2) convenience (experts' warnings are ignored for various, often political, reasons); (3) the adoption of too narrow an outlook (it is not enough to consider technical feasibility and economic acceptability). In practice, engineering or economic planning objectives may determine the character of a scheme with the result that hydrological, geographical, biological, or socioeconomic difficulties may overshadow the successful achievement of the primary project goals (Farvar and Milton 1972; Farvar 1976; Sachs 1976; Dasmann *et al*. 1973:182–232).

While there is an abundance of literature on post-project appraisals of water resource development, most are subjective and lacking in empirical evidence. Few studies, if any, have had adequate breadth, enough detail, or have continued for a sufficient length of time (including pre-development studies) to give a complete picture of a given project. Authors have tended to review environmental and other impacts either for single projects or, in a few cases, from one region or nation. Each river system is unique; hence the production of generalized impact prediction guidelines awaits further studies and detailed synthesis of hindsight knowledge (World Bank 1974).

Problems of inadequate food supplies can be resolved more rapidly through improved supplies of water and improved water management than through attempts at technological innovation (Pearse 1980:223). Agricultural development, particularly the "Green Revolution" innovations, demands input of fertilizer, pesticides, and above all irrigation water (James 1978). Higher yields may be achieved by simply improving water supply and using good quality seed. However, modern, high-yielding improved crop varieties may require two to three times more water than traditional varieties.

According to Lvovich (Beaumont 1978:38), exploitation of water has only just begun, with only 21 per cent of the African continent's runoff being regulated by dams. The percentage is even lower in Australasia (6.1 per cent) and South America (4.1 per cent). Agriculture makes great demands on water resources and is therefore heavily dependent on artificial supplies. The 1974 World Food Conference estimated that 200 million ha depend on irrigation and, according to Obeng (1980a:118), 80 per cent of the world's available water is being used for irrigation. By the year 2000, two-thirds of the total world streamflow will be regulated, with much of that development taking place in tropical nations (Szesztay 1972 in Petr 1978:368). In addition to irrigated crop production, shallow aquatic ecosystems in the tropics offer tremendous potential for fresh water fish production.

The rising world oil prices are increasingly prompting nations to develop hydro-electric resources to conserve scarce foreign exchange. Furthermore, recession among developed nations prompts engineering and planning consultants to promote large-scale water resource development among tropical countries for by no means purely altruistic reasons (Johnson 1979). Increased domestic and industrial water consumption accompanies development. As per capita incomes rise, water demands increase, sometimes by as much as fifty times (from about 14 litres/day/capita to about 682 litres/day/capita). Consumption in the urban communities of developed nations may reach 7,000 litres/day/capita (Obeng 1980a:118).

The provision of safe water plays a vital role in improving a nation's agriculture and economy, because improving people's health, especially in tropical nations where water-borne diseases are more prevalent, increases productivity (Biswas 1979c). With improved water supplies, much less time would be spent in fetching water and contact with infectious diseases, such as schistosomiasis, guinea worm, and trypanosomiasis, would be reduced if people no longer visited overcrowded water collection points where flies and contaminated soil pose health risks (Biswas 1980). The provision of improved domestic water supplies would enable people to devote more time to agricultural labour. The World Health Organization estimates that about 25 million people die every year from diseases caused by unclean or inadequate water and by lack of sanitation. To remedy this situation, the 1980s have been designated by the United Nations as the International Drinking Water Supply and Sanitation Decade (Agarwal 1980; Report of the Independent Commission on International Development Issues 1980:56).

Water is thus a key resource for developing nations. Attempts to rectify the disparity between resource potential and availability will probably cause more environmental, social, and economic change and stress than the development of any other resource (Obeng 1980b). The demand for environmental quality is closely related to per capita income, and while people are denied access to safe drinking water, adequate food, and secure employment, environmental problems generated by water resource development are unlikely to receive much attention (Walker and Uglow 1979:103; Shapiro and Smith 1979:1147). Environmental attitudes to water resource development in tropical nations thus reflect circumstances (Farmer 1971; Long 1971; Walker 1978). If environmental interest exists at all in many developing nations, it is mainly at the level of concern for water quality in relation to human health. Health ministries may be involved in water resources management, but few nations have progressed to serious environmental protection. For example, in reviewing the South American situation, Biswas (1979d:32) cited only two countries, Colombia and Venezuela, which had progressed to environmental protection measures. International agencies, for example, FAO, UNESCO, WMO, and WHO, are increasingly active in water resources management. In the view of the Independent Commission on International Development Issues ("Brandt Report" 1980:94), the biggest single amount of development investment required during the 1980s to aid the Third World will be for irrigation and water management.

Modification of Tropical Water Resources

The hydrological cycle may be modified in broadly three ways: diversion, regulation, and groundwater extraction. For about seventy-five years, the focus of water resource development has been the regulation of rivers by large dams (Smith 1972). Large dams are often linked to large irrigation schemes which may generate environmental, health, and socio-economic difficulties, the latter including upset to traditional economies by surplus crops depressing market prices.

It is only during the past thirty years or so that earth-moving techniques and concrete technology have evolved in temperate developed nations to make possible the construction of inexpensive large impoundments (larger than 100 km^2 surface area) and canals. Since the Second World War, there has been a massive transfer of the technology of water resource development from temperate developed nations to tropical developing nations. This transfer process has been largely aided by intragovernmental development assistance agencies and multinational corporations (Milton 1975:207). Motorized pumps, tubewells, and borehole technology have also spread, and within the humid tropics exploratory attempts have been made to modify natural rainfall through cloud-seeding. Technological innovation has also taken place within the tropics, with peasant farming communities evolving cheap bamboo tubewells and bullock cart-mounted diesel pumps from more expensive Western factory-produced models. Such innovations will greatly modify water usage in areas served by aquifers. Milton (1975: 208) commented that "river basin development based on the construction of large dams and reservoirs has been perhaps the single, most ecologically-significant engineering modification of the environment in recent years". The features which characterize many recent dam and irrigation schemes are: (1) their large scale, (2) the speed with which they are implemented (not infrequently ahead of schedule), (3) their demands for infrastructure for construction, optimum development, and maintenance, (4) their tendency to accelerate destruction of biological resources and cultural patterns, and (5) the high risks that such schemes will tie up a substantial part of a developing nation's resources.

The ill-considered transplantation of techniques successful in one tropical region to another may result in problems, as much as the application of techniques developed in temperate environments to the tropics (Almeyra 1979). Many major projects have seriously disrupted ecosystems and peoples, which cautions that the "idolisation of the big dam" (Smith 1972) is dangerous. What worked for the Tennessee Valley Authority, or even in tropical Latin America, may not function well in tropical Asia or Africa, and vice versa.

The preoccupation with large-scale engineering approaches to water resource development has resulted in the relative neglect of alternative techniques. Flood control may be achieved not only by building a large dam but also by good flood plain and watershed management. It is difficult to separate water management from land use, as processes in the catchment determine both yield and quality of water. Road bridges designed for flows from a forested catchment may fail if that catchment is cleared and developed (Jackson 1977). With this type of problem in mind, Dasmann *et al.* (1973:208–10) gave a list of tropical watershed management considerations. Hydro-electricity generation may be obtained from environmentally benign, run-of-water axial tube turbines, and with such cheap, low-maintenance, easily manufactured installation, the need for a large impoundment and a high dam to regulate a river is dispensed with in many localities (provided flows are not too erratic). A review of the environmental impacts of hydro-power projects has been given by Ash (in Gangstad 1978:21–32).

River Basin Development and Planning

The bio-geophysical characteristics of a river basin tend to form a discrete hydrological and ecological system, and because of this river basins are often used as units for development planning (UN 1971:16; Organization of American States 1978:5; Faniran 1980). Increasingly, development involves the complex optimum development of entire river basins, often with multinational cooperation (UN 1970). Early water resources planning dealt with specific aspects of water development within a river basin. Multi-purpose planning developed later, but declined from favour because of the difficulties generated through the various competitive uses of water. Integrated basin planning sought to coordinate, develop, and harmonize water uses within a river basin with other development pressures, both within and outside the basin (UN 1970). Comprehensive river basin planning is an extension of integrated development planning, and goes beyond water resources to cover many aspects of regional and socio-economic planning. The character of river basin planning has thus evolved and expanded, with agencies today frequently having powers ranging beyond the hydrological basin, concerned with far more than simple water resource development, or even inter-basin transfer.

Already a number of authors have adopted an approach to the study and development of river basins which includes the environment. For example, Williams and Howard (1977) undertook an inter-disciplinary study of the development of the Kafue Basin, Zambia (developed primarily for hydro-electric power); El-Hinnawi (1980) has attempted to elucidate an environmentally sound approach to the development of the Nile Basin, and Lubin (1977) examined the environmental impact of developing the Senegal basin. River basin development is especially important to nations which have only one or a few major rivers, particularly when a substantial percentage of their total population may be concentrated along these rivers, as in Egypt and Zambia (Beaumont 1977).

The development of a basin's water resources may involve the regulation of flow of rivers serving more than one nation. The management of such water resources has been the subject of a UN report (UN 1975). Well-coordinated plans covering whole basins, rather than piecemeal developments are rare — "integrated river basin development is more often an aim than a reality in many developing countries" (Odingo 1977:148). River basin development in the tropics could greatly benefit if a multi-purpose approach was adopted more often, particularly if more reliable data were collected and used in the planning process.

River Basin Development: Large Man-made Lakes

The tropics, especially Southeast Asia, are poor in natural lakes (Fernando 1976). The science capable of monitoring and predicting environmental problems in lakes originated in the northern temperate latitudes from studies of northern temperate lakes. The monitoring and prediction of problems in man-made lakes in the tropics is more difficult for a variety of reasons: more rapid and complex bio-geochemical activity, more prevalent human and livestock diseases with no cold season to reduce the activity of vector organisms or to check growth of weed which may harbour them, rainfall which may be very heavy resulting in massive silt loads in streams, poor soils which may force peasant farmers to adopt practices which increase erosion and siltation of water courses, and high rates of evapotranspiration.

As water resource development impacts may be transmitted by complex, multiple routes, prediction and remedial action depend on a sound knowledge of ecosystem function and reactions to disturbances. Ill-planned changes may result in costly, possibly irreversible problems. Such changes include: urbanization which cannot easily be reversed once initiated, natural forest which cannot be replaced once cleared, soils which may be virtually impossible to rehabilitate once damaged, and fish species which are disrupted by construction may become extinct.

Decisions in water resource developments have been, and still are, dominated by engineers and economists. A range of disciplines has not been effectively introduced into water resource management — reports submitted by soil scientists, ecologists, and fisheries experts once a project is underway are not enough. A framework is needed whereby decision-makers can properly consider such contributions. Many environmental problems of water development only manifest themselves after a delay, by which time consultants would have left and development funds may be exhausted.

Each successive man-made lake has yielded data. Since the early 1970s, a number of descriptions of tropical waters, especially African, have appeared (e.g., Beadle 1974; Allanson 1973). Predictions of volumes and flows of water for hydro-electric generation, flood control, or irrigation are adequate, but aquatic biologists still face difficulties in predicting how water, as a medium for plants and animals (including disease organisms), will react to impoundment. Ecological models for prediction with the precision of physical or engineering models are rare (Benson 1973:588).

The first large impoundment in the tropics was the Kariba Lake which was completed in 1958. Many large impoundments were constructed between 1958 and 1968, so many that it has been dubbed the "decade of large dams" (Obeng 1978). Unlike most products of modern technology, large impoundments are as common per unit area in many developing countries as in high income countries: more than 40 impoundments exceed 1,000 km^2 surface area, and at least 315 are 100–1,000 km^2 in surface area (SCOPE 1972; Fels and Keller 1973; Freeman 1974:3; Mermel 1978; Gangstad 1978:45–46). Biswas (1979d) noted that in Latin America the dams constructed in the 1960s were almost five times the previous average size; those built in the early 1970s were almost five times the size of those built in the 1960s (see Table 13.1).

TABLE 13.1 SOME DETAILS OF LARGE TROPICAL IMPOUNDMENTS

Scheme	Country	Approximate Date Full	Approximate Surface Area (km²)	General Vegetation Cover	Approximate Maximum Drawdown (m)	Ratio of Inflow to Outflow	Original, Planned Purpose (If Known)
Volta	Ghana	1969	8482–9000	Savanna/woodland	3–4	1:4	Hydro-electric power
Kariba	Zambia/S. Rhodesia	1963	4300–5250	Savanna/woodland	9–14	1:9	Hydro-electtic power
Kainji	Nigeria	1968	1500	Savanna/woodland	10–11	4:1	Hydro-electric power, flood control, fisheries, navigation
Kossou	Ivory Coast	1978	1600	Savanna	3		
Aswan High	Sudan/UAR	1974	4000–6216	Desert	10–20	1:2	
Cabora Bassa	Mozambique	1975	2700	Savanna/forest	31	—	
Nam Pong	Thailand	1965	—	Monsoon forest	—	—	
Lam Pao (11 projects)	Thailand	1963–71	—	Monsoon forest	—	—	
Pa Mong	Thailand	Projected	4000	Monsoon forest	—	—	
Nam Nguam	Thailand	1971	—	Monsoon forest	—	—	
Upper Pampanga	Philippines	1973	—	Tropical rainforest	—	—	
Ord	Australia	1972	3089	Savanna/monsoon forest	Considerable	—	Irrigation
Brokopondo	Surinam	1974	1560	Tropical rainforest	3	4:3	Hydro-electric power
Sobradinho	Brazil	Projected	4500	Tropical rainforest	36	—	

Source: Petr 1978; Freeman 1974; Fels and Keller 1973; Lowe McConnell 1969.

The primary purpose of virtually all the large dams of the 1960s and early 1970s was hydro-electric generation. Sri Lanka and Brazil already generate nearly all of their electricity from hydro-electric sources. Recent schemes, especially in Africa, are also aimed at increasing food production (Odingo 1977).

The only large man-made lake to be created in a developed tropical nation was Lake Argyle in Australia, created by the closure of the Ord River Dam in 1972. The opportunity for solid scientific research on what was both a pioneering scheme for tropical Australia and a scheme devoid of many of the financial, political, and logistic problems of developing nations was missed. Petr (1978:381–82) regarded the environmental studies made as sparse and inferior to those conducted in tropical African nations. The scheme, in his view, was a "monument to political expendience".

Dam construction is not a new practice, for small and medium-sized dams abound. In some countries, the total area of small dams is considerable; for example, the N'cema Dam project in the Limpopo basin, Africa is reported to comprise 464 dams with a total area of 209 km^2 (Obeng 1977:46). There are probably 10,000 or more reservoirs in Sri Lanka covering about 70,000 km^2 (Fernando 1980, pers. comm.). While small dams may generate public health problems or affect ground water distribution, it is generally accepted that the larger the reservoir the greater and more complex are the upsets to environmental and socio-economic systems, and the more difficult the rectification of such problems. Small dams, provided siltation problems can be overcome, will probably play a central role in aiding in Third World's small farmers to increase crop production. Massive new dam-building and river control programmes are projected for the not-too-distant future in some parts of the tropics, and environmental problems are thus likely to continue to be generated.

Although biological systems are more complex and unpredictable than physical systems, the major hindrance to the development of predictive models has been the fact that hardly any large dam has been studied in sufficient detail, or for long enough (even in the United States) for long-term implications to be clearly established, as "no tropical reservoir effort has yet incorporated into the project a system of comprehensive monitoring of social and natural ecosystem change" (Milton 1975:208). The development of ecologically sound alternatives to current engineering practices will continue to be hampered until comprehensive, integrated environmental research is incorporated into all phases of river basin projects from initial pre-development surveys to final post-development evaluations.

Tropical and sub tropical dam projects are frequently situated in remote, inhospitable terrain, which makes environmental studies both difficult and costly. The proposals for environmental studies for the Parari Hydropower

Scheme in Papua New Guinea, for example, was to cost US$2.3 million — far too large a sum for a developing nation to afford unless aided by international assistance. Political difficulties, such as the break-up of the Rhodesian Federation after the completion of the Kariba Lake, or the political unrest when the Cabora Bassa Dam was filling, make the collection of data and monitoring of reservoir developments difficult. To date there have been numerous partial analyses but few analyses of the complete dam construction – reservoir stabilization system. It is also difficult to construct generalized guidelines from such research as each new project is virtually unique with conditions which alter from one river basin to another and often involve human activities within the catchment, which are especially unpredictable (World Bank 1974; Obeng 1976). The 1972 SCOPE Report summarized the present situation: "information so far collected has not been synthesized in a way to enable formulation, without further analysis, of a useful guide for setting detailed policy for new reservoirs" (p. 17).

Damming a river rapidly alters riverine and valley terrestrial ecosystems. Conditions in the man-made lake are unstable, transient, and difficult to predict for at least ten years (Margalef 1975:1842; SCOPE 1972:62; McConnel and Worthington 1965). Stability is never fully assured as changes in water management, lack of rainfall, excessive drawdown, the introduction of an exotic species of plant or animal may all upset the reservoir's equilibrium (Bandler 1975:580).

Five stages may be recognized during the creation of a man-made lake: (1) pre-impoundment, feasibility study stage (particularly subject to political influences), (2) a final design and construction stage, (3) dam closure, initiating a period of gross stability lasting for perhaps six to ten years, (4) a "stabilization" phase, between the decline of phase 3 disruptions and imbalances and the attainment of "stability", and (5) the final phase "stabilization" which is largely controlled by project design (determined during phases 1 and 2 and subsequently difficult to alter), management strategy, and processes within the river basin.

Although no detailed guidelines exist, researchers have produced graphical representations of the physical, chemical, and biological changes which result from impoundment of tropical rivers (Freeman 1974; Petr 1978) in an attempt to provide general guidelines for policy-makers and planners. Such guidelines are vital if environmental and human problems are to be avoided and if the life of a man-made lake is to be extended to its maximum.

Those creating large man-made lakes in Africa, South America, Asia, and Southeast Asia during the 1960s and 1970s had little hindsight knowledge on which to draw as the great African impoundments — Kariba, Aswan, the Volta, and Kainji — were pioneer efforts. Several symposia have subsequently presented planners with information (Lowe-McConnel

1966b; Rubin and Warren 1968; Obeng 1969; Ackermann *et al.* 1975; Stanley and Alpers 1975), and attempts have been made to list or review the experiences gained in creating large lakes (Edgcomb 1965; Baxter 1977; UN 1970; Obeng 1978) and even to produce reservoir planning guidelines (e.g., Lagler 1969).

Perhaps the best general review of the socio-economic, physical, and biological impacts of tropical impoundments and irrigation schemes is that given by Dasmann *et al.* (1973). The bulk of information was generated covering the 1960s and early 1970s in West Africa (Bardach and White 1969; Lawson 1970; Visser 1970; SCOPE 1972), with the Aswan Scheme adding to the flow of knowledge. Information from outside the African continent has been provided by the Brokopondo Project in Suriname (Leentvar 1966; 1975), while Milton (1975:209–12; 214–15) and more recently Biswas (1979) have summarized the environmental impacts of man-made lakes in the American humid tropics. Information from Asia and Southeast Asia, apart from the Mekong basin (e.g., Bardach 1972) is more scarce. Petr (1979) has examined the environmental impacts of water resource development in Papua New Guinea, and Pantulu (1979) presented a case study of the environmental aspects of the Pa Mong Project, Thailand (see Table 13.2). A general review of development impacts on water resources in Peninsular Malaysia has been given by the author (Barrow 1980).

With the notable exception of the Volta Preparatory Commission (HMSO 1956a; 1956b),[1] most information has resulted from studies commissioned only when problems have become manifestly apparent, or has resulted from university, government, or international agency research, funded only while construction was in progress or completed. In general, research has started too late and on too small a scale. However, international and national organizations, especially in the United Nations, have now begun to co-ordinate research on man-made lakes (Petr 1978).

The SCOPE Report of 1972 has so far been the most comprehensive summation of knowledge on tropical man-made lakes. Disjointed and fragmentary research, plus the fact that man-made lakes may still be unstable ten years after filling, has prevented a "holistic picture of the total impact of a reservoir in a general geographical context". This has resulted initially in largely speculative management, which has been generally changing to systems which introduce corrective measures on the basis of firm scientific and socio-economic knowledge. Rationalization is thus replacing adventurism (Petr 1978).

---

[1]There are still considerable gaps in knowledge about the Volta Lake mineral cycles, sedimentation, and the effects of the scheme on groundwater (Freeman 1974).

TABLE 13.2 SOME ENVIRONMENTAL PROBLEMS ASSOCIATED WITH MAJOR TROPICAL DAM PROJECTS

Volta	Kainji	Aswan High	Kariba	Cabora Bassa	Brokopondo	Mekong Basin
*increased weed caused increased schistosomiasis until 1968 *sea water incursion upset clam fishery *lateral fish migration stopped *loss of fertile silt left by floods *flow too fast for women to work clam fishery *river blindness increased initially due to weed (solved by controlled release of water). †algal blooms +weed growth: (*Pistia*; *Salvinia*; *Azolla*) ‡marginal vegetation growth favours tsetse ‡seismic activity	*decreased fish catch *migration hindered †schistosomiasis, increased incidence †costly clearance of vegetation before flooding; not much value †algal bloom †weed growth (*Pistia*; *Salvinia*) ‡schistosomiasis	*nutrient loss affects marine and marine fisheries *loss of silt for agriculture *erosion of Nile Delta *damage to bridge foundations, etc. *nutrients leached from irrigation schemes *raised water-table †evaporation greater than expected; lake unlikely to fill to projected level †stratification ‡raised regional water-table ‡schistosomiasis ‡seismic activity ‡malaria increased	*fish migration upset *water-table downstream reduced affecting valley grazing †stratification †weed growth (*Salvinia*) †deep water deoxygenation ‡schistosomiasis increased markedly ‡trypanosomiasis increase on south shore ‡seismic activity	*fish migration hindered *water-table downstream reduced affecting valley grazing *erosion of channel & estuary †siltation of reservoir	*hydrogen sulphide-rich water †fall in Ph from 6.5 to 5.5 + weed growth (water hyacinth) †deoxygenation †stratification ‡schistosomiasis ‡malaria increased incidence (following use of herbicide in water hyacinth, dead plants provided mosquito habitat)	*fish migration †Nam Pong Dam: weed growth ‡loss of wildlife ‡liver fluke disease, increased incidence.

Notes: *downstream impact
          †impoundment problem
          ‡surrounding region

Environmental Problems Generated by Large Reservoirs

Water quality. In the humid tropics, high rainfall, nutrient-poor crystalline rocks, and "tight" nutrient cycling under forest vegetation may result in nutrient-poor (oligotrophic) runoff. Acidity may range from high pH (rich in cations) to low pH flows. Streams of pH 7 on less (pH 3.8–4.9 is not unknown) may be rich in humic compounds — the "blackwaters" of tropical areas. The productivity of nutrient-poor tropical rivers is heavily dependent on detritus from forest vegetation. Such productivity may be reduced if the forest cover is removed. Eutrophic waters are those well supplied with nutrients.[2] Intermediate between these extremes are mesotrophic flows of medium nutrient content. The character of a natural river flow may thus be altered by forest clearance, leaching of agricultural chemicals, fertilizers, or pollution from agro-industries or human sewage. The river basin developer should be aware of the quality of the water which is to be developed. It is quite clear that lakes formed in regions of tropical forest differ from those in desert or savanna woodland zones (Petr 1979). Allanson (1973) stresses that differences are inevitable between reservoirs in regions of high rainfall with constant inflow and those seasonally fed by variable river flow.

During the unstable transition period following the closure of a dam, biological productivity in these man-made lakes rises (as flooded vegetation and soil release nutrients into the water) and then declines to a more "stable" state. How long this process takes and the degree to which the lake water nutrient content increases depends on a range of factors: (1) the nutrient status of inflowing water, (2) the ratio of inflow to outflow, (3) depth, (4) circulation, (5) the quantity and rate of release of nutrients from drowned vegetation (a process accelerated by burning vegetation before flooding to reduce risks to fishing gear and navigation), and (6) the flora and fauna able to colonize the suddenly created lake.

Oligotrophic "blackwaters" like the Caroni River, Venezuela (impounded by the Guri Dam), are less likely to suffer weed, algae, and snail-borne disease problems than might richer flows. The calcium-deficient water hinders shell formation and is thus unfavourable to aquatic snails — for this reason some tropical impoundments have escaped schistosomiasis problems. Nevertheless, it is rash to generalize, as some reservoirs in Africa, for example, the Volta have a more or less oligotrophic inflow but are still not snail-free (Freeman 1974:63). The rate of through-flow (ratio of inflow to outflow) greatly determines the character of a reservoir — surface water storage essentially postpones runoff, sometimes for more than a year. A rapid throughflow tends to flush out nutrients from the impoundment;

---

[2]The term is frequently misused to imply excess nutrients resulting in pollution; water may be well supplied without ecological problems arising in every case.

however, if the inflow and outflow are not balanced and result in drawdown (the seasonal exposure of marginal shallows), game and cattle may graze these areas and their dung may rapidly return nutrients to the lake upon flooding (Baxter 1977). To increase inflow to reservoirs in some humid tropical countries, for example, in Colombia and Malaysia, experiments with cloud-seeding have been tried. The results are inconclusive although there would appear to be a risk of generating sudden, intense storms which may bring silt-laden water.

Abundant nutrients are not the sole cause of excessive weed or algae growth. Lake currents, areas of shallows, and the anchorage provided by drowned trees also determine the location and the nature of problems of biological productivity in man-made lakes. Planners without well-defined guidelines can only bear in mind the nutrient status of a stream and adjust monitoring programmes accordingly. Generally, oligotrophic rivers in the tropics are best for hydro-electric impoundment (turbines are less liable to suffer from corrosion or electrolytic deposition and weed growth is also less likely). Mesotrophic and eutrophic rivers, while offering more risk of excessive weed growth (and indirectly, disease) if impounded, are suitable for diversion for irrigation (Beauchamp 1969:4; Goodland 1980).

The time taken for man-made lakes to "stabilize", according to research in the Soviet Union, is probably of the order of six to ten years for reservoirs in latitudes of less than 50° and 25 to 30 years in higher latitudes (McConnel and Worthington 1965). In the case of Lake Kariba (the oldest of the large man-made lakes), unstable eutrophic conditions persisted from 1958 to about 1967 (Petr 1979:374). Petr (1978) felt that changes continued to take place in tropical man-made lakes eight or more years after impoundment, but that such changes became slower with time.

Stratification. Stratification in tropical reservoirs is inadequately understood (Goldman 1974; Beadle 1974) despite attempts to clarify the process in both natural and man-made lakes (Beauchamp 1969; Henderson 1973). A tropical lake, lacking the seasonal turnover which occurs in lakes in more temperate climates, may become stratified if inflow and outflow rates are such that the impounded water changes only slowly. In a stratified lake very little vertical mixing of the dissolved gases or nutrients occurs, so that the lake depths become depleted in oxygen and often rich in products of decomposition. In short, the lake becomes stagnant.

Whether or not a tropical lake becomes markedly stratified depends on altitude, wind exposure, and particularly on depth. Steep-sided and sheltered reservoirs seem especially prone to stratification as are deeper bodies of water. Cold inflows from mountain catchments, or water cooled in shallows at night, may also determine the degree of stratification. When the process is fully understood, it may be possible to plan tropical reservoirs to utilize natural features which aid mixing (Goldman 1979:5), thereby avoiding the creation of an ecologically depauperate lake (releases of water from which may kill downstream life). Mixing of water can be so poor that below even only a few metres depth, anaerobic conditions kill fish and generate hydrogen sulphide in solution which may precipitate on turbines as sulphides, and may render the water unpalatable or poisonous downstream from the dam. The Siranamum Dam near Port Moresby, Papua New Guinea has developed just such a high hydrogen sulphide content through anaerobic decomposition under stratified conditions (Goldman 1979); the waters of Lake Ayame, Ivory Coast, have damaged hydro-electric turbines through corrosion and sulphide deposition (SCOPE 1972:58). If stratification exists at depth in an impoundment, there is always a risk that marked changes in wind or other weather conditions will mix the deoxygenated bottom waters with adequately oxygenated surface layers causing massive kills of fish (Baxter 1977:270; Petr 1979:179). Trees drowned when an impoundment is created may remain sound in deoxygenated conditions and therefore pose a danger to shipping or the nets of fishermen for decades. This is a problem in the Gatun Lake, Panama (Addo-Ashong 1969; Lawson 1970). Of the large African reservoirs, Kariba exhibits some degree of stratification, but the Volta has so far remained unstratified (Obeng 1977).

Weed problems.   The generally high level of nutrients in a reservoir during its early stages of development will cause aquatic weed and algae growth to increase. Similar growth will result if development around the lake increases the nutrient output. During the "unstable" phase, blooms of algae and excessive weed growth may lead to deoxygenation of the water, hinder fishing and transport or harbour disease vector organisms. Algal blooms (e.g., *Microcystis spp.*) may poison fish and species like *Anabaena* may even cause gastroenteritis in children, adults, and cattle which drink from the lake or from its outlets (Beadle 1974:305). Plants unnoticed during pre-impoundment survey may spread from remote parts of the catchment or may be introduced by wild life or fishermen. Typically, in the newly created environment invading species tend to increase rapidly. Little (1969), Obeng (1973), and Hall *et al.* (1969) have provided considerable insight into weed problems in African reservoirs. Gangstad (1978) has also examined the benefits and problems which aquatic weeds present to the river basin manager.

The greatest problems have been caused mainly by free-floating species such as the water fern (*Salvinia auriculata* and *S. molesta*), water lettuce (*Pistia stratoides*), water hyacinth (*Eichhornia crassipes*), or various filamentous algae. On the Bangweulu Lake, Zambia, a floating grass *Vossia* has proved a nuisance (Gaudet 1979) and on the Volta Lake the water fern *Azolla africana* has proved troublesome. *Salvinia* (like *Eichhornia*, a native of South America) has caused difficulties on Lake Kariba (Boughey 1962) and also in India (Thomas 1979). *Eichhornia*, presently the world's greatest nuisance weed, is a problem in Argentina, Australia, Bangladesh, Brazil, China, Egypt, India, Indonesia, Nicaragua, Pakistan, the Philippines, Puerto Rico, the Sudan, Thailand, Zaire, and many other countries. The weed deoxygenates water, reduces nutrient content and hence fish stocks, provides a suitable habitat for disease-carrying mosquitoes and snails, and increases evaporation losses from reservoirs and irrigation channels, which it also chokes. Estimates by the Mekong Committee have suggested a cover of *Eichhornia* increases evapotranspiration losses from a reservoir by 1.1 to 3.5 times or more that of an unvegetated surface. This is a major problem in small reservoirs and canal systems. Biswas (1980) has estimated that transpiration losses of water from *Eichhornia* at the Aswan High Dam amounted to US$216 million. *Eichhornia* may also increase siltation when it grows at lake inlets or outlet channels or in irrigation supply channels because the plant slows down the rate of flow. Within two years of its invasion, 410 km² of Lake Brokopondo, Surinam, were covered by the weed. In the 1950s it invaded the Congo system, seriously affecting navigation and fishing (Biswas 1980; Milton 1975).

Floating "sudd" vegetation (mats of species such as *Cyperus papyrus* and *Scirpus cubensis*) may hinder traffic and turbines and may enable other plant species to colonize the lake and also displace water, so reducing the storage capacity of the reservoir. This "sudd" vegetation may become anchored by submerged trees making clearance very difficult.

Submerged and semi-submerged weed species such as water lilies (*Nymphaea* spp.), "pond weeds" (*Ceratophyllum demersum* or *Elodea* spp.) are a world-wide nuisance when they multiply in reservoir shallows and downstream channels. They provide cover for disease vectors, as do the floating weeds, especially in Africa.

However, where the weeds are present in limited quantities they provide anchorage for invertebrates on which fish may feed and in which fry may seek refuge and adult fish spawn. Weeds in excessive quantities may be costly and difficult to control. The herbicide 2, 4-D will kill water hyacinth (Turner 1971), but the rotting remains may pollute the lake and cause blue-green algae blooms, while the herbicide itself is a threat to wild life. Efforts are being made to find a suitable biological control, such as the release of the manatee, grass carp (*Ctenopharyngodon idella*), or *Tilapia rendalli*, or suitable insects. The main spur is the need to control human schistosomiasis. *Eichhornia* has been blamed not only for harbouring the aquatic snail vectors of schistosomiasis but also for extending the northern limit of the two most important snail vectors beyond their former range on the Nile (Hammerton 1972:201).

In time the weed may even become a resource rather than a nuisance. *Eichhornia's* rapid, prolific growth and capacity to remove pollutants, especially heavy metals, from water slowly flowing past it in shallow channels may provide a cheap treatment for sewage and industrial pollution, as well as a source of methane, alcohol, fertilizer, charcoal, fish culture, cattle and pig food. In China and Singapore, the weed has already been used for pig feed (Milton 1975:211; Land 1980), and in America NASA is investigating the weed's potential (Wolverton 1980).

Siltation. Of major importance to the longevity and ecology (including fisheries yield) of a reservoir, and below the reservoir to riverine and inshore marine fisheries and seasonal flood-agriculture, is the silt load (Goldman 1979; Glymph 1973; Christiansson 1979). Excessive silt deposition reduces the storage capacity of reservoirs and may choke irrigation and navigation channels. Silting problems are encountered through the miscalculations of siltation rates or because of altered agricultural practices. Perhaps half of mankind is affected by the way watershed lands are used. Yet those generating silt by their practices are generally unaware of the link between say upland deforestation and lowland flooding and other silt-caused problems. Excessive silt may reach streams in the humid tropics as a result of logging and logging road construction or through the activities of shifting cultivators or squatter farmers. In the semi-arid and arid tropics, savanna grazing may increase silt loads, as in Argentina where livestock has increased dredging costs in the Plate Estuary, and industrial or extractive development may choke rivers. Malaysia, for example, has suffered greatly from tin tailing waste. The Anchicaya Hydro-electric Project, Colombia (1955), built to serve the city of Cali, relied on erroneous estimates of silt loads. These were aggravated by nearby highway construction, which under humid tropical conditions resulted in severe erosion and also attracted (by aiding communications) agricultural settlers into the forest, to the extent that by 1967 80 per cent of the reservoir's storage capacity had been lost despite costly dredging (Milton 1975). Worries have also been voiced about the rate of siltation in the Cabora Bassa Dam, which may exceed pre-impoundment projections (Boulton 1978). The reduction of mainstream flow as a consequence of dam construction may result in deposition and reduction of channel depth and width where tributaries join a river (Petts 1980).

Excessive silt damages aquatic ecosystems, altering or killing fish populations, weed, and invertebrates. A sudden influx of silt may induce weed or algae growth in a reservoir if the silt is rich in nutrients. A man-made lake affects a river's siltation. The river flow upstream is slowed by the lake and silt may accumulate to the extent that flooding may result. With the lake acting as a silt trap, downstream flow may more actively erode banks, bridge supports, and stream channels.

In the humid tropics, landslips may enter a reservoir and cause a catastrophic breaching of the dam. The smaller reservoirs are particularly vulnerable to this danger. In 1963 the Viant Dam in Italy suffered such a landslip, killing 2,000 people. Such a disaster, although in a Mediterranean climate, is a warning to humid tropical nations where landslips are a year-round risk. In many developing nations with densely populated river valleys the death toll would be much higher.

Drawdown.　Release of water during the dry season to power turbines, to maintain downstream flow, for irrigation, or to make room for incoming flood water in the next wet season results in drawdown, particularly in arid and semi-arid regions. There is some indication that drawdown reduces the diversity of benthic lake organisms. However, the grass and herbs that may seasonally flourish provide extensive grazing for game and habitats for wild animals and birds. If the lake level changes abruptly or unpredictably, some of the wild life may suffer, others may actually depend upon inundation areas for spawning and feeding (Baxter 1979). Drawdown areas may provide breeding areas for insect pests, particularly locust species which may then devastate surrounding vegetation.

Climatic change caused by large lakes.　A large body of water may affect local climates, especially in cooler climatic regions. However, in most cases accurate pre-impoundment data are lacking, so that a clear picture has not yet emerged. Local air temperatures may be altered, in particular the diurnal range of air temperature may be moderated in the vicinity of a large lake. Onshore and offshore breezes, increased fog, cloud and slight changes in precipitation may also occur but these are difficult to assess (Bardach and Dussart 1973; Baxter 1977; DeHeer Amissah 1969).

Seismic changes.　The filling of a lake undoubtedly results in some seismic activity. At Kariba and Kayna (India) activity has reached or exceeded Richter six (Biswas 1980). It is unlikely that the activity results from the sheer weight of water; it would appear more likely to be the result of increased groundwater pressure acting on rock fissures aiding the release of existing geophysical stresses or adding enough pressure to trigger them off. Deeper lakes may pose more risk compared with large but shallow impoundments (Gupt and Rostogi 1976; Nace 1974). Whatever the cause of seismic activity, the effects are traumatic as seen in the 1967 Kayna Dam disaster, the epicentre of which coincided with the dam site (and was probably caused by the reservoir). The collapse of the dam resulted in more than 200 deaths (Biswas 1980).

Effect on ground water.   Following impoundment, local water-tables may well rise within a river basin. The Aswan High Dam is reported to have considerably raised regional water-tables (Biswas 1980), while in the Nile Valley ground water once 14 m or more below the surface is now 0.5 to 8 m deep and still rising, thereby endangering archaeological relics. Irrigation may be hindered by regional water-logging and salinization caused by raised ground water levels.

Wild life.   Every attempt should be made, before developing a basin, to weigh the benefits of resource development against the loss of wild life, among which there may be species of genetic potential or tourist attraction value. River valleys are frequently among the richest environments for the wild life of a region and, therefore, flooding such an area will have particular impact on flora and fauna.

Dams may all too easily eliminate unique wild life and have, as a result, been the subject of conservation debates in many nations. In Southeast Asia the development of the Lower Mekong basin will doubtless eliminate many species. While dam construction has been halted for conservation reasons in the United States, it is unrealistic to expect tropical developing nations to halt much-needed schemes or to invest much in pre-impoundment inventories of wild life. In some cases, a reservoir may provide increased cover and grazing and may, by providing water-sport attractions, draw tourists which may help make the development of a lakeshore game reserve economically attractive.

There are many varieties of migratory species of fish in tropical rivers. As species react differently to environmental stimuli, conditions and time of migration spawning vary. It is thus inevitable that dam construction will hinder the passage, or alter the environmental triggers necessary for spawning of some species (even if fish-passes or ladders are installed), resulting in some extinctions. The provision of fish-passes is a waste of time if during construction a dam prevents a species from spawning; by the time facilities are ready the species may be so reduced in numbers that it may not recover. Many migratory fish rely upon their sense of smell to return to their spawning grounds and may be disorientated by interbasin transfer or changes in water quality as a result of the development of water resources. There is much still to be learnt about the fishery potential of man-made lakes — catch returns in particular have often been unreliable when migrant fishermen camp in remote areas.

Several fish species have been deliberately introduced to new habitats, man-made lakes, or irrigation systems to try to control mosquitoes, or to utilize vacant niches which the pre-impoundment riverine fauna could not adapt to (Lezek and El-Zarka 1973, Ridley and Steel 1975). Not all introductions have, however, been intentional. Water development may enable some species to extend their ranges. Inevitably, water resource development will have great effect upon the fish faunas of Africa, South America, Asia, and Australasia. The evolution of fresh water fish species in South America and Africa owes much to complex past-history of isolation and occasional linkage of centres of fresh water fish evolution. Increased contact between river basins by water transfer and the introduction of species to new waters will break down the degree of isolation that has prevailed in the past. The dual interest of controlling algal blooms and of utilizing vacant, unproductive niches (generally the surface layers of large lakes well away from the shore) has resulted in the introduction of pelagic, plankton-feeding species like *Limnothrissa miodon* (Lake Tanganyika "sardine") and *Pellonula afzelius* to African impoundments and *Tilapia* spp. (especially *T. mossambica*). *Tilapia* (e.g., *Tilapia nilotica*) and the Nile perch (*Heterotis niloticus*) have been introduced into Southeast Asia to improve the fish productivity of certain reservoirs (Fernando 1976). There is need for special caution against the introduction of predatory "sport" fish such as the Nile perch into reservoirs. Unfortunately, introductions can occur by accident, carried by wild life, flood, or through the actions of misguided persons (as Florida has learnt to its cost, having acquired a lengthy list of unwanted exotic species). In all probability, the range of introduced species will expand as governments experiment with fish species for weed, snail, or insect control. The "mosquito fish" *Embusia affinis* has been especially widely employed to control mosquito larvae. Diamant (1979:54) notes that more than 265 species of fish have been used against 35 mosquito species in more than forty countries. Some of the species will become more a nuisance than a benefit unless great care is taken in screening such species before release. Not all "biological control" introductions need be predatory; for example, the snail *Marisa cornuarietis* may be released to compete with and replace snails which harbour schistosomiasis (Diamant 1979).

The plight of terrestrial wild life during dam construction has attracted much attention. As Balon (1978) has commented, "the plight of Lake Kariba's (50,000-odd) refugees caused less concern than the much publicized "Operation Noah" (which was a costly scheme to rescue wild life from the rising waters of Lake Kariba). However, not much appears to be known about what happens to animals forced (or rescued) from a river basin impoundment. In all probability, conflict will arise over finding new territory (SCOPE 1972:58).

Health impacts.   Alterations to natural flora and fauna as a result of impoundment or irrigation development has resulted in changed patterns of human and livestock disease, which in turn may, by discouraging settlement or grazing, affect soil erosion rates and vegetation cover. The health implications of damming rivers and of creating large irrigation schemes has been widely researched (Stanley and Alpers 1975; Gangstad 1978:36–38). The health and settlement consequences of large river impoundments in developing nations has also been reviewed recently (Barrow 1981).

Downstream impact of large dams.   According to Milton (1975), lake fisheries seldom compensate for downriver or even marine catch reductions caused by the construction of a large impoundment, for even if the lake yields are good they are often under-utilized. Many riverine species depend on seasonal variations of flow to stimulate spawning, or to make food available by flooding riverine land. The regulation of flow by a dam upsets such flows. The release of water to meet short-term fluctuations in hydro-electricity demand may upset downstream fish populations. The sudden release of water from a dam may remove weed, disease-carrying snails or blackfly larvae from downstream channels. However, blackfly (*Simulium*) may breed in poorly designed outlet channels if the water becomes turbulent enough to be well-oxygenated.

While a single dam, if provided with fish-passes, may not hinder migrant species too greatly, a series of dams compounds the effects, which may then be catastrophic. For example, the Kariba Dam failed to halt migrating eels, but the addition of the Cabora Bassa Dam to the Volta halted their migration.

Many of the effects of a dam on riverine environments downstream are roughly the reverse of the effect of a dam on upstream environments; the lake may slow flow and oxygenate upstream and by retaining nutrients, silt, and heat make downstream flow more erosive, richer in oxygen (depending on the depth of withdrawal), poor in nutrients and colder.

The downstream effects of the Aswan dams of the Nile have attracted considerable attention. The Nile is hydrologically the best-known river in the world, with records of flows going back to A.D. 660 (Oglesby *et al.* 1972:176; Jackson 1975; Abul-Ata 1979; Kaises 1973). The reduction of the Nile's silt load through the impoundments acting as silt traps has denied Egyptian farmers fertile silt for their land. Seasonal inundation agriculture was affected by impoundment, as the reduction of annual floods has removed the means of delivering the free natural fertilizer. Erosion and marine incursion is affecting the Nile Delta and nearby coast and fish production, even well away from the Nile in the south-eastern Mediterranean, has suffered, as seen in the smaller catches of sardine, scombroid, and crustacea catches (George 1972). The south-eastern Mediterranean sardine, *Sardinus maderensis*, catch is reputed to have fallen from 18,000 to 500 t a year, and according to Turner (1971), the largest annual catch Lake Nasser is likely to yield is 5,000 to 5,500 t.[3] The foundations of old bridges and barrages, which for decades have been stable, have suffered structural damage since the Nile was dammed, because of reduced silt loads and the increased erosive capacity of the river. Turner (1971) claims the Aswan Scheme could have been planned to avoid trapping silt (thereby avoiding many of the environmental problems) by using the Wadi Rayan Depression as a separate storage reservoir linked to the Nile by a canal.

During the construction of the Volta dam, flow was interrupted allowing sea water to penetrate further inland than normal up the Volta River. This changed the usually slightly brackish conditions required for the commercially important clam *Egeria radiata* to reproduce. When flows were finally restored, they had been so altered as to affect the catches of the clams (Hilton and Kowu-Tsri 1970).

While the regional water-table in the vicinity of a large dam may be raised, fears have been voiced that the construction of large dams such as Kariba or the Cabora Bassa will lower downstream valley water-tables so much that the capacity of the flood plains vegetation to sustain even the grazing of wild game will be severely reduced (Petr 1978:377). Petr also warns against the risks of damming rivers which feed inlands lakes without outlets, such as Lake Chad in Africa. The Caspian Sea, even though it does not experience tropical evaporation conditions, has suffered a marked increase in salinity as a result of the regulation of the Volga, and this has drastically affected biota. With greater evaporation in the tropics, lakes like Chad could be rapidly altered should their inflows be dammed.

Water drawn from deep in a reservoir may be supersaturated with gases; air may also be mixed into water by passage through turbines or if the water tumbles down spillways into a deep basin. Fish exposed to such water may develop "gas bubble diseases" and be killed for considerable distances downstream of dams. In the depths of a tropical reservoir, water may be significantly cooler than is normal in a flowing river at such latitudes. If such water is released, it may chill the river below a dam, which will result in damage to the biota. These problems are difficult to overcome once a dam is completed and, therefore, it is desirable to consider them at the design stage. With suitable monitoring equipment and better design of outlets, water quality of dams outflow could be controlled (Bardach and Dussart 1973:815; Allanson 1973:484).

---

[3]There is a risk that the rich fishery potential of the Gulf of New Guinea could be similarly reduced by the development of the Purari River for hydro-electricity (Goldman 1979).

Environmental Problems Generated by Irrigation

Often linked with large dams or barrages are large irrigation schemes. An example is the Gezira Scheme in the Sudan. Over the past few decades, irrigation has spread rapidly and will probably continue to do so in the coming decades.

After a brief period of increased crop yields, irrigated agriculture all-too-often encounters problems of salinization or alkalinization. Millions of hectares of productive land have become barren mainly because of inadequate drainage. Excessive ground water extraction has also exhausted ground water supplies which are then only slowly replenished. In some cases, such supplies — relics of past, wetter climates — could not be replaced. On a global scale, Biswas (1979:243) estimates that at least 200 to 300 thousand ha of irrigated land are lost every year to salinization or water-logging, and he suggests that in all 20 to 25 million ha of once productive land is now saline. A large proportion of that world total is in Pakistan, Saudi Arabia, Peru, Mexico, Israel, South Africa, and India (Michel 1972).

Irrigation water released to a river downstream of a scheme is usually of increased salinity and hardness, and may contain additional amounts of nitrogen, phosphorus, and nutrients leached from fertile soil, which may then cause algal or weed blooms downstream. Pesticide contamination is an ever-present risk in runoff from irrigated land. There is also the risk, in areas where waters are naturally too acidic for the snail vectors of human schistosomiasis, that irrigation water returned to the streams may have leached lime and carbonates applied to agricultural land, thereby rendering downstream flows more hospitable to snails and so assisting the spread of the disease.

Water consumption is at present very wasteful — it is common to find that 80 to 85 per cent of the water delivered to an irrigation system never reaches the crops (Biswas 1979c). The history of exploitation of ground water in developing nations is often one of a rate of exploitation exceeding the rate of natural renewal, and it is inevitable that over-depletion of ground water will affect some tropical nations.

Modern, permanent irrigation schemes have altered the natural seasonally dry environment of many parts of the tropics in such a way that many water-related diseases are becoming more prevalent (Egbuniwi 1976). Aquatic environments are suitable not only for snails but also for insects such as mosquitoes. The nature of modern irrigation is such that seasonal water fluctuations which formerly held pests in check are often replaced with regulated year-round flows. Irrigation may provide new habitats suitable for plant diseases and insect pests; the former may particularly be encouraged by overhead sprinkler-irrigation. Large areas of irrigated crops may favour the development of large populations of rodents or birds, which in turn may devastate surrounding vegetation or necessitate the heavy use of pesticides. Irrigation has greatly increased the prevalence (in addition to schistosomiasis) of malarial and filarial infections, meningitis, and various parasitic worms (Diamont 1980).

Attention to lining irrigation canals and ensuring that their gradient is sufficient to ensure a flow of at least 30 cm³ per second should reduce infiltration leading to raised local water-tables, and should reduce snail and some mosquito species breeding. Like the appropriate design of dam spillways to reduce blackfly breeding and snail screens in distribution canals, such "environmental engineering" cannot always be effectively applied as an afterthought. Engineers must be aware of the environmental problems that can be generated by water development and should design to avoid them from the initial planning stages. Unfortunately, many governments assign the administration of irrigation schemes to ministries of agriculture or economic development because the goal is to increase food production or to aid rural poor, with the result that awareness, let alone consideration of health and environmental problems, may be lacking.

River Diversion, Channel Improvement and Canals

Among the many river diversion schemes that have actually been implemented (many proposals have never been put into practice) since the Second World War, the Jonglei Canal ranks as one of the most ambitious. Since the early part of this century, there have been plans to construct a 350 km canal through the sudd swamp region in southern Sudan from near Jonglei (Bor) to Malakal, initially to divert 20 million m³ per day of Nile water that would otherwise evaporate in the marshes, and to make this water available for irrigation. Work on this equatorial Nile project started in the mid-1970s and will result in vast changes to sudd ecology and to the traditional economy of at least 700,000 Nilotic peoples. In particular, changes in the regional water regime will probably affect soil quality, vegetation, wild life, fish, game, and livestock movement (with crossing points being necessary to reduce hindrance to migration) and may cause tension between tribes. The canal may also open up a region, at present protected by its remoteness, to game poaching.[4]

Low altitude intercoastal ship canals, provided salinity changes and navigation locks are not a barrier, may allow the passage of species from one sea to another. There have been reports that such passage has in recent years increased through the Suez Canal as the salinity of the Great Salt Lake has fallen. Navigation locks on the present Panama Canal do much to

---

[4]An extensive review and bibliography of the Jonglei Canal is provided by Platenkamp (1978); more recently, Tahir and El Sammani (1980) have written on the environmental and socio-economic impacts.

restrict migration, but such barriers would disappear if a new lower altitude canal were ever constructed.

Channel improvement to reduce flood damage and to aid navigation and land drainage generally involves straightening, dredging, and embankment of the formerly sluggish, meandering, lower reaches of a river. The surrounding land is thus deprived of seasonal flood, and old channel swamps and oxbow lakes are no longer refilled. The river channel generally provides a more uniform, less diverse range of habitats for aquatic species and wild life along the banks, and on the flood plains the flow regime may also become erratic.

The development of deltaic areas in the tropics has been the subject of considerable research. A useful introduction to such developments has been provided for southern Asia, the Pacific, and Southeast Asia by the United Nations (1978). Widespread flood control, land reclamation and irrigation, despite commonly acidic, saline, or sulphatic soils, by means of barrages, tide-gates, and embanking, will increasingly affect nutrient supply to inshore marine fisheries and will have an impact in many regions, especially coastal mangrove swamps and estuaries which are among the most productive of natural tropical ecosystems.

Problems of Water Resource Development

The problems of water resource development in the tropics may be summed up as follows:

1. Lack of an established body of knowledge. There are fewer natural lakes in the tropics than in other latitudes. Tropical man-made lakes were created only after the 1950s. Limnological knowledge in the tropics drew heavily on temperate latitude research until the mid-1970s.

2. The transfer of ideas and expertise from temperate to tropical water resource development has so far not been particularly successful.

3. Biogeochemical processes in tropical latitudes are more complex, more rapid, and are not moderated by cool seasonal conditions. Understanding of the structure and function of tropical ecosystems improved greatly with the launching of the International Biological Programme during the 1960s.

4. Many environmental problems are manifest only after consultants or funds become unavailable on the completion of a project.

5. Water resource needs in many tropical nations are so pressing that development may be hurried, ruthless, and divorced from consideration of long-term environmental costs.

6. Many tropical developing nations are too poor to invest in adequate environmental surveys and continued monitoring, and face political and administrative difficulties which hinder such research. The richer tropical

nations have largely neglected the few opportunities they have had to study the impact of man-made lakes.

7. As a result of the problems mentioned above, virtually no tropical lake has been studied in enough detail continuously, for long enough to yield enough insight to enable the synthesis of general planning guidelines. The same is probably true for large tropical rivers, although there have been small-scale studies of tropical catchments.

8. Each river basin, each lake or canal is to some degree unique, and to generalize from the results of study in one or a few cases is dangerous.

9. Water resource development affects human populations, even in sparsely populated areas. The Kariba, Volta, Aswan, and Kainji schemes alone probably necessitated the evacuation and resettlement of about 1.5 million people. Disturbed populations may resort to environmentally damaging practices such as shifting cultivation, or may be settled in localities unable to support agriculture without environmental damage.

10. Development planning teams have been dominated by engineers and economists whose tendency is to under-emphasize the human and environmental aspects of water resource development.

11. Indigenous politics, foreign aid, investment, or foreign consultants may determine water resources policies and in some cases difficulties are generated. For example, Smith (1972:iv) felt that the Aswan scheme was "one of the greatest river mis-management follies of all times . . . a result of political rivalry between the United States and the Soviet Union".

12. Water resource development impacts may be indirect: hydroelectric power generated by a dam may support environmentally damaging extractive industry or smelting, sometimes at some considerable distance from the actual dam.

13. As tropical metropolitan communities develop new sewage systems, as existing systems are expanded, and as agro and manufacturing industries spread pollution, this will increasingly affect river, lake, and coastal marine environments.

Discussion

The adverse environmental problems faced by water resource development in the tropics may be grouped into three general categories:

1. Environmental disruption by inadvertance (the result of lack of knowledge, lack of expertise, or inadequate expertise).
2. Environmental disruption from decisions because of convenience, or for expediency. Experts' warnings are ignored or are not even sought for a variety of reasons such as lack of finance, or political or personal factors.

3. Environmental disruption as a result of too narrow an outlook, that is, planners have considered technical, economic, and engineering feasibility, but not social or environmental implications (Sachs 1976).

Category 1 above should decline as specialist bodies, some international (like the FAO) and some indigenous to tropical nations, invest effort in water resource development studies (Hoon 1976). At present, although channels for exchange of information are well-developed in and among the temperate, developed nations and, to a lesser extent, between North and South America, to a lesser degree again between Europe and Africa or Asia, there is a poorly developed exchange between tropical nations despite their shared socio-economic and environmental problems. A review of the problem is given by di Castri and Hadley (1979).

Category 3 has attracted the attention of planners and researchers and should decline in importance. However, ecological modelling is unlikely to generate forecasts as reliable as those in the physical sciences. Ecological and human problems will continue to arise but on fewer occasions.

The problems which result from human decision-making (category 2) are likely to prove more persistent and unpredictable.

The persistence of the problems of accurate ecological forecasting and of monitoring and administering large developments in tropical nations suggests that a more cautious approach to water resource development should be adopted. Programmes of gradual change should be adopted whenever possible, the implementation of large-scale developments must be structured to be flexible enough to adapt quickly and cheaply to meet unforeseen problems. The most commonly favoured development scenario to date has been the large-scale dam, irrigation or drainage project in which development authorities adopt, at one extreme, an approach whereby they hasten to satisfy primary project goals (hydro-electric generation, flood control, etc.) and then hope to deal with secondary, particularly environmental problems, as they are perceived, in the hope these will not be too intractable or expensive. The other extreme is for a development authority to delay construction until careful research and establishment of adequate monitoring facilities enable confident prediction and measurement of impacts — a costly, slow, and often imperfect approach. Between these two extremes should be a flexible approach suitable for large-scale, medium to high technology water resource development schemes. Such an approach should include: (1) some pre-development study well ahead of construction to identify possible problems; (2) the research and evaluation of relevant hindsight experience; and (3) consultation with people likely to be affected by the development to ensure that they are aware of the situation and to identify human reactions which might impinge on the environment.

There is special need to ensure that there is real multi-disciplinary cooperation in pre-development research and in monitoring. Throughout the literature there are phrases to the effect: "past unnecessary mistakes could have been avoided by better planning and closer collaboration between engineers, social and life scientists". A vital requirement should be that water resources schemes be managed by a single, strong authority, with adequate jurisdiction to control the whole region affected by the development. Barkin and King (1970) felt the cooperation of various agencies and interests could best be obtained by establishing a river basin development agency with broad powers to cut across administrative, disciplinary, and even national boundaries (see Fig. 13.1).

Unless advances are made in combating siltation by means of suitable silt traps or watershed management, leakage, and evaporation, smaller-size irrigation or reservoir development seems unlikely to oust the large-scale schemes. A large scheme composed of numerous small reservoir or irrigation plots should be both more adaptable and responsive to environmental difficulties than a single large project. Assuming component units of such a composite irrigation scheme all vary in management, the scheme should in theory be more environmentally stable; not all plots would succumb to disease or climatic damage as would a large mono-culture irrigation scheme. Smaller irrigation schemes would demand less funding at any one time, and an area could therefore be developed gradually. Small schemes would also better promote rural employment, although water losses are generally greater on smaller holdings (but are often improved by land consolidation). Medium-sized impoundments seem to be nearly as vulnerable to siltation and other problems as small man-made lakes (Johnson 1979); however, medium-sized irrigation projects may generate less impacts than large-scale schemes.

On larger rivers, provided flow does not fluctuate excessively, dams may be unnecessary. Instead "low-head" axial tube turbines installed in low weirs or barrages would generate hydro-electricity with less environmental disruption than established engineering technology (Goodland 1980). Fashion, financing, and the previously discussed politico-economic factors determine the development scenario. Schemes may be initiated and financed by various bodies such as an indigenous government department, indigenous river basin development authorities which have wider powers, or multi-national corporations, aid agencies, or communist bloc agencies which may dictate more or less appropriate plans and employ expatriate staff to implement and monitor projects. In general, aid agencies and national governments in developing nations today do not emphasize small-scale schemes aimed at peasant farmers. The level of technology may range from low to highly sophisticated and varies greatly in the degree of environmental impact it generates.

Whatever scenario of development is pursued, whether it be indigenously organized or influenced and paid for by external bodies, every effort should be made to avoid irreversible environmental change (the extinction of fauna, flora, or the loss of soil). Long-term benefits should be weighed against short-term gains, as far as this is practicable in view of the fact that environmental concern is largely related to a nation's per capita income (Shapiro *et al.* 1979; Walker and Uglow 1979). Unfortunately, the excesses of the conservation movement have in many nations alienated many decision-makers who play a large part in water resource development. Gradually, however, environmentally conscious planning and legislation is being adopted in the Third World. Malaysia, the Philippines, and Papua New Guinea are among the first to do so.

Acknowledgements

I am grateful to Prof. C.M. Elliott and colleagues of the Centre for Development Studies and to the University College of Swansea for their support. I would like to thank the University College of Cardiff for use of library facilities.

REFERENCES

Abul-Ata, A.A. 1979. "After the Aswan". *Mazingira: The World Forum for Environment and Development*, no. 11:21–26.

Ackermann, W.C.; White, G.F.; and Worthington, E.B. eds. 1973. *Man-Made Lakes: Their Problems and Environmental Effects*. American Geophysical Union, Geophysical Monograph no. 17. Washington.

Allanson, B.R. 1973. "Summary: Physical Limnology of Man-Made Lakes". In American Geophysical Union, *Man-Made Lakes: Their Problems and Environmental Effects*. Washington.

Agarwal, A. 1980. "A Decade of Clean Water". *New Scientist* 88:356–59.

Balon, E.K. 1978. "Kariba: The Dubious Benefits of Large Dams". *Ambio* 7:40–48.

Bandler, H. 1975. "Environmental Implications of Dams and Man-Made Lakes". In International Association of Hydrological Sciences, *Hydrological Characteristics of River Basins: Papers Presented at International Association of Hydrological Sciences Symposium*, Dec. 1975. Tokyo.

Bardach, J.E. 1972. "Some Ecological Implications of Mekong River Development Plans". In Farvar, M.T., and Milton, J.P., eds., *The Careless Technology*. New York.

______, and Dussart, B. 1973. "Effects of Man-Made Lakes on Ecosystems". In American Geophysical Union, *Man-Made Lakes: Their Problems and Environmental Effects*. Washington.

Barkin, D., and King, T. 1970. *Regional Economic Development: The River Basin Approach in Mexico*. London.

Barrow, C.J. 1981. "The Health and Resettlement Consequences and Opportunities Created as a Result of River Impoundment in Developing Countries". *Water Supply and Management* 5:135–50.

Baxter, R.M. 1977. "Environmental Effects of Dams and Impoundments", *Annual Review of Ecology and Systematics* 8, 255–83.

Beadle, L.C. 1974. *The Inland Waters of Tropical Africa: An Introduction to Tropical Limnology*. London.

Beauchamp, R.S. 1969. "Hydrological Factors Affecting Biological Productivity: A Comparison Between the Great Lakes in Africa and the New Man-Made Lakes". In Obeng, L.E., ed., *Man-Made Lakes: The Accra Symposium*, Nov. 1966, Accra.

Beaumont, P. 1977. "Resource Management: Case Study of Water". *Progress in Physical Geography* 1:528–36.

———. 1978. "Man's Impact On River System: A World-wide View". *Area* 10:38–41.

Benson, N.G. 1973. "Summary: Man-Made Lakes as Aquatic Ecosystems". In American Geophysical Union, *Man-Made Lakes: Their Problems and Environmental Effects*. Washington.

Biswas, A.K. 1978. "Environmental Implications of Water Development for Developing Countries". *Water Supply and Management* 2:283–97.

———. 1979a. "Climate, Agriculture and Economic Development". In Biswas, M.R., and Biswas, A.K., eds., *Food, Climate and Man*. New York.

———. 1979b. "Water: A Perspective on Global Issues and Politics". *Journal of the Water Resources Planning and Management Division, ASCE* 105, WR2:205–22.

———. 1979c. "Water in Developing Countries: Problems and Prospects". *Geojournal* 3:445–56.

———, ed. 1979d. *Water Development, Supply and Management. Volume 12: Water Management and Environment in Latin America*. Oxford.

———. 1980. "Environment and Water Development in Third World". *Journal of the Water Resources Planning and Management Division, ASCE* 106, WR1:319–32.

Biswas, M.R. 1979. "Agriculture and Development". *Technical Memoir, No. 3, International Commission on Irrigation and Drainage*. New Delhi.

Boughey, A.S. 1962. "The Explosive Development of a Floating Weed Vegetation on Lake Kariba". *Adansonia* 3:49–61.

Boulton, P. 1978. *The Control of Water Resources in the Zambezi Basin and Its Implications for Mozambique*. University of Edinburgh School of Engineering Science Occasional Papers on Appropriate Technology. Edinburgh.

Bowmaker, A. 1970. "A Prospect of Lake Kariba". *Optima* 20:68–74.

Brennan, J.P. 1973. "Introduction". In Meggers, B.J.; Ayensu, E.S.; and Duckworth, W.D., eds., *Tropical Forest Ecosystems in Africa and South America*. Washington.

Christiansson, C. 1979. "Imagi Dam: A Study of Soil Erosion, Reservoir Sedimentation and Water Supply at Dodoma, Central Tanzania". *Geographiska Annaler* 61, Ser. A:113–45.

Dasmann, R.F.; Milton, J.P., and Freeman, P.H. 1973. *Ecological Principles for Economic Development*. London.

De-Heer Amissah, A.N. 1969. "Some Possible Climatic Changes that May Be Caused by the Volta Lake". In Obeng, L.E., ed., *Man-Made Lakes: The Accra Symposium*, Nov. 1966.

Diamant, B.Z. 1980. "Environmental Repercussions of Irrigation Development in Hot Climates". *Environmental Conservation* 7:53–58.

Di Castri, F., and Hadley, M. 1979. "A Typology of Scientific Bottlenecks to Natural Resources Development". *Geojournal* 3:513–22.

Edgcomb, G., ed. 1965. *Man-Made Lakes: A Selected Guide to the Literature (An Aid to Planning and Multidisciplinary Research on New African Reservoirs)*. National Academy of Sciences/National Research Council. Washington.

Egbuniwi, H. 1976. "Public Health Aspects of Tropical Water Resource Development". *Water Resources Bulletin* 12:393–98.

El-Hinnawi, E. 1980. "The State of the Nile Environment: An Overview". *Water Supply and Management* 4:1–11.

Faniran, A. 1980. "On the Definition of Planning Regions: The Case for River Basins in Developing Countries". *Singapore Journal of Tropical Geography* 1, no. 1:9–15.

Farmer, B.H. 1971. "The Environmental Sciences and Economic Development". *Journal of Development Studies* 7:257–69.

Farvar, M.T. 1976. "The Interaction of Ecological and Social Systems". In Mathews, W.H., ed., *Outer Limits and Human Needs: Resource and Environment Issues of Development Strategies*. Uppsala.

Fels, E., and Keller, R. 1973. "World Register on Man-Made Lakes". In American Geophysical Union, *Man-Made Lakes: Their Problems and Environmental Effects*. Washington.

Fernando, C.H. 1976. "Reservoir Fisheries in South East Asia: Past, Present and Future". *Symposium on the Development and Utilization of Inland Fisheries Resources, 27–29 October, Colombo, Sri Lanka (11th Session UN Indo-Pacific Fisheries Council)*. FAO. Rome.

Freeman, P.H. 1974. *The Environmental Impact of a Large Tropical Reservoir: Guidelines for Policy and Planning (Based on a Case Study of Lake Volta, Ghana in 1973 and 1975)*. Office of International and Environmental Programmes, Smithsonian Institution. Washington.

Ganstad, E.O. 1978. *Weed Control Methods for River Basin Management*. CRC Press, Palm Beach, Florida.

Gaudet, J.J. 1979. "Aquatic Weeds in African Man-Made Lakes". *PANS* 25:279–86.

George, C.J. 1972. "The Role of the Aswan Dam in Changing Fisheries of the South Eastern Mediterranean". In Farvar, M.T., and Milton, J.P., eds., *The Careless Technology*. New York.

Glymph, L.M. 1973. "Summary: Sedimentaiton of Reservoirs". In American Geophysical Union, *Man-Made Lakes: Their Problems and Environmental Effects*. Washington.

Goldman, C.R. 1979. "Ecological Aspects of Water Impoundment in the Tropics". *Unasylva* 31:2–11.

Goodland, R.J. 1980. "Environmental Ranking of Amazonian Development Projects in Brazil". *Environmental Conservation* 7:9–26.

Gupta, H.K., and Rastogi, B.K. 1976. *Dams and Earthquakes*. Amsterdam.

Hall, J.B.; Laing, E.; Hassain, M.; and Lawson, G.W. 1969. "Observations on Aquatic Weeds in Volta Basin". In Obeng, L.E. ed., *Man-Made Lakes: The Accra Symposium*, Nov. 1966.

Hammerton, D. 1972. "The Nile River: A Case History". In Oglesby, R.T.; Carlson, C.A.; and McCann, J.A. eds., *River Ecology and Man*. New York.

Harrison, P. 1979. "The Curse of the Tropics". *New Scientist* 84:602–4.

Henderson, F. 1973. "Stratification and Circulation in Kainji Lake". In American Geophysical Union, *Man-Made Lakes: Their Problems and Environmental Effects*. Washington.

Hilton, T.E., and Kowu-Tsri, J.Y. 1970. "The Impact of the Volta Scheme on the Lower Volta Floodplain". *Journal of Tropical Geography* 30:29–37.

Hoon, R.N. 1976. "Utilizing Reservoirs for Optimum Benefits". *Water Power and Dam Construction* 28:39–42.

Jackson, P.B. 1975. "Fish". In Stanley, N.F., and Alpers, M.P., eds., *Man-Made Lakes and Human Health*. London.

Jackson, I.J. 1977. *Climate, Water and Agriculture in the Tropics*. London.

Johnson, B. 1979. "Dammed Alternatives". *Development Forum* no. 7:7.

Kassas, M. 1973. "Impact of River Control Schemes on the Structure of the Nile Delta". In Farvar, M.T., and Milton, J.P. eds., *The Careless Technology*. New York.

Kneese, A.V. 1979. "Development and Environment". *Third World Quarterly* 1:84 90.

Lagler, K.F., ed. 1969. *Man-Made Lakes: Planning and Development*. UNDP/FAO. Rome.

Lanley, J.P., and Clement, J. 1979. "Present and Future Natural Forest and Plantation Areas in the Tropics". *Unasylva* 31:12–20.

Land, T. 1980. "Profits from a Beautiful Pest". *Development Forum* 8, no. 7:10.

Lawson, G.W. 1970. "Lessons of the Volta: A New Man-Made Lake in Tropical Africa". *Biological Conservation* 2:90–96.

Leentvar, P. 1966. "The Brockopondo Research Project, Surinam". In Lowe-McConnel, R.H., ed., *Man-Made Lakes: Proceedings of the Royal Geographical Society Symposium*, Sept./Oct. 1965. London.

———. 1975. "The Brockopondo Research Project". In Stanley, N.F., and Alpers, M.P., eds., *Man-Made Lakes and Human Health*. London.

Lezek, A., and El-Zarka, S. 1973. "Ecological Comparison of the Pre-Impoundment and Post-Impoundment Fish Populations of the River Niger and Kainji Lake, Nigeria". In American Geophysical Union, *Man-Made Lakes: Their Problems and Environmental Effects*. Washington.

Little, E.C. 1969. "Weeds and Man-Made Lakes". In Obeng, L.E. ed., *Man-Made Lakes: The Accra Symposium*, Nov. 1966.

Lowe-McConnell, R.H. 1966. *Man-Made Lakes: Proceedings of Royal Geographyical Society Symposium*, Sept./Oct. 1965. London.

Lubin, S. 1977. "Environmental Impact of the Senegal Basin Project". *Kidma* 3:36–39.

Margalef, R. 1975. "Typology of Reservoirs". *Verh. Internat. Verein. Limnol.* 19:1841–48.

McConnel, R.L., and Worthington, E.B. 1965. "Man-Made Lakes". *Nature* 208:1039–42.

Mermel, T.W. 1978. "Major Dams of the World". *Water Power and Dam Construction* 30:43–53.

Michel, A.A. 1972. "The Impact of Modern Irrigation Technology in the Indus and Helmand Basins of Southwest Asia". In Farvar, M.T., and Milton, J.P., eds., *The Careless Technology*. New York.

Milton, J.P. 1975. "The Ecological Effects of Major Engineering Projects". In *IUCN Publication (New Series), No. 31: Proc. Meeting on the Use of Ecological Guidelines for Development in the American Humid Tropics, 20–22 February, 1974, Caracas, Venezuela (Session VI, No. 13)*. International Union for Conservation of Nature and Natural Resources. Paris.

Nace, R.L. 1974. "Environmental Hazards of Large-Scale Water Developments". In Leversedge, F.M., ed., *Priorities in Water Management (Western Geophysical Series Vol. 8)*. Department of Geography, University of Victoria. Victoria, British Columbia.

Obeng, L.E., ed. 1969. *Man-Made Lakes: Planning and Development*. UN Development Programme/FAO. Rome.

———. 1973. "Volta Lake: Physical and Biological Aspects". In American Geophysical Union, *Man-Made Lakes: Their Problems and Environmental Effects*. Washington.

———. 1976. "Man-Made Lakes and Problems of Human Settlement". In Owens-Jones, J., and Rogers, P. eds., *Human Ecology and the Development of Settlements*. London.

———. 1977. "Should Dams be Built? The Volta Experience". *Ambio* 6:46–50.

———. 1978. "Environmental Impacts of Four African Impoundments". In Gunnerson, C.G., and Kalbermatten, J.M., eds., *Environmental Impacts of International Civil Engineering Projects and Practices (Proc. ASCE National Convention, 17–21 October, 1978, San Francisco)*. New York.

———. 1980a. "Some Environmental Issues in Water for Development". *Water Supply and Management* 4:115–28.

———. 1980b. "Water in the Environment: The UNEP Experience". *Water Supply and Management* 4:155–70.

Odingo, R.S. 1977. "African Experience: Some Observations from Kenya". In White, G.F., ed., *Environmental Effects of Complex River Development*. Boulder, Colorado.

Oglesby, R.T.; Carlson, C.A.; and McCann, J.A., eds. 1972. *River Ecology and Man*. New York.

Organisation of American States. 1978. *Environmental Quality and River Basin Development: A Model for Integrated Analysis and Planning.* Government of Argentina /OAS/UNEP. O.A.S. Washington.

Pantulu, V.R. 1979. "A Case Study of the Pa Mong Project: Environmental Aspects". In *UNAPDI: The Environmental Dimensions in Development (Proc. Consultative Meeting on Methodology and Techniques for Identification and Incorporation of Environmental Dimensions in Development Planning, 14–16 February, 1979).* Bangkok.

Pearse, A. 1980. *Seeds of Plenty, Seeds of Want: Social and Economic Implications of the Green Revolution.* Oxford.

Petr, T. 1978. "Tropical Man-Made Lakes: The Ecological Impact". *Arch. fur Hydrobiol.* 81:368–85.

________. 1979. "Possible Environmental Impact on Inland Waters of Two Planned Major Engineering Projects in Papua New Guinea". *Environmental Conservation* 6:281–86.

Petts, G.E. 1980. "Long-term Consequences of Upstream Impoundment". *Environmental Conservation* 7, no. 4:325–32.

Plattenkamp, J.D. 1978. *The Jonglei Canal: Its Impact on an Integrated System in the Southern Sudan.* Institute of Cultural Anthropology, and Sociology of Non-Western Peoples. Publication no. 26. Leiden.

Report of the Independent Commission on International Development Issues (under Chairmanship of Brandt) (1980). *North-South: A Programme for Survival.* London: Pan Books.

Ridley, J.E., and Steel, J.E. 1975. "Ecological Aspects of River Impoundments". In Whitton, B.A., ed., *River Ecology (Studies in Ecology vol. 2).* Oxford.

Rubin, N.N. 1976. "Hindsight Evaluation of Dams in Africca". In *UN River Basin Development: Policies and Planning, Vol. 1 (Proc. UN Seminar on Basin and Interbasin Development.* Sept. 1975. Budapest.

________, and Warren, W.M., eds. 1968. *Dams in Africa: An Interdisciplinary Study of Man-Made Lakes in Africa.* London.

Sachs, I. 1976. "Environmental and Strategies for Development". In Matthews, W.H., ed., *Outer Limits and Human Needs: Resources and Environment Issues of Development Strategies.* Uppsala.

SCOPE. 1972. *Man-Made Lakes as Modified Ecosystems (SCOPE Report 2).* International Council of Scientific Unions Working Group on Man-Made Lakes. Paris.

Shapiro, P.T.; Miyuo, T.; and Smith, T.R. 1979. "Environmental Quality and Economic Development". *Environment and Planning* A 11:1147–56.

Smith, A.W. 1972. "Ecological River Basin Development". *National Parks and Conservation Magazine* 46:I–IV.

Stanley, N.F., and Alpers, M.P., eds. 1975. *Man-Made Lakes and Human Health.* London.

Tahir, A.A., and El Sammani, M.D. 1980. "Environmental and Socio-Economic Impact of Jonglei Canal Project". *Water Supply and Management* 4:45–51.

Thomas, K.J. 1979. "The Extent of Salvinia Infestation in Kerala (S. India): Its Impact and Suggested Methods of Control". *Environmental Conservation* 6:63–69.

Toran, J. 1973. "The Consequences of Building Dams on the Environment". *Proc. First World Congress on Water Resources*, Sept. 1973. Chicago.

Troll, C. 1963. "Landscape, Ecology and Land Development, with Special Reference to the Tropics". *Journal of Tropical Geography* 17:1–11.

Turner, D.J. 1971. "Dams and Ecology: Can They Be Made Compatible?" *Civil Engineering — ASCE* 41:76–80.

U.N. 1970. *Integrated River Basin Development: Report of a Panel of Experts* (revised edn. annex V). New York.

________. 1972. *Development and Environment: Report of the Working Party and Papers of a Panel of Experts Convened by the Secretary General, UN Conference on Human Environment*, June 1971, Founex, Switzerland.

________. 1975. *Management of International Water Resources: Institutional and Legal Aspects* (*Report of the Panel of Experts on the Legal and Institutional Aspects of International Water Resource Development*. UN Department Economic and Social Affairs. Natural Resources/Water Series no. 1. New York.

________. 1978. *Proceedings of the Third Regional Symposium on the Development of Deltaic Areas, UN Economic and Social Commission for Asia and the Pacific, Bangkok*. Natural Resources/Water Series no. 50. New York.

Visser, S.A. 1970. *Kainji: A Nigerian Man-Made Lake* (*Kainji Lake Studies*, vol 1). Nigerian Institute of Social and Economic Research. Ibadan: Ibadan University Press.

Walker, I. 1978. "Environmental Attitudes in Less Developed Countries". *Resource Policy*, Sept. 1978, pp. 200–4.

________, and Uglow, J. 1979. "Environmental Policies in Developing Countries". *Ambio* 8: 102–9.

White, E. 1969. "Man-Made Lakes in Tropical Africa and Their Biological Potentialities". *Biological Conservation* 1:219–24.

Williams, G.J., and Howard, G.W., eds. 1977. *Development and Ecology in the Lower Kafue Basin in the Nineteen Seventies* (Papers from the Kafue Basin Research Committee of the University of Zambia).

Wolverton, B.C. 1979. "The Water-Hyacinth". *Mazingira: The World Forum for Environment and Development*, no. 11:59–65.

World Bank. 1974. *Environment, Health and Human Ecologic Considerations in Economic Development Projects*. Washington.

14

Natural Resources and Development Planning: An Analysis of Divergence and Convergence in the Guyana Economy

DERYCK BERNARD

Natural resources may be defined as those economic factors in production or consumption which owe their origin and existence, either dominantly or totally, to natural phenomena or to processes that occur autonomously in nature. In discussions of development planning, the role, utilization, and potential of the resource base of poor countries has been given major attention. There is a widely held assumption that the natural resources of a poor country indicate the levels of economic development possible. This assumption remains popular even though the evidence of a comparative analysis of poverty has demonstrated that the rate of economic change and the level of material well-being are not necessarily related to the resource base (Thomas 1974).

The utilization of natural resources is intricately linked with the nature of economic and political structures which exist in the international economic system and the patterns of ownership and distribution of wealth which exist within poor countries. It is essential that we understand the structures which exist and which determine what natural resources are obtained and for whose ultimate benefit they are used. As geographers, we would ignore such considerations at the risk of impoverishing our contribution to studies on the nature of development.

A natural resource is not an objective description of a clearly recognizable phenomenon. A resource becomes a resource only because someone is willing to invest in its exploitation, and the decisions about such an investment may not rest in the hands of the state wherein the phenomenon is located. Second, if the resource is owned and controlled by an external agency and operated for its benefit, then the resource is not a resource of the poor country but part of the capital of the owner. There are numerous expositions on this issue, including analyses of the role of multinationals in the exploitation of the resources of various Caribbean territories (Girvan 1967; Omawale 1971; Thomas 1974; Brewster 1967; Sears 1969). As Girvan has noted,

> Dependence on foreign demand, prices and decision making, the large amount
> of profit repatriation and its reflection, the low share of the industry's value
> which is "returned" to the national economy; are the surface manifestations
> of the institutionalized relationships between subsidiaries and their parent
> firms (Girvan 1967).

The economic and political contribution to the study of natural resources is reinforced by a more fundamental and unashamedly psychological factor — the right to economic self-determination. As one Caribbean politician writes:

> without the right of peoples and nations to enjoy the principle of complete
> freedom of action in determining the use of their resources, they cannot
> develop that basic attribute of self-reliance which is inherent in genuine devel-
> opment. In the exercise of self-reliance, the foundation is properly laid for
> peoples and nations to exercise authority over their wealth and resources . . .
> in the interest of their national development (Reid 1980).

It is within the context of the political economy of natural resource utilization that we will analyse the contribution of the convergence model of development planning and its application to the Guyana economy. Convergence and its antipole, divergence, are most clearly stated in the seminal works of two economists — Thomas and Brewster (Brewster and Thomas 1967; Thomas 1974). Dynamic divergence has been defined by Thomas (1974:59) as:

> The lack of an organic link rooted in an indigenous science and technology,
> between *the pattern and growth of domestic resource use* on the one hand and
> the pattern and growth of domestic demand, and . . . the divergence between
> domestic demand and the needs of the broad mass of the population (my
> italics).

Dynamic convergence seeks to reverse these relationships. As a planning strategy, it seeks to effect a convergence between the pattern and rate of domestic resource use and domestic demand. Of crucial importance is the adjustment of the resource use and the pattern of production to meet a defined set of basic needs on the one hand, and the reorientation of domestic demand to reflect domestic capabilities. The key to the concept of convergence planning is the notion of "totality". This describes a political and economic situation in which there exists the will and the machinery to control and adjust the patterns of domestic production, investment, foreign trade, and domestic consumption. The policies of a government attempting to generate convergence within the framework of totality will be: (1) to minimize dependence upon the importation of basic consumer goods; (2) to eliminate the importation of consumer goods for which there are acceptable substitutes; (3) to eliminate the importation of non-basic consumer goods; (4) to generate the growth and utilization of indigenous technology in the productive sectors.

The adaptation of a convergence strategy for development represents a distinctive ideological position and political style. Major priority must be given to disengagement from international capitalist systems and the generation of "self-reliance". It also implies a rejection of development based on consumer goods industries which cater for a small section of the population and which are based on imported technology or on management controlled by external institutions. Further, the state abrogates a wide range of consumer decisions, especially, but not exclusively, in relation to imported consumer goods. Convergence is, therefore, only feasible under conditions where the allocation of foreign exchange is strictly controlled and where vital economic decisions regarding what is produced and what is consumed are not made by external agencies.

The political basis of natural resource development can be described as a choice between "totality", with its emphasis on disengagement, and "dependency", which implies that the major decisions on the development of resources are taken outside of the developing country in the international centres of economic power. The implementation of totality as a means of effecting convergence requires a drastic alteration in life styles and consumer behaviour. The introduction of restrictions in consumer freedom in a society whose tastes and habits have been determined by values imported from contact with larger and more developed societies may not even be always politically feasible in a political system dependent upon "Westminister"-style political institutions for support. Influential groups in the economic sectors and the more articulate groups in the society are likely to resist the alteration or inhibition of consumption and production patterns to which they are accustomed. As a result, totality in practice means the control of not only the economic apparatus but eventually the restriction of critical activities. As a result, we may assert that only entrenched and stable governments can successfully introduce a policy of convergence which insists on short-term sacrificial and irksome restrictions in return for possible future long-term benefits.

The People's National Congress (PNC) Strategy

From 1970, the Guyana government has gradually moved towards a broad "totality" policy in the internal and external functioning of the economy. The most coherent expression of this policy of convergence was the 1972-76 Development Plan and its accompanying self-sufficiency campaign (Feed, Clothe and House Ourselves campaign). Subsequent years have seen the strengthening of the expression of "totality" in economic management as the state gradually acquired control of all the major importing and exporting organizations. The aspects of totality outlined are evidence of policy goals or de facto elements in the country's economic organization.

There are two elements of the political aspect of convergence. First, there is the need for local control over industries which depend upon the exploitation of local resources. This has been adopted as a basic policy of the Guyana government (Burnham 1971). Though there are emotive and ideological bases for this policy, its major justification is the need to control the volume and location of production. In addition, there is the presumed ability to generate or create linkages between natural resources. The investment policies of transnational enterprises which control the operation of resource-based industries has been well documented (Girvan 1967). For example, the refusal of the Demerara Bauxite Company to use suitable domestic input and its reluctance to examine the feasibility of tapping co-existent kaolin resources were major factors that led to the nationalization of that company (Burnham 1971).

During the period under discussion, there has been a consistent implementation of a convergence policy. The government has nationalized the bauxite and sugar industries, bought or invested heavily in the forest industry, and the operations of all locally export-based resource industries have had their export activities centralized and brought under state control. Along with the implementation of control of domestic industry, the government has adopted a policy towards resource investment which emphasizes the use or substitution of local materials and the attainment of self-sufficiency. Aspects of this policy include the encouragement of wood in construction, the production and use of domestic commodities, and the adaptation of domestic technologies and materials in industrial processes.

The second aspect in the implementation of convergence is the control of external trade and external transactions. During the period under review, the Guyana government has gradually eroded the freedom of enterprises and individuals to import commodities which may compete with domestic production or for which the production of a local substitute is feasible. This strategy is aimed especially at protecting the food, forest, and related industries. One inevitable result of these policies has been the development of stricter control over private foreign exchange transactions and the distribution of basic commodities.

Quantifying Divergence

Discussions on divergence and its impact on West Indian economics have focused on those contributions which have analysed the role of export-based mineral industries, particularly bauxite in Jamaica and oil in Trinidad and Tobago. Brewster and Thomas have identified two measures of divergence in an under-developed economy — the import domestic expenditure coefficient and the coefficient of divergence. The former is defined by the equation:

$$\lambda = \frac{Mu}{E} = \frac{M}{O} - \frac{Mx}{X} \qquad (1)$$

when Mu is imports for domestic use, E is domestic expenditures, M is total imports, X is exports, Mx is the import content of exports, and O is gross national product. This measure seeks to relate the value of imports for domestic use to domestic expenditure. The more recent coefficient of divergence is defined by the equation:

$$D = \tfrac{1}{2} \left(\frac{\pi - \phi}{E} + \frac{x^{1} - \phi}{o} \right) \qquad (2)$$

where π is the volume of imports which cannot be produced at home, x^{1} are exports which do not compete with domestic expenditures, and ϕ is the import content of exports. Conceptually, equation 2 represents a clearer and more advanced level of understanding of the nature of dynamic divergence. However, the values proved difficult to quantify, particularly since the identification of x^{1} would rest upon unsatisfactory subjective criteria. As a result, equation 1 has been used as the statistical index of divergence. Since this is the more widely used index, it has the further advantage of comparability with related work and the original contributions of Thomas. Table 14.1 shows values of λ for 1967 to 1979. Values of λ approaching unity indicate increasing convergence, whereas values approaching o indicate divergence.

The estimation of λ as shown in Table 14.1 for 1967 to 1979 uses the official estimates of the Bank of Guyana. Imports, GNP, and exports are easily extracted from such statistics, but estimates of the import component of export earning (MX) are based on the classification of imports. This is particularly difficult in the case of capital imports for sectors which had only a small export earning for the years under discussion. Values of λ by this calculation are in the same range but somewhat higher than the range of values obtained by Brewster and Thomas. Nevertheless, the values, though indicating divergence, are far too subject to higher values in the returns from exports in particular years. In particular, the high values of λ for 1974 and 1975 and the sudden rise from 1978 to 1979 seem to be related to fluctuations in sugar prices rather than to basic structural changes in the national economy.

Intransigence of Divergence

The evidence of this analysis is that the policies of the Feed, Clothe and House Ourselves campaign, the 1972–76 Development Plan, and the basic policy of self-reliance and resource control have not so far been a material success. Not only are the indices of divergence pointing to an increasingly unpredictable situation but there is no real evidence that the structures which support economic divergence have been effected even in the face of

TABLE 14.1 DIVERGENCE IN THE GUYANA ECONOMY

Year	Imports	GNP (million G$)	Exports (million G$)	MX1	$\frac{M}{o}$	$\frac{MX}{x}$	λ
1967	219.0	397.0	211.7	59.2	0.55	0.27	0.28
1968	212.9	428.7	229.0	62.9	0.50	0.27	0.23
1969	234.4	456.8	252.9	91.7	0.51	0.36	0.15
1970	266.3	490.5	264.8	78.8	0.54	0.29	0.25
1971	266.0	524.9	290.9	83.2	0.51	0.28	0.23
1972	297.9	576.0	300.0	92.8	0.52	0.31	0.21
1973	365.0	617.4	285.0	140.6	0.59	0.49	0.10
1974	567.0	905.4	600.0	319.0	0.62	0.53	0.90
1975	810.6	1,139.1	858.0	226.3	0.71	0.26	0.55
1976	927.4	1,057.0	711.3	456.9	0.87	0.64	0.23
1977	800.9	1,053.2	661.2	510.2	0.76	0.73	0.03
1978	711.1	1,193.0	753.8	430.0	0.59	0.56	0.03
1979	785.0	1,250.0	737.5	409.8	0.63	0.55	0.80

the totality of state control. Guyana's inability to halt the impact of divergence may be attributed to a number of major factors.

First, the ownership of resources does not in itself determine the volume of production and the pattern of distribution of produce. One of the structural factors which remains even after control of natural resources is achieved is that there may be a negligible domestic market for the product of major resource industries. In Guyana, the export of bauxite, sugar, and their by-products remain important pillars of the Guyana economy. The small size of Guyana's population and the historical fact that these industries are tied to external markets cannot be ignored. Since these industries are major earners of domestic revenue as well as for foreign exchange, the importation of the capital and input requirements of these industries must be given priority, whatever the balance of payments situation in the country.

Second, the creation of linkage between domestic resources is not accomplished by mere ownership and control. Though domestic resources may exist which may provide the bases for linked agricultural and industrial development, the development of the linked industries may be dependent upon external financing through agencies and mechanisms which increase rather than decrease dependence on the international economic system. The clearest example is the bauxite industry where linkages are being pursued in the direction of hydro-power development and ceramic engineering, both of which will be dependent upon imported technology and financing.

Third, the management of consumer behaviour in the absence of totalitarianism has proved unsuccessful. The tastes and patterns of consumer behaviour in Guyana are often a function of cultural values with ethnic significance or of values and tastes perceived to be superior because of their associations with the dominant aspects of Caribbean culture and life style. As a result, the exclusion and elimination of imported commodities which are not replaceable by domestic substitutes is only achieved in the face of considerable opposition from political groups and other articulate elements in the society. In Guyana there is the particular problem that consumers evaluate their own material well-being in terms of conditions existing in other CARICOM countries where the majority of governments exercise minimal control over consumer commodity preferences. As a result, the population has become preoccupied with consumer preferences and commodity availability. This preoccupation with basic commodity supply and the concomitant market reactions such as hoarding, black market, smuggling, and the bazaar economy have prejudiced the perceptions of the population to the government's economic strategies.

Fourth, convergence in a neo-colonial economy was closely connected with the development of indigenous technology. Development programmes have remained dependent upon imported technology, and efforts to evolve local technology have been neither significant nor numerous. In all the major sectors of the economy, there has been increasing dependence upon imported technology, and the evolution of a truly indigenous technological base is not a short-term strategy as is nationalization or import controls. Since the evolution and implementation of industrial technologies emanates from investment in scientific research or innovations from within the production process, it can be argued that there are limitations of scale and financial resources which inhibit the possible levels of technological change. The significant point is that as long as Guyana depends upon exogenous technology, it will remain tied to the capital and skills from the developed world. The need to earn hard currency to pay for capital and repay foreign investment necessitates the continued earning of hard currencies from traditional markets. Indeed, the volume and value of imports and the propensity to import have been higher during the period under review (Table 14.2).

One can indeed argue that during a period of prolonged emphasis on natural resource development and in the absence of indigenous capital and technology, divergence must increase in the form of a pronounced difference between basic consumer needs and the importation of capital.

TABLE 14.2 IMPORT PROPENSITY IN GUYANA, 1969–79

Year	GDP (million G$)	Imports (million G$)	Import Propensity
1969	438.0	234.4	0.53
1970	467.0	266.3	0.57
1971	495.2	266.0	0.54
1972	529.3	297.9	0.56
1973	576.4	365.0	0.63
1974	865.0	567.0	0.65
1975	1,096.4	810.6	0.73
1976	1,024.5	927.4	0.90
1977	1,011.5	800.9	0.79
1978	1,126.0	711.1	0.63
1979	1,170.0	785.0	0.67

Finally, the reorganization of the economy on the basis of a concept of dynamic convergence needs a thorough consideration of spatial issues. Because economic planning and implementation remain firmly in the hands of actors who think primarily in sectoral terms and conditions to be limited by the confines of a professional specialization, the utilization of natural resources in the reorganization of the economy has not been soundly conceived. The localization or regionalization of investment and other economic decisions is subsequent to the allocation of sectoral priorities. Regional plans in Guyana represent a regionalization of national economic decisions rather than a truly regional planning process. As a result, there is a conspicuous absence of the benefits of spatial linkage between economic activities and the spatial links possible between activities which utilize resources that are coexistent or contiguous. The most common example is the frequent inability of land development and forestry sectors to be adequately linked in the development of virgin land.

The difficulties in conceptualizing and implementing spatial linkages, particularly in the 1972–76 National Development Plan and its related regional plans, are made particularly serious by the changing sectoral emphases common among political decision-makers. Key personnel respond to perceived or expressed national priorities rather than to the objective evaluation of what a particular region can produce. Indeed, one of the weaknesses of convergence theory is that it is conceived primarily at a macro-level and the spatial/regional aspect of the national resource base is given little attention. Few states, however, are so homogeneous in the distribution of natural resources or population that an economic strategy which is optimal for the whole economy can be optimal for the particular combinations of natural and human resources to be found in the regions. The intractability of the coastal orientation of the economy is one of the more conspicuous aspects of the failure to generate convergence. The 1972 development plan identified the development of the interior or hinterland as a specific major objective:

> Our hinterland has been neglected. The concentration of social amenities on the coast makes it difficult to achieve the desired mobility of labour from the comparative shelter of our coastlands to the potentially more rewarding hinterland. Regional development is therefore one of the strategies that will be employed as a means of affecting this settlement in the hinterland where new economic activities will contribute to an acceleration of the total development process (King 1972).

This emphasis was reflected in a concentration of investment on land development and similar projects in hinterland locations between 1972 and 1978. Major locations include the north west region, the intermediate savannas, and the Upper Mazaruni area. Since 1979, however, the focus of investment has been on the established sectors and in established or contiguous locations. Priority projects include drainage and irrigation projects on the Demerara and Essequibo areas such as the Mahaica-Mahaicony-Abary and the Tapakuma drainage and irrigation schemes (Table 14.3).

TABLE 14.3 REGIONAL INVESTMENT ALLOCATIONS
IN RESOURCE-BASED SECTORS,
GUYANA, 1979–80

	(million G$)
Region I North West	3.12
Region II Essequibo	42.85
Region III Demerara	81.09
Region IV Berbice	28.39
Region V Mazaruni-Potaro	3.38
Region VI Rupununi	1.08
Georgetown	76.49
Linden	32.51
New Amsterdam	7.00

Source: State Planning Commission, Guyana

Conclusion

It must be emphasized that the restricting of the economy of a dependent export-based economy is to be approached not merely from the more superficial macro-economic policy instruments but also from a more thorough appreciation of the problems of spatial and other forms of linkage. Without a successful reorganization of the spatial relationship between resources, and between resources to the pre-existent divergence economy, the privations and sacrifices of economic "totality" may not result in tangible economic benefits. It is also necessary to question the validity of dependency as a realizable policy objective in the short run, particularly in economies where the infrastructural and other forms of inertia tie export-based economies to a reliance upon foreign exchange earnings to finance investment. Perhaps there is a vicious circle: one cannot diversify and broaden one's resource base without the admittedly poor returns of export earnings, yet dependence upon limited export sectors may hinder diversification.

REFERENCES

Brewster, H., and Thomas, C.Y. 1967. *The Dynamics of West Indian Economic Integration.* Institute of Social and Economic Research, Mona.

Burnham, L.F.S. 1972. *Towards Progress and Change.* Extracts of address during 1971 Budget Debate, Ministry of Information.

Girvan, N. 1967, 1976. *The Caribbean Bauxite Industry.* Institute of Social and Economic Research, Kingston and New York.

______. 1967. *Corporate Imperialism.* Monthly Review Press, New York.

King, K.F.S. 1973. *National Development Plan, 1972–1976.* Georgetown.

Omawale. 1972. *Conservation and Natural Resources.* Georgetown.

Reid, P.A. 1976. "Consolidating Self-Reliance". *New Nation*, 14 Sept. 1976, Georgetown.

Sears, D. 1964. "The Mechanism of an Open Petroleum Economy". *Social and Economic Studies* 13:243–53.

Thomas, C.Y. 1976. *Dependence and Transportation.* Monthly Review Press, New York.

15

Population Change and Natural Resources in Africa South of the Sahara

ADETOYE FANIRAN

That this is a period of widespread concern about the nature of the relationship between man and nature is no more a matter for debate.[1] The problem is concerned essentially with our manner of utilizing the stock resources within an area such that they can continue to serve us rather than be wasted and lost. Much has been written about resource utilization and conservation, especially in North America and Europe, but very few studies have focused specifically on Africa south of the Sahara. This work is therefore exploratory in nature. It is hoped that the issues raised will not only arouse discussion among experts but stimulate further studies on this subject.

Some Basic Concepts

To focus attention on the real issues and avoid ambiguities and confusion in terminology, some key words and phrases need to be defined.

Resources

The dictionary offers three definitions of the term resources: a means of satisfying some want or deficiency; a stock or reserve upon which one can draw when necessary; the collective means possessed by any country for its own support or defence. However, the tendency among scholars working in this general area is to emphasize the functional relationship which exists between man's wants, his abilities, and his appraisal of his environment (Zimmermann 1951; Hunker 1964). This is different from the classical concept in which resources are thought of in terms of tangible (physical) objects such as water, land, capital, and labour. The term *resource* is currently defined as an abstract concept which hinges upon man's perception of the means of attaining certain socially and economically valued goals, within

---

[1] I am aware of the philosophical problem involved in the concepts of man and of nature (Yi-Fu Tuan 1971) and have deliberately chosen to separate them here for purposes of the argument in this paper, i.e., man (population) making use of nature (natural resources).

the limits of the biophysical environment (O'Riordan 1971:60). But although this is the current view, the emphasis in this paper is on the tangible objects, especially the naturally occurring ones, which a country has available for its use, that is, its natural resources.

Natural Resources

A natural resource is often defined broadly to include any object or feature in the God-given (cosmic, natural) environment that can be used by man. The environment here includes the atmosphere and outer space, the lithosphere, the hydrosphere, and the biosphere, excluding man. The terms *renewable*, *reusable*, *non-renewable*, *potential*, and *developed* often used in relation to natural resources are widely discussed in the literature (O'Riordan 1971; Pryde 1972), and so need not be defined here. However, ideas change according to the amount and nature of available information, and soils, forests, water, wild life — though all theoretically renewable — have been "killed" in many parts of the world owing to indiscriminate use, based perhaps on the notion that they can renew themselves without man's efforts.

Population Change

The concept of population change in connection with resource utilization implies increase or decrease in number, and/or demand function; changes in patterns of distribution; and improvement in technology and skills among other things. Although it is not easy, because of inadequate and unreliable census figures, to determine the rate of population increase in sub-Saharan Africa, estimates, especially from detailed sample censuses, give reasonable figures which vary from 1.0 to 3.5 per cent a year (Table 15.1).

A very important cause of population change in Africa is European colonization. Starting from the fifteenth century, European influence in Africa eventually led to the partitioning of the entire area (except Liberia and Ethiopia) into European spheres of influence. This resulted, among other things, in the establishment of educational institutions, health facilities, trade, factory-type industries, and a new-style government system, all of which were geared towards the development of both local and overseas trading. Overseas trading started with a few commodities, notably human beings (as slaves) and a few local products including gold, ivory, and spices. Later, after the abolition of the slave trade, the demand for local products increased substantially. The result was an increase in the range of locally produced crops and intensified mineral exploitation.

Resource Utilization, Conservation, and Management

Classical economic theory lists four resource inputs — land, labour, capital, and entrepreneurship — as vital to the production process and hence necessary factors in generating and stimulating economic development. In the classification, land approximates what is here referred to as natural resources. Just as land is important to the entire process, the ability to exploit (utilize) it varies with culture, time, space, accessibility, and other factors. Also, man's attitude towards natural resources exploitation (utilization) varies between cultures and with time. Therefore, while the possession of a sizeable and diversified stock of natural resources is a major factor in the satisfaction of wants and desires, it is not a sufficient factor. It is true that the resources possess latent utility values, but these will have to be realized by man. The real key to the successful development of the natural resources of a country, therefore, lies in the full awareness of the nature, extent, and possible values of the resources, as well as of the full appreciation of the methods of and approaches to their exploitation.

Biophysical Systems

The earth system is the highest ranking example of a system divisible into a number of sub-systems, including the physical (environmental) systems, the biophysical systems, and the human/biological systems. Of these, the biophysical systems are the most relevant here, referring as they do to the interphase between the human/biological and the physical (environmental) systems, and underlining the significance of the links and interactions between the physical systems or natural resources (air, water, rocks and minerals, plants, wild life, and land forms) and the human systems involving the use of tools, techniques, symbols, and the like.

The main argument here is that the entire resource wealth of a place is an interrelated whole, exhibiting the main attributes of systemic entities or organizations including the connectivity, interconnection, interdependence, and interrelatedness of the various components (Chorley and Kennedy 1971). Because resources are interrelated, their exploitation must take cognizance of possible consequences, particularly the adverse ones, to other resources. To do this, at least two types of information are essential: (1) a proper survey inventory and evaluation of the various stocks of natural resources that exist (Areola and Faniran 1977); and (2) a proper assessment of the side effects of the exploitation of one resource, initially on the resource world generally and eventually on man. The ideal approach is that in which all the resources are planned for in a comprehensive development plan within a functional, definable region (Faniran 1972; 1980).

TABLE 15.1 POPULATION CHARACTERISTICS OF SELECTED AFRICAN COUNTRIES, SOUTH OF THE SAHARA

Region/Country	Population Estimates Mid-1972 (million)	Annual Births per 1,000 Population	Annual Deaths per 1,000 Population	Annual Rate of Popula-tion Growth (per '000)	Number of Years to Double Population	Population Projections to 1985 (millions)	Annual Deaths to Infants under One Year of Age per 1,000 Live Births	Percent of Population in Cities of 1,000,000
Angola	5.9	50	30	0.21	33	8.1	192	6
Botswana	0.7	44	23	0.22	32	0.9	175	—
Burundi	3.8	48	25	0.23	30	5.3	150	—
Cameroon	6.0	43	23	0.20	35	8.4	137	6
Cape Verde Islands	0.3	39	14	0.25	28	0.3	121	—
Central African Republic	1.6	46	25	0.21	33	2.2	163	11
Chad	3.9	48	25	0.23	30	5.5	160	—
Congo (People's Republic of)	1.0	44	23	0.21	33	1.4	148	21
Ethiopia	26.2	46	25	0.21	33	35.7	—	3
Equatorial Guinea	0.3	35	22	0.14	50	0.4	—	—
Gabon	0.5	33	25	0.80	87	0.6	184	—
Gambia	0.4	42	23	0.19	37	0.5	125	—
Ghana	9.6	47	18	0.29	24	14.9	122	18
Guinea	4.1	47	25	0.23	30	5.7	216	6
Ivory Coast	4.5	46	23	0.24	29	6.4	138	12
Kenya	11.6	48	18	0.30	23	17.9	—	7
Lesotho	1.1	39	21	0.18	39	1.4	181	—
Liberia	1.2	50	23	0.27	26	1.6	137	—
Malagasy Republic	7.3	46	25	0.21	33	10.8	102	6
Malawi	4.7	49	25	0.25	28	6.8	119	4

TABLE 15.1 (*con't*)

Region/Country	Population Estimates Mid-1972 (million)	Annual Births per 1,000 Population	Annual Deaths per 1,000 Population	Annual Rate of Population growth (per '000)	Number of Years to Double Population	Population Projections to 1985 (millions)	Annual Deaths to Infants under One Year of Age per 1,000 Live Births	Percent of Population in Cities of 1,000,000
Mali	5.3	50	27	0.23	30	7.6	190	5
Mauritania	1.2	44	23	0.21	33	1.7	137	—
Mauritius	0.9	27	8	0.19	37	1.2	58	17
Mozamb que	8.1	43	23	0.21	33	11.1	—	4
Namibia (Southwest Africa)	0.7	44	25	0.20	35	0.9	—	—
Niger	4.1	52	23	0.29	24	6.2	148	—
Nigeria	58.0	50	25	0.26	27	84.7	—	7
Portuguese Guinea	0.6	41	30	0.11	63	0.7	—	—
Rhodsia	5.4	48	14	0.34	21	8.6	122	14
Rwanda	3.8	52	23	0.29	24	5.7	124	—
Senegal	4.1	46	22	0.24	29	5.8	—	15
Sierra Leone	2.8	45	22	0.23	30	3.9	136	7
Somalia	2.9	46	24	0.22	32	4.2	190	7
South Africa	21.1	41	17	0.24	29	29.7	138	32
Tanzania	14.0	47	22	0.26	27	20.3	162	3
Togo	2.0	51	26	0.25	28	2.8	163	10
Uganda	9.1	43	18	0.26	27	13.1	160	4
Upper Volta	5.6	49	29	0.20	35	7.7	182	—
Zaire (Dem. Rep. Congo)	18.3	44	23	0.21	33	25.8	115	7
Zambia	4.6	50	21	0.29	24	7.0	159	11

Environment as Resources and Constraints in Sub-Saharan Africa

Sub-Saharan Africa is unique in that it is the land of the Negroes. Although a few million whites might be counted, and although Negroes also live in other parts of the world, Africa south of the Sahara cannot be rivalled by any other region in the "purity" and "homogeneity" of the black population.

The Negroes have many distinctive characteristics apart from the colour of their skin. They have not until very recently been known for achievements in the literary and technological fields. On the contrary, they have been used as slaves in the Middle East, the Maghreb, and the Americas. Their territory has been colonized, emerging as fragmented, and in many cases, insolvent nations. However, it is the physical environment which usually provides the basis for the pattern of settlement and the economic activities of a given area. This environment consists of the rocks (structure), relief, drainage, climate, soils, and vegetation, which separately and together constitute both resources and resistances (limitations) to development.

Physical Environment

Structure, relief, and drainage.   In terms of relief and geological structure, sub-Saharan Africa may be divided by a line drawn from northern Angola to western Ethiopia into two provinces (Fig. 15.1). To the west is what is popularly referred to as Low Africa and to the east is High Africa. The greater part of Low Africa consists of old pre-Cambrian rocks, mainly metamorphic and igneous rocks with a few old volcanic rocks. In addition, sedimentary rocks, ranging in age from the Palaeozoic to Recent, outcrop very widely. The Cretaceous rocks are probably the most significant because they contain the only economic coal deposits of the sub-region (for example, in southeastern Nigeria) as well as the oil and gas deposits (e.g., in the Niger delta, Gabon, and Angola). The geology has influenced the development and present-day distribution of the important resources of the region, including the rocks and minerals, land forms, and soils (Figs. 15.2 and 15.11).

An outstanding feature of the landscape of Low Africa is the existence of extensive plains, both depositional and erosional, at varying elevations. Apart from the low plains along the coast, there are high plains and plateaus, such as the Jos Plateau, which have extensive areas of grass cover and are important for animal husbandry (Faniran 1968). The surfaces are usually separated by scarps or by narrow zones of broken country. Other major land forms are the hills, particularly the randomly distributed rock hills (inselbergs) of the crystalline base-complex rock areas (Faniran 1974a) and the residual hills of the sedimentary (Cretaceous) rock areas, the incised

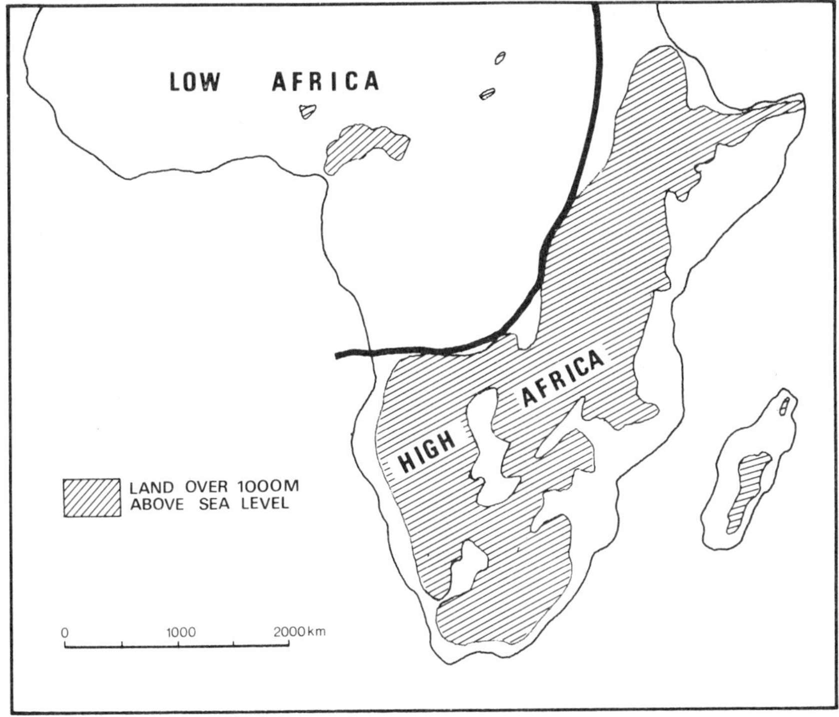

Fig. 15.1 High and Low Africa

river valleys, and the highlands or mountains such as the Cameroon-Ba-menda highlands. There are also basins (notably the Chad basin and the Congo basin) which, in many ways, resemble the plains, especially the low plains. A common sequence of land forms consists of a narrow rejuvenated (incised) valley and an extensive plain (pediment) backed by a hill which may be an inselberg, a quartzite ridge, or a residual hill (Burke 1967).

In contrast, nearly all High Africa rises above 500 m (Fig. 15.3). Even the Kalahari basin lies more than 500 m above sea level, while in East Africa the surface of Lake Victoria stands more than 1,000 m above sea level. In this region are a number of lofty mountains and peaks, including the Drakens-berg (South Africa) which has an average height of 3,000 m and extends north into the highlands of Zimbabwe, Zambia, and Malawi. Also in the region are the volcanic peaks of East Africa, namely, Kilimanjaro (5,971 m), Kenya (5,194 m), Ruwenzori (5,118 m), Dashan (4,572 m), Elgon (4,310 m), and many others.

There are two drainage patterns in sub-Saharan Africa: that in the low-land highland plains and depressions of Low Africa, and that in the high-

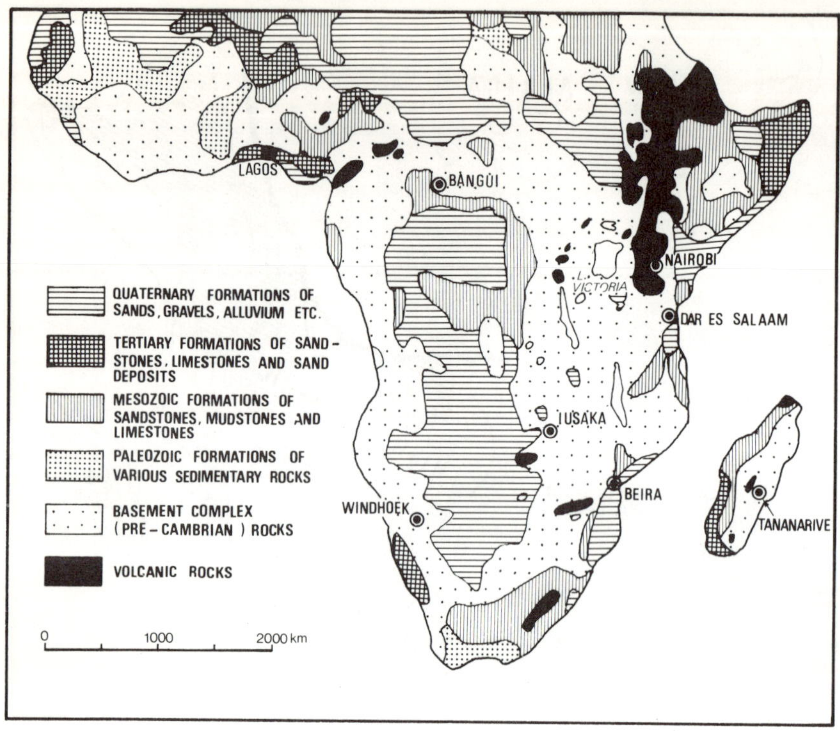

Fig. 15.2 Geology

relief areas of High Africa. River courses over sections of the plains and depressions are not permanent. It has been observed, for example in the Chad Basin, that a fisherman once caused the permanent diversion of river water over a 2,600 km² area just by placing a fish trap along the river channel (Morgan and Pugh 1969:231). This is because the river flood plains are usually quite extensive and of low gradient.

The wide, almost flat plains provide ideal conditions for deep weathering and duricrusting (Faniran 1974b). Agriculture cannot be practised where the duricrusts outcrop at the surface, as they do over large areas. The extensive plains also experience a high rate of infiltration, reduced runoff, and a substantial ground water recharge (Asseez 1972; Omorinbola 1979). In contrast, rivers in the high-relief areas are fast and short and marked by waterfalls; many of them have already been dammed and harnessed for hydro-electricity (Fig. 15.3).

Climate and water supply. The main features of the climatic environment of sub-Saharan Africa (Fig. 15.4) can be summarized as follows:

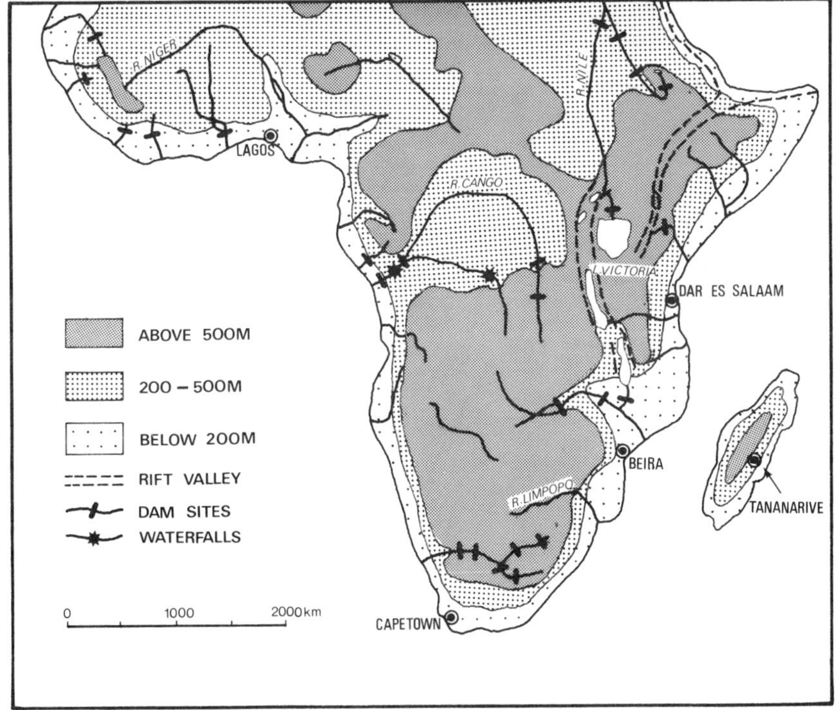

Fig. 15.3 Relief and drainage

1. Virtually the entire area lies within 35 °N and S of the equator, so that the climate is typically tropical. Indeed, Africa is the most tropical of all the continents.

2. The basic pattern of air movement is generally towards the equator. This pattern is modified by the N-S movement of the Inter-Tropical Discontinuity (ITD). Because this zone of discontinuity is not always parallel to the equator — for instance it is pushed far south in the east and south during the southern summer, apparently under the influence of relief — the wind pattern is complicated, as is the weather pattern.

3. Rainfall varies markedly in amount and distribution over the area. Generally speaking, the western half is wetter than the eastern half, except in southern Africa. The areas of low and unreliable rainfall are badly affected by recurrent periods of drought and flood. The Sudan Zone, for example, was affected by a drought which lasted from 1969 to 1973 and caused substantial losses to the region (Fed. Dept. Water Resources 1977).

4. The reliability of rainfall depends on the effective temperature and its evaporative power. Although the southeastern part is considerably cooler

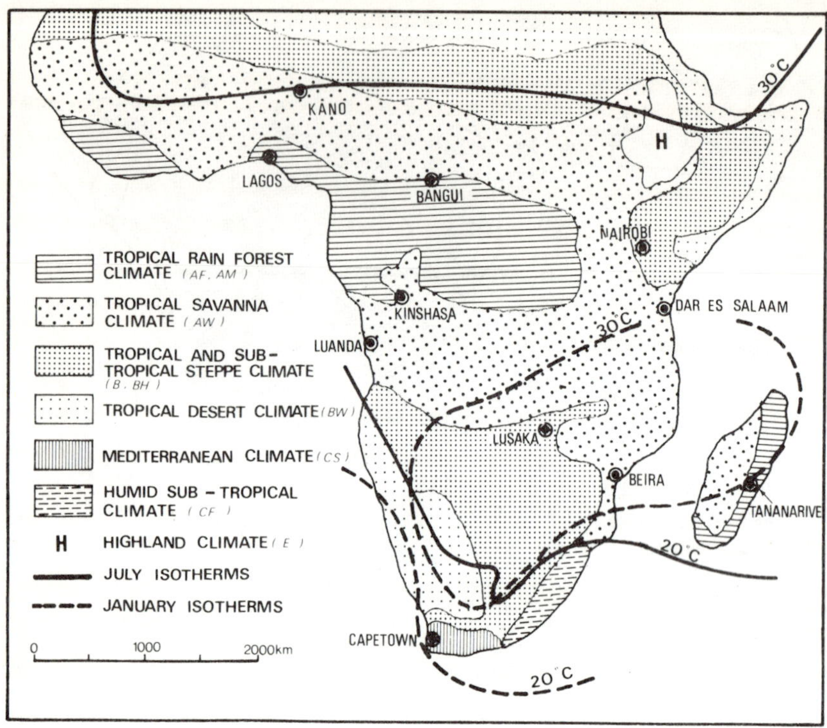

Fig. 15.4 Climate

than the northwestern part, mainly on account of greater altitude, the rate of evaporation is still high over the entire area as the lowest mean monthly temperatures are still of the order of about 18 °C; while above 30 °C figures are common (Table 15.2; Fig. 15.4).

5. The tempo of economic activities is dictated mainly by rainfall, both in terms of its absolute annual totals and, more important, in terms of its seasonal distribution. Plant growth ceases with the onset of the dry season, when temperatures are high and humidities low. Traditional African life, based on a subsistence (arable crop) economy, is adversely affected by a long dry season, which means a long period of under-employment. Also regional differences in river regimes, vegetation, and land use can be explained primarily in terms of the variation, from place to place, in the amounts and seasonal distribution of the rainfall. It is only recently that large-scale irrigation projects are being undertaken, the effects of which are still salutory (Wall 1976).

6. Large parts of sub-Saharan Africa suffers from aridity — permanent, seasonal, or occasional — in one form or another. In the view of some

TABLE 15.2 MEAN ANNUAL TEMPERATURES (T°C) AND RAINFALL (Rmm) FOR SELECTED AFRICAN STATIONS

Station		Jan.	Feb.	Mar.	Apr.	May	June	July	Aug.	Sept.	Oct.	Nov.	Dec.
Freetown (Sierra Leone)	T(°C)	17/81	18/3	18.6	17.9	17.6	16.6	14.8	15.3	16.6	18.4	17.9	17.0
	R(mm)	48.3	91.4	106.7	226.1	142.2	55.9	22.9	27.9	30.5	58.4	134.6	71.1
Goree (Senegal)	T(°C)	20.4	19.2	20.3	20.7	16.6	25.9	27.5	27.7	28.2	28.0	25.9	22.4
	R(mm)	0	0	0	0	0	124.5	91.4	251.5	132.1	17.8	2.5	0
Mongalla (Sudan)	T(°C)	25.3	27.3	31.8	34.8	36.1	32.8	29.0	27.8	28.1	29.4	28.6	25.3
	R(mm)	0	0	0	0	15.2	99.1	210.8	201.8	142.2	48.3	7.6	5.1
Lulaabourq (Zaire)	T(°C)	27.1	27.8	28.3	27.4	26.3	25.4	24.6	24.5	25.3	25.8	26.3	26.4
	R(mm)	2.5	17.8	38.1	106.7	137.2	116.8	132.1	147.3	124.4	109.2	45.7	7.6
Swakopound (Namibia)	T(°C)	17.4	17.4	17.8	15.6	16.0	14.8	13.7	12.8	13.5	14.6	14.9	16.5
	R(mm)	0	2.5	5.1	0	0	0	0	0	0	2.5	0	5.1
Pietermaritzburg (South Africa)	T(°C)	23.1	23.2	22.1	20.1	16.7	14.3	14.9	17.0	18.6	19.8	20.8	22.3
	R(mm)	129.5	157.5	129.5	66.0	27.9	7.6	2.5	20.3	45.7	63.5	196.6	127.0
Do W'ala (Cameroun)	T(°C)	26.5	26.8	26.4	26.2	25.9	19.5	23.9	23.8	24.4	24.6	25.7	26.1
	R(mm)	203.2	276.9	434.3	434.3	629.9	1,516.4	635.8	1,465.6	1,656.1	1,148.1	1,675.6	383.5

(con't overleaf)

TABLE 15.2 (*con't*)

Station		Jan.	Feb.	Mar.	Apr.	May	June	July	Aug.	Sept.	Oct.	Nov.	Dec.
Entebbe (Uganda)	T(°C)	21.9	21.9	22.0	21.5	21.8	20.9	20.5	20.5	20.9	21.3	21.3	21.4
	R(mm)	66.0	914.4	147.3	246.4	215.9	129.5	73.7	78.7	78.7	88.9	127.0	129.5
Nairobi (Kenya)	T(°C)	17.8	18.3	18.6	17.9	17.6	16.6	14.8	15.3	16.6	18.4	17.9	17.0
	R(mm)	48.3	91.4	106.7	226.1	142.2	55.9	22.9	27.9	30.5	58.4	134.6	71.1
Lagos (Nigeria)	T(°C)	27.4	16.9	28.7	28.3	27.9	26.5	25.8	25.6	26.0	26.6	27.7	27.7
	R(mm)	27.9	53.3	94.0	144.8	266.7	475.0	271.8	71.1	134.6	198.1	66.0	20.3
Cape Town (South Africa)	T(°C)	21.2	21.5	20.2	17.5	15.1	13.3	12.7	13.2	14.5	16.4	18.1	20.1
	R(mm)	17.8	15.2	22.9	48.3	96.5	114.3	94.0	86.4	58.4	40.6	27.9	20.3

writers, no drop of rain falling on the surface should be allowed to escape to the sea, at least not until it has fulfilled some function or brought some benefit to mankind (Stamp 1959:88). Others have suggested that a comprehensive approach to the continent's water resource development be taken (Faniran 1971; 1974c). At present, large quantities of water go waste via the major and minor rivers to the adjoining seas. In contrast to the peoples of the Nile basin, western Europe, and Southeast Asia where wells, shadoof, water wheels, flumes, tanks, irrigation ditches and channels, and terraces among other devices are available for reaching, storing, and using water, the inhabitants of sub-Saharan Africa still adhere in the main to traditional sources and techniques. With few exceptions, their crops depend upon the rains and their domestic supplies are from rivers and streams (when they flow), lakes, shallow hand-dug wells, pits, and from direct collection of rain water in earthernware pots and calabashes (Akintola *et al.* 1979; Faniran *et al.* 1980).

Soils.   The soils of sub-Saharan Africa are affected by the climatic and drainage conditions. Most soils show the effects of the intense pedogenic process commonly termed lateritization or duricrusting. The high temperatures and humidities, the abundant rainfall over large areas, and the seasonality of the rainfall accelerate the process of deep weathering which, especially in the low relief areas, has resulted in the widespread occurrence of duricrusts or ferrugionous crusts and related soils (Faniran 1974b:Fig. 15.5). These soils, except in isolated places, are mostly infertile, and have a low base status and cation exchange capacity. Even where the ferruginous crusts are not exposed and fertility is maintained by the vegetation, clearing for cultivation quickly deprives the soils of their humus supply and exposes them to surface wash. The result is that the soil is rendered infertile within at most five years, after which the plot has to be abandoned.

Pockets of quite fertile soils occur in the east and south, where volcanic rocks have weathered into very rich soils such as those of the "White Highlands" of Kenya and scattered other places in the sub-region.

Vegetation.   The vegetation consists of three major formations — forest, grassland, or savanna and desert (Fig. 15.6) — each of which has been markedly modified by man. The tropical rainforest has the most imposing structure in the plant world. It covers large parts of west and central Africa. It is made up of numerous species of trees of different heights and ages, and is the storehouse of virtually all the valuable timber in Africa. The tropical savanna consists of the sub-humid tropical woodland and grassland occupying the plateau country of south, central, and east Africa as well as the northern margins of the Congo basin to the Atlantic coast, south of the Gambia River. They supply fewer export commodities than the

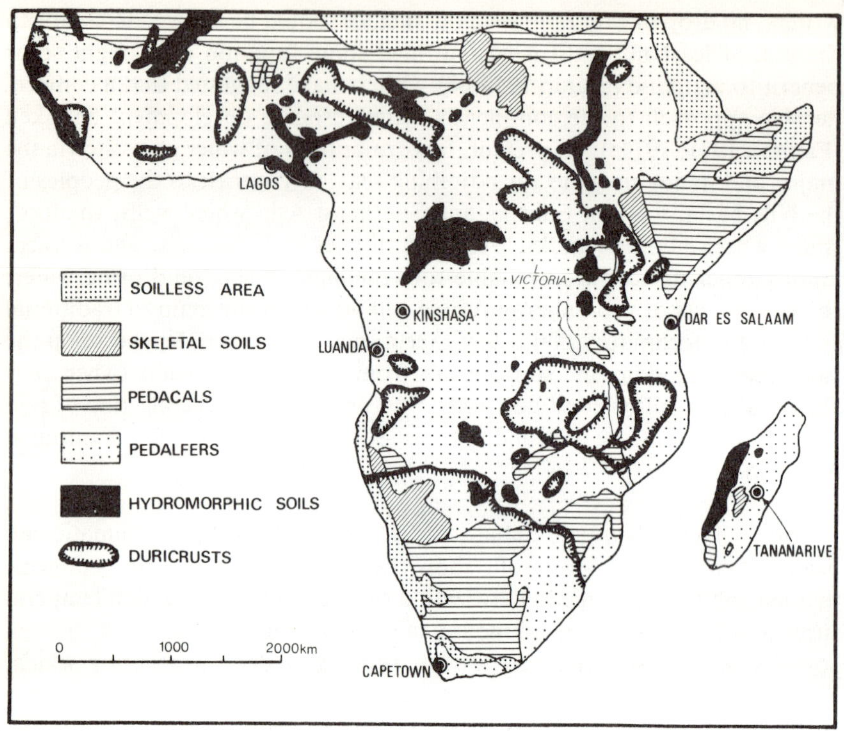

Fig. 15.5 Soils

forests, the most important being shea butter and gum arabic. They also play important roles in the local economies as sources of wild fruits, medicaments, dyes, timber, and firewood. In West Africa, the savanna merges, southwards and northwards respectively, into wooded steppe, thorn scrub, and finally desertic vegetation. In southern Africa, the succession is influenced by the occurrence of Mediterranean and humid temperate climates along the southern and southeastern coasts respectively.

Wild life.   Africa's mammallian fauna is remarkably rich, consisting of about forty families. Fossil remains indicate that many of the animals now confined mainly to central and east Africa were much more widespread in the past, and were distributed between the Atlas Mountains and the basin of the Vaal River. Ancient rock engravings and paintings of hippopotamus, giraffe, and other large beasts have been found in remote desert regions from where they had long since disappeared. The fauna left is of very great interest to scientists and tourists alike. National parks have been demarcated where hunting is prohibited and conscious efforts are being made to protect the fauna population (Fig. 15.7).

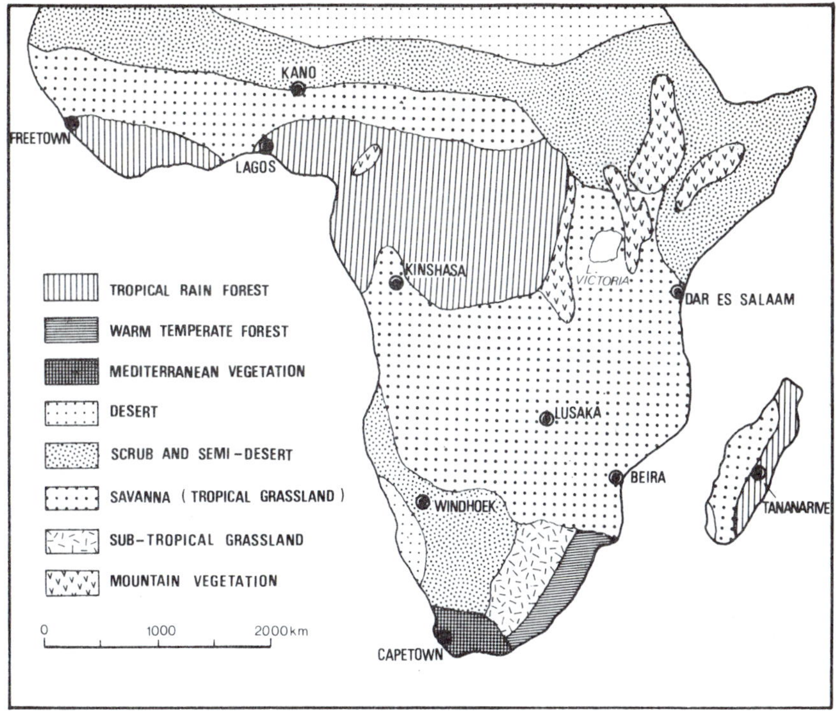

Fig. 15.6 Vegetation

The human environment. The success of any development effort (resource utilization) is a function of both the physical and the human environment, the latter consisting, in the nomenclature of Zimmermann (1951), of the social and institutional environments. However, while the aspects of the physical environment are more or less obvious and straight-forward, those of the human environment are less so, particularly in sub-Saharan Africa where both the social and institutional environments cannot be seen as being directly relevant to modern economic development and progress. Nevertheless, the following topics are often discussed in this connection: population characteristics, history, and institutional arrangements.

Population. The population of Africa south of the Sahara stood at more than 220 million in 1968. Except for perhaps the Republic of South Africa, censuses are usually unreliable and estimates may be wrong by a margin of up to 15 per cent. Consequently, rates of population increase, important for planning purposes, are seldom available, and crude approximations are being used. These range, as mentioned earlier, between 1.0 and

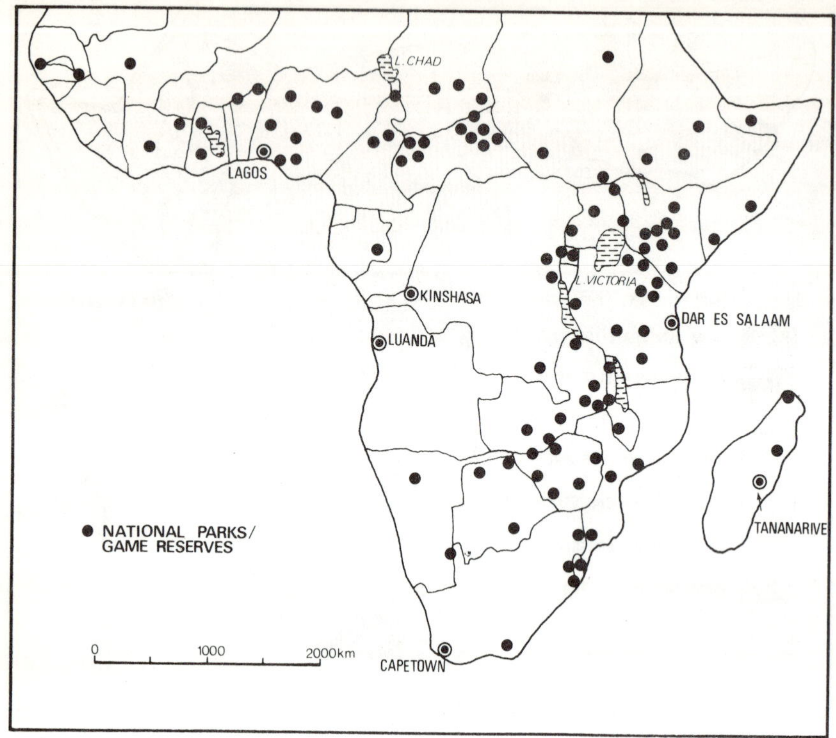

Fig. 15.7 National parks and game reserves

3.5 per cent annually, (Table 15.1), meaning that the population in some parts of the sub-region may double by the end of the century. The following are the salient points about the population and demography of this sub-region, which have relevance for the utilization of its natural resources:

1. Large areas are still sparsely populated with fewer than 25 persons per km² (Fig. 15.8). These areas, except those with valuable mineral resources, are likely to remain poor and backward.

2. The densely settled regions are few and far apart, but are nevertheless of great economic importance. These, with exception of the mineral-rich areas of southern and central Africa, are essentially agricultural regions with population densities of more than 50 persons per km².

3. Of the largest densely settled regions, two are in West Africa — one in the coastal forest belt from Nigeria to Ghana and the other in the Sudan belt in the Kano and Mossi regions of Nigeria and Upper Volta. Another densely settled area is in East Africa, in the Lake Victoria basin. Other densely settled areas include enclaves in southern Africa (between East London and

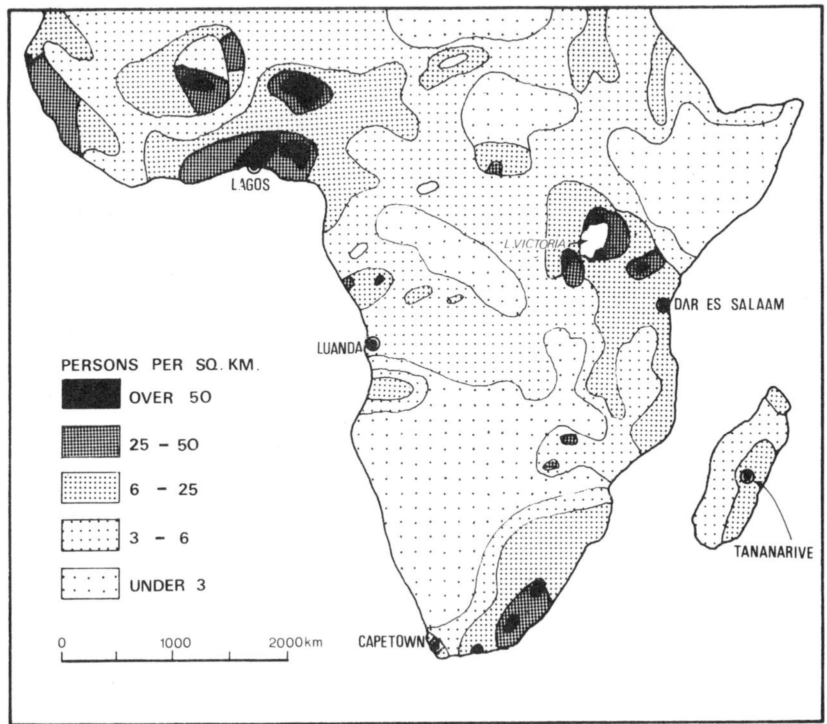

Fig. 15.8 Population density

Durban), the hinterlands of Freetown (Sierra Leone) and Dakar (Senegal), and the borders of the rift valleys in Ethiopia and Malawi (Fig. 15.8).

4. Nearly all the densely populated areas are important producers of one cash crop or another. The populous belt near the Gulf of Guinea is among the world's greatest sources of cocoa. Hausaland produces considerable quantities of groundnuts, cotton, hides, and skins, while cotton, coffee, tobacco, and tea are valuable exports from the country bordering Lake Victoria (Fig. 15.9).

5. Africa is the least urbanized of the continents, with only about 10 per cent of its population living in towns with more than 5,000 inhabitants. In fact, until the end of the nineteenth century, nearly all Africans lived in small rural settlements, practising subsistence agriculture. The exceptions are the caravan route termini in Sudan West Africa; Arab ports along the east coast; national capitals such as Kampala, Addis Ababa, and Kumasi; and the towns of Yorubaland in southwestern Nigeria (Fig. 15.10).

6. The colonial period saw the establishment of new towns as administrative, commercial, mining, or industrial centres.

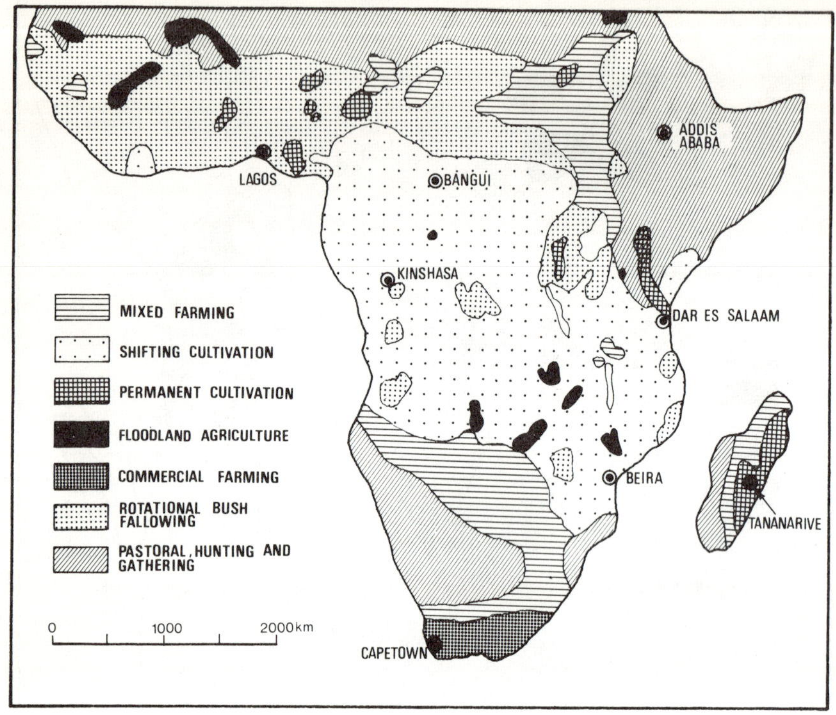

Fig. 15.9 Agricultural systems

7. The history of population in Africa is marked by large-scale movements of population over hundreds of kilometres. Although movements of whole tribes have since ended, these have been replaced by labour migrants to mines, factories, and large-scale farming concerns or by skilled labour and other elite to the coastal and other towns (Fig. 15.10).

8. There is marked contrast in manpower development between areas. Generally speaking, manpower development seems to have progressed faster in the former British territories (including the Union of South Africa and Zimbabwe) than in the other countries. Nigeria, with thirteen existing and seven proposed universities is a case in point. Nevertheless, there is a general shortage of technical and professional people which necessitates the importation of expatriates, formerly mainly from Europe but now from virtually all over the world.

History and ethnography.   To fully understand the nature of the relationship between man and his environment in Africa, we must identify a few events or periods, namely, the indigenous, proto-colonial, colonial, and post-colonial — all of which have left indelible marks on the present-day

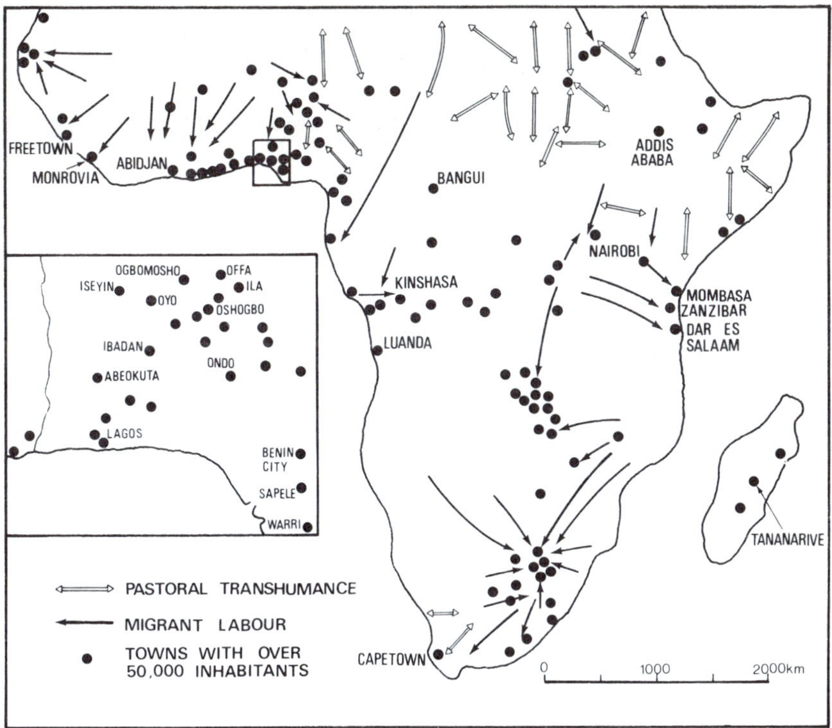

Fig. 15.10 Urbanization and migration

society. The indigenous period is taken as the period before A.D. 1450, when there was little or no contact with the outside world. Although there were considerable developments during the mediaeval and early modern periods, these were exclusively in connection with trade, upon which many of the early African empires such as Ghana, Mali, and Songhay flourished. Agriculture was basically subsistence since the trade items were essentially non-agricultural.

General economic development based on intensive natural resource utilization only started as a result of Europe penetration and settlement. The first permanent European trading station of Goree Island (Dakar, Senegal) was established in A.D. 1445. Another was built at Elmina (Ghana) in A.D. 1482. These early contacts were made by Portuguese explorers who were basically concerned, apart from trade, with finding the legendary kingdom of Prester John. They therefore directed their attention mainly towards the Indian Ocean.

The incidental features of the Portuguese effort, especially the introduction of new crops from the New and Old Worlds, had the most far-

reaching consequences. Bananas and yams from Asia and maize, groundnuts, cassava, and sweet potatoes from the Americas have since altered the
social life of the people (Fig. 15.9). Other notable events of the proto-
colonial period are the slave trade, the white settlement of South Africa,
and the nineteenth-century explorations which brought the British, French,
Dutch, German, and other colonial powers, as well as the missionaries, into
the continent.

The partitioning of Africa in 1885 into states under various colonial
powers marked the real beginnings of the period of "organized" utilization
of Africa's natural resources. Therefore, the period 1450–1885 may be
regarded as a transitional phase, which extended into the colonial period to
about 1945 when the tempo of national economic development increased
markedly. The year 1960 saw the "wind of change" blowing through Africa
with the granting of independence to African states. Today all the nations
of Africa are very conscious of their natural endowments as well as the need
to develop them for their own benefit. The various national development
plans formulated at periodic intervals are an indicator of this consciousness.

Institutional arrangements.   The indigenous administrative set-up is, in
most communities, organized on the basis of a subsistence economy. The
main occupation is subsistence agriculture so that the major item of property is land. This is commonly held as an inalienable right. Superimposed
on this is the European system of commercial farming, private ownership,
and advanced technology. The method of imposition varies between the
different colonial powers. The indigenous societies were also not homogeneous. The impact of the changes in the traditional system is therefore
different from place to place.

First, it is possible to differentiate between the different colonial administrations. Whereas the indigenous population in many former British territories transformed their economies into semi-commercial ones by producing
coffee, cocoa, oil palm, and groundnut for export, the production of these
crops in some former French, Belgian, Portuguese, and German territories
was, and in some cases still is, largely in the hands of expatriate farmers.
But this is not to say that the situation is similar in all the countries. In
Nigeria, for example, it is possible to distinguish between the cash croppers
of the forest south and Sudan north from the predominantly food croppers,
many of them subsistence farmers, of the Middle Belt. Also the difference
between, say, a Yoruba cocoa farmer and a Biron on the Jos Plateau is such
that it is difficult to imagine that they both belong to the same country.

The level of economic attainment thus differs widely within sub-Saharan
Africa, just as there are wide disparities in the distribution of their natural
resources.

Natural Resource Utilization in Sub-Saharan Africa

The natural resources which are being exploited in sub-Saharan Africa are the rocks and minerals, soils, water, land forms, and forests (or plants) and faunal resources. The main activities connected with this exploitation are hunting, gathering, mining, agriculture, pastoralism, fishing, manufacturing, and tourism.

The tendency so far has been to emphasize the impact of resource utilization on the economy. Thus White and Gleave (1971) identified two major themes in a discussion of the West African economy. First, there is continued improvement and modernization of the traditional economy consisting essentially of farming and craft industries. Second, perhaps because of the variation in the distribution of the resources and also because of some locational advantages and disadvantages, the rate and degree of modernization so far achieved vary greatly among the various parts of the region as well as among the various sectors of the economy.

In contrast, the approach taken in this paper underscores the effects of population change on the utilization of natural resources. The following themes are considered relevant in such an approach: population/resource ratio, optimum population, and resources utilization.

Population/Resource Ratio

Many models, some of them very simplistic, have been used in the discussion of the concept of population/resource ratio. The early models are based on largely discredited concepts, especially of resource. One such model, as mentioned earlier in this paper, is that of the inexhaustibility of resources, particularly the renewable resources.

Among the best examples of population/resource ratio analysis in sub-Saharan Africa is that of the population carrying capacity of the land, elaborated for West Africa by Morgan and Pugh (1969) and illustrated for the various cultivation methods. The idea is based on the assumption that people with the same cultivation technique in the same environment should have the same coefficient of density (Urvoy 1942). The result of the application of this idea to the agricultural regions of West Africa is summarized in Table 15.3. The density figures are those of *cercles* and divisions which are large enough to include wasteland. Only divisions or *cercles* with at least two-thirds of their area in the appropriate region are used, while those which are markedly affected by commercial cultivation have been excluded.

Table 15.3 shows that the traditional systems of agriculture will support (in round figures) about 8–20 persons per km^2 in the cereals region; about 12–35 per km^2 in the mixed roots and cereals region, and about 27–77 per km^2 in the roots region.

**TABLE 15.3 CULTIVATION SYSTEMS AND POPULATION
DENSITY IN WEST AFRICA**

Cultivation System and Region	Population Density
	(per km²)
Shifting Cultivation	1.5 – 9
Rotational Bush Following:	
Northern cereals region	8 – 20
Bulrush millet cereals region	2.3 – 7
Bornu	9 – 12
Maradi Dosso	7 – 10
Tilla bery	5 – 12
All	
Guinea corn dominant subregion	2 – 32
The S.W. upland rice region	5 – 36
Mixed cereals and roots region	3 – 71
	11 – 35
The roots region	45 – 170
N. Iboland (est. max)	116 – 154
S. Iboland (est. max)	147 – 190
Overall	27 – 77
Nigeria — Cameroon border	
(Cocoyam — Plantain region)	
Kumba	13
Victoria	13
Mixed Farming — Kaolack (Serer)	20
Labe (Sedentary Fulani)	19
Dikwa (Shuwa)	20
Permanent Cultivation	93
Kano Division	240 – 337
Ibo and Ibibio	66
Jos Plateau	91
Kabrai (Togo)	
Floodland Cultivation	
Niger Valley/Tosaye-Say	19 – 50
Gao	50

Source: Morgan and Pugh 1969.

Optimum Population

Nash (1941) estimated that the minimum density that will keep a level of clearing and control tsetse infestation in the Anchau area of Kaduna State (Nigeria) was 27 per km², and Stamp (1938) gave the maximum figures for savannas as 34 and for forest as 56. Research conducted in Africa shows that there are close links between population characteristics and systems of land use. It is apparent from Table 15.3 that the listed cultivation systems reflect, among other things, differences in population density. Indeed, it appears that man in sub-Saharan Africa, as in other parts of the developing world, has reacted to increases in number by changing his cultivation system to meet his food needs. That he has not succeeded in all cases to adjust to increasing presence of population on land is reflected by such environmental problems as soil degradation, forest depletion, and outmigration.

One example of areas showing the impact of accelerated population growth is the Kigezi district of Uganda (Manshard 1965; Greenzebach 1980). Greenzebach (1980:35) summarizes the situation in these words:

> as a result of the population explosion . . . the former abundant supply of land which was freely available to everyone according to their needs has recently become a subject of speculation . . . , is barely sufficient and, most recently, has become a negotiable asset. . . .
>
> The general drift from the overpopulated agricultural areas of southern Kigezi, . . . over a period of no more than ten years must be considered a portentous process in the general development of the cultural landscape and society of this problem region of Africa.

Greenzebach (1980) also made the following points about the region:

1. Owing to shortage of land, 50 per cent of the families studied had no chance of earning their living from their own farm land, even if they used intensive cultivation methods.

2. Only about 7 per cent of the farms in the area studied were more than 1 ha in size. The largest farm (3.4 ha) belonged to a staff member of the Agriculture Ministry, and was located in a newly developed, formerly swampy valley land.

3. Almost everywhere the soil's fertility was at risk as a result of permanent use, inadequate input of fertilizers, and complete failure to allow for a fallow period.

4. The general shortage of land has led to a marked drop in the numbers of small animals being kept since the small food surplus is needed in the field. The result — unbalanced diets, shortage of protein and malnutrition.

5. The general shortage of land greatly limited the cultivation of supplementary cash crops since many holdings were not large enough for self-sufficiency in food.

6. There was a consistent increase in the number of land litigations, which totalled about 10,000 cases between 1960 and 1969.

7. Few non-agricultural alternative sources of employment and income were available. Consequently, migration was the only, in many cases inevitable, alternative for many inhabitants of the southern Kigezi district. The target area was the nearby, less densely populated northern Kigezi district.

These various points apply equally to pockets of overpopulated and ill-adjusted areas of sub-Saharan Africa.

Resource Utilization

The utilization of the natural resources of an area is often revealed through the various activities of man. These activities, as listed earlier, include hunting and gathering, fishing, farming, stock rearing, mining, manufacturing, and commerce. In terms of the related natural resources, hunting, gathering, collecting, fishing, and stock rearing involve the exploitation of the biospheric (plant and animal) resources; farming depends on the soil and water resources; mining on the rocks and minerals, while manufacturing and commerce do not directly involve natural resources and so need not be discussed here.

Rocks and Minerals

The rock and mineral resources of sub-Saharan Africa, which are being exploited for developmental purposes, belong to two major categories — metallic and non-metallic. The Ghana Empire of mediaeval times obtained gold from the Faleme River and the adjacent Bambouk mountains. It is also on record that the Mali (Manding) Empire in the fourteenth century probably sent its merchants into what is now the Ivory Coast and modern Ghana to buy gold among other articles of trade (ivory, kolanuts, and slaves). Indeed, gold was one of the minerals which attracted Europeans to Africa. Besides gold, several metals were smelted and worked into objects of everyday use by the indigenous population, long before the European era. Examples are tin on the Jos Plateau (Nigeria), brass in Ashanti (Ghana), brass and bronze in Benin and Bida (Nigeria), and iron ore almost everywhere. The working of both natural and prepared rock products was even more widespread, as shown by the widespread occurrence of the traditional clay and pottery industry throughout the continent.

However, it is the organized and commercial aspects of the mining industry which are of interest here, and these began with the establishment of colonial rule in the late nineteenth century. The first of such mining operations started in Ghana in 1878 and led to the establishment of rail transport and the opening up of the country.

In fact most of the railways in Africa were built to transport minerals to external markets, for example, the lines in Sierra Leone, Liberia, Guinea, and Mauritania, and the Eastern Railway of Nigeria which was built to join the Jos Plateau tin fields and the Enugu coal fields with the port city of Port Harcourt. Other examples of such railway networks are those of Zaire, Zambia, Zimbabwe, Mozambique, Angola, and the Union of South Africa (Fig. 15.11). A few examples of these rock and minerals are discussed below to further illustrate the population-change factor.

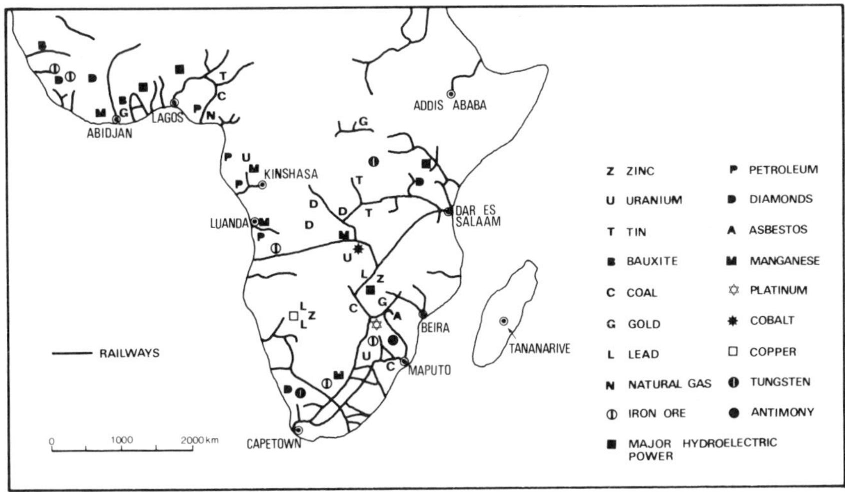

Fig. 15.11 Minerals

Rocks. Africa has a great abundance of suitable rocks for building material. Some of these have been worked for millennia for various direct uses. These materials provide perhaps the best examples of the effect of population change on the exploitation of natural resources. Until very recently, workings have been on very small and localized scales, mainly around the big towns. But in many cases the demand for these materials has grown so much as to cause concern.

A recent study in the Lagos State of Nigeria shows that large areas in the interior are being dug up for earth-fill materials, while the mining of river sand is so widespread and intensive as to be causing the fishing communities, including the Government Fishery Division, very great concern (Faniran *et al.* 1976). Moreover, the establishment of factory-type industries, utilizing local clay deposits (e.g., the defunct Ikorodu Ceramic Industries established in 1953 and the Clay Minerals [Nigeria] Limited, Iregun), sand (for the glass industry at Badagry and Remo), limestone (there are more than ten cement factories in Nigeria), and so on, has led to the intensified exploitation of existing deposits and stimulated a search for other sources. The best example here is perhaps the burnt bricks industry, several of which are being established by both the Federal and State governments. Oyo State has two such industries under construction and plans to establish at least five more.

The greatest demand is from the building industry, and this is to be expected not only because of the rapid, in fact unprecedented, rise in demand for residential, industrial, commercial, and other buildings in the rapidly growing urban areas, but also because of the change in the type of material being used in building houses and in road and similar constructions. The cement block, clay brick, and rock flabs have replaced mud as building material, while gravel and crushed stone are widely used in road and pavement construction.

Minerals. The exploitation of the mineral resources of Africa has been similarly influenced by changes in the structure of its population. The main areas of change in this respect relate to the establishment of modern-style government during the period of colonial administration and the development of overseas trade as well as local industries. However, for a long time the manner of mineral exploitation was little different from the traditional systems, except in its scale and its overseas markets target. Hand labour and head porterage were used for all operations, and workings were mostly on the sites of existing mines or quarries.

But today many mines have been mechanized to a large extent so that local miners, using traditional techniques, are rapidly being displaced. This is particularly true of the mining operations in southern Africa as well as those of central and west Africa. The petroleum industry is different in that it had to be mechanized as well as operate on a large scale right from its inception. Also, unlike the other minerals, mineral oil was never a locally exploited mineral.

The most important mineralized regions in sub-Saharan Africa lie south of the equator. They include the Katanga (Zaire); the Zambia copper region, the Zimbabwean coal fields south of the Zambezi; the manganese occurrences of Zaire, Gabon, and Angola; the gold, diamond, and other deposits of South Africa, including Southwest Africa (Namibia) and the hot steam areas in the rift-valley zones of Uganda and Kenya. Countries such as Zambia and Zaire depend on minerals, just as Nigeria and Gabon now depend on petroleum (Fig. 15.11).

Soil Resources

The importance of soil resources emanates from their role as the anchorage for plant roots and the medium in which the roots develop; as a habitat for the micro and macro fauna which play active roles in soil formation; and as a store of the water and nutrients needed by the plants for metabolism and chemo-synthesis. These and other functions are not independent of each other, for the plants themselves affect the ability of the soil to perform the functions. It is evident that there exists a complex system of interrelationships, interconnections, and interdependence of soil-plant symbiosis, commonly considered in ecosystemic terms. The utilization of soil resources also involves the entire field of soil management in relation to the various agricultural land uses. This will necessitate at least a brief consideration of aspects of indigenous agriculture, especially the controversy which pertains to its relationship with the environment in which it operates.

The African cultivator, particularly in the tropical regions, is often portrayed as a wasteful and destructive exploiter of soil and forest resources. However, it may be difficult to justify this view of early observers from the known features of African indigenous agriculture, many of which are conservative of biological resources. The most important of these features are the following:

1. Only small patches are cultivated, partly because of the amount of labour involved in forest clearing and also because of the limited technology. This introduces only minimal disturbance of the soil-plant system and so minimizes runoff and attendant soil loss.

2. Cultivation lasts for only two to five years, after which follows a period of natural fallow, partly due to loss of fertility and partly to weed problems. In the forest belt, in particular, weed encroachment can be very rapid.

3. Many large trees are not cleared while the stumps of smaller trees are left in the ground. The roots are rarely disturbed. These supply considerable quantities of organic matter while cultivation continues; they also allow for quick regeneration of the woody vegetation on abandonment to fallow.

4. A considerable number of adventitious plants are usually left to grow with the crops. Many of these plants are very useful to the farmer as medicinal and/or as food items. Mulching and the cultivation of cover crops are other ways in which the indigenous farmer ensures the conservation of soil moisture, the interception of the usually heavy rainfall which is potentially destructive of soils, and the ready supply of organic matter to the soil.

5. Tree crop plantations of cocoa, rubber, and oil palm are widely distributed in the humid tropical regions. These, apart from being sources of the cash indispensable in an increasingly commercialized economy, are crops which cause little disruption to the plant-soil ecosystem in the humid tropical environment.

6. In the savanna areas, the use of animal droppings and household manure helps considerably in maintaining soil fertility.

In short, in large parts of tropical Africa the system of agriculture shows evidence of a seemingly high degree of sophistication marked by the use of many soil-plant conservational devices. Consequently, soil erosion was not a general problem.

This situation has, however, changed in many places. The main factor of change, as seen earlier, is population pressure. Table 15.3 shows tolerable population-density levels for different cultivation systems identified in West Africa. These levels have been exceeded in many areas, resulting in the breakdown of the indigenous system. In particular, the fallow period has shortened appreciably in many places leading to impoverished soils and vegetation in some areas and accelerated soil erosion and gullying in others (Ofomata 1965; Ologe 1970; 1972; Faniran and Areola 1974; 1978).

Topographic Resources

Landforms and other topographic features constitute a resource when they are used as amenities. From time immemorial, man has derived satisfaction in different forms from landforms of various types. Examples include people worshipping mountains; mountaineers deriving satisfaction from either climbing the highest peaks or taking the most difficult and often very dangerous approaches to the tops of mountains; hills providing safer sites for settlements; waterfalls and the development of hydro-electricity; tourists visiting unique and spectacular landforms and other sites; suitable coasts developed into busy harbours and ports; beaches providing recreational facilities; and coral reefs serving as attractions to both tourists and scientists (Faniran and Ojo 1980; Faniran, forthcoming).

The use of rivers for the development of hydro-electricity is well known, typified by such dams as the Kariba, Kainji, Volta, and a host of other smaller dams all over Africa (Fig. 15.3). The potential is particularly high in east and central Africa.

Water Resources

Africa can be divided, on the basis of the water supply situation and use, into (1) areas lacking sufficient moisture for cultivation except from rivers and underground sources; (2) areas having sufficient water supplies for cultivation and other uses only in the wet season; and (3) areas having sufficient water for cultivation and other uses all the year round. The first two situations are the most common as there are few settlements in Africa without a water supply problem, whether on account of marked seasonality and scarcity during the dry season, on account of poor water quality, or on account of location far away from water sources. Some settlements are known to be up to 16 km away from their regular sources of water, so that there is a great loss of labour time in fetching water (Faniran *et al.* 1980).

The solutions to the problem of water supply in this sub-region vary in the different zones, but all are aimed at improving the supply situation. Wells have been sunk in areas of proved aquifers while many dams have been built especially for urban water supply. The wells are concentrated in the sedimentary rock areas, where extensive aquifers exist; such areas are common in Senegal, southeastern Mauritania, western Mali and the Sokoto, Chad and Kalahari basins.

Nevertheless, there is still much to be done, while the environmental and other effects of the various schemes are yet to be appraised. In the first place, only a few large centres have supplies with a daily capacity of more than 140 litres per head. Many large towns have less than 45 litres per head a day. And with the very rapid rate of population increase in these centres, many of the schemes are already obsolete by the time they are completed (Faniran 1975). This situation is best illustrated by Lagos, Nigeria's federal capital. In 1959, a reservoir with capacity of 55,000 m³ could supply 200 litres per head a day. But by 1963 one with a capacity of 82,000 m³ could only supply 122 litres per head a day. The situation has since worsened as the population has passed the 2 million mark, while the rate of industrialization continues to accelerate without any substantial improvement in the supply of water.

Wild Life

The activities connected with the exploitation (utilization) of the plant and animal (wild life) resources of sub-Saharan Africa include collecting, gathering, hunting, fishing, lumbering, animal husbandry (livestock rearing), and tourism, including game viewing.

Man's earliest known activities of collecting, gathering, and hunting still persist in the present-day economies of many African communities. Indeed, until very recently, gathering provided the main items of overseas trade, and hunting and fishing the bulk of the animal protein. In Senegal and Mauritania, for example, the trade in gum arabic, gathered from acacias, led to the founding of French trading posts on the coast and along the Senegal river, notably St. Louis, Rufisque, and Portendik. Rubber, because of its commercial value, was gathered and exported from numerous locations, particularly from areas of dense forest. Trade in elephant ivory has been important in equatorial Africa for a long time, being much sought after by the early Portuguese traders.

Piassava is another important wild product exported from sub-Saharan Africa. It is from the sheaths of the rafia palm, and used chiefly in making brushes, brooms, and mats. The piassava industry is particularly important in the coastal swamp lands of Sierra Leone, the best quality coming from the Bouthe district. Other produce gathered for export include kolanuts, shea butter and nuts, locust beans, gum, opal, various species, beeswax, kapok, dyewoods, and barwood.

A related industry in the forest belt is lumbering. The timbers for export are almost entirely obtained from self-sown (wild) species, the planted species such as teak and cassia being cut chiefly for the internal market. But apart from the export market, the internal market for woodfuel (firewood) and for building makes heavy demands on African timber resources. In spite of attempts by governments and private concerns alike to establish timber or woodfuel lots, and also in spite of the growing use of other fuel such as kerosene, gas, and electricity, most of the people still rely on wild tree species, so much so that many forested areas have been cut and turned into treeless grasslands.

The utilization of the wild life resources of an area includes game hunting for valuable food and game viewing (tourism). In sub-Saharan Africa, the hunting of wild animals has been and still is either the main or subsidiary occupation of a large proportion of the indigenous population. In the forest regions, where cattle cannot be reared, game meat or bush meat is important in the diet of the rural people, which make up about 80 per cent of the population. Riney (1967) estimated that bush meat, including game, fish, insects, snails, maggots, and the like constituted more than 80 per cent of the fresh meat consumed in Ghana, a figure which is very close to the 75 per cent estimated for Africa south of the Sahara (Asibey 1972) and 79 per cent for southern Nigeria (Charter 1970). The total value of bush meat in Nigeria was put at ₦30 million and that of naturally produced protein food at ₦100 million (FAO 1970).

The gathering industry in Africa has reacted in at least two ways to changes in population. First, some plantations, for example, rubber and oil palm, were established under the various colonial administration. Second, the intensification of tapping and gathering led, particularly in the case of rubber, to overtapping and destruction of trees and vines. By 1899, it was estimated that 75 per cent of the rubber plants in the forests of southwestern Nigeria had been destroyed.

Consequences of Resource Utilization

Resource utilization takes place in the process of development, which invariably involves costs and benefits (Faniran and Areola 1975). A development project is judged as successful if its envisaged benefits are realized at tolerable social and other costs. In connection with natural resources, the costs are reflected mainly in the physical environment, in which case the increasing public concern with environmental quality as a legitimate goal of resource allocation and use provides a basis for evaluating resource utilization.

The benefits of natural resources utilization in sub-Saharan Africa are in some cases so pervasive and overwhelming that the costs and implications are often discounted. The contributions of mineral exploitation in the development and economic advancement of countries such as Guinea, Sierra Leone, Liberia, Ghana, Nigeria, Zaire, Zambia, and South Africa are cases in point. But while these benefits have all been documented and widely discussed, the side effects have not been so emphasized. These effects relate mainly to environmental pollution.

Environmental Pollution

The term *pollution* has a very broad meaning. According to the dictionary, to pollute means to make foul or unclean, to defile, profane, dirty, or desecrate. Here the term is used in the physical sense, pertaining to the adulteration or contamination of water, air, and land generally. Thus the Committee on Pollution of the United States National Research Council defines pollution as an undesirable change in physical, chemical, or biological characteristics of air, land, and water that may or will harmfully affect human life or that of other desirable species, our industrial processes, living conditions, and cultural assets; or that may or will waste or deteriorate our raw material resources. In other words, pollution is wanton and indiscriminate use of natural resources with the danger of rendering them unavailable for continued use.

Pollutants are often classified, on the basis of their origins, as biological, chemical, or physical. Biological pollutants include the human parasites such as certain water-borne viruses, bacteria, parasitic fungi, protozoa, and plant spores and pollen grains. The chemical pollutants are either organic or inorganic. The biggest source of organic pollution is domestic sewage. Other sources are from industrial wastes. Agricultural land use and animal husbandry can also result in organic pollution from pesticides, fertilizers, and manure. Inorganic pollutants originate from industrial activities, transportation, and agricultural land use.

Physical pollutants originate from changes brought about by the physical alteration of the properties of the environment. They consist of particulates in water (mainly sediment caused by soil erosion) and in air (smoke from industrial plants, buildings, vehicles, and incinerators). Wind is another cause of air pollution, as is evidenced by the dust storms of world-wide occurrence and the harmattan dust of West Africa. Other pollutants include solid wastes on land, thermal pollutants, radioactive pollutants, and noise pollution.

To begin with, pollution is not regarded as a serious problem in large parts of sub-Saharan Africa, at least not as serious as in many developed countries. Also, by its very nature development is usually accompanied by some form of pollution. Agriculture involves clearing the land and applying fertilizers, insecticides, and pesticides, all of which change the natural environment and may even defile not only the soil but the water in the streams into which the farm residues (particularly the noxious parts of the fertilizers, insecticides, and pesticides) are washed. Similarly, in the case of an industrial plant, the air, water, and land in its vicinity are polluted. Yet development is necessary. The problem is to balance the benefits of development against the costs of environmental pollution. The examples below are meant to call attention to the extent of environmental pollution, and so solicit efforts to solve the problem before it gets out of hand. They all arise mainly as a result of population change in both quantitative and qualitative terms.

Land Pollution

This is a very broad area which embraces all the changes on the earth's surface. It includes pollution from solid and other wastes in urban areas, and in rural areas effected by industries, by soil erosion and soil degradation, by forest clearing and deforestation, and by despoilation of beach and other coastal environments.

Urban wastes. Most of the large population centres in sub-Saharan Africa are now faced with the problem of efficient waste disposal. The rapid rate of urbanization has been accompanied in the native areas by increased dumping and accumulation of solid waste. This problem has aroused public concern in many countries. The most common avenues of complaint are the daily papers and periodicals. The *Daily Times* of Nigeria has on several occasions called attention to this problem, using Ibadan as an example (see Faniran 1980b). The *Renaissance*, another Nigerian daily, on 26 September 1972, said this about Enugu:

> It is rather difficult to believe; but here is a street in the heart of Enugu which has automatically been rendered impassable by an evergrowing heap of rubbish. . . . The congestion in the Enugu motor park spells out the need for another modern park to serve the enlarged township (quoted by Inyang 1973).

Noye-Nortey (1974) has this to say of towns in Ghana:

> Lack of proper drainage, unsatisfactory waste disposal services and lack of adequate water supply which have turned these communities into sanitary eyesores in the cities.

Soil erosion and soil degradation. Soil is a renewable resource in that its fertility can be renewed. And as observed earlier, traditional agriculture achieves this through a combination of farming processes during cultivation and by the use of the fallow. Modern agriculture also has this as a basic requirement and uses a number of conservational and soil management procedures to guarantee continued soil fertility (Faniran and Areola 1978). The plight of agriculture in large parts of sub-Saharan Africa today is that the traditional system has broken down while modern agriculture is yet to make any significant impact. The gullies of southeastern Nigeria and of Zaria are examples of the consequences of this breakdown. Soil erosion and degradation are serious in the high-density rural population centres of northern Ghana, central Upper Volta, southern Uganda, Rwanda, and Burundi.

Soil degradation is also caused by other factors. Mining is a particularly important cause, mainly because it involves, in most cases, surface diggings and other forms of soil destruction. The worst-affected parts are those with open-cast mining, such as the tin mining areas of the Jos Plateau and the oil mining areas of the Niger delta areas, both of Nigeria, and the iron and bauxite mines of Liberia, Sierra Leone, Guinea, and Mauritania, to quote a few examples.

Water Pollution

The location of water bodies at the lowest points in an area makes them natural depositories for waste. Rivers, lakes, ponds, and the ocean have been and will continue to be the depository for all the materials removed from dry land, while underground water resources are sometimes impaired through seepage of polluted water. Sediments entering lakes, rivers, or oceans owing to natural erosion may modify the water to a degree that it no longer can serve consumptive purposes. And if the river load and other particulates are sufficiently fine and widely dispersed, the settlement of the fine silt may take quite some time and so render the water unusable. In southwestern Nigeria, for example, a number of dams constructed about thirty years ago are now rapidly silting up (Alavi 1979). The silting of harbours is another example of the trouble caused by natural erosion and sedimentation but accelerated by man's activities.

Another important aspect of water quality which is relevant here is the ability of streams of purify themselves. This process is recognized by many communities in Africa which do not believe that water in flowing streams is ever polluted to the extent that it can be a health hazard. This viewpoint, which holds to some extent in traditional rural life, is no longer valid. Many rivers, lakes, lagoons, and ponds have been polluted to the extent that no one will use them for anything. Examples of polluted streams are the Kudeti and Ogunpa in Ibadan, the Asata in Enugu, and the Calabar (all in Nigeria), and the Odaw in Accra and the Subin in Kumasi (both in Ghana). The lagoons, especially those near urban centres, are the most polluted as they take all the waste from the rivers. Human and other waste are also dumped into the lagoons. Boateng (1974) wrote as follows of the Odaw river in Accra area:

> Wastes from the Brewery are poured into the river and as the river moves along domestic wastes and other wastes from the city are also dumped into it. The stinking Odaw river eventually discharges into the Korle lagoon which is now behaving as a large cesspool and the people living around this area can bear testimony to the foul odours which exist there.

The effects of water pollution are all well known. Water-borne diseases are now rampant even in the rural areas. Living organisms are also adversely affected. A recent survey in the Ibarapa division confirms the widespread occurrence of dysentery among other diseases (Faniran *et al.* 1980). Indeed, the University of Ibadan campus at Igbo-Ora has as its major aim the study and eradication of these diseases from the area. Another study, this time in the Lagos area, shows that the fish population in parts of the Lagos Lagoon has not only declined in number but also in quality because of water pollution. The situation is aptly described by Inyang (1973) who wrote as follows of the pollution of the Calabar river.

> It can be identified by less and less catches of fish, by the turbid nature of the water at all times, by the layer of oil spreading over considerable areas of the river and by the general absence of swimmers along the beaches.

Oil Pollution

The oil industry in Nigeria has expanded at a phenomenal rate since commercial production started in 1957. Today it is the dominant sector of the country's economy. A number of African countries are now also prospecting for oil. It is therefore necessary to consider pollution as pertains to the oil industry. The following sources of oil pollution have been identified: (1) discharge of oil slugs by visiting tankers, (2) disposal of crude oil in oil fields during production tests, (3) disposal of drilling mud and chemicals, (4) crude oil spillage from pipelines and flow stations, (5) well blowouts, (6) spillage of crude oil due to burst hoses during loading operations, (7) refinery effluents, and (8) thermal pollution caused by gas flaring. Reports of oil pollution are becoming more and more frequent, both in the oil-producing areas and elsewhere. Tankers have discharged slugs at seaside resorts such as Tarkwa Bay (Lagos) and the Ilashe and Ilado beaches (Okoloko 1974). The Nigerian agricultural, fish, and shrimp industries have been seriously threatened by several oil spills in oil-producing areas.

Discussion and Conclusion

That Africans, like their counterparts in other parts of the world, are concerned about these and other environmental consequences of population change is not in dispute. Concern for the environment has been expressed as much on the pages of local newspapers and other mass media as in contributions at conferences, seminars, and workshops. In the past few years, the environment has featured prominently within the sub-region. Examples from West Africa include: (1) the Accra meeting of the West African Regional Conference of the Commonwealth Geographical Bureau, in September 1970, with its theme: Environment and Development in West Africa, (2) the conference on Environmental Resource Management in Nigeria, July 1973, organized by the Department of Geography, University of Ife, Ile-Ife, (3) Scientific Committee on the Problems of the Environment (SCOPE) Conference I (1974) and II (1980), Accra, organized by the Ghana Academy of Arts and Sciences for the Ghana National Committee for SCOPE, and (4) AASA[2] workshop on "Utilization of Agricultural, Forestry and Fisheries Waste Products" in Doula, Cameroun, November 1980.

At these meetings, resolutions are often passed spelling out the problem and advocating specific lines of action. A portion of the resolution passed at the end of the September 1974 Accra meeting referred to above reads as follows:

> *Bearing in mind*, the pace of industrial development, the rate of urbanization and the general growth of population in the subregion
>
> *Being conscious* of the need for the monitoring of environmental pollutants, seismic phenomena, drought potentialities and the incidence of epidemic diseases
>
> *Taking into account* the need for closer co-operation between African States in environmental matters and for the sharing of information and experience relating thereto, (the Conference) *Urges* all African States, as a matter of immediate importance, to set up Environmental Protection Agencies or Councils with the express purpose of —
>
> (i) creating awareness of the importance of protecting the environment;
>
> (ii) advising the respective governments on measures to maintain the quality of life;
>
> (iii) ensuring the judicious exploitation of natural resources in order to secure a proper balance between population growth and economic developments;
>
> (iv) creating data banks for the collection and retrieval of essential environmental information at the national and inter-national levels;
>
> (v) ensuring the preservation of genetic pools in plants and animals and the conservation of endangered species; and
>
> (vi) promoting environmental education at all levels of formal education and in the home.

Various governments have set up bodies responsible for environmental problems. These bodies in Ghana include the National Committee of SCOPE, the Environmental Protection Council, and the Ghana Standards

---

[2]Association for the Advancement of Agricultural Sciences in Africa, with headquarters in Addis Ababa, Ethiopia.

Board. Similar bodies exist elsewhere. There is therefore sufficient evidence that at least at the government level, and also among the press men and academics, there is definite awareness of the problem of the environment.

To conclude, I shall draw the various factors together and offer some concrete suggestions to solve the problems. These problems have arisen from unplanned growth, both of population and of the economy. Everywhere, the traditional order has been upset, without being replaced by any definite new order. Thus, as Mabogunje (1974) has observed, the problem of African urbanization has arisen because the traditional principle of social integration was violated by the colonial administration, partly on account of ignorance and partly on account of the necessity of modernizing the economy.

The governments have taken the initiative by establishing the machinery for environmental management, but for this machinery to operate successfully the right type of personnel as well as community response and co-operation are needed. This will depend on education and public involvement. As a strategy for successful environmental conservation programmes in sub-Saharan Africa, it is necessary to combine, on the one hand, an original high level of research and government with public education on the quality of life.

Much more research is needed about the sources and effects of the various types of pollution. Information is also needed on the stock of natural resources, the rate at which they are being removed, and perhaps, more important, the impact of the utilization of one resource item on other related resources and the total environment.

The question of public education is fundamental. It is one thing to make laws to protect a group of people, but another thing entirely for the people to be convinced that they need such laws. Traditional African life is tolerably clean and free from serious pollution problems. It is therefore understandable that not all Africans believe that flowing water, for instance, can be dangerous, or that disease and death are caused by dirt, germs, and pathogens. In other words, it is necessary to educate the public on the connection between environmental sanitation and clean environments on the one hand and quality of life on the other.

In this connection, one may take a cue from the Soviet Union which has an elaborate programme of engendering public awareness of the problems of the environment (Pryde 1967). It involves a number of national and local groups which plan the course of environmental and natural resources conservation. The task of maintaining an overall view of conservation efforts and of coordinating the activities of the various research and development agencies is vested in a central commission — the Central Directorate for the Conservation of Nature, Natural Preserves and the Hunting Economy —

which is directly responsible to Cabinet or the Council of Ministers. The government also recognizes and financially supports a number of public conservation organizations or societies, such as the All-Russian Society for the Conservation of Nature, and the Moscow Society of Naturalists. In addition, Soviet citizens are actively engaged in conservation work by passing resolutions, writing articles, or sending letters to newspapers and periodicals on both general and specific environmental issues. The conservation of natural resources forms a major part of the programmes in Russian institutions of higher learning, while the Ministry of Education arranges a broad range of relevant extra-curricular activities on this topic.

The political environment in Africa is different from that of the Soviet Union. Nevertheless, with the will on the part of the governments and peoples of all states, it should be possible to start organizing conservation education at state levels and later to encourage cooperation among the various states, as envisaged in one of the resolutions quoted above.

REFERENCES

Akintola, F.O.; Areola, O.; and Faniran, A. 1979. "The Elements of Quality and Social Costs in Rural Water Supply and Utilization". *Water Supply and Management* 4:1–8.

Alavi, N.A. 1979. "Water Storage and Abstraction from Basement Complex Drainage Basins of Oyo and Kwara States: A Simulation Study". Ph.D. thesis, University of Ibadan.

Apeldoorn, G. Jau Van 1978. *Drought in Nigeria*, vols. 1 and 2. Centre for Social and Economic Research, Ahmadu Bello University, Zaria.

Areola, O., and Faniran, A. 1977. "A Framework for Land Resource Evaluation in Nigeria". In Mabogunje, A.L., and Faniran, A, eds. *Regional Planning and National Development in Tropical Africa*. Ibadan University Press.

Asibey, E.O.A. 1972. "Wildlife as a Source of Protein in Africa, South of the Sahara". *F.A.O. African Foresty Commission*. Nairobi, Kenya.

Assez, L.O. 1972. "Rural Water Supply in the Basement Complex of Western State, Nigeria". *Bulletin International Association of Hydrological Sciences* 18:97–110.

Boateng, J.W. 1974. "Oxygen Economy of Polluted Streams". *Conference Environment and Development in West Africa*. Ghana Academy of Arts and Sciences, Accra.

Burke, K.C. 1967. *The Scenery of Ibadan*. Mimeo. Department of Geology, University of Ibadan.

Carter, J.D. 1956. "The Rise in the Water-Table in Parts of Potishum Division, Bornu Province". *Geological Survey Nigeria, Records*, pp. 5–13.

Charter, J.R. 1970. "The Economic Value of Wildlife in Nigeria". *1st Annual Conference, Forestry Association of Nigeria*.

Chorley, R.J., and Kennedy, B.A. 1971. *Physical Geography: A Systems Approach*. London: Prentice-Hall.

Faniran, A. 1968. "Creating a Commercial Diarying Industry in a Nomadic Pastoral Economy". *Australian Geographer* 10, no. 5:392–401.

_______. 1971. "The Development of International Drainage Basin". In *Hydrology and Hydrometeorology in Africa*, vol. 2. WMO, Geneva.

————. 1972. "River Basins as Planning Units". In Barbour, K.M., ed., *Planning for Nigeria, the Geographical Approach*. Ibadan University Press.

———— . 1974a. "Nearest-neighbour Analysis of Inter-Inselberg Distances: A Case Study of the Inselbergs of Southwestern Nigeria". *Zeits fur. Geomorph. Suppl. Bd. 20*, pp. 150–69.

————. 1974b. "The Extent, Profile and Significance of Deep Weathering in Nigeria". *Journal of Tropical Geography* 38:19–30.

————. 1974c. "Drainage Basin and Political Boundaries in Africa". *Nigerian Journal of Economic and Social Studies* 16, no. 3:445–59.

————. 1975. "Rural Water Supply in Nigeria's Basement Complex: A Study in Alternatives". *Proceedings, Second World Congress, International Water Research Association, New Delhi*. vol. 3.

————. 1980a. "On the Definition of Planning Regions: The Case for River Basins in Developing Countries". *Singapore Journal of Tropical Geography* 1, no. 1:9–15.

————. 1980b. "Economic and Ecological Aspects of Solid Waste Disposal in Developing Agricultural Economies: A Case Study of Ibadan, Nigeria". *Association for the Advancement of Agricultural Sciences in Africa, Workshop on Appropriate Technologies for the Development of Agriculture in Africa*. Duala, Cameroun.

————. 1981. *African Landforms*. Forthcoming.

————, and Areola, O. 1974. "Landform Examples from Nigeria, No. 7: A Gully". *Nigerian Geographical Journal* 16, no. 1:57–60.

————. 1975. "Ecology and Economics in the Lagos State, Nigeria". *Nigerian Journal of Economic and Social Studies* 17, no. 1.

————. 1977. "A Framework for Land Resource Evaluation". In Mabogunje, A.L., and Faniran, A., eds., *Regional Planning and National Development in Tropical Africa*. Ibadan University Press.

————. 1978. *Essentials of Soil Study*. London: Heinemann.

Faniran, A., and Ojo, O. 1980. *Man's Physical Environment*. London: Heinemann.

Faniran, A.; Sada, P.O.; and Areola, O. 1976. "Riversand Mining: Its Organization and the Building Industry in the Lagos Area". *Quart. Jour. Admin.* 10, no. 4:423–35.

Faniran, A.; Filani, M.O.; Akintola, F.O.; and Acho-Chi, C. 1980. *Improvement of Water Supplies for Small Nucleated Settlements in Rural Nigeria*. Department, University of Ibadan/IDRC, Ottawa.

Federal Department, Water Resources Lagos/Centre for Social and Economic Research. 1977. *The Aftermath of the 1972–74 Drought in Nigeria*. Ahmadu Bello University, Zaria.

Federal Government of Nigeria. 1962. *1st National Development Plan 1962–70*. Lagos.

————. 1970. *Second National Development Plan 1970–74*. Lagos.

————. 1975. *Third National Development Plan 1975–80*. Lagos.

Greenzebachs, K. 1980. "Methods of Regional Landuse Analysis in Southwest Uganda: A Precondition for Agricultural Regional Planning". In *Applied Geography and Development*. Institute for Scientific Co-operation.

Harrison-Church, R.J. 1974. *West Africa*. 7th ed. London: Longman.

Hilton, T.E. 1956. "The Population of the Gold Coast". *I.G.U. Symposium*. Makerere.

Hunker, H.L., ed. 1964. *Introduction to World Resources*. London: Harper & Row.

Inyang, P.E.B. 1973. "Pollution of the Most Valuable Environmental Resources (Land, Water and Air) in Parts of the Eastern States of Nigeria". *Conference of Environmental Resource Management in Nigeria, University of Ife, Ile-Ife*.

Mabogunje, A.L. 1974. "Value-Systems and the Urbanized Environment". *18th Annual Conference of Nigerian Geographical Association, University of Nigeria, Nsukka*.

Manshard, W. 1965. "Kigezi (Southwest Uganda): The Geographical Structure of an East African Mountain Region". *Erkunde* 19:192–210.

Morgan, W.B., and Pugh, J.C. 1969. *West Africa*. London: Methuen.

Nash, T.A.M. 1941. "The Anchau Settlement Scheme". *Farm and Forest* 2, no. 2:77.

Noye-Nortey , H. 1974. "Sanitation and the Urban Environment in Ghana". *Conference of Environment and Development in West Africa, Ghana Academy of Arts and Sciences, Accra.*

Ofomata, G.E.K. 1965. "Factors of Soil Erosion in Enugu Area of Nigeria". *Nigerian Geographical Journal* 8, no. 1:45–59.

Okoloko, G.E. 1974. "The Problem of Oil Pollution in Nigeria". *Conference of Environment and Development in West Africa, Ghana Academy of Arts and Sciences, Accra.*

Ologe, K.O. 1972. "Gullies in the Zaira area: A Preliminary Study of Headscarp Erosion". *Savana* 1, no. 1:55–66.

Omorinbola, E.O. 1979. "A Quantitative Evaluation of Groundwater Storage in Basement Complex Regoliths in Southwestern Nigeria". Ph.D. thesis, University of Ibadan.

O'Riordan, T. 1971. *Perspectives on Resource Management*. London: Pion Ltd.

Pryde, T. 1967. *Conservation in the Soviet Union*. Cambridge: Cambridge University Press.

Riney, T. 1967. *Conservation and Management of African Wildlife*. FAO/IUCN, Rome.

Stamp, L.D. 1938. "Land Utilization and Soil Erosion in Nigeria". *Geographical Review* 28:32–45.

————. 1959. *Africa: A Study in Tropical Development*. New York: John Wiley.

Wallace, Tina. 1979. *Rural Development through Irrigation Studies in a Town on the Kano River Project*. CSER, ABU, Zaria.

Urvoy, Y. 1942. "Petit atlas ethnographique du Soudan, Memoires". *L'Institut Francais d'Afrique Noire* 5:14.

White, H.P., and Gleave, M.B. 1971. *An Economic Geography of West Africa*. London: Bell.

Yi-Fu Tuan. 1971. "Man and Nature". *Commission on College Geography Resource Paper no. 10*. Association of American Geography, Washington, D.C.

Zimmermann, E.W. 1951. *World Resources and Industry*. New York: Harper & Bros.

16
Natural Resources and Economic Development in Developing Countries*

RICHARD S. ODINGO

In recent years there has been a new awareness and a reawakening of interest in natural resources throughout the world and especially in developing countries. This has come about largely because of the realization that the process of economic development in all countries is closely linked to the resources, internal or external, which are available for that development.

The great interest in natural resources has arisen from the general assumption that no meaningful economic development can be achieved without natural resources. The need to manage and use resources wisely has arisen from the assumed close linkage between natural resources and development. Agriculture and food production, for example, are seen as resource development using the soil, water, and climatic resources of a given country. "Agriculture relies on biological processes and on the productive capacity of renewable natural resources" (FAO 1977:5). But the concept of natural resources is somewhat illusive and changes much with time (UNESCO 1979:3). It thus acquires an essentially dynamic character which changes with the needs of man, with the evolution of his technology, and with the choices he makes to achieve his aims. This is important in considering the role of natural resources in the development process. Today many countries consider the existence of significant quantities of non-renewable natural resources as the best assurance for the success of their initial development and the creation of wealth sufficient to enable them to derive the full benefit of improvements in agricultural and industrial production.

It is a truism that a country which is endowed with adequate natural resources and which manages those resources carefully is able to achieve development provided the correct technology for the use of those resources and the appropriate cultural environment are available. In this paper, an attempt is made to look at the natural resource situation in developing countries and how these countries are beginning to view and use their

---

*The views expressed in this paper are those of the author alone and do not necessarily represent those of the United Nations University.

resources in their effort to achieve economic development. The paper examines the recent trends in resource utilization and looks at some of the difficulties in economic, technological, and environmental fields which have stood in the way of many developing countries and the realization of their aim of the proper use and management of their natural resources to bring about economic development. The question in natural resources is so vast that a short paper can only touch on some of the basic points under consideration. For this reason, this paper neither lays claim to being complete or exhaustive, but points at some of the directions now being taken on resource considerations in developing countries.

Use and Management of Natural Resources in Developing Countries

Until recently, most developing countries were not fully aware of their natural resources and did not even have concrete plans to use them in their efforts to achieve economic development. Today this picture is rapidly changing, and there is a realization that there is an important link between resources and development. One aspect of this, to be examined later, is the move by the developing countries to assert their claims to full sovereignty over their natural resources. The resources were previously either squandered and exported, or exploited for very small returns by transnational enterprises. But today things are beginning to take a different shape. Most developing countries now agree that they need more detailed scientific knowledge and data for the better management of their natural resources. If the available resources in the developing countries can be used carefully, and if the concept of management can not only enter but also be accepted as the method of dealing with the problems facing those countries, a long-term solution will have been found. It is accepted that there are many constraining factors to the better management of natural resources in the form of physical, biological, social, economic, technological, and political considerations. However, the proper use of management techniques, with special reference to natural resources, is intended to overcome these very problems.

The role of population growth relative to resource availability and utilization is still improperly understood, and there are many conflicting theories. Many well-meaning scholars can see the restriction of population growth as being the short-cut to resource development. They suggest such an approach even before the existing resources are explored, recorded, and documented. In this context, the concept of carrying capacity should be borne in mind. This concept, which tries to put into proper focus the lack of balance between population and resources and the factors hindering resource development and realization, is more scientific than those which study crude population and land characteristics without fully understanding the relationships.

Another area which needs greater in-depth studies concerns technology and resource utilization. The relationships between technology and resources, especially in relation to the full realization of land, water, and mineral resources, are fundamental to the understanding of the value of the use and management of natural resources to bring about economic development in the developing countries. The technology problem is well illustrated in the management of energy resources in developing countries. Today many of these countries are acutely aware of the role of energy in their continued economic development. However, because of the absence of adequate technology they are finding the problem very perplexing and are still unable to begin to tackle it meaningfully. Exceptions such as China and India exist, where efforts at developing alternative sources of energy are already advanced and operational. However, with the exception of China which is considered a leading developing country in the field of alternative energy, most other countries have yet to find answers as to what are the best societal organizational arrangements required to bring the full benefits of these new sources of energy to their populations.

The case of energy clearly illustrates the fact that it is only through the wise use and management of their natural resources that the developing countries can hope to meet the basic needs of their people, to maintain political stability, and to obtain an equitable share of the world's wealth. "Economic advances of these countries will depend on: (1) the wise exploitation of natural resources, (2) sustained productivity from ecosystems, (3) observation of the carrying capacity of land and water and, (4) avoidance of degradation of environmental quality" (Carpenter 1980). Unfortunately, all these will not come about unless these countries pay even greater attention to the integrated survey and exploration of their natural resources followed by well-coordinated management schemes. The surveys will pinpoint the location and characteristics of the resources, and the management schemes will suggest how best they can be used to bring about economic change in the countries concerned. This can be brought about by incorporating scientific and technological information in the national planning process, and focusing that planning on resource development and conservation.

Data Needs for Resource Development

The availability of resources for development may not be as serious as the lack of reliable planning data for that development. Developing countries cannot be expected to incorporate elaborate plans for resource development in the absence of detailed surveys and reliable data on their natural resources. Economic planning can only be meaningful if the data used for that planning are adequate and reliable. The data needs for planning the development of natural resources cover many fields including soils and land use, agriculture and livestock, forestry and fisheries, wild life resources, water resources, mineral resources, and energy. The main problem to be faced by most countries is that such planning as exists is based on either sketchy or very inadequate data. The data, if available, are not often in the form most suitable for decision-making at policy level. To complicate the problem, with the recent realization of the importance of energy as a sector of natural resources needing government attention, policy decisions in many countries have had to be made in the complete absence of data. Second, the data systems used are not only often inadequate but also quite unreliable. They nevertheless continue to form the basis of planning.

It is true that the availability of data on natural resources for planning varies considerably from one developing country to the next. Traditional methods of data collection are the most common and also the most popular. But recently, through aid programmes, many developing countries are gaining access to modern data collection methods. The technology available for providing data on natural resources is developing rapidly. Even more important is the fact that the availability of data is increasing more rapidly than the ability of many countries to select and analyse it. For these reasons, many developing countries need assistance with data and information systems for the use and management of their natural resources. In the process of collecting information on natural resources, efforts need to be made to integrate traditional methods, such as surveys and aerial photography, with new methods such as remote sensing. But even more important is what happens to the data after it has been collected. To ensure correct and well-informed development planning, therefore, developing countries will need to pay a much greater attention on their data requirements and even more important storage and retrieval for policy formulations.

The use of satellite images as one of the methods of data collection is quite promising as a possible solution to the data problems of developing countries (National Academy of Sciences 1977). Whereas it is true that ground surveys for map making and for detailed resource inventories are still essential, airborne (aerial photographs) and satellite images could cut the time required to obtain operational data. Satellite imagery enables countries to obtain reliable maps at scales of 1:250,000 which can be used for topographical, geological, and soil surveys, for forest mapping, for crop inventories, and for the general mapping of agricultural land use through many of the stages required for planning purposes. But developing countries which are acquiring this technology are finding difficulties in using it because of lack of standardization and because the older and more well-tried technologies of data collection are still preferred. Efforts should therefore be made to use remote sensing technology to solve the data problems of the developing countries. This can only be done if the developed countries with the technology do not concentrate too much on selling the equipment instead of training the developing countries on how to use the new technology to improve their planning data requirements.

Basic Resources Required for Development

There are certain basic natural resources on which economic development can be based and which most developing countries possess. These include soil, air, water, forests, minerals, and the ocean. Not all countries have uniform quantities of these basic resources and it is therefore to be expected that development problems will be related to the abundance or absence of any one or all of these resources. It has often been said that most developing countries have at least the land and the land-based resources in plenty as a basis for their development. In most peoples' minds, developing countries are always associated with primary agriculture and, before the end of the colonial period, with the exploitative form of agriculture known as the plantation system. Second, the economies of several developing countries such as Botswana, Bolivia, Brazil, India, Jamaica, Malaysia, Mexico, the Philippines, and Zaire are greatly dependent on minerals and mining-related industrial activities. But even in these cases, land still remains a major resource, and the careful management of this particular resource will spell success or failure in their efforts to meet the basic needs of their populations. In many of these countries, the land resource is still under-utilized, both because of lack of capital and because of the lack of technical skills. The development of land resources entails the organization of land use in such a manner as to bring about optimum productivity in selected areas with regard to the biological resources, water resources, as well as social and economic attributes of the region. The correct use of land resources also means creating a system of land use which is sustainable and conservation-based so as to maintain growth and productivity well into the future. Land resource development can be aided by good topographical maps as well as soil maps which are often lacking in many developing countries. Such maps are also relevant for the development of other resources, for example, the geological surveys and mining activities, irrigation and agricultural development, forestry inventory and development, as well as for the development of water resources and even wild life resources.

The continued development of mineral resources in developing countries, especially through the activities of transnational enterprises, has become highly politically charged. Their contribution to economic development will largely depend on the agreed national philosophy in each case, which will set out the conditions under which mineral exploration and development may operate. The aim in each case should be a fair and equitable arrangement which ensures that the mineral resources of a given country contribute to the creation of employment through primary and secondary processing, through the encouragement of related industries, and finally, through earnings from sales in the world market.

The development of forest, water, and fisheries resources is also closely linked with the efforts to raise standards of living in general in many developing countries. In the development of these as well as the other resources already outlined, there are numerous economic, cultural, and social constraints. Efforts have been made to study and isolate some of these constraints so as to assist the developing countries to achieve orderly and unhindered development.

Science and technology are needed for resource management and development, and many developing countries have failed to realize the full benefits of their resources because of lack of both. Four main areas in resource development are directly affected by science and technology: (1) resource inventory, (2) analysis of resources, (3) management decisions, and (4) management policies. In short, resource development must be based on sound scientific and technological knowledge. The experience of the developed world has shown that, as a result of scientific and technological advances, the value of certain resources has been upgraded. This is because technology has made it easy to exploit available new resources. Second, because of better inventory techniques and new technological discoveries, the full value of resources can be realized. Because this is often seriously lacking in developing countries, these countries are asking for scientific and technological assistance programmes to help with exploration, inventory, development, management, and the orderly exploitation of their natural resources, including the initiation and execution of plans for their conservation. With reference to technology, however, it is important to sound a word of warning on technological transfer: ecological know-how is often site specific and it is impossible to transfer information on how to deal with natural resources from one ecological zone to the next. It is also important to remember that the characteristics of resource management are likely to differ markedly from one area to the next.

With reference to strengthening the position of the developing countries relative to resource use, Ignacy Sachs (1979:29) has this to say:

> What is needed then for a meaningful dialogue on T (technology) between North and South? The initial bargaining position of the South must be strengthened. . . . The single most important step in this direction would be to initiate in deeds, not in words, a policy of collective self-reliance in S and T among Third World countries.

One of his suggestions for doing this is by way of the following example (1979:30):

> Tropical countries in Asia, Africa, and Latin America have a common interest in jointly organising a research programme on new industrial uses of renewable resources, with special emphasis on biomass energy and biochemistry as well as solar energy at large.

However, most developing countries will be obliged to start by bringing about improvements in their agricultural sector first before they look at the other sectors, because agriculture still provides employment in each case for more than 60 per cent of the total population.

Development of Agricultural Resources

The rationale for agricultural development is that it is the mainstay of the economy of many developing countries. Agriculture remains the foundation of the entire economies of these countries. Its capability for production growth, labour absorption, and the supplying of food and industrial raw materials will continue to govern the rate and pattern of expansion in industry and other sectors (FAO 1978). In many of these countries agricultural development is necessary (1) for supplying food and other essential products in sufficient quantity and satisfactory quality (UN Conference on the Human Environment 1972:16), (2) for ensuring the conservation of a large part of the environment, (3) for employment opportunities, and (4) for maintaining and enhancing the quality and attractiveness of rural areas for recreation and as buffer zones between urban areas.

Therefore, one of the greatest challenges for the developing countries is to achieve agricultural development first before concentrating on other forms of resource development. Agriculture must continue to supply food in the context of fast growing populations in many of these countries. In a typical developing country, population is growing at an annual rate of 2.5 to 3 per cent (FAO 1970:10). Agriculture must also hope to cater for those populations which can only be reached through marketed output, and it must develop in such a way as to reduce the dietary deficiencies, common in many developing countries. Second, resource developments in other sectors will mirror developments in agriculture. Where minerals are absent, agriculture remains the largest earner of foreign exchange. This is one of the most serious bottlenecks to economic growth in most developing countries. Third, within individual countries, agriculture is expected to contribute significantly to the savings needed to finance development. In summary, the agricultural sector is clearly the most important in the economies of most developing countries and will continue to be so for some time to come.

One of the problems which must be dealt with to ensure agricultural development is that of data availability. More data are needed for planning balanced agricultural development. Governments of developing countries are very much aware of the problem, and they are committed to strengthening agricultural research geared at improving ecological understanding, as well as the establishment and the strengthening of national programmes for the conservation of soil resources. Not only is there a commitment to the expansion of agricultural production in general but there are efforts to improve crop fields, to upgrade soils to use purchased inputs wherever possible, and to use all available technology to increase output. Similar improvements are necessary in livestock rearing and fisheries to increase the output of beef, dairy products, and fish.

One of the problems in agricultural development in developing countries concerns cash crops. In many countries of Africa, for example, non-food crops have traditionally occupied the best land, continuing the colonial pattern which favoured industrial crops for export. Much of the newly developed land is being used for such crops, and indeed in many countries the proportion of land used for domestic food production may even decline. "In countries with an acute shortage of cultivable land, the choice between food and industrial crops may be a difficult one. However, it is clearly desirable that where possible, the better land should be reserved for food crops" (FAO 1978). In Africa, to increase food production, the cropped areas must be increased by about one-third between 1975 and 1990, and the ratio of cropped area to arable area from 52 per cent in 1975 to 60 per cent in 1990. The largest increases in area would be for maize, millet, sorghum, and pulses. A high proportion of the increase would come from converted rangeland (FAO 1978).

In Asia, according to the FAO (1978), many of the problems of food and agriculture concern rice, which provides 40 per cent of the dietary energy supply of the 1,220 million people who live in this densely populated part of the world. In Latin America, the process of agricultural modernization has probably gone further than in the other developing regions. However, the expansion of the modern sector appears to have been accompanied by a breakdown of the traditional sector, so that rural socio-economic disequilibria have been accentuated (FAO 1978). In Brazil, for example, widespread conversion of biomass from agriculture (sugar cane and cassava) into methanol could be regarded as a long-term threat to agriculture in general and to food production in particular.

In general, it must be pointed out that the greatest need for agricultural development in the developing countries is to encourage systems which will be renewable, economic, and well-adapted, with a high added biological value. Marcel Mazoyer (1979) emphasizes that this can only be done through the mobilization and steady improvement of the means and methods of the peasant economy. This agricultural research in the developing countries should be sensitive to environmental requirements while drawing upon the wealth and diversity of the world agricultural heritage. This would involve being sensitive to the social norms under which new agricultural methods are to be introduced. Every effort should be made to pick out relevant adaptive solutions where they have been identified by the people themselves; there should, therefore, be room for a diversity of crops, farming communities, and use of livestock; a diversity of management of cultivated ecosystems and ways of renewing production capacity (soil fertility); and diversity of tools, cultural practices, and know-how.

Today most developing countries have realized that they must move from the position of being mere producers of one crop (monoculture) to a more diversified system of agriculture geared to satisfying their domestic food requirements as well as meeting their industrial needs. This is bringing about changes in areas formerly affected by the plantation system of agriculture (tea, coffee, cocoa, banana, pineapple, groundnut, cotton, and oil palm) which largely catered for the needs and markets of the developed countries. Ways in which some of these changes are coming about include more research directed at food crops rather than at export crops, at livestock rearing as well as local farming systems in general.

Development of Other Resources

Developing countries have been striving to increase the returns from the exploitation of their mineral resources. An increase in the share of raw material output that is locally processed to, for example, refined metal, semi-manufactures, or even fabricated products, has for a long time been a primary objective of developing countries (UN Committee on Natural Resources 1979). Today, some rapidly industrializing developing countries (e.g., Brazil) are importing significant amounts of unprocessed raw materials, usually from other developing countries, for processing for the local market. Other developing countries, such as the oil exporting countries of the Middle East, are establishing mineral processing facilities which will use local energy supplies and imported raw materials, with the final products largely intended for export. But the most important development is the gradual increase both in the share of developing countries' mineral products which are processed locally, and in the proportion of developing countries' mineral output which is exported in processed form, as opposed to concentrates or unprocessed ore.

In general terms, the increased processing of minerals in developing countries is expected to generate additional employment, additional government revenue, higher export earnings, and expected general improvements in the economies of the particular countries. Unfortunately, such improvements are not always realized because of the high cost of imported technology for mineral processing. As for the present, while the less developed countries produced one-third of the world output of minerals in 1970, they consumed only 6 per cent. Four-fifths of their production was exported to the West (Harrison 1979). There is the danger of the exhaustion of the minerals in the developing countries to supply the needs of the developed world.

The energy resources issue in developing countries has assumed overwhelming significance in the past seven years. This is particularly true for those, in the majority, who are neither oil-producing nor oil-exporting countries. But even in the oil-producing developing countries, the supply of energy to the rural population is seriously deficient. In the words of the Advisory Committee on the Programme on the Use and Management of the Natural Resources of the United Nations University (1977):

> For millions of people in villages and other small rural communities, the availability of energy resources is a continuing problem.

Efforts are now being made to resolve this problem. The United Nations University, for example, is concentrating its limited resources for energy work almost entirely on rural energy needs. Management requirements for energy in the rural environment need to be worked out by most developing countries. This sector still relies to a great extent on non-commercial fuels such as firewood, dung, and agricultural wastes. This is true for many countries and it is only in the past few years that any serious thought has been given to this problem. Only in India and China have steps been taken to develop alternative sources of energy. Elsewhere, not enough priority has been given to tackling this problem (Morgan *et al.* 1980). The per capita fuel demands in the rural environment are extremely small (Table 16.1), but they are entirely necessary for survival. By comparison with these very low figures, the average person in an industrialized country consumes about 105,000 Kcal, and considerable quantities of energy are incorporated into the commodities imported from the developing countries (McKillop 1980).

One way of solving the rural energy problem is by using the local biomass sources which are easily available. Of these, firewood is already being used in those countries whose forestry resources have not yet been depleted. But this is quickly leading to a crisis situation as rural population increases. In China and India, biomass sources are already being used for the manufacture of rural biogas plants (also called gobar gas plants in India) (UNU 1979). To date, it is estimated that there are about 7.5 million biogas plants in China and 70,000 in India. According to Ramachandra and Gururaja (1977), 12 million rural households in India own enough animals to install

TABLE 16.1 ESTIMATED PER CAPITA DAILY ENERGY USE IN SELECTED RURAL AREAS, 1977

Use/Source	Rural India	Hunan, China	N. Nigeria	Rural Bangladesh
		(in '000 Kcal)		
Human labour	0.67	0.64	0.61	0.67
Animal labour	1.00	0.92	0.13	1.00
Fuel wood	2.86		10.27	0.93
Crop residues	1.16	13.69		1.65
Dung	0.67			0.57
Total non-commercial	6.36	15.25	11.01	4.82
Coal, oil, gas, and electricity	0.53	2.05	0.02	0.27
Chemical fertilizers	0.22	0.34	0.05	0.10
Total from all sources	7.11	17.64	11.08	5.19

Source: Revelle 1978.

the currently available family-sized plants of 2 m³ to 3 m³ capacity, which can provide the domestic energy needs of 90 million people and significantly reduce the demand for non-commercial fuels. The cost of a family-sized biogas plant ranges from US$200 to US$300, still high for extensive rural applications. Other countries are beginning to take an interest in this renewable energy technology from China and India. It is likely that following the United Nations Conference on New and Renewable Sources of Energy held in Nairobi in August 1981, many developing nations will launch national programmes for biogas development or related developments in the energy field.

Population in Relation to Resource Development

Much has been written on population and resource development in the developing countries. It is maintained that the general increase in population has led to an increased demand for natural resources and hence increased pressure on available natural resources. The authors of the much-debated book *The Limits to Growth* put it in the most extreme form as follows:

> If the present growth trends in world population, industrialisation, pollution, food production and resource depletion continue unchanged, the limits to growth on this planet will be reached sometime within the next one hundred years. The most probable result will be a rather sudden and uncontrollable decline in both population and industrial capacity (Meadows *et al.* 1972:23).

There have been vehement responses to this statement. What is not disputed, however, is that the debate did expose ignorance about the link between resources and population. Even today the place of natural resources in the debate on development is not fully understood. It is agreed that in the developing countries (1) there is a need to study that uneven geographical distribution of population relative to available land, (2) there is also a need to pay attention to the more efficient use of natural resources, (3) there is a need to study resource depletion and accompanying environmental degradation, (4) population growth is a matter of general concern, (5) there is a need to study resource distribution and transfer, and finally (6) there is a need to study the value of environmental management and to improve resources use. Developing countries are well aware of all these needs, and the spate of UN Conferences in the 1970s has helped to focus on nearly all these aspects in an effort to get action started.

It is true that many developing countries are experiencing rapidly increasing population, while productivity and per capita incomes remain low. Therefore, natural resource use is one way to raise the level of per capita productivity and the incomes of individual families, particularly in the poor sectors of the society. One should bear in mind that population growth and productivity increase are not independent forces running a race. Rather, additional persons cause technological advances by inventing, adapting, and diffusing new productive knowledge. Additional persons, instead of being a permanent drag, lead to an increase in per worker output starting thirty to seventy years after birth — that is, ten to fifty years after entering the labour force. Offsetting the negative capital-dilution force of more people in developing countries are the positive forces of increased work done by parents, extra stimulus to agricultural and industrial investment, increased social infrastructure, and other economies of scale (Simon 1980).

Recent Trends in Natural Resources Development

Developing countries have for many years been concerned about the status of natural resource development in the world and the share which they command even of their own natural resources for their development. Arising from this there have been moves by the developing countries in various international meetings to exercise more sovereignty over their natural resources, and to stake a claim to those other resources which have not yet been "annexed" by the developed countries.

During the colonial era, the resources of many now independent countries were controlled by metropolitan powers. Even after independence, the control of their own natural resources did not automatically pass on to the newly independent countries because of capital implications and the fact that exploitation started by the former colonial masters had continued into the new era. More often than not, management of some of these natural resources, for example, agricultural plantations and mining activities, passed into the hands of multinational corporations. Therefore, the newly independent countries find that they are unable to make direct decisions on how best to develop some of these resources.

The importance of national control over natural resources has become a question of sovereignty and is now a subject of fierce debate. In the view of many developing countries, the issue of sovereignty over natural resources cannot be abstracted from the concept of economic independence. In this respect, the activities of certain transnational corporations or enterprises are regarded by many developing countries as obstacles to the full implementation of permanent sovereignty (UN Committee on Natural Resources 1979). But the problem of sovereignty over natural resources is closely linked to the lack of knowledge about these resources and the scarcity of data for planning purposes which have been discussed in the earlier parts of this paper. Second, developing countries are concerned that through the use of new technologies such as satellite photography, developed countries may get advance information about a developing country's natural resources and compromise any local plans for the development of these resources. To help deal with some of these problems, the view has been expressed that greater economic and technical cooperation among developing countries should be encouraged as a means of strengthening the capacity of states to exercise sovereignty over their own natural resources.

Whereas they have found it difficult at home to have full sovereignty over their own natural resources, the developing countries have taken the opportunity through concerted action to take important stands over those resources which have still been held loosely by many nations. The best examples of this can be found in the discussions which have now been going on for several years in the Law of the Sea Conference. Before this is discussed, a few points should be made about the oceans and their importance as a resource not only for developed but also for the developing countries.

The oceans are by far the largest and most valuable parts of the earth which still await full utilization. They cover more than two-thirds of the planet. They comprise the surface of the seas, the water column, the seabed, and the subsoil, and have all the features of emerged land — mountains, plains and valleys, a varied flora and fauna, and mineral resources. Ocean pace is of vital importance for the following reasons:

1. It contains more than 95 per cent of the world's water, probably more hydrocarbons and certainly vastly greater quantities of a wide range of other minerals than are found on land; it also contains vast living resources which can make a far greater contribution to world food needs than at present.

Some of these resources such as krill and marine plants, are still virtually unexploited.

2. It is an immense potential source of energy which awaits exploitation.

3. It is not merely the last and greatest resource reserve of our planet; it also offers space for a variety of activities which are at present land-based, and is essential for international trade, and for the maintenance of national security as perceived today.

4. It is of fundamental importance to climates, indeed to life on earth, and it is the ultimate sink of the enormous, growing, and increasingly toxic wastes produced by our expanding industrial society (Pardo and Borges 1977).

Since the manner in which ocean space will be used and exploited affects the perceived natural interests of every nation in the world, it is vitally important in the creation of any new international order. This paper focuses on the ocean as a resource. It also attempts to show how, through the activities of the developing countries, a more rational approach is being taken by the whole world to this, the last of the major resources.

One of the major outcomes of the Law of the Sea negotiations which are still going on is the virtual acceptance of the seabed as a *common heritage of mankind*. The concept of the common heritage of mankind, if finally accepted, will supersede the traditional freedom of the sea. It would mean no nation could appropriate it; it would also mean it can be used but not owned; that there would be a system of common management of this great resource, and that all the benefits from the seabed would be shared by all nations including the land-locked states. These are important developments in the field of natural resources, and if they become international law they would change drastically many concepts about resources. The conference is still working out a system by which all nations, coastal and land-locked, developing and developed, will participate in decision-making concerning ocean use and the management of the seabed resources. Involving more than 150 countries, it is the largest and most complicated participatory conference on law ever held.

Many new concepts affecting natural resources have been discussed and some unexpected national actions intended to expand and protect national sovereignty over some parts of the sea have come about. One of these is the rights by states over the contiguous zones of the sea. Another is the concept of *exclusive economic zone*. This economic zone forms the baseline from which the breadth of the territorial sea is measured 200 nautical miles from the coastline. Many countries have been quick to declare their sovereignty over it. The exclusive economic zone is a new concept which conveniently consolidates a variety of claims to exclusive access, by coastal states, to resources such as fisheries and marine mineral deposits, and to the control of activities in the marine environment advanced by coastal states with increasing frequency in recent years. It means coastal states will be able to explore and exploit, conserve and manage the natural resources, renewable or non-renewable, within this zone, including the sub-soil and the superjacent waters. But the problems of definition and counter-claims to resources even within this zone are so complex that one must await the final outcome of the full conference before categorical statements can be made. One thing, though, is clear — that is the developing countries, like the developed, stand a much better chance of getting a fair deal over these resources compared with other areas which should have been common but were annexed by the various powers with the required technology. The best example of this is the continent of Antarctica, which should have been used for the benefit of all humanity but is increasingly being used for the national ends of a few developed countries.

In conclusion, it can be said that the whole field of natural resources is extremely complex and the developing countries have only recently become aware of the complexity of the problem, and have started to take a keen interest in it. In all cases, their priorities should be with the use and management of their own natural resources to endeavour to raise living standards for their population. To succeed, they will need assistance with technology for data collection and for the conversion of some natural resources into more useful forms. Even more important are the surveys which have still to be done to reveal the location and value of these resources. Discovery of new resources could mean a change from poverty to riches overnight. But the most important need is the need to protect the sovereignty of developing countries over their resources so that they can be used for their own development.

REFERENCES

Carpenter, R.A. 1980. "Using Ecological Knowledge for Development Planning". *Environmental Management* 4, no. 1:13–20.

FAO. 1976. *Food and Environment*. Rome

———. 1978. *The State of Food and Agriculture*. Rome

Harrison, P. 1979. *Inside the Third World*. London: Penguin.

Mazoyer, M. 1979. *Science and Technology for Agricultural Development: Impasses and Future Prospects*. United Nations Conference on Science and Technology for Development; Science Technology and the Future. A/Conf. 81/5 Add. 2. Vienna.

McKillop, A. 1980. "Economic Considerations for Solar and Renewable Energy in Developing Countries". *Natural Resources Forum* 4, no. 2:165–79.

Meadows, D.L.; Meadows, D.H.; Randers, J.; and Behrens III, W.W. 1972. *The Limits to Growth*. Report for the Club of Rome's Project on the Predicament of Mankind. New York.

Morgan, W.B.; Moss, R.P.; and Ojo, G.J.A. 1980. "Rural Energy Systems in the Humid Tropics". *Proceedings of the First Workshop of the United Nations University Rural*

Energy Systems Project. Ife, Nigeria, 10–12 August 1978, United Nations University, Tokyo. (NRTS-4/UNUP-93).

National Academy of Sciences. 1977. *Resources Sensing from Space: Prospects for Developing Countries*. Washington National Academy of Sciences.

Pardo, A., and Borgese, E.M. 1977. *The New International Economic Order and the Law of the Sea*. International Ocean Institute, Malta.

Ramachandra, A., and Gururaja, J. 1977. "Perspectives on Energy in India". *Annual Review of Energy*, no. 2:365–86.

Revelle, R. 1978. "Renewable Energy Resources and Rural Applications in the Developing World". In Brown, N.L., ed., *Renewable Energy Resources and Rural Applications in the Developing World*. Boulder, Colorado: Westview Press, 1978.

Sachs, I. 1979. "Controlling Technology for Developmnet". *Development Dialogue*, no. 1:24–32.

Simon, J.L. 1980. "Resources, Population Environment: An Oversupply of False Bad News". *Science* 208.

UNESCO. 1978. *New Perspectives in International Scientific and Technological Cooperation*. Paris. A/Conf. 81/BP UNESCO.

United Nations. 1970. *Natural Resources of Developing Countries: Investigation, Development and Rational Utilization*. Report of the Advisory Committee on the Application of Science and Technology to Development, Department of Economic and Social Affairs, New York.

United Nations Conference on the Human Environment. 1972. *Environmental Aspects of Natural Resources Management*. A/Conf. 48/7. Stockholm.

United Nations. 1979. *ECOSOC Committee on Natural Resources*. E/C 7.97. United Nations, New York.

______. 1979. "Some Energy Problems and Issues in Developing Countries". *ECOSOC Committee on Natural Resources*. New York.

United Nations University. 1977. *Report of the Ad Hoc Advisory Committee Meeting on the Use and Management of Natural Resources*. NRR-2/UNUP-1E. Tokyo.

______. 1979. "Bioconversion of Organic Residues for Rural Communities". Paper presented at the Conference on the State of the Art of Bioconversion of Organic Residues for Rural Communities, held at the Institute of Nutrition of Central America and Panama, Guatemala City, Guatemala, 13–15 Nov. 1978. *Food and Nutrition Bulletin*, Supplement 2.

Index